E:CO:E

Emergence, Complexity & Organization
E:CO ANNUAL
Volume 11

ISBN: 978-0-9842165-6-7

Edited by:

Peter M. Allen (Cranfield University, UK)
Kurt A. Richardson (Exploratory Solutions, US)
Jeffrey A. Goldstein (Adelphi University, US)

Published and distributed by:

EMERGENT™
PUBLICATIONS

3810 North 188th Avenue
Litchfield Park, AZ 85340

Emergence, Complexity and Organization: E:CO Annual Volume 11
Edited by: Peter M. Allen, Kurt A. Richardson & Jeffrey A. Goldstein
Library of Congress Control Number: 2005906985
ISBN: 978-0-9842165-6-7

Printed in the United States of America.

Emergence: Complexity & Organization
E:CO Annual VOLUME 11 2009

CONTENTS

Academic Paper Section

Practitioner Paper Section

Philosophy Section

Classical Papers Section

Forum Section

Appendices

Editorial

Peter M. Allen
Cranfield University, ENG

This collection of articles from Volume 11 demonstrates the amazing breadth and depth of thinking today about complexity, emergence and organization. I well remember in the 1970s overhearing pompous professors at an illustrious English university, commenting on a talk by Ilya Prigogine, saying—"nothing useful comes out of this stuff, at best it is about some minor transitory phenomena"! Prigogine had chosen to plough an unusual, and what was then, a rather lonely path by looking at open systems that were not at equilibrium. Many chemists, physicists and later economists initially opposed these ideas, mystified by the idea that there could be anything of interest or importance arising from the study of open systems out of equilibrium. For many of them, the very possibility of 'solving' a problem scientifically relied on the assumption of simultaneous equations arising from an equilibrium condition. For them, whatever happened to the system on its way to equilibrium was really transitory, probably rapid and of no consequence. And yet, they themselves, biologically (with a metabolism driven perhaps by college dinners and vintage port), socially and culturally were the product of on-going multi-level evolutionary processes over millions of years. This fact alone should give us serious pause for thought in supposing that evolutionary complexity leads either to an 'optimal state' or to any kind of monotonic 'optimization' process, yet this was the traditional scientific view.

Of course, Prigogine was not alone in signalling the dawn of a new era in science, as people such as Joseph Schumpeter, Herbert Simon, Gregory Bateson, Hermann Haken, Rene Thom, Francisco Varela, Humberto Maturana and others all helped to illuminate different parts of the new landscape. In the natural sciences and engineering the study of non-linear dynamics and complexity has become dominant, and the pompous professors have long since disappeared. The social sciences have in general also embraced these ideas, as they seem to provide a more relevant basis for understanding than any simple mechanical or equilibrium paradigm. In economics the opposition has been the toughest to break down, as neo-classical economists retreated into their pseudo-physics theologies, and used their power and positions to exclude 'heretics'. However, there had already been economists such as Thorstein Veblen and Joseph Schumpeter who had long ago proposed dynamic and evolutionary views. There was also an Austrian school of economics which viewed economic systems as complex self-organizing systems, particularly by Nobel laureate Friedrich Hayek. They generally hold the view that the complexity of human behavior makes mathematical modelling of an evolving market extremely difficult and therefore advocate a 'laissez faire' approach to the economy. In 1981 Kenneth Boulding published a small book called "Evolutionary Economics" and in 1982, Richard Nelson and Sidney Winter published "The Evolutionary Theory of Economic Change", and with this emerged a school of 'evolutionary economists'—such as Stan Metcalfe, John Foster, Brian Arthur, Giovanni Dosi and Paolo Saviotti to name but a few. Now many others have helped create an evolutionary, disequilibrium view of economics and stressed the importance of innovation and change. Institutional and political economists, such as Geoff Hodgson have also embraced evolutionary views and books by authors such as Paul Ormerod have gained a wide audience for their non neo-classical views. Today hundreds of economists attend the annual 'Workshop on heterogeneous, interacting agents' conferences, which are about modelling and about complexity and evolution.

So, complexity itself demonstrates its own paradigm—starting off small, driven by a few, non-average individuals and eventually changing the world, and the understanding we have of it. Its development was not simply quantitative—with greater numbers of chemists and physicists getting involved, but was qualitative—with ecologists, biologists, sociologists, psychologists, urban and regional modelers and others, opening new dimensions and topics for the new ideas. By luck, I happened to be part of Prigogine's group during the crucial years of the 70s and 80s where part of this revolution was born, and the new, exciting ideas were conceived and started to emerge and spread at meetings such as the AAAS. I well remember the day in 1977 when Prigogine was awarded the Nobel Prize for Chemistry—with a special note that expressed the importance of the ideas for a wide range of subjects including social systems. However as with any new idea, particularly with a new scientific paradigm, the overall, long term impacts depend on the breadth of the phenomena affected and whether the effects are considered beneficial. The real impact on society depends on the long term advantages of the new ideas, and to what extent they can improve our technology, resilience, actions and interventions.

This remarkable collection of papers in this Annual shows how much progress we are making in so many different fields now. The range of papers is extraordinary: organizational adaptation, geographical patterning, social network development, power law distributions, fractals, patterns in rugby playing, learning about complexity from Ibsen, sustainable networks, logistics, supply chains, complexity and soccer are among the subjects explored in this annual. Instead of complexity being a small domain concerning transitory effects, we see that it is the scientific underpinning of evolutionary thinking, and this is in fact the basis of physical, ecological, cognitive, psychological, social, cultural and economic systems.

Complexity is perhaps the most important advance in scientific thinking since the enlightenment, as its content is particularly relevant for society. It changes the whole basis of our 'understanding' away from the mechanical to a creative, multi-level coevolutionary process whose future is uncertain. In some ways it exchanges certainty about simple systems or complicated ones at equilibrium, for the knowledge of the limits to our knowledge about real, living systems. For people either seeking or familiar with the 'old' certainties of scientific prediction, this may seem frustrating. However, with reflection it should be obvious that if precise prediction were actually possible in the real world, then life would not be worth living. There would be no place for creativity, for novelty, for surprise, for learning or for humanity, since choices and free will would not exist. This would correspond to the dead world of 'equilibrium'—with no qualitative changes or real development ever occurring. If anyone were awake in such a world, they would soon get completely bored with the fixed routines and predictable course of events.

The real world is unpredictable, uncertain, dangerous and as a result thrilling. The amazing thing is the multiple ever-expanding paths that are open to us, as complexification builds upon itself. If we think of a piece of music that the composer has written up to a certain point, then if we ask what happens next—there are thousands of possible, diverging pathways. A genius such as Mozart could pick, with great rapidity and seemingly with no doubts or vacillation, 'the' path into the future reflecting and affected by the preceding passages, yet still holding both some surprise and also some fulfilment of our expectations. So each step we take both reduces our options in some directions, but opens other opportunities and vistas. Our experience is one of 'becoming' as we live and struggle in a changing world rather than simply 'being'. There can be enormous consequences for seemingly small differences or events, and at other times the constraints of existence can close our choices right down.

So in a matter of a few decades we have moved from a situation in which the pillars of science were clearly modern physics and Darwin's explanation of biological evolution—but one in which nobody could say exactly how one followed from the other. With the study of open, non-equilibrium systems the ideas of complexity science emerged and uncovered a unifying evolutionary view of the universe. The price, however, was that we lost the simplicity of structurally stable systems with given variables and dimensions, and moved to a much more uncertain and problematic one with emergent behaviors and changing dimensions. This new view does not fit well with the traditional academic disciplines based on a reductionist view of the world, in which understanding and knowledge was obtained by looking deeper and deeper into ever lower levels of structure, with expertise and reputation being in ever narrower channels. We were inspired by 'analysis' and tried not to think about 'synthesis' and interdependence. Complexity has blown this away. We have moved from an ontology of isolated elements to one of 'connectedness'. We are shaped by and inhabit multi-level coevolving systems and must try to find out how to do better than we otherwise would—

how to make learning, adaptable organizations and systems, and how to find adequate values and goals with which to evaluate them.

In this new view, a journal such as *E:CO* is a valuable element. It links many areas of study and reflection and in so doing also provides a place where work on complexity can be published and read by other individuals who seek the new way. A glance at the list of editorial associates shows the remarkable diversity that lies behind it, but as one of the Editors-in-Chief I can use this opportunity to say that, apart from the avid readers and contributors, the continued development and growth of *E:CO* is really down to the managing editor Kurt Richardson. Without his indefatigable energy and activity in dealing with contributors, potential contributions, reviewers and printers then *E:CO* would not have made it into its second decade.

This Annual is a remarkable tribute to the spreading importance of complexity ideas and their ability to help both understanding and practical experience. In the end, these ideas will persist if they are essentially 'therapeutic' in that we feel better for knowing and using them. It is not always nice to be told that we face uncertainty, but it can lead to real improvements in resilience and in preparedness. Complexity recognizes the importance of the subjective understandings and views of people that are interdependent. There is not necessarily a simple, objective truth about what is going on, and what 'should' be done. Indeed, any models developed will in reality be part of the 'politics' of the system, but nevertheless understanding better our connectedness and the interconnectedness of things, can help us identify moves that lead towards more sustained improvements.

Academic and Practitioner Paper Section

In this volume our aim is to bring together academics, practitioners and also articles with news and items of general interest. This is because we recognize that in a complex world, where structures, organizations, people and their intentions evolve and change over time, then knowledge is no longer what we thought it was. We find that knowledge and ignorance trace out a delicate dance which involves the abandonment and the birth of some knowledge, possibly requiring new terms and concepts, and inhabiting some new dimensions. This fundamental creativity is not what traditional science could deal with, nor what it prepared us for, and so we must move into new territory in which continual learning and experiments occur, as actions lead on to emergent strategies and intentions. We are fundamentally in the domain of open systems interacting with their environments, and indeed participating in a coevolutionary process of mutually exploring transformations. The difference between the 'academic' and 'practitioner' papers is simply that we expect the former to be nested in some particular literature, and to advance the understanding, or limits to understanding of that particular domain. The insights may or may not be seen as immediately useful, as we are not faced here with needing to give a response to the 'so what?' question. Rather we are interested in the behavior and surprising possibilities that complex social and organizational systems are capable of and in the way that we can understand and perhaps deal with them. We take from science the idea that we are interested in knowing more, for its own sake, but do not expect knowledge to be cumulative as it was in the 'simple' sciences, nor that we shall see the utility of some new phenomenon. For the practitioners, we are interested in their reporting of experiments, in the hands-on implementation in reality of some ideas or suggestions, and in the noting of phenomena that do not seem readily explicable.

By bringing together academic and practitioner papers, and also by having papers reviewed by both types of referee, then we hope to avoid pointless 'academic' speculation on the one hand, and uninteresting but practical description on the other. The papers in the next two sections cover a wide range of issues, from interdisciplinarity, cognition, networking and organizational change, and give us an excellent start in moving toward a deeper, although always to be incomplete, understanding of complex systems.

Peter M. Allen

Academic paper section

1. Derivation of Factors Facilitating Organizational Emergence Based On Complex Adaptive Systems and Social Autopoiesis Theories

Ghada Alaa
Information Technology Institute, Ministry of Communications & Information Technology, Egypt

Modern turbulent business environments are characterized by rapid change that make businesses unpredictable, which brings emergence to the core of modern organizations. Deriving factors facilitating organizational emergence has been undertaken by drawing on complex adaptive systems (CAS) and social autopoiesis theories. Social autopoiesis was particularly chosen as it focuses on social elements, such as communication, morale, trust, etc. and their relation to social emergence, whereas CAS theory concentrates more on adaptive mechanisms that make a CAS produce emergent order, such as inter-relations, interactions, edge of chaos, feedback, etc. This led to the identification of various factors facilitating emergence and the development of a framework for utilizing these factors that were organized into two dimensions. First the factors are classified as either tangible or intangible. Second, the factors are classified as either dynamic, i.e., realize emergent properties, or they are concerned with the enabling infrastructure, i.e., enable the dynamic factors to become effective, or they are controlling factors, i.e., they attempt to balance excessive change with stability to prevent descent into chaos. The framework was applied to an Information Systems Development (ISD) project which showed that it is applicable to any type of business sector. This framework is argued to be a step forward to realize organizational emergence based on complexity principles derived from literature. The split between factors facilitating emergence and generic principles of CAS is not clear in the complexity literature and it is argued to be an important contribution of the paper.

Introduction

In turbulent business environments organizations need to react quickly and creatively to make the most of new opportunities and business models. These new imperatives of business practice require organizations to self-organise and become more flexible to handle change (Goldman *et al.*, 1995). Of key importance to organizations in responding successfully to change is the concept of emergence. Complexity science, it has been argued, is a way of addressing and improving such capabilities in organizations, as it is concerned with the role of chance, emergence and contingency in the face of frequent and continuous change (Montuori, 2003). McKelvey (1997), Stacey *et al.* (2000) and Mitleton-Kelly (2003) illustrate the growing interest in understanding organizations and new management practices in terms of theories of complexity that seek to provide new ways of thinking and reasoning in relation to emergent behavior.

In this paper factors facilitating organizational emergence have been identified by interpreting complex adaptive systems (CAS) and social autopoiesis theories with the aim of identifying mechanisms or strategies that raise the emergent properties of social business enterprises. Social autopoiesis was chosen as it focuses on social elements of emergence, such as communication, collaboration, morale, trust, etc., whereas CAS theory concentrates more on adaptive mechanisms that make a CAS produce emergent order, such as inter-relations, interconnectivity, edge of chaos, feedback, etc. A thorough literature review of management-related contributions in the field of complexity and social autopoiesis theories was undertaken to extract mechanisms or strategies that are argued will facilitate the emergence of new work arrangements in the face of frequent change. Based on this a framework has been derived that summarizes the so-called factors that facilitate organizational emergence. The framework classifies factors as tangible and intangible, and it differentiates between dynamics, enabling infrastructure and controls, amongst emergence factors. Preliminary validation of the framework was carried out through its empirical application in the context of an information systems development (ISD) project, a business to business e-commerce portal. Results show that the framework is generalizable and can be applied to any kind of business sector. It provides insight into the various elements of emergence realized or missing in the business enterprise. By enforcing factors facilitating emergence and avoiding factors prohibiting emergence, it is argued that organizational emergence will be leveraged leaving space to project teams to innovate and continuously evolve appropriate solutions in order to adapt to an ever-changing business environment.

Complex Adaptive Systems Theory and Organizational Emergence

Complexity science seeks to explain the process of self-organization, emergence of new properties and the spontaneous creation of new order. The behavior of complex adaptive systems (CAS) is typically unpredictable, but exhibits various forms of order and regulation. Complexity principles emphasise that emergence of properties and creation of new order are not explicable from a purely reductionist viewpoint, but the whole is greater than the sum of the parts (Kaufman, 1993). Therefore the focus of attention shifted from understanding the parts or entities of which the whole was composed to the interaction of subsystems (agents) to form a system (Lewin, 2000). Heylighen (2001) defines CAS as a system composed of interacting agents, which undergo constant change, both autonomously and in interaction with their environment.

Heterogeneous agents exhibit various agent behaviors that can be defined in terms of "simple rules" where they adapt and evolve through their interactions and by changing their rules through learning as experience accumulates (Holland, 1996).

Many contributions define CASs and describe their main characteristics and principles. While a uniform description or interpretation of CAS is still not provided, several key aspects that characterise CAS were suggested in the literature. Different authors surveyed various complexity literature (e.g., Axelrod & Cohen, 2001; Dooley, 1997; Holland, 1996; Kauffman, 1993; Lewin, 2000; McKelvey, 1997, 1999) and based on that characteristics/principles of CAS were outlined. For example, Benbya & McKelvey (2006) derived eight main characteristics of CAS to include large number of components, variation and diversity, self-organization, dynamism and liveliness, adaptation to their environment, interactions, non-linearity and selection. Mitleton-Kelly (2003) identified ten generic principles that characterise CAS, these include connectivity and interdependence, coevolution, far-from equilibrium/edge of chaos, space of possibilities, feedback, historicity and time, path-dependence, self-organization, emergence and creation of new order. Whereas, Webb & Lettice (2005) suggested six pertinent complexity science principles that include self-organization, edge of chaos, diversity, historicity and time, unpredictability and pattern recognition. The different characteristics/principles of CAS are summarized in Table 1, which illustrates how the literature overlap and also complement in order to form a broad overview of the major characteristics of CAS. From the comparison (Table 1) it is apparent that Benbeya & McKelvey (2006) were more concerned with the agent-based interpretation of CAS, where CAS consists of large number of agents that interact and re-structure their inter-relations and behavior as response to change (American view of CAS), whereas Mitleton-Kelly (2003) & Webb & Lettice (2005) interpret CAS as energy-driven actors residing at the edge of chaos/ the far from equilibrium transition phase (European view of CAS).

Organizations can be defined as complex adaptive social systems that evolve and produce emergent behavior in an unpredictable way (McKelvey, 1997; Mitleton-Kelly, 2003). As according to Baskerville *et al.* (1992) organizational emergence is characterized as the organization being in continual change, following no predefined pattern and never reaching a steady state. Although CAS theory is traditionally associated with natural sciences, Kim & Kaplan (2006) argue that CAS in management domains can be interpreted as actors, interacting with their environment and other actors within neighborhoods, and employ a variety of context-bound strategies in steering the behavior of actors towards maximizing the perceived interests or benefits of the organization. CAS in management contexts marks a fundamental paradigm shift from a mechanistic perception of an organization towards a self-organizing, autonomous understanding (Stacey *et al.*, 2000). The past view of organizations is seen as a hierarchy of components, whereas organizations are now seen as social networks of autono-

	Benbya & McKelvey (2006)	Mitleton-Kelly (2003)	Webb & Lettice (2005)
Large number of components	x		
Variation and diversity	x		x
Connectivity and interdependence		x	
Far-from equilibrium/edge of chaos		x	x
Unpredictability and non-linearity	x		x
Space of possibilities / adaptation to environment (context)	x	x	
Interactions	x		
Feedback		x	
Pattern recognition/learning			x
Historicity and path-dependence	x	x	x
Selection and Self-organization	x	x	x
Coevolution		x	

Table 1 *Characteristics/ Principles of Complex Adaptive Systems.*

mous components or actors that are organized by themselves and exhibit the previously discussed CAS characteristics (Table 1).

Social Autopoiesis and the Autopoietic Enterprise

Luhmann (1986) redefines social systems as entities of communications that recursively produce and reproduce, which he refers to as social autopoiesis. The word autopoiesis means "self-production", so a living, biological system is called autopoietic as it produces and is produced by itself (Mingers, 1994). Essential features of autopoietic systems are that they are autonomous, bounded and self-making. They consist of components that have specific properties where the overall behavior is generated through interactions between the components (Mingers, 1994). Human societies also possess features of autopoietic systems, as they are capable to survive by their own and change their entire structure and appearance (Beer, 1995). Considering human societies as autopoietic systems is useful in analyzing enterprises in terms of the form and structure they take and how they transform and reform over time (Whitaker, 1995).

The issue of how autopoiesis can be applied to social systems pointed out difficulties in the transfer of the natural sciences' concepts of autopoiesis to social domains. As autopoiesis implies the production and reproduction of the constituent components, so for social systems it is difficult to specify what do they exactly produce and reproduce (Mingers, 1994). There are two primary approaches in applying autopoiesis to social systems (Whitaker, 1995):

Enterprises as Autopoietic Systems

This view advocates applying concepts of autopoiesis theory, such as autonomy, self-making and self-reference, on the social system. It is criticized in terms of its straight forward applicability on social system's constituents that are human beings and that need according to autopoiesis theory to re-make themselves (Varela, 1981). In addition social systems do not exhibit topological boundary that is another attribute of autopoietic systems (Mingers, 1994). Therefore, Luhmann (1986) redefines social systems as entities of communications, or in other words the constituent elements of social systems are communications. He argues that autopoiesis of social systems is realized through production of communications, i.e., in social systems communications recursively produce and reproduce. Another interpretation characterizes social systems as a system of concepts, ideas, descriptions or messages that interact and self-produce (Varela, 1981).

Enterprises as a Medium for Autopoiesis

Maturana (1988) argues that social systems are not themselves autopoietic but constitute the medium for autopoiesis, i.e., social systems interact in a way that autopoiesis of actions and interactions contribute to the survival of the social system. Thus, this view advocates that the social system is constitutively emergent from interactivity among its participants. This generates networks of interactions and relations through structural coupling between participants and the social system. In that regard Küppers (1999) suggests the analysis of social interactions and their dynamics of reformation.

Regardless of the two different interpretations of social autopoiesis it is important to identify and understand the emergence of new behavior within social contexts and what social mechanisms may drive social interactions and their reformation. Thus, the application of autopoiesis in social contexts needs to take into consideration the dynamics of social interactions and inter-relationships, as well as the network of communications that develops between the actors. For this paper we will explore social autopoiesis literature and review their interpretations and arguments towards improving the emergent properties of a social enterprise. Based on the analysis carried out in the rest of the paper it is argued that communication, collaboration, interactions, trust and morale are the driving force of social autopoiesis. Furthermore, it is important to develop an understanding of the intersection between the literatures of social autopoiesis and CAS, as this area is still nascent and under research.

Why Derive Factors of Emergence?

Mitleton-Kelly (2003) suggests that the complexity approach to management is about fostering and creating enabling conditions, which will permit an organization to explore the space of possibilities and facilitate the creation of new organizational forms that will be sustainable in a constantly changing environment. This means it is important to identify elements/factors of emergence that improve the ability to respond to change and let work arrangements emerge according to what the situation implies, rather than being dictated in advance. Alaa and Fitzgerald (2004) support the same view and emphasise the need to instill into organizations habits and dynamics that improve their emergent properties instead of leaving them happen by chance. It is important to identify factors that will encourage and reinforce emergence and avoid those that suppress or inhibit such activities. Once inhibitors are removed and enablers put in place, new behavior emerges enabling the business enterprise to quickly adapt to change.

Generic principles/characteristics of CAS, i.e., conditions and attributes that characterise CAS were widely identified in the literature (see Table1). But Holland (1996) outlined more specific elements of CAS, i.e., mechanisms they adopt to realize such characteristics. These included aggregation of interactions, flows, internal building blocks and structure hierarchies, which ensure free propagation of change, adaptation and emergence of new order. Thus, Holland (1996) outlined constituent elements of CAS and mechanisms they adopt to realize emergent behavior rather than characteristics/properties of CAS. This split between mechanisms/factors that facilitate emergence and generic characteristics/principles of CAS is not clear in the complexity literature and it is argued to be an important contribution of this paper.

It is argued that in order to harness complexity concepts in business contexts an explanatory framework that identifies enabling factors of emergence is required to put generic characteristics/principles of CAS (Table 1) into action. Therefore this research aims at identifying factors of enabling conditions that could improve the emergence of new order in organizations. Various factors facilitating emergence were suggested in scattered literature that will be surveyed and collated into a framework in the following section. This will show how the different factors that would facilitate organizational emergence can be integrated and interlinked. The framework provides a roadmap that guides based on complexity principles managers and employees to harness emergent properties in their organizations.

Derivation of Factors Facilitating Organizational Emergence

Most of the work on complexity and the development of complexity theories have been undertaken in the context of the natural sciences and there has been relatively little work on developing or applying such theories in the social sciences. The literature generally suggests that there is a fundamental difficulty in attempting to do this because of the nature of human beings and human interactions, and specifically the notions of emotion, conflict and cultural elements. However more recently a number of attempts have been made to apply complexity principles in social and management disciplines. A thorough review of complexity and social autopoiesis literatures is undertaken in this section with special focus on management-related contributions to extract mechanisms or groupings of factors that are argued will facilitate emergence in social and management contexts. The analysis resulted in a classification into several groupings; dynamics (social construction factors/intangible dynamics and adaptive factors/tangible dynamics), enabling infrastructure (tangible and intangible), and control factors (tangible and intangible). Dynamics are factors that realize emergent properties, the enabling infrastructure include elements that enable the dynamics to become effective, whereas controlling factors attempt to ensure balance of dynamics to prevent descent into chaos.

Social Construction Factors/Intangible Dynamics

The review of the literature resulted in the identification of social drivers and stimulators that have been suggested as important in facilitating emergent social behavior. These are identified in bold and are of course frequently overlapping, as follows:

- The development of autopoietic society requires *communication*, meaning and consciousness that form an essential driver of emergent behavior (Luhmann, 1986);
- *Constant dialogue* is an essential social driver that creates a willingness to *communicate* with a growing level of *trust*, both of which enable coevolution of a social enterprise. For example trust facilitates better communication, which in turn enables the formation of activities and processes as response to problem situations (Lewin & Regine, 2003);
- Facilitation of *interaction* in the development of social organizations put *co-operative interaction* and relationships at the centre of organizational emergence, which can be achieved through *participation*, *collaboration* and *team working* (Stacey *et al.*, 2000);
- *Interactions* are important in the business ecosystem in order to respond to the problem situation and take appropriate actions. Local interactions are responsible for new order creation and emergence of global structures (Lewin, 2000);
- The quality of *interactions* between human agents is a function of the diversity, density, and intensity of those relations. These may be formal or informal, designed or un-designed, implicit or explicit (Mitleton-Kelly, 2003);
- Individual *motives* or intentions and individual emotions and *morale* act as driving forces for social autopoietic systems influenced by interests, social

context and forms of *cooperation* and *collective behavior* towards achieving a specific goal (Küppers, 1999).

Thus, the important social construction factors are *communication, collaboration, interaction, trust* and *morale*. These appear to be the important elements of complex social systems as they are responsible for social interactions and stimulation of creative thinking that will lead to human empowerment and leveraging self-organization. Next we look at more mechanistic or adaptive factors/tangible dynamics.

Adaptive Factors/Tangible Dynamics

According to Whitaker (1995) the dynamic of an evolving social entity is determined by inter-component relationships that outline its form and internal arrangements. Complex behavior arises from the inter-relationship and interconnectivity of elements within the social system and between it and its environment (Mitleton-Kelly, 2003). Thus, adaptive factors are required to improve the ability of the social system to re-arrange, re-form its structure and quickly respond to change, these include the following elements:

- Complexity theory implies that the internal dynamics and form of a system play a major role in determining its behavior. In a social context each individual belongs to many groups and different contexts and the contribution depends partially on the other individuals within that group and the way they *inter-relate* (Lewin, 2000; McKelvey, 1997);
- Propagation of influence through an ecosystem depends on the *degree of connectivity, interdependence* and *strength of coupling* (Heylighen, 2001);
- The *degree of interdependence* between entities may not always have beneficial effects and lead to *flexibility* because as one entity tries to evolve or adapt and improve its fitness other entities may respond by hindering this process and impose additional effort or cost (Küppers, 1999);
- In human systems, *connectivity* between individuals or groups is not a constant or uniform relationship, but varies over time (Mitleton-Kelly, 2003);
- Complexity thinking is about wholes and complex *inter-relationships*. This requires boundaries to be drawn around issues and the *break-down* of the problem under consideration into manageable wholes for better understanding of the problem situation (Holland, 1996);
- Difficulties created by the unpredictability of complex human processes and interdependencies are problematic, therefore *short-term orientation* and *simple solutions* (*simplicity*) are likely to result in better outcomes and more predictable developments whereas long-term solutions are likely to fail, as requirements and conditions can only be articulated and understood as events evolve (Stacey *et al.*, 2000);

- Conditions for experimentation and exploration of possibilities need to be provided as complexity theory suggests that several different chances and attractors will be possible and need to be explored. This also usually implies *small-scale orientation* in order to *quickly* try out various options and get *quick* feedback (*rapidity*) without requiring large scale resources and time (McMillan, 2004).

Thus, the more mechanistic, adaptive factors reflect the degree of interdependence, connectivity, structural coupling and quick re-formation of internal arrangements represented in *flexibility*, *short-term* and *small-scale* orientation, *simplicity* and *rapidity*. These elements help facilitate fast response and quick, internal adaptation and re-formation of system components.

Enabling Infrastructure

In order to facilitate emergent response an enabling infrastructure is required that allows emergence to occur, without such an infrastructure dynamic and emergent behavior will be limited or repressed. Aspects of an enabling infrastructure that facilitates emergence in social contexts include:

- *Hierarchy* and *structure* are pre-conditions that enable or inhibit the emergence of new behaviors and working ways (Heylighen, 2001);
- Action of organization members is shaped to a high degree by the existence of specific *organizational* form and *structures* (Levinthal, 1997);
- Conditions that facilitate the day-to-day management of an organization, for example, *management style*, good *leadership* and the provision of psychological space and freedom are necessary for learning and emergence to occur (Lewin & Regine, 2003);
- Enabling infrastructures also include *cultural conditions* that facilitate new work habits and intensions, such as manoeuvrability and risk-taking in ambiguous, uncertain situations (Kelly, 1995);
- Established research in complexity suggests that rigid procedures and bureaucratic *regulations* hamper emergent behavior. (Carlisle & McMillan, 2006);
- Analysis of the influence of *external factors* like power, money and control *regulations* like contracts and conventions is also important as these often act as constraints that limit social dynamics in complex situations (Stacey *et al.*, 2000).

Control Factors

Complexity theory in social contexts is designed to enable creativity, spontaneity and emergence but it also requires some kind of moderating or control mechanisms, which seeks to balance excessive change with stability, possibilities with constraints, innovation with tradition, etc. (Lewin, 2000; Montuori, 2003). In order to adapt to a changing environment, the system needs a variety of stable states that is large enough to react to all changes but not so large as to make its evolution uncontrollable or chaotic (Küppers, 1999).

- Stability is sustained by the property of the *edge of chaos* that limits the spread of destruction through the system. Change and stability are thus balanced and the edge of chaos is a critical point of the system, where a small change can either push the system into chaotic behavior or tip the system back into a more stable state (McKelvey, 1997);
- *Edge of chaos* is controlled by equilibrium models which attempt to bound a system to ensure that the system is always pushed back to stable conditions and will not result in absolute chaos (Whitaker, 1995). Also termed stability dynamics (Heylighen, 2001) which counteract excessive change before it endangers the essential organization;
- The mechanisms by which complex systems maintain control and achieve certain goals is by *feedback*, *learning* and *frequent small adjustments* to counteract any excessive tendencies to change (McMillan, 2004);
- Continuous *reflection*, *learning* and *circular causality* mutually reinforce social relationships and interactions (Küppers, 1999);
- Simple *high-level rules* are a way to achieve a balance between dictation and freedom enabling team members to interact with each other guided by these rules (Stacey *et al.*, 2000).

Based on the above analysis of literature we identify the first grouping of factors facilitating emergence, i.e., *dynamics* that include those factors that operationalize the emergent behavior. The factors of a complex social system are also classified into *intangibles* and *tangibles*. Intangibles represent the social factors that uniquely characterise social human systems, as opposed to natural systems, whereas tangibles represent the mechanistic/adaptive factors, those elements responsible for the internal connectivity of system components. Social construction factors include communication, collaboration, interaction and trust. These are important to raise human empowerment and self-organization. On the other hand adaptive/mechanistic factors reflect the degree of interdependence, connectivity and quick internal adaptation of the social system components. This is represented in flexibility strategies, short-term and small-scale orientation, simplicity and rapidity mechanisms that will facilitate fast response and quick re-arrangement of the entities within the enterprise.

The second grouping is the *enabling infrastructure* that enables or allows the social and adaptive elements to either be effective or inhibited. This includes organizational structure, hierarchies, management style, work culture, leadership, etc. These elements can also be *tangible*, such as structures, hierarchies and external factors or *intangible*, such as culture, management style and leadership. The third grouping is *control*, as in order to facilitate emergent behavior without complete chaos or anarchy, controls need to be in place and maintained, but they need not to be too restrictive. This will ensure a balance between excessive change and stability (edge of chaos). For example, reflection, circular causality and learning are *intangible* controls, whereas feedback, continuous adjustment, and generative rules are *tangible* controls. The different groups and elements of each category are illustrated in Figure 1 that collates the various factors extracted from literature. It is argued that this forms a useful framework for identifying and understanding factors that facilitate organizational emergence.

A Case Study: An e-Marketplace Portal

We now move on utilizing the "Organizational Emergence" framework (Figure 1) to interpret an information systems development (ISD) project in modern, turbulent environments. Ciranet is a B-B (business-business) e-marketplace relevant to the pharmaceutical sector in Egypt. It offers electronic trading tools for the pharmaceutical industry covering drugs, cosmetics, medical supplies, personnel and childcare products. A portal application had already been developed before the fieldwork was carried out. The development approach the company adopted covered planning and management, analysis and design, and development and maintenance activities towards the implementation of the IT-based marketplace application. The framework (Figure 1) will be applied to the different stages of the development process which identifies factors that facilitated and/or inhibited emergence in the ISD project. In so doing it is possible to identify if this project possessed elements of emergence or not.

Planning and Management

The management team adopted a *long-term* planning approach that caused some problems, as it froze the organization with a set of assumptions that were likely to change, due to rapid changes in the environment. Problems were also identified with the *hierarchical organizational structure* and the *command-control* style of *management* adopted. For example, one individual, the business consultant, was dominant in defining the requirements, often ignoring the ideas and contributions of other team members that might have made the project more successful. This *limited collaboration* and *reduced trust* among team members and project stakeholders. The researchers, in an Action Research mode tried to intervene and introduce a more agile approach into the organization but this was resisted by the management who were used to the *optimization-oriented work culture.*

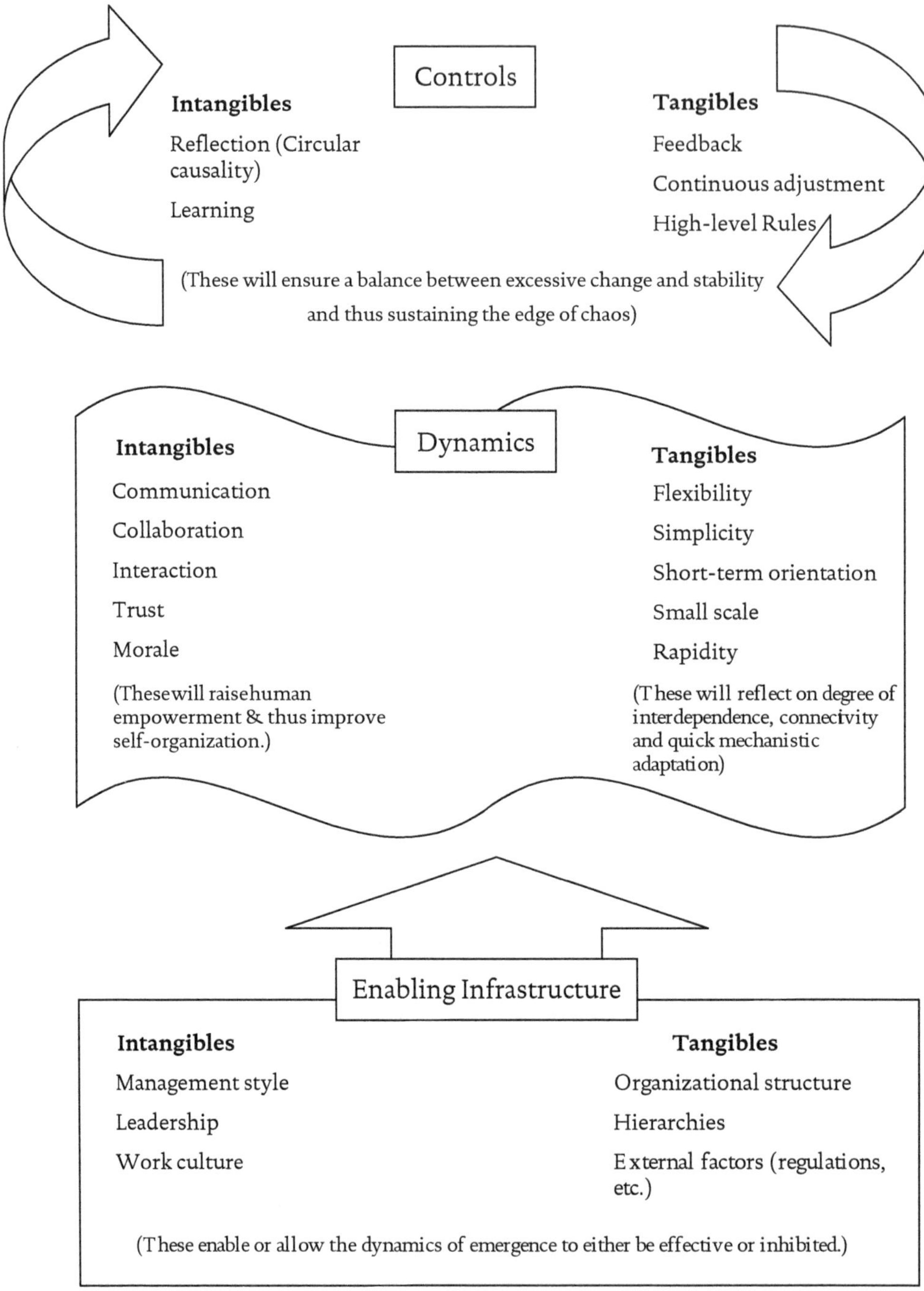

Figure 1 *Framework of Factors Facilitating Organizational Emergence (Organizational Emerge Framework)*

Analysis and Design

The original analysis and design approach focused on business process modelling of the pharmaceutical trading supply chain. For example, the order lifecycle was decomposed into order preparation, order placement, order fulfilment, etc., also the 'return of expired products' lifecycle was identified and was extensively analyzed. During the fieldwork, over a period of five months, problems were identified with the use of the *heavy-weight development* approach (*no high-level/agile rules*). It was systematic, comprehensive and plan-driven, which tended to limit and inhibit emergence in turbulent business environments, such as e-commerce. For example, the researchers identified problems and drawbacks resulting from the developers becoming overwhelmed with too much detail from the comprehensive analysis of the e-commerce workflows. This made the team produce *complex,* rather than simple *designs* and led the developers to ignore requests for changes. As the project progressed it became clear that the requirements needed to be re-thought and that some degree of innovation, as the business environment shifted, was required and yet this did not happen, at least partly because of the work already invested in the complex designs and a reluctance to abandon this.

Development and Maintenance

The IT application was outsourced and in contrast to Ciranet and its management, the development team of the outsourcing company adopted a more *incremental, iterative development* approach, where the systems modules were incrementally implemented and more flexibly improved with new features and add-ons as deemed appropriate. These represent a kind of *small-scale* and *feedback* mechanisms. The team also used *prototyping* techniques that facilitated *small scale and incremental development*, this also helped and enabled some *reflection and learning*. For example, the first prototype covered the basic functions, such as the catalogue and shopping cart, and then the prototype was refined through *feedback* from the clients. *Rapidity* was achieved through *short development cycles* and *parallel development* that quickened the development cycle. Furthermore, the portal was implemented with *component-based* technology that provided some degree of *flexibility* as it utilized *re-usable components as the technical architecture.* Then the modules were *tested* to fix errors and *customer reviews* were arranged to identify any flaws or problems with the business rules, which helped assure the *quality* of the software product (*quality controls*). But after launch of the application there were *no* attempts to continuously *adjust* and *re-develop* it to cover updates and deal with the changing necessities of the business context which prohibited the emergence of the IT application.

The different factors that facilitated or prohibited emergence at each stage of the development process in Ciranet are illustrated in Table 2 that distinguishes between the factor of emergence and the way/strategy to realize it

	Dynamics	Realize Dynamics in ISD	Enabling Infrastructure	Realize Enabling Infrastructure in ISD	Controls	Realize Controls in ISD
Planning & Management	*No Collaboration* *No Trust* *Long-term orientation*		Management style Work culture Organizational structure	*Command-control management* *Optimization-oriented* *Hierarchical*		
Analysis & Design	*No Simplicity*	*Complex designs*			*No Generative Rules*	*Heavy-weight development methods*
Development & Maintenance	Small scale Flexibility Rapidity	Incremental development, prototyping Component-based development Short development cycles & parallel development	**Technical architecture**	Flexible, re-usable architecture (e.g., component-based)	Reflection & Learning Feedback *No continuous adjustment* **Quality controls**	Prototyping Iterative development *No continuous re-development* Testing & Reviews

Table 2 *Analysis of Factors Facilitating Emergence and Ways to Realize Them in Ciranet/ ISD Project (Inhibitors in Italic)*

in ISD discipline. The factor represents the factor that will facilitate emergence, whereas the strategy is the way that will operationalize this factor in the particular domain under consideration, i.e., ISD for this case (see Table 2).

Thus, from the analysis of the case above it is found that social construction elements, such as collaboration, trust, etc. were missing in Ciranet. For example, the lack of collaboration between the business consultant and the development team, the command-control culture, etc., and this inhibited emergent response. But some tangible dynamics of emergence existed as introduced by the outsourcing company, such as rapidity, realized by short development cycles and parallel development. Small scale aspects were realized by incremental development, and flexibility was achieved by component-based development. However, other dynamics were found that inhibited emergence, e.g., the long-term orientation and complex designs. Also, the enabling infrastructure was found to be mainly inhibiting emergence. For example, the command-control management style, hierarchical organizational structure and optimization oriented work culture. Whereas, the technical architecture was found to contribute to the facilitation of emergence, as it was based on a component-based, reusable architecture. Additionally, the control mechanisms in the case were too rigid as Ciranet relied on heavy-weight development methods that produced rigid designs and thus limiting the developers' freedom, despite the more agile approach adopted by the outsourcing company. So, based on the analysis of Ciranet and the application of the "Organizational Emergence" framework (Figure 1), many factors were found that potentially inhibited emergence (according to the framework) and yet there were also a number of other factors that should have facilitated emergence. Thus, complete emergent properties were not fully-realized in the case but this is perhaps not surprising as most projects are probably somewhat mixed in relation to emergent factors. Figure 2 summarizes the analysis that was undertaken as illustrated in Table 2. Factors helping to encourage emergence are in plain text and factors inhibiting emergence are in italic, whereas between brackets are the strategies that were taken to operationalize the factors in the ISD domain.

Reflection

The proposed framework that outlines factors facilitating organizational emergence (Figure 1) was utilized to analyze an information systems development (ISD) project in turbulent e-commerce environments. It is found that the "Organizational Emergence" framework is useful as it represents an *interpretive analytical tool* that provides insight into the elements that facilitate or prohibit emergence in real case projects (Figure 2). The case is an example of an unclear situation where the tangible factors, such as flexibility and rapidity might have been expected to enable emergence more than they did, while ignoring intangible factors, such as communication, collaboration and trust have dramatically prohibited human stimulation and self-organization, thus limited

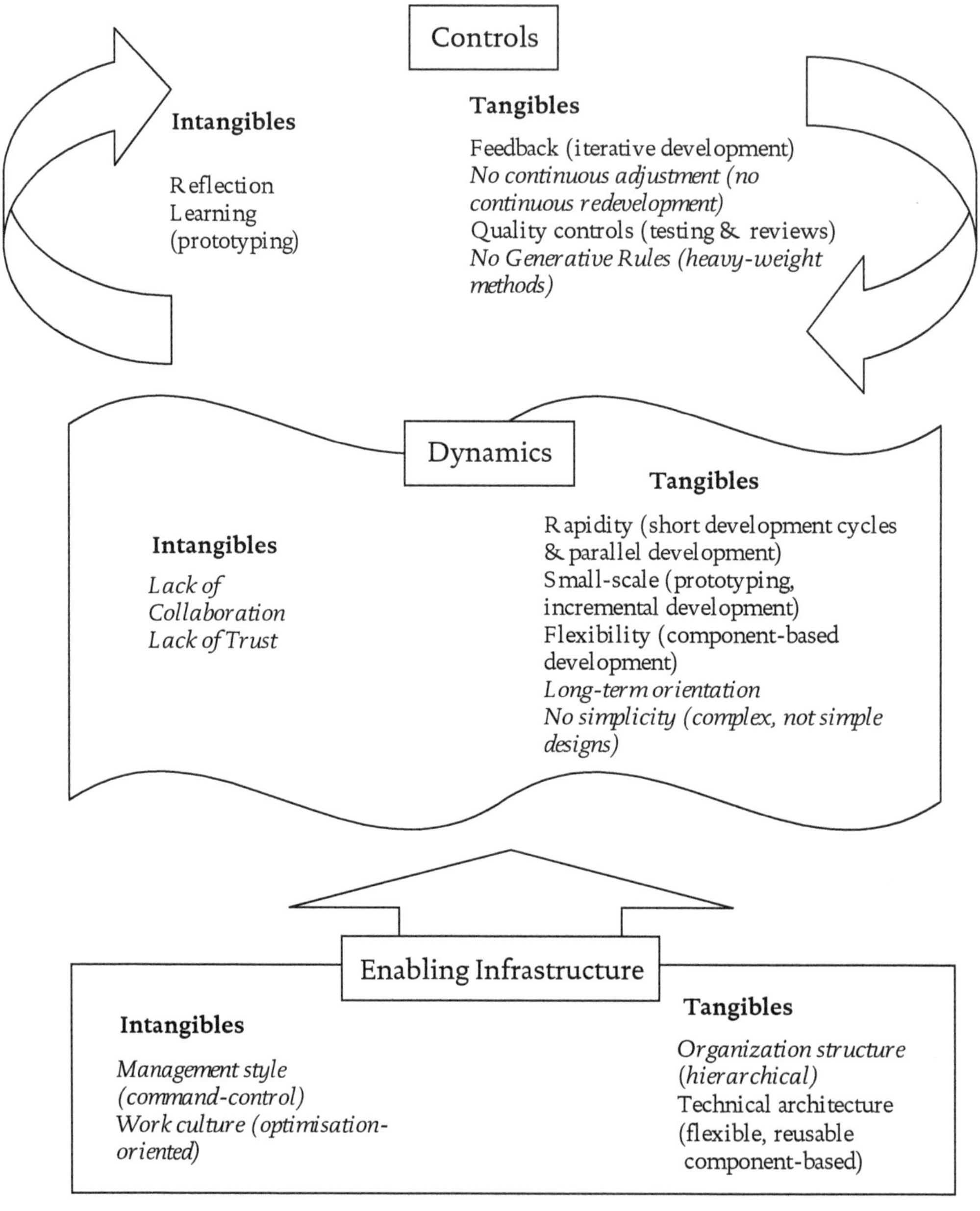

Figure 2 *Organizational Emergence Framework for Ciranet (Inhibiting Factors of Emergence in Italic)*

emergent response. It also shows the importance of the enabling infrastructure and the controls in determining whether emergence will occur and without anarchy or not.

As the "Organizational Emergence" framework was derived from contributions in social and management contexts that it is applicable to ISD projects provides support to the framework and its generalizability. Thus, the identified factors facilitating emergence; e.g., flexibility, small-scale approaches, short-term orientation, communication, collaboration, etc. are generic to any social enterprise/business sector. But the strategies that realize the factors are specific to the business domain under consideration. As shown in Table 2 different strategies were adopted in the ISD domain to operationalize the generic emergence factors identified in Figure 1. For example, prototyping technique operationalizes both a small-scale approach, as well as an experimentation tool to support reflection and learning, component-based development realizes flexibility, iterative development represents a feedback mechanism and leads to adjustment of the IT application, etc.

Thus, Figure 1 addresses generic emergence factors in social and management contexts, whereas Figure 2 provides the same framework but with special attention to ISD requirements and targeting strategies that will operationalize these factors in the ISD domain. But there are some factors specific to ISD, as derived from the above analysis of the case (Figure 2). For example, technical architecture and quality controls were not addressed directly in the management literature, these factors reflect enabling infrastructure and control elements. The technical architecture is a key element of Information Technology (IT) infrastructure whereas quality controls are used as a way to control the development process and ensure the quality of the software product. Furthermore, it is found that unlike in management discipline ISD focuses more on the internal project elements, e.g., technical architecture, development methodology and techniques, etc., not paying much attention to external factors that affect the emergence of the project.

In addition it is important to notice that the introduced framework is holistic where the identified factors do overlap and sometimes produce counteracting tension. For example, it is unlikely to collaborate, unless you communicate and interact, there is also no way to reflect and learn unless you have a kind of feedback mechanism, etc. Other factors produce counter-acting effect, e.g., extensive analysis of flexibility and producing re-usable designs might be time consuming and therefore will counter-act the rapidity factor, etc. The project team needs to manage and balance this uncertainty and tension as a way of surviving in a turbulent environment.

In light of this discussion it is concluded that the "Organizational Emergence" framework represents a foundation to realize social/organizational emergence by enforcing factors that facilitate emergence instead of leaving them happen by chance. The framework is a step forward on generic characteristics and principles of CAS (see Table 1) as it provides a way to realize these principles through factors and strategies in action. It is also important to note that the framework does not provide pre-planned business solutions or so called all-inclusive rules that model all possible business settings. Instead it raises the emergent properties of organizations so that project teams can freely innovate and evolve suitable business solutions to address the particular business context. The framework provides the following uses and benefits to business and industry:

Creates a way to articulate intangibles of organizational emergence

The framework articulates the importance of intangible factors of emergence in social contexts, such as communication, collaboration, morale, etc. These have been previously missed out or partially acknowledged intuitively.

Creates an emergence mirror of the enterprise

It provides a mirror for the company to identify what factors of emergence have been stipulated and what other factors have been undermined or completely missed out.

Creates a way to manage and align emergence factors

Emergence is a holistic phenomenon where the identified factors are intertwined and sometimes produce counteracting tension. As the framework portrays the different factors under consideration it is possible to determine the importance of each of them and how to align them and balance their effect.

Creates a way to balance emergence and avoid anarchy

It offers a new lens in balancing between enabling conditions of emergence and complete chaos. This is achieved through the control factors that need to be not too restrictive.

Conclusion

In this paper complex adaptive systems theory (CAS) and social autopoiesis have been interpreted with the aim to identify factors realizing emergent properties in organizations. Social construction elements, such as communication, collaboration, interaction, trust, etc. are argued to be critical drivers of human empowerment and thus self-organization, whereas mechanistic, adaptive dynamics like flexibility, short-term orientation, small scale approaches, simplicity and rapidity will ensure fast response and quick adaptation to the problem situation. However, emergence cannot be fully realized without the necessary enabling infrastructure that will allow the dynamics of emergence to become effective, e.g., management style, work culture, organizational structure etc. Also appropriate control mechanisms, such as feedback, reflection, learning, etc., need to be in place in order to balance absolute freedom and restrictive structure to ensure emergence to happen without descent into anarchy. The elements or factors in each category have been identified and related in a framework, to help understand and analyze the phenomenon of emergence in social organizations.

It is important to point to the difficulty we faced in identifying factors facilitating organizational emergence. As complexity principles emphasise that emergence is a spontaneous process, so how can a framework represent this phenomenon? In that regard it is important to note that the introduced framework guides project teams with factors or strategies that will enhance their free response to problem situations rather than dictating them with pre-planned business solutions. The framework can be seen as a significant improvement on generic complexity principles suggested in literature, such as diversity, large number of agents, interactions, edge of chaos, etc. that refer to emergence characteristics but without providing a clue on how to realize these concepts in action. A case, a business-business e-marketplace application, has been used to show how the framework can be used and how the factors can enhance the emergent properties of the project.

Especially, it is important to notice that the framework represents a holistic approach where the various identified factors are intertwined and some of them may produce counteracting effect. This uncertainty and tension are major characteristics of the emergence phenomenon and if managed properly it is argued will determine the success of organizations in modern turbulent environments. Future research will focus on further validation of the framework through other empirical applications. Especially of interest is to test if the framework does help companies better understand and manage the emergence phenomenon and put forth intentionally factors that raise the emergence of new work arrangements, or not? Generalizability and completeness of the framework are also important to test in future work.

References

Alaa G. and Fitzgerald G. (2004). "Evolving self-organizing activities: Addressing innovative and unpredictable environments," in B. Fitzgerald and E. Wynn (eds.), IT *Innovation for Adaptability and Competitiveness*, ISBN 9781402079993, pp. 3-19.

Axelrod, R. and Cohen, M.D. (2001). *Harnessing Complexity: Organizational Implications of a Scientific Frontier*, ISBN 9780465005505.

Baskerville R., Travis J., Truex D. (1992). "Systems without method: The impact of new technologies on information systems development projects," *The Impact of Computer Supported Technologies on Information Systems Development*, ISBN 9780444896490, pp. 241-269.

Beer S. (1995). *The Heart of Enterprise*, ISBN 9780471948377.

Benbya, H. and McKelvey, B. (2006). "Toward a complexity theory of information systems development," *Information Technology & People*, ISSN 0959-3845, 19(1) 12-34.

Carlisle, Y. and McMillan, E (2006). "Innovation in organizations from a complex adaptive systems perspective," *Emergence: Complexity and Organizations*, ISSN 1521-3250, 8(1): 2-9.

Dooley, K. (1997). "A complex adaptive systems model of organization change," *Nonlinear Dynamics, Psychology & Life Science*, ISSN 1090-0578, 1(1): 69-97.

Goldman, Steven. L., Roger N. Nagel and Kenneth Preiss (1995). *Agile Competitors and Virtual Organizations: Strategies for Enriching the Customer*, ISBN 9780442019037.

Heylighen F. (2001). "The science of self-organization and adaptivity," in: Knowledge Management, Organizational Intelligence and Learning, and Complexity, in: The Encyclopedia of Life Support Systems (EOLSS), Publishers Co. Ltd, pp. 253-280.

Holland, J.H. (1996). *Hidden Order: How Adaptation Build Complexity*, ISBN 9780201442304.

Kelly K. (1995). *Out of Control: The New Biology of Machines, Social Systems and the Economic World*, ISBN 9780201483406.

Kim, R. and Kaplan, S. (2006). "Interpreting socio-technical coevolution: Applying complex adaptive systems to IS engagement," *Information Technology & People*, ISSN 0959-3845, 19(1): 35-54.

Küppers. G., (1999). "Self-organization: The emergence of order from local interactions to global structures," Simulating Self-organizing Innovation Networks project, http://www.societyofcontrol.com/library/reform/paperno_selforganisation.pdf.

Levinthal, D. (1997). "Adaptation on rugged landscapes," *Management Science*, ISSN 0025-1909, 43(7): 934-950.

Lewin, R. (2000). *Complexity: Life at the Edge of Chaos*, ISBN 9780226476551.

Lewin, R. and Regine, B. (2003). "The core of adaptive organizations," in E. Mitleton-Kelly (ed.), *Complex Systems and Evolutionary Perspectives on Organizations: The Application of Complexity Theory to Organizations*, ISBN 9780080439570, pp. 167-183.

Luhmann, N. (1986). "The Autopoiesis of Social Systems," in F. Geyer and J. van der Zouwen (eds.), *Sociocybernetic Paradoxes: Complexity, Autopoiesis, and Observation of Social Systems*, ISBN 9780313314186, pp. 172-192.

McKelvey, B. (1997). "Quasi-natural organization science," *Organization Science*, ISSN 1047-7039, 8: 351-381.

McKelvey, B. (1999). "Self-organization, complexity catastrophe, and microstate models at the edge of chaos," in J. A. C. Baum and B. McKelvey (eds.), *Variations in Organization Science: In Honor of Donald T. Campbell*, ISBN 9780761911265, pp. 279-307.

McMillan E. (2004). *Complexity, Organizations and Change*, ISBN 9780415314473.

Mingers, J. (1994). *Self-Producing Systems: Implications and Applications of Autopoiesis*, ISBN 9780306447976.

Mitleton-Kelly E. (2003). "Ten principles of complexity and enabling infrastructures," in E. Mitleton-Kelly (ed.), *Complex Systems and Evolutionary Perspectives on Organizations: The Application of Complexity Theory to Organizations*, ISBN 9780080439570.

Mitleton-Kelly, E. (2004) "An integrated methodology to facilitate the emergence of new ways of organizing," *Conference Proceedings of the 3rd European Conference on Research Methodology for Business and Management Studies*, ISBN 9780954709617.

Montuori, A. (2003). "The complexity of improvisation and the improvisation of complexity: Social science, art and creativity," *Human Relations*, ISSN 0018-7267, 56(2): 237-255.

Stacey, R., Griffin, D. and Shaw, P. (2000). *Complexity and Management: Fad or Radical Challenge to Systems Thinking?* ISBN 9780415247610.

Webb, C. and Lettice, F. (2005). "Facilitating learning and sense-making with complexity science principles in organizations by means of a complexity starter kit," *Proceedings of the Complexity, Science and Society Conference*, 11th-14th September, Centre for Complexity Research, University of Liverpool, UK.

Whitaker R. (1995), "Overview of autopoietic theory," http://www710.univ-lyon1.fr/~jmathon/autopoesis/ATReview.html#Background.

Ghada Alaa holds a PhD in Information Systems, Brunel University, UK where her research focused on strategic requirements modelling for e-commerce systems and improving information systems development methodologies in fast-changing environments by building on complex adaptive systems theory. Her research interests are in the areas of e-commerce, web-based development, agile development methodologies and complex adaptive systems theory. Prior to the PhD studies, she worked as web developer and systems analyst in the Egyptian Cabinet's Information and Decision Support Centre. She also holds an MSc. in Business Information Technology and a BSc. in Biomedical Engineering. Her PhD work was presented as one of best 40 international information systems research projects for 2003, ICIS (International Conference of Information Systems) Doctoral Consortium, Seattle, USA, 2003.

2. Studying Urban Development in Tel Aviv, Based on Its Morphology Alone

Efrat Blumenfeld-Lieberthal[1], Iris Aravot[2] & Daniel Czamanski[2]
1 The Porter School of Environmental Studies and the Faculty of Arts, Tel-Aviv University, ISRAEL
2 Faculty of Architecture and Town Planning, Technion-Israel Institute of Technology, ISRAEL

It is possible to understanding the spatial behavior and structure of cities based on urban morphology alone. The units of analysis are urban clusters, defined as contiguous built-up urban areas instead of municipalities defined by politically determined boundaries. By means of historic data of the Tel-Aviv metropolis we present analyses of urban cluster statistics from 1935 to 2000. We focus on the largest cluster which includes the city of Tel-Aviv and several surrounding municipalities. The results suggest anomalies in the years 1964 and 1985. Based on the character of cities as self organizing systems, our study suggests that the analysis of urban cluster dynamics is an efficient tool to study urban phenomena.

Introduction

The spatial evolution of cities has been the subject of many studies for over a century. Previous work attempted to find explanations for the way cities evolve and arrange themselves in space, based on data collected at the municipal level (Jefferson, 1939; Zipf, 1941; Anas *et al.*, 1998). In reality, however, the spatial evolution of cities extends beyond municipal boundaries. The nature of the data used in past studies created distortions in the description of urban phenomena and in tests of hypotheses. Furthermore, the models used to study urban dynamics, whether they originated in fields of cultural studies, politics, sociology, or economics, utilized unrealistic mathematical functions (Christaller, 1933; Alonso, 1964). Thus, no existing model provides a complete and precise explanation of the physical similarity found among urban structures (Batty & Longley, 1994).

Recent research has illustrated that it is useful to think of cities as self-organizing (Batty & Longley, 1994; Portugali, 2000). The presented work relies on this conception of cities and the fact that urban morphology is amenable to mathematical description and analysis. It utilizes concepts and tools from theories of complexity, self-organization, chaos, and fractal geometry. Previous work has shown that urban areas can be characterized as fractals (Batty & Longley, 1987, 1994; Batty *et al.*, 1989; Frankhauser, 1990, 1994; Batty & Xie, 1996; Shen, 1997, 2002; White & Engelen, 1993; White *et al.*, 2001). Makse *et al.* (1995) proposed a percolation model that relates the morphology of built areas to the location decisions of its population. Benguigui *et al.* (2001a) developed a cellular automaton model that simulated the development of the footprint of the built areas in one city within the Tel-Aviv metropolis.

Patterns of urban morphology exhibit similarity to phenomena known in physics as cluster formation. This led to the hypothesis that there may be a general explanation of the arrangement of these structures in space (Schweitzer, 1997). In a previous work we presented an approach to understanding the spatial distribution and structure of cities based on morphology alone (Benguigui *et al.*, 2006). The novelty in this approach is the replacement of municipalities as units of examination by urban clusters, defined as contiguous built areas. This enabled us to ignore changes of the cities' boundaries that are related to political decisions rather than to real changes in the city spatial distribution.

In this study we focus on the largest cluster in the Tel-Aviv metropolis and show how changes in its morphology might be used to identify socio-economic changes. This is as the largest cluster is compounded of the core city of the metropolis (i.e., Tel Aviv) and its surrounding municipalities. This agglomeration of cities works as the financial, political, and cultural center of the metropolis and in many terms of the entire country.

The next section "Methodology" provides a methodological discussion at the backdrop of the research. "The Largest Cluster in the Metropolitan Context" focuses on the analysis of the morphology of the largest cluster. The "Discussion" section that follows relates the mathematical results with socio-economic phenomena in the history of the Tel-Aviv metropolis. Based on this analysis several conclusions and recommendations are presented in the final section.

Methodology

We identified urban clusters in the Tel-Aviv metropolis based on a series of maps of the Tel-Aviv region representing the years 1935 and 2000. The boundaries of the area analyzed remained fixed throughout the study period. Thus, the boundaries were determined independently of the boundary of the metropolitan area. The latter changed with time. The area studied is bounded by the city of Natanya in the north and by the city of Yavne in the south. These outer borders of the Tel-Aviv metropolis are widely accepted and include the metropolis in all the periods included in the study.

We digitized the historic maps and copied the built areas into new maps, suppressing non-built areas such as roads and railroads. The new maps are presented in Figures 1a to 1h. We used a black and white reduction of the maps. A black pixel represents an occupied site (a built area) while a white pixel an empty one. We defined a cluster by contiguity properties of the black pixels. The contiguity was determined as the sequence created by the touch of each pixel with its eight closest neighbors on a square lattice. Small clusters that had no

	(dN/dt)	(dA/dt)	α	α, Largest Cluster	Rank Size Slope
1935-1964	> 0	increases	0.75	1	-1.00 ±0.04
1964	0	Max	0.75	1	-1.00 ±0.06
1964-1985	< 0	Varies	0.75	1	-1.15 ±0.08
1985	0	Max	0.80	?	-1.50 ±0.13
1985-2000	0	decreases	0.70	?	-1.54 ±0.15

Table 1 *Summary of the Analysis of Clusters in the Tel-Aviv Metropolis*

urban characteristics (e.g., military facilities) were defined as noise. Such clusters were deleted. Each of the deleted clusters was smaller than 45,000 m^2.

In order to conduct a precise analysis of the clusters, we calculated the values of the area and the perimeter of each using the MATLAB computer software. The basic unit of the measurement was 1 pixel which represents a length of 30 meters.

Several different analyses of the data for each of the clusters were carried out at two levels: the first for the entire Tel-Aviv metropolis and the second examined the largest cluster alone. In Benguigui *et al.* (2006) we focused on the morphology of the entire metropolis and Table 1 presents the summary of these results.

One can distinguish three stages in the city's evolution, based on the changes presented in Table 1. The first stage occurred between the 1930s and the 1960s. The second was between the 1960s and the 1980s. And the third took place in the years between the 1980s and the 2000s. The changes observed in the 1980s were found in all the examined parameters. Thus, we assumed the processes that led to these changes were more varied and powerful. As a result when we tried to associate the socio-economic trends and processes we have decided to focus on these years.

The next section presents the analysis of the data for the largest cluster comprised of the city of Tel-Aviv and its adjacent towns.

The Largest Cluster in the Metropolitan Context

Cluster size

The area of the cluster was used as its main indicative characteristic. Figure 2a presents the growth of the area A(t) of the largest cluster and of all the clusters in the metropolitan area over time. It can be seen that the area of largest cluster behaved similarly to the total area of the other clusters; it experienced two significant increases—the first in the 1960s and the second in the 1980s. These increases are visible by examining the derivative of the area growth (Figure 3).

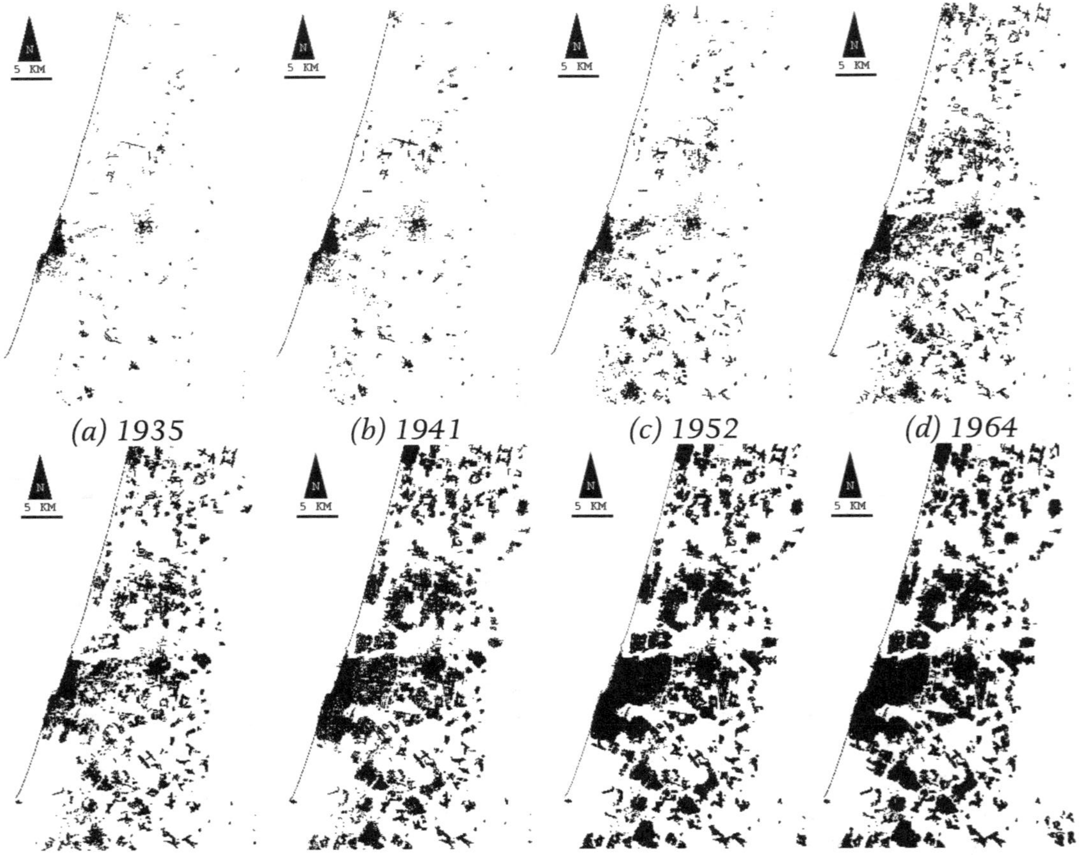

Figure 1 *Maps of the Tel-Aviv Metropolis (Built Area).*

A comparison between the growth of area of the largest cluster and the growth of the population of the Tel-Aviv and its surrounding municipalities did not explain the anomalies in the curve A(t) (see Figure 3a). The total population of the municipalities that create the largest cluster experienced linear growth from the 1950s onward. The population of the individual cities, however, Bat Yam, Qiryat Ono, Holon, and Petah Tiqwa, that are adjacent to the city of Tel-Aviv, developed rapidly between the 1960s and the 1980s (see Figure 4). These cities became part of the largest cluster in 1985. Thus the intensive development they experienced previous to the 1980s influenced the area of the largest cluster from that time onward.

From 1935 to 1974 the largest cluster was surrounded by many small clusters (see Figure 5). In 1985 the largest cluster merged with 11 of its neighboring clusters and thus its area increased significantly. These clusters were compounded of the municipalities, adjacent to the core city—Tel Aviv. This agglomeration occurred due to the continuous growth of each cluster, i.e., each cluster developed towards its neighboring clusters until they merged together.

Consequently, the largest cluster, which was compounded of the city of Tel Aviv only until the 1970s, included additional municipalities from the 1980s onward.

Area and Perimeter

As seen in the Figure 1, the urban clusters in the Tel-Aviv metropolis have irregular shapes. We used the box-counting method on a random sample of 20 clusters (including the largest cluster), and found that all of the examined clusters displayed self-similarity. The clusters exhibited the characteristic of a fractal, both in terms of their areas and perimeters. The average fractal dimension of the clusters area was found to be 1.8 and the average fractal dimension of the clusters perimeter was 1.3.

We used the fractal characterization of the clusters to calculate the value of α which indicates the shape, or the morphology, of the clusters:

(1) Perimeter ~ Area$^{\alpha}$,

(where $\alpha = DL/DA$).

Where DL represents the fractal dimension of the perimeter and DA represents the fractal dimension of the area of the clusters. The values of α are bounded between 0.5 and 1 (as it is reasonable to assume that DA>DL). Values that are close to 0.5 indicate a quasi-circular cluster while values close to 1 indicate a highly irregular shape.

The values of α for the largest cluster were constant from 1935 to 1985 and equal to approximately 1 (see Figure 6). This value is higher than the average value of α for the rest of the clusters that was 0.75. The value 1 represents

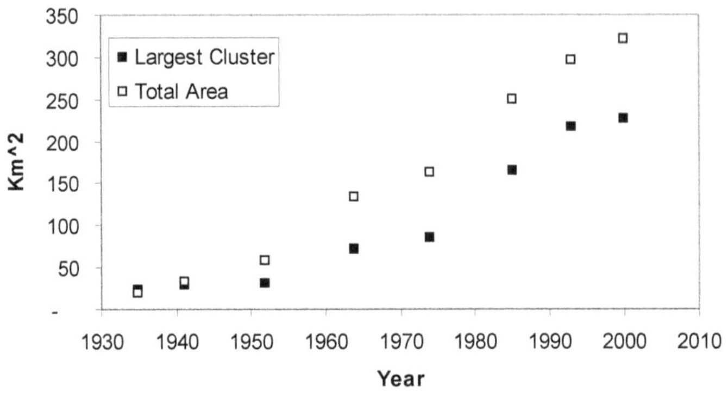

Figure 2a *The Growth of the Total Area and of the Largest Cluster Area between the Years 1935-2000 (the values are normalized).*

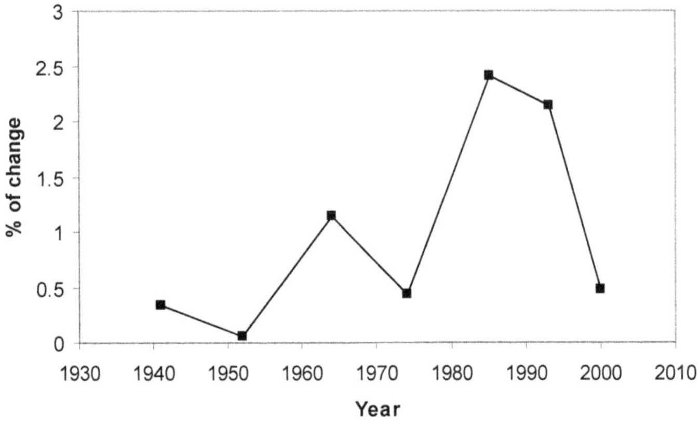

Figure 2b *The area growth rate of change of the largest cluster.*

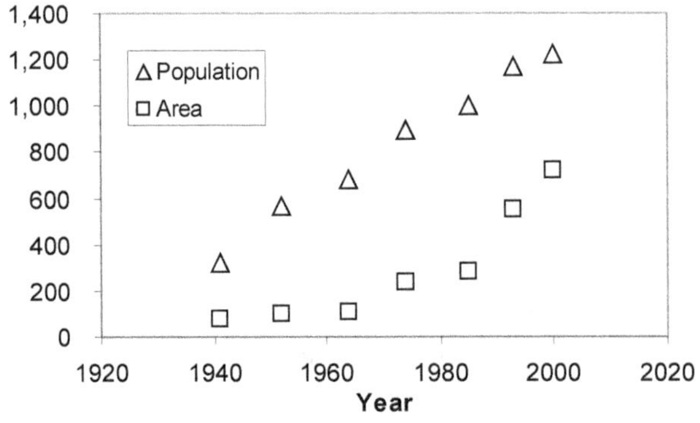

Figure 3 *The Growth of Area and population of the largest cluster and its municipalities between the Years 1935-2000 (the values of the area are normalizes).*

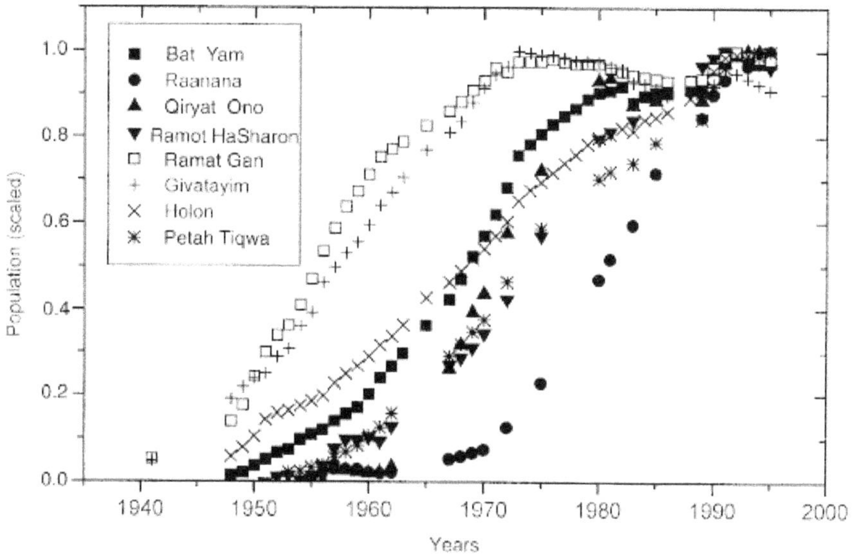

Figure 4 *The population growth in cities adjacent to Tel-Aviv (Benguigui et al., 2001).*

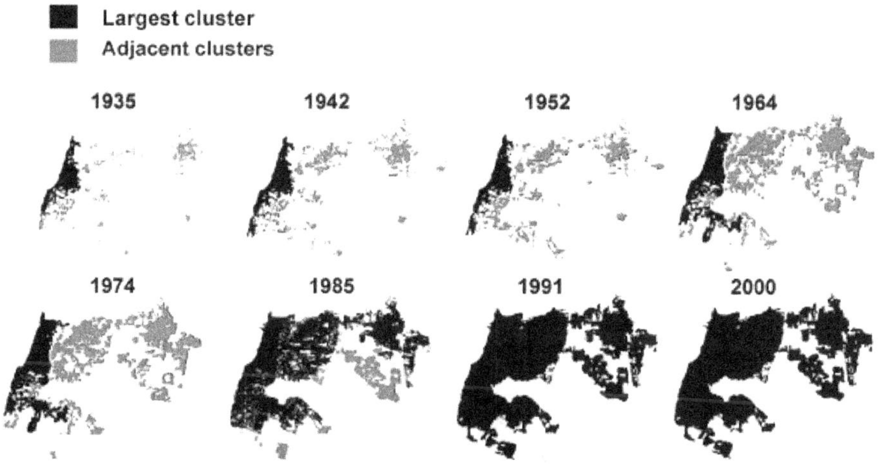

Figure 5 *The development of the largest cluster and its adjacent clusters from 1935-2000.*

an irregular morphology and indeed, the largest cluster had a very complicated shape during these years (see Figure 5). From 1985 onwards the values of α did not align with the rest of the data. There was not enough data for these years to perform a new regression, thus we cannot determine conclusively the values of α from 1985 onwards. We can assume, however, with high probability that the relationship between the area and perimeter of the largest cluster is always valid but with α smaller than 1.

The Rank Size Distribution

From Auerbach (1913) and onwards, it has been widely accepted that the Pareto distribution represents the city size distribution accurately (i.e., the relationship between the ranks of cities and their sizes). The Pareto distribution is also known as the "rank size rule". "Zipf's law" (Zipf, 1941) is the case in which the Pareto exponent is unity which means the largest city is twice the size of the second largest city, and *N* times the size of the city with rank *N*.

Schweitzer (1997) was the first to examine the Pareto relationship between urban areas based on their morphology instead of their population. In Benguigui *et al.* (2006) we followed Schweitzer in calculating the rank size distribution of urban clusters in the Tel-Aviv metropolitan area in different years (Figure 7).

Figure 7 shows that the largest cluster did not align with the rest of the clusters in the entire period studied (the circled dots represent the largest cluster). This resembles the phenomenon of a primate city, i.e., the largest cluster was and is more than twice as big as the next cluster in size. In other words, the largest cluster was the most influential cluster throughout the entire period studied (see further discussion in the next section).

Discussion

The cluster approach to the study of urban dynamics revealed a number of major changes in the history of the morphology of the largest cluster in the Tel-Aviv metropolis.

The major changes in the morphology occurred twice during the city's history—in the 1960s and in the 1980s. The changes in these years were revealed both in the morphology of the largest cluster and in the morphology of the entire metropolis. As a result, there were three distinct stages in the city's history. The first stage occurred between the 1930s and the 1960s. The second stage was between the 1960s and the 1980s. And the third stage took place in the years between the 1980s and the 2000s. The changes observed in the 1980s were more explicit than those of the 1960s and are revealed in all of the parameters examined. We therefore assumed that the processes that led to these changes were more varied and powerful. By examining several socio-economic phenomena that affected the development of the largest cluster, we suggest some explanations for these findings.

The significant increase in the area of the largest cluster that occurred in the 1980s can be explained by two main factors. The first is the inclusion of residential areas, as well as business and industrial area, in the largest cluster. The latter were affected by the construction of the main highway in the metropolis, the Ayalon Highway (see Figure 8, the Ayalon highway is represented by the gray line). The Ayalon Highway, located on the banks of the Ayalon stream, improved

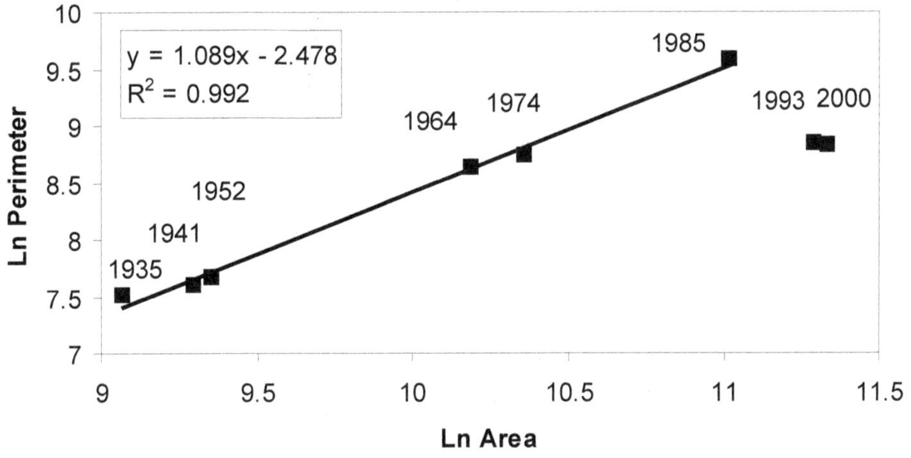

Figure 6 *The Logarithmic Relation between the Area and Perimeter of the Largest Clusters with Time.*

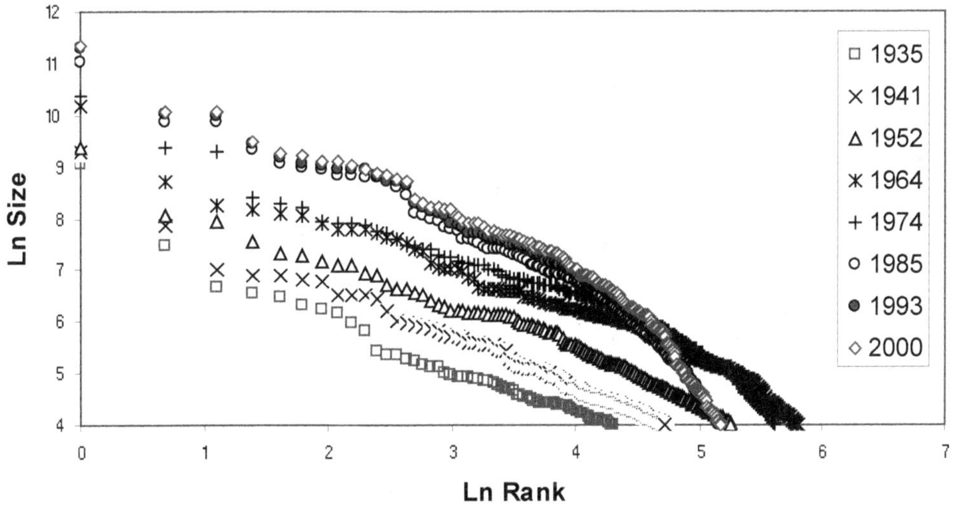

Figure 7 *Rank Size Distribution of Clusters in the Tel-Aviv Metropolitan Area in 1935-2000.*

the accessibility at regional and national scales, causing many post-industrial firms to move to this area and creating an increased demand for office space on the banks of the Ayalon Highway. This demand was further compounded with the adoption of an urban plan named the Mazor plan by the municipality of Tel-Aviv. One of the guidelines of the Mazor plan was the relocation of non-residential usages that were in residential buildings. Thus, many firms had to find a new base for their offices. Although the major impact of the Mazor plan was in the late 1980s, its influence on the city began early in that decade.

As a result of these changes, the CBD (Central Business District) of Tel-Aviv grew to span the entire area from the traditional CBD to the opposite side of the Ayalon Highway (see Figure 9). The massive building on the banks

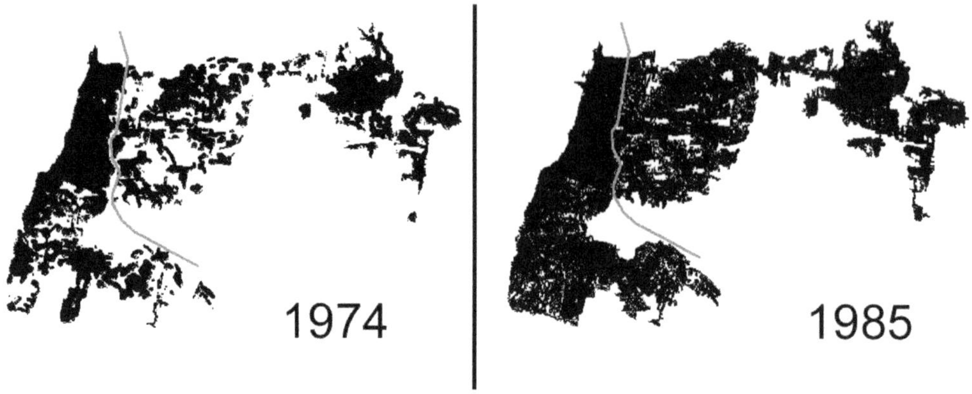

Figure 8 *The largest cluster in 1974 and in 1985.*

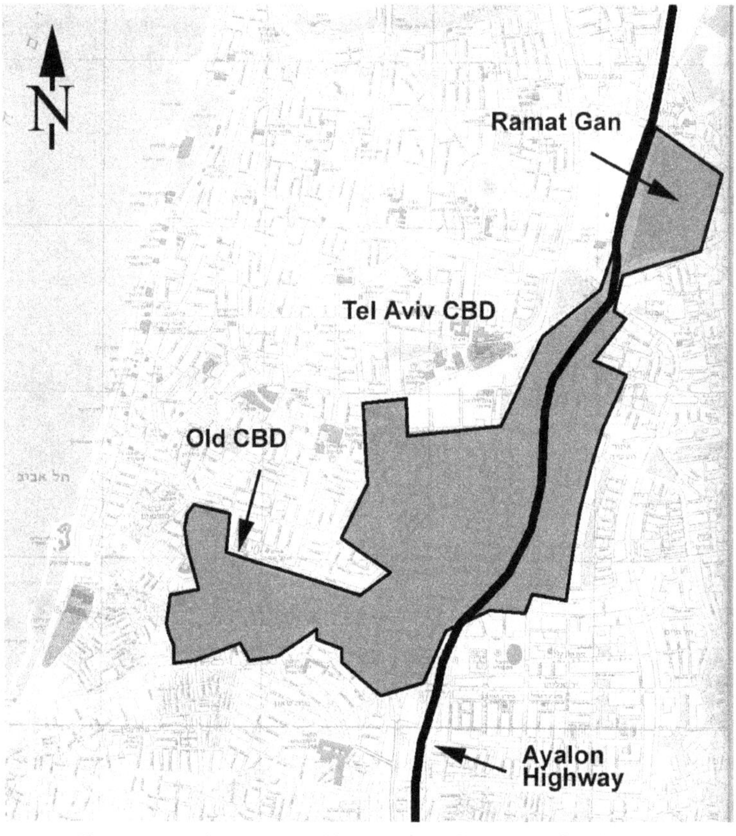

Figure 9 *The central business district of Tel-Aviv (based on Tuchler-Aharoni, 1997).*

of the Ayalon changed the morphology of the built environment as it united both sides of the Ayalon into a unique cluster. Thus, the largest cluster, which in 1975 was composed of the cities Tel-Aviv, Bat Yam, Holon, and part of Rishon LeZion agglomerated with additional municipalities that formed clusters on the eastern bank of the Ayalon (Figure 8). The result was a significantly large cluster containing more than 25% of the total built area of the metropolis. In 1974, in comparison, the largest cluster occupied only 12.2% of the total built area of the metropolis.

The second factor that influenced the morphology of the largest cluster was the increase in the population of cities adjacent to Tel-Aviv (Figure 4). The development of each of the individual cities within the metropolis took place in four stages (Benguigui *et al.*, 2001). This form of development is known as "leap frogging". In stage 1 there is a constant increase in the population in each cluster. In the second the rate of population increase reaches its maximum. Stage 3 defines population saturation and in stage 4 the population of the city oscillates.

Most of the cities in the Tel-Aviv metropolis, (presented in Figure 4,) reached the third stage by the 1980s, i.e., they had reached population saturation by that time. The cities, Bat Yam, Qiryat Ono, Holon, and Petah Tiqwa, went through the second stage (significant increase in population) between the 1960s and the 1980s. Thus, prior to the mid 1980s, all these cities faced an intensive development which influenced their built area.

There are economic underpinnings to this rapid and significant development. In the years following the Six-Day War (1967), Israel's real per capita GDP increased substantially. This can be seen in the steep slope after 1967 in Figure 10. In addition, a political upheaval in 1977 led to the formation of the first right-winged liberal government in Israel's history. The new government established a new economic policy which was based on the principles of a capitalistic market. The resulting economic prosperity, along with the liberalization of the market, led to an improvement in the standard of living and in the standard of residences (Carmon & Czamanski, 1992). As a result, the demand for one-floor housing in the central district and in the Tel-Aviv district increased significantly. The demand surplus created an atmosphere of lack of land and increased the political pressures to change land usage from agricultural to residential. These pressures were fruitful and in the late 1970s Israel Land Administrator increased the number of tenders in one-floor housing neighborhoods by 130%. The city's population migrated to new and more spacious settlements beyond the geographical core of the metropolis (Benguigui *et al.*, 2000). This extension of urban activities to suburbs in the rural area as well as to dormitory towns, adjacent to the city of Tel-Aviv, led to the creation of new urban centers and to a significant change in the Tel-Aviv morphology and in the largest cluster area.

The development of the cities near Tel-Aviv can be divided into two processes. At first there was enough space to allow leap frogging, but when each of the cities came close to saturation, the only way the cities could continue to grow was by filling vacant locations within their boundaries. Clusters within each of these municipalities agglomerated with their neighboring clusters and created new continuous clusters. As the boundaries of these municipalities often overlapped, clusters located at the boundaries of the municipalities agglomerated with clusters of adjacent municipalities. Thus, in 1985 most of the

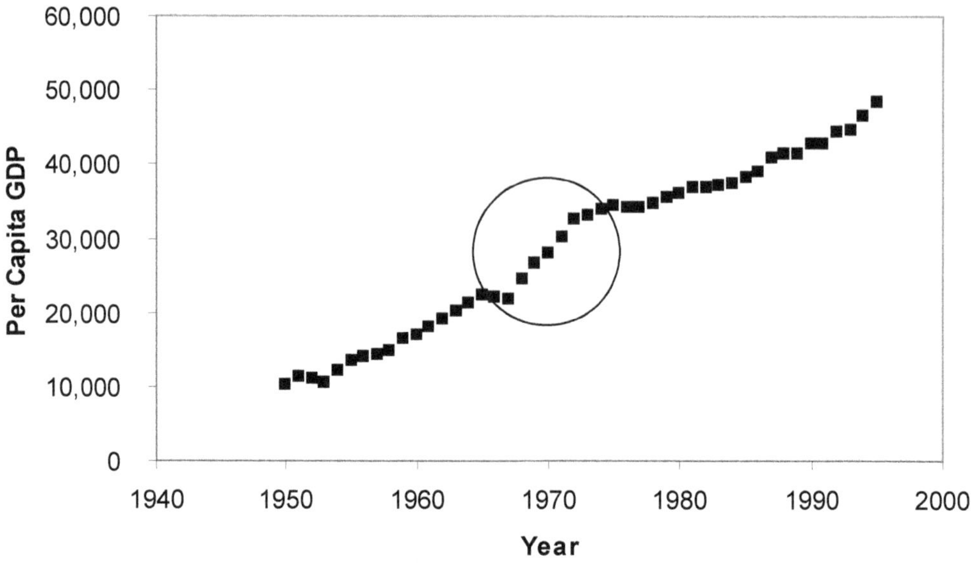

Figure 10 *Israel's real per capita GDP.*

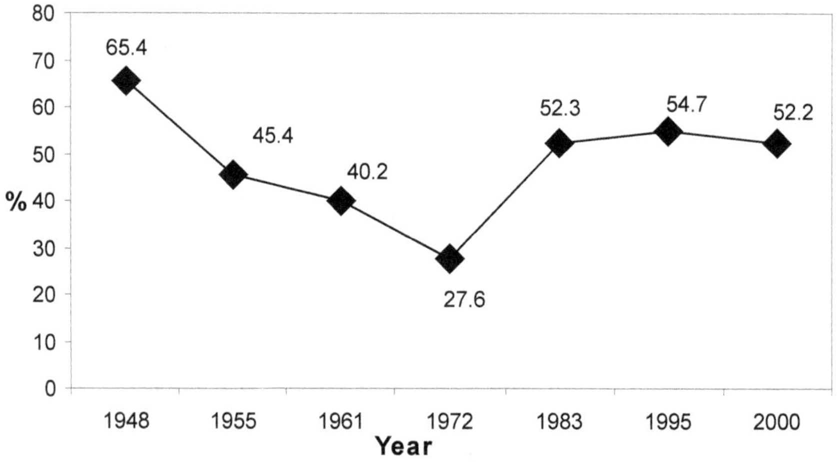

Figure 11 *The population of the largest cluster as a percentage of the population of the metropolis.*

clusters, adjacent to the largest cluster, agglomerated with it and the intensive development of these cities affected the area of the largest cluster in that year. From 1990s onward the population of these cities hardly changed as did the area of the largest cluster.

The leap frogging process influenced the development of the municipalities adjacent to Tel-Aviv until 1974. Each of these municipalities developed at a different rate, but by the 1980s most of them had reached saturation. Thus, from 1985 onwards the continuous growth process dominated the development of the clusters within these municipalities. In other words, from 1985 onwards the creation of new clusters within the area of the largest cluster stopped. From 1993 onwards the agglomeration of the largest cluster with its neighboring clusters ceased as well. These findings are not with agreement with Schweitzer's (1997) conclusion about the development of the largest cluster, which is based on his analysis of the development of urban clusters in Berlin. He claimed that the largest cluster did not grow proportionally to its size, but only through agglomeration with its neighboring clusters. In the case of the Tel-Aviv metropolis the largest cluster stopped growing as a result of agglomeration with its neighboring clusters, and from 1993 onwards it grew only as a result of self-filling, i.e., filling the empty location within the cluster's area.

These changes in the dominate growth process can also explain the change in the values of α for the largest cluster. There are three processes that yield a decrease in the value of α: First, the fractal dimension of the perimeter (DL) decreases. Second, the fractal dimension of the area (DA) increases. And third, both the first and the second processes occur at the same time. In the Tel-Aviv metropolis the largest cluster became more compact and less irregular with time mainly due to a self-filling process. This process affected the fractal dimension of both the area and the perimeter of the largest cluster. The self-filling of locations within the cluster's area caused the increase in its DA, while the self-filling of locations on the cluster's perimeter caused the decrease of its DL. Thus, there was a decrease in the value of α (which equaled 1 until 1985).

The rank size distribution of the clusters in the Tel-Aviv metropolis shows a resemblance of the largest cluster to a primate city. This can be explained both by socio-economic activities and by the size of the population. As opposed to the population of its neighboring clusters the population of the city of Tel-Aviv decreased constantly from the 1960s until the mid 1980s. Until the mid 1970s, the largest cluster was composed of the built area within the city of Tel-Aviv alone, but nevertheless, it was the dominant cluster and resembled a primate city. This phenomenon can be explained by socio economic activities within the city of Tel-Aviv. Tel-Aviv was declared an independent municipality in 1921. From its early days it was both the core of and the dominant city in the metropolis in terms of business and culture activities. During the 1930s and the 1940s Tel-Aviv matured to be a central city in terms of employment and services. After Israel's declaration of independence in 1948 the economic growth which characterized the country at the time had a positive influence on the economic activitiy in Tel-Aviv and the city became a national center of banking and finance. In 1950 Jaffa was formally appended to Tel Aviv and most of the immigrants who came to Tel Aviv after 1948 were dwelled in the deserted houses of Jaffa and the Arab villages near Tel Aviv. During the 1960s and 1970s, Tel-Aviv became an employment base for a growing number of commuters from the entire metropolis. In addition, the city developed as the major cultural center of Israel, and expanded its cultural activities and facilities. As mentioned earlier, clusters were defined as a contiguity of built-up area, without distinguishing between different land usages. Thus, the fact that the population of the city of Tel Aviv decreased did not affect the growth of the largest cluster. It was due to the development of the city as a business and cultural center rather than the development of new neighborhoods which influenced its development in the early years. Primate cities dominate the country in terms of influence, and are the national focal-point. Thus, it is not surprising that the largest cluster, which until the 1980s was composed mainly of the city of Tel-Aviv, had the characteristics of a primate city.

From the 1980s onward the largest cluster agglomerated with other clusters and contained the cities of Tel-Aviv-Jaffa, Ramat-Gan, Givataim, Bney-Brak, Holon, Rishon LeZion, Petah Tiqwa, Qiryat Ono, and Bat-Yam. As a result, the population of the largest cluster as a percentage of the population of the metropolis increased from 27.6% in 1972 to 52.3% in 1983 (see Figure 11). The characteristic of the largest cluster as a primate city strengthened with the increase in the centrality of Tel-Aviv and its surrounding cities in terms of business facilities and cultural activities.

To summarize, the largest cluster characteristic as a primate city was influenced by both the fact that Tel-Aviv has always been a national business and cultural center, and the fact that from 1985 onwards, the population of the largest cluster included more than 50% of the metropolis population.

An additional reason for the growth of the largest cluster lies in the new settlement policy established by the new government, elected in 1977. This policy legitimized Jewish settlements in the west areas of Judah and Samaria (The West Bank). Most of these settlements were essentially suburbs and relied on nearby urban centers for services and occupation. The eastern cities of the Tel-Aviv metropolis therefore evolved rapidly from dormitory towns to developed urban centers. This process caused the agglomeration of small clusters within the area of these cities into large clusters, an example for such a city is Petah Tiqwa (see Figure 12). The agglomeration of small clusters within the municipality of Petah Tiqwa into one large cluster caused the agglomeration of the latter with the largest cluster in the metropolis in 1985.

The morphology of the largest cluster in the years from the mid 1980s onward can be viewed as a "quasi stationary state". During this period the morphology of the largest cluster hardly changed. We suggest that this is due to the fact that there were no substantive parcels of land left to be developed. Protected agricultural land and national nature reserves occupy almost all of the available land. This situation is likely not to be sustainable over time. It is probable that market forces and pressure from private developers on the government will create additional developable land. In the third stage of the metropolis development, the only possibilities of development was to fill unoccupied land inside of the clusters (see Figure 5) and to construct buildings with significant heights (Czamanski & Roth, 2007).

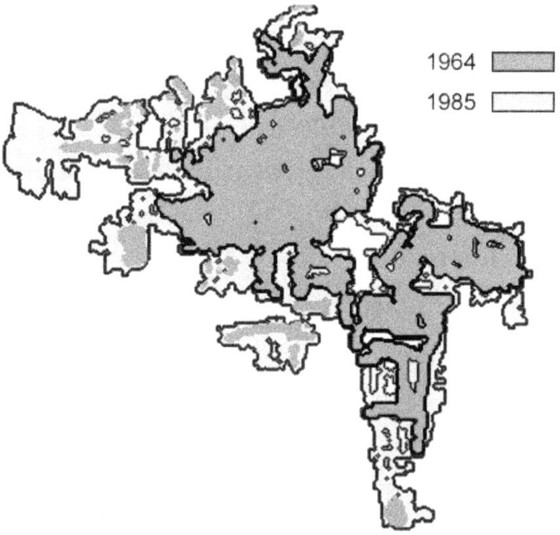

Figure 12 *The Agglomerated cluster of the Petah Tiqwa area.*

Conclusions

To conclude, this work considers the urban area as a physical object, disregarding municipal boundaries. It is based on the assumption that the city is a complex system with self organizing characteristics. It reflects patterns of consistency and homogeneity in the perceptions, constraints and desires of people, as expressed by their behavior and decisions. This homogeneity results in great similarity, i.e., the urban clusters in different locations evolve in a similar way, which can be described mathematically. In addition, when clusters are fractals, their fractal dimension can be used in order to discover phenomena connected with their evolution.

From this study we suggest three important conclusions:

1. The built-up area of the largest cluster (and its components) in the Tel-Aviv metropolis is indeed a self organizing system; although it was not affected by political and planning decisions (as clusters are not subject to municipal boundaries), it presented some common characteristics. This result was found in the morphology of the largest cluster through time as well as in the behavior of all the clusters in the Tel-Aviv metropolis.
2. The development of the morphology of the largest cluster in the Tel-Aviv metropolis can be divided into 3 stages: 1935-1964, 1964-1985 and 1985-2000. These division correlates with the development of the entire metropolis morphology and can be associated with socio-economic trends and processes. Thus the cluster analysis has a significant potential to be used as a research tool to investigate urban development.

3. The largest cluster in the Tel-Aviv metropolis has come to saturation in terms of land availability, i.e., there is no more available land for development within its boundaries. To enable further development the following options should be considered by the planning authorities:
    a. Continue to construct buildings with increasing heights.
    b. Regeneration of old neighborhoods that are characterized by houses with one or two floors.
    c. Relocation activities that contain land uses that are not related to urban activities—such as military facilities, from the metropolis core (the largest cluster).

References

Anas, A., Arnott, R. and Small, K.A. (1998). "Urban spatial structure," *Journal of Economic Literature*, ISSN 0022-0515, 36: 1426-1464.

Alonso, W. (1964). *Location and Land Use*, Cambridge, MA: Harvard University Press.

Auerbach, F. (1913). "Das gesetz der Bevolkerungskoncentration," *Petermanns Geographische Mitteilungen*, ISSN 0031-6229, 59: 74-76.

Batty, M. and Longley, P.A. (1987). "Urban shapes as fractals," *Area*, ISSN 0004-0894, 19: 215-221.

Batty, M., Longley, P. and Fortheringham, S. (1989). "Urban growth and form: Scaling, fractal geometry, and diffusion limited aggregation," *Environment & Planning A*, ISSN 0308-518X, 21: 1447-1472.

Batty, M. and Longley, P.A. (1994). *Fractal Cities: A Geometry of Form and Function*, ISBN 9780124555709.

Batty, M. and Xie, Y. (1996). "Preliminary evidence for a theory of the fractal city," *Environment & Planning A*, ISSN 0308-518X, 28: 1745-1762.

Benguigui, L., Czamanski, D., Marinov, M. and Portugali, Y. (2000). "When and where is a city fractal?" *Environment and Planning B: Planning and Design*, ISSN 0265-8135, 27: 507-519.

Benguigui, L., Czamanski, D. and Marinov, M. (2001a) "Urban spatial dynamics: The case of Petah Tikvah," *Environment and Planning B: Planning and Design*, ISSN 0265-8135, 28: 447-460.

Benguigui, L., Marinov, M. and Czamanski, D. (2001b). "City growth as a leap-frogging process: An application to the Tel-Aviv metropolis," *Urban Studies*, ISSN 0042-0980, 38(10): 1819-1839.

Benguigui, L., Blumenfeld-Lieberthal, E. and Czamanski, D. (2006). "The dynamics of the Tel Aviv morphology," *Environment and Planning B: Planning and Design*, ISSN 0265-8135, 33(2): 269-284.

Blumenfeld-Lieberthal, E. (2005). The Dynamics of Urban Morphology in the Tel-Aviv Metropolis, http://techunix.technion.ac.il/~efratb5/index.html.

Carmon, N. and Czamanski, D. (1992). "Housing in Israel: From planned economy to semi-free market management," *International Journal for Housing Science and its Applications*, ISSN 0146-6518, 16(1): 47-60.

Christaller, W. (1966). *Central Places in Southern Germany*, ISBN 9780131226302.

Czamanski, D. and Roth, R. (2007). "Do planners cause sprawl? The impact of characteristic time on developers' behavior," http://sadapt.inapg.inra.fr/ersa2007/papers/111.pdf.

Frankhauser, P. (1990). "Aspects fractals des structures urbaines," Espace Géographique, ISSN 0046-2497, 19: 45-69.

Frankhauser, P. (1994). *La Fractalite des Structures Urbanies*, Collection Villes, Anthropos, Paris, France.

Imre, A.R. and Bogaert, J. (2004). "The fractal dimension as a measure of the quality of habitat," *Acta Biotheoretica*, ISSN 0001-5342, 52: 41-56.

Israel Central Bureau of Statistics, http://www.cbs.gov.il/engindex.htm.

Jefferson, M. (1939). "The law of the primate city," *The Geographical Review*, ISSN 0016-7428, 29: 226-232.

Mandelbrot, B. (1982). *Fractal Geometry of Nature*, ISBN 9780716711865.

Makse, H.A., Havlin, S. and Stanley, H. (1995). "Modeling urban growth patterns," *Nature*, ISSN 0028-0836, 377: 608-612.

Portugali, J. (2000). *Self-Organization and the City*, ISBN 9783540654834.

Shen, G. (1997). "A fractal dimension analysis of urban transportation networks," *Geographical & Environmental Modelling*, ISSN 1361-5939, 1: 221-236.

Shen, G. (2002). "Fractal dimension and fractal growth of urbanized areas," *International Journal of Geographical Information Science*, ISSN 1365-8816, 16: 419-437.

Schweitzer, F. (1997). "Urban cluster growth: Analysis and computer simulations of urban aggregations," in F. Schweitzer (ed.), *Self-Organization of Complex Structures: From Individual to Collective Dynamics*, ISBN 9789056990275, pp 501-518.

Schweitzer, F. and Steinbrink, J. (1998). "Estimation of megacity growth: Simple rules versus complex phenomena," *Applied Geography*, ISSN 0143-6228, 18(1): 69-81.

Tuchler-Aharoni, M. (1997). *The Impact of the Ayalon Throughways on the Structure of the CBD in the Tel-Aviv Metropolitan Area*, unpublished MA thesis, Technion IIT.

White, R. and Engelen, G. (1993). "Cellular automata and fractal urban form: A cellular modeling approach to the evolution of urban land use patterns," *Environment & Planning A*, ISSN 0308-518X, 25: 1175-1199.

White, R., Luo, W. and Hatna, E. (2001). "Fractal structures in land use patterns of European cities: Form and process," *12th European Colloquium on Quantitative and Theoretical Geography*, Saint-Valérie-en-Caux, France.

Zipf, G.K. (1941). *National Unity and Disunity: The Nation As a Bio-social Organism*, Bloomington, In: The Principia Press Inc.

Efrat Blumenfeld-Lieberthal is an architect and a lecturer of Architecture at the David Azrieli School of Architecture Tel Aviv University. She received her B. Arch, and PhD at the Faculty of Architecture and Town Planning, at the Technion – Israel Institute of Technology. Her first Post doctorate research was at the Center for Advanced Spatial Analysis (CASA), University College London. She has recently completed her second post doctorate research as a Porter Fellow at the Environmental Simulation Laboratory (ESLab), Tel Aviv University. Her research interests are applying theories of complexity to urban environments; urban morphology; size distribution of entities in complex systems; and complex networks in urban systems.

Iris Aravot, B. Arch., M.Sc., Ph.D., Technion, I..I.T., Israel. Philosophy studies at Haifa University. Postdoctoral studies at the A.A. Graduate School, London. Associate Professor in Urban Design and Urban Design Theory at the Faculty of Architecture and Town Planning. Research interest: Urban design with special emphasis on new-towns. Phenomenology of Architecture and Urbanism; Imagining and conception; Organizer and chair of the 1st International Conference on Architecture and Phenomenology, Haifa 2007. Head of The Research Centre for Architectural R&D and former Vice Dean for Research and Graduate Studies. Published numerous articles in Hebrew and English. Visiting scholar in European and North American schools of architecture.

Daniel Czamanski has three degrees in economics. He teaches economics in the city and regional planning program and serves as associate dean of architecture faculty at Technion – Israel Institute of Technology. He is the director of the city complexity research lab. He has published a number of books and tens of academic papers. In the past, Czamanski served as advisor to three ministers of energy in Israel and was the head of a task force concerning the re-organization of Israel Electricity Corp. During 2003-2005 he was the senior vice president for research at the Canadian Energy Research Institute.

3. A Dynamical Model for Information Retrieval and Emergence of Scale-Free Clusters in a Long Term Memory Network

Ignazio Licata
Institute for Scientific Methodology and Ethics, ITA
Associazione Italiana per la Ricerca sui Sistemi, ITA

The classical forms of knowledge representation fail when a strong dynamical interconnection between system and environment comes into play. We propose here a model of information retrieval derived from the Kintsch-Ericsson scheme, based upon a long term memory (LTM) associative net whose structure changes in time according to the textual content of the analyzed documents. Both the theoretical analysis carried out by using simple statistical tools and the tests show the appearing of typical power-laws and the net configuration as a scale-free graph. The information retrieval from LTM shows that the entire system can be considered to be an information amplifier which leads to the emergence of new cognitive structures. It has to be underlined that the expanding of the semantic domain regards the user-network as a whole system. It hints an epistemological shifting from the ontological models to the ontogenetic ones in describing knowledge dynamical representation.

Introduction

The concept of an intelligent agent involves a world description by which the agent makes its choices, activating some data evaluation strategies and selecting the most significant elements on the basis of a given objective and the interaction with the environment.

The old strong AI approaches were based on formal logic tools and heuristic rules for achieving a sufficiently exhaustive world description.

Expert Systems used in various fields—chess, military and economic strategies, clinical diagnosis etc.—even though they used different formal tools for conceptualization, they were all classifiable within the strong AI and they shared a static knowledge representation form.

A world description made up by a set of well defined and ever accessible production rules inevitably leads to a limited efficacy as the semantic domain is expanding.

Connectionistic approaches based on neural networks and parallel distributed processing systems have opened new prospects characterized by a different and closer relation between the system and the environment.

The new systemic-cybernetic approach by N. Wiener, L. Von Bertalanffy, R. Ashby, H. Von Foerster, H. Maturana and F. Varela makes a world description depending on the observer. This is considered to be an interacting system that selects knowledge according to its inner structure and its dynamic history.

Between the observer (the intelligent system) and the observed (the world) there is no longer a linear and deductive relationship defined by a formal model, but there is a circular function based on a continuous adaptation and coevolution strategy. This implies at least two important differences from the approaches adopted in strong AI.

A relationship of thermodynamic openness between the system and the environment must be considered. Information and energy make up a flow that crosses and continuously modifies their relationships and their structure.

This dissipative feature of the system is only a necessary but not sufficient condition. The system must also have a logic openness i.e., it must show emergent behaviors which depend on the dynamic state of the relationship between the system and the environment. These emergent behaviors should lead the system to more complex adaptive situations by the production of new knowledge.

So there are no longer definite and independent knowledge representation forms of the agent-observer world description. But the agent continuously generates adaptive processes that produce at any moment a world description bound to the structure and the objectives of the agent.

Classic AI representations conform to the adoption of models with a low logic openness. They can be seen as samples of a knowledge structure that is always dynamic in the natural systems.

Being aware of these limits and problems we are searching for common aspects between the symbolic and the connectionistic approaches.

Firstly we consider an ontology as a linguistic theory of beings. This implies two strong assumptions:

1. The "neutrality" of the ontological representation. This is formalized by the second order logic and it must guarantee the system portability i.e., the possibility to make different conceptualizations;
2. The possibility to describe the world by the discrete units of the second order logic at any moment.

Furthermore a knowledge acquisition system must exhibit coherence, operational closure and must produce a world by intrinsic emergent processes based on the dialogic relationship between the system and the user.

So we must leave the ontological models for the ontogenetic models of knowledge representation.

A Dynamic Knowledge Acquisition System

During the discourse comprehension it has been demonstrated that the human mind generates dynamic structures that are adapted to the particular context of use. In particular Kintsch (1998) noticed that human knowledge in these cases can be modelled by networks of propositions.

This particular model extends and combines the advantages of the classic static forms of representation such as "associative networks" (Meyer & Schvaneveldt, 1971), "semantic networks" (Collins & Quillian, 1969), "frames" (Minsky, 1975) and "scripts" (Schank & Abelson, 1977). They are all based on the predicate-argument schema, i.e., their atomic parts are propositions linked to each other by weighted and not labelled arcs.

This particular representation form presents great advantages in comparison to classic formalisms. While semantic networks, frames and scripts organize knowledge in a more ordered and logical way, the networks of propositions are disorganized and chaotic, but their structure can change dynamically in time on the basis of cumulated experiences and the knowledge can be retrieved with particular procedures that consider the perceived context.

The meaning of a node, i.e., its correspondent concept, is given by its position in the network because it is constrained by the close nodes. But during the comprehension process only a part of the neighborhood is considered. This subset of nodes is "activated" by the particular context of use represented by other nodes.

So the meaning of every concept in the network is not static, i.e., fixed and permanent, but it is always built in a particular structure called "working memory" by the activation of a particular section of the network.

To specify the activation modalities Kintsch and Ericsson (Kintsch, *et al.*, 1999) introduced the concept of "Long Term Working Memory" (LTWM). This is a part of the "Long Term Memory" (LTM) that is the entire network of propositions. LTWM is generated by the short term part of the working memory (STWM). This process is allowed by fixed and stable memory structures called "retrieval cues" that link the objects present in the STWM to other objects present in the LTM.

After the definition of their model Kintsch and Ericsson have tried to implement it. Two particular problems had to be solved, the creation of the LTWM and the formation of the retrieval cues.

For the definition of the LTWM Kintsch developed two methods. The first, defined with Van Dijk (van Dijk & Kintsch, 1983), is a manual technique that starts from the propositions present in the text (micropropositions) and using some organizing rules arrives to the definition of "macropropositions" and "macrostructures" and even to the definition of LTWM.

The second is based on the "Latent Semantic Analysis" (LSA) (Landauer, Foltz, and Laham, 1998). This technique can infer, from the matrix of co-occurrence rates of the words, a semantic space that reflects the semantic relations between words and phrases. This space has typically 300-400 dimensions and allows to represent words, phrases and entire texts in a vectorial form. In this way the semantic relation between two vectors can be estimated by their cosine (a measure that according to Kintsch can be interpreted as a correlation coefficient).

In the second solution the modalities of the information retrieval from the semantic space are not clearly specified. When the "textbase", i.e., the representation obtained directly from the text, is sufficiently expressed, the retrieval of knowledge from the LTM is not necessary. In other cases a correct comprehension of the text (or the relative "situation model") requires the retrieval of knowledge from the LTM.

After the creation of the LTWM the integration process begins, i.e., the activation of the nodes correspondent to the meaning of the phrase. Kintsch uses a diffusion of activation procedure that is a simplified version of the one developed by McClelland and Rumelhart (1986). Firstly an activation vector is defined whose elements are indexed over the nodes of LTWM. Any element's value is "1" or "0" depending on the presence or the absence of the corresponding node in the analyzed phrase (i.e., in the STWM). This vector is multiplied by the matrix of the correlation rates (the weights of the links of the LTWM) and the resulting vector is normalized. This becomes the new activation vector that must be multiplied again by the matrix of the correlation rates. This procedure goes on until the activation vector becomes stable. After the integration process, the irrelevant nodes are deactivated and only those that represent the situation model remain activated.

There is also another problem that must be considered. Theoretically the position occupied by a word in the LTWM is determined by a lifetime experience, i.e., by the continuous use that is made of it. Obviously this kind of knowledge cannot be reached practically and Kintsch build his semantic space using information taken from a dictionary. It is important to note that this operation is done only one time and the semantic space is not further updated. So this kind of implementation loses some of its dynamic properties.

An Alternative Implementation of the LTWM Model

We think that all the previous problems can be fully solved only by dropping the intermediate representation of the semantic space and trying to find a method that define directly the retrieval cues. Unfortunately the lack of adequate textual parsers able convert the paragraphs of a text on the correspondent atomic propositions has driven us to develop simple dynamic models of associative networks that are based on the LTWM model of Kintsch and Ericsson.

Our associative network can be extended, in a second phase, using the relations provided by an existing general ontology. In this way we think that can be derived a network of propositions.

We are considering a particular ontology built over WordNet (Gangemi *et al.*, 2002). Each concept is represented by a synset and is linked to other concepts by relations (propositions). A synset is a set of synonims that represent the same concept (Miller, 1993).

There are many systems for the ontology extraction (Navigli *et al.*, 2003) or for the semantic disambiguation that are based on WordNet. In particular we are considering a technique (Ide & Veronis, 1998) that is based on the extension of the nodes representing the terms of a text (phrase) with all the correspondent synsets of WordNet.

Once the correct synsets have been activated with the diffusion of an activation signal, the relative relations can be added to the structure.

In this way can arise a dynamic structure made by propositions and literals that can be represented in an ontological way, maybe using the formalisms of the Semantic Web Languages (RDF, RDF Schema).

This ontology is created dynamically considering the context, i.e., the content of the analyzed text and the nodes representing the goals of the user (query).

The coherence of the ontology is guaranteed by the coherence of the upper level ontology that is used during its assemblage. Now we are facing the problem of the creation of the associative network that is the first step towards the automatic generation of ontologies.

Figure 1 describes the system that we used for the extraction of knowledge from textual documents. The part of the document that is analyzed—a section, a paragraph or a simple group of words enclosed in a window—is stored temporarily in a buffer. Its content must be codified on the basis of the context before being elaborated by the working memory block. This has been implemented by a simple scale-free graph model (Albert & Barabási, 2001), due to the fact that recently it has been found that human knowledge seems to be structured in this way (Steyvers & Tenenbaum, 2001).

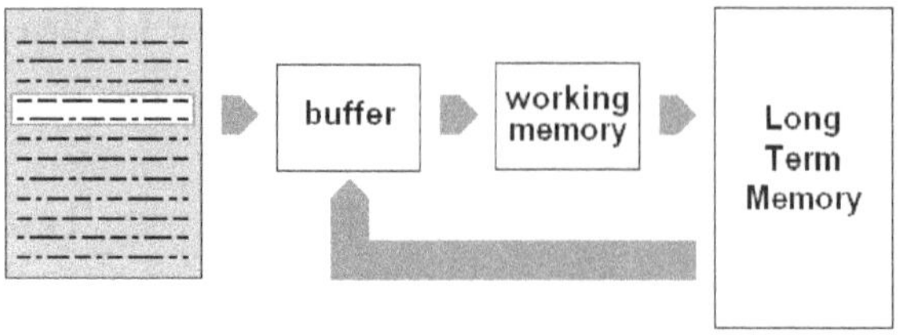

Figure 1 *A possible architecture of a system for the dynamical acquisition of knowledge from a repository of documents.*

The analysis is performed over all the paragraphs transferring their content in the buffer. This structure contains not only the words of the analyzed paragraph, but also words retrieved from the LTM by the diffusion of an activation signal starting from the LTM nodes that represent words in the buffer. Theoretically the buffer should contain also words activated during the analysis of the previous paragraph, but this aspect of Kintsch model has not been considered for its computational complexity.

The buffer, the working memory and the activated part of the LTM block can be compared (but they are not the same structure) to the LTWM defined by Kintsch and Ericsson.

During the acquisition of the paragraph is used a stoplist of words that must not be considered (as articles, pronouns etc.). For any word in the text, the paragraphs where it has appeared (or where it has been inserted after the retrieval procedure) are stored. When the entire text has been parsed and the data of all the N not filtered words have been memorized, the formation of the network of concepts in the working memory begins.

The model adopted is similar to the one defined by Bianconi and Barabasi (2001). The process starts with a net consisting of N disconnected nodes. At every step t=1..N each node (associated with one of the N words) establishes a link with other M units (M=5). If j is the selected unit, the probability that this node establishes a link with the unit i is:

$$P_i = \frac{U_i k_i}{U_1 k_1 + \ldots + U_N k_N}$$

where k_i is the degree of the unit i[1], i.e., the number of links established by it, while U_i is the fitness value associated to the node that can be computed as the ratio between the number of paragraphs that contain both words i and j and the number of paragraphs that contain either i or j.

LTM is an associative network that is updated with the content of the WM. Whenever a link of the WM corresponds to a link present in the LTM, the weight of this one is increased by "1". For example if the WM links "market" to

1. Each node is connected to itself by a loop.

"economy" and in the LTM "market" is linked to "economy" with weight "7" and to "stock" with weight "4", in the updated LTM "market" is linked to "economy" with weight "8" and to "stock" with weight "4" (unchanged). To perform the diffusion of the activation signal all the weights must be normalized. In this case "market" must be linked to "economy" with weight 8/(8+4) and to "stock" with weight 4/(8+4).

Since the scale free network that represents the content of the WM is used to update the content of LTM, this associative networks should take the form of a scale free graph. Unfortunately the modalities of evolution of the LTM does not allow the definition of a simple equivalent mathematic model, that is necessary to make useful previsions about its evolution.

While in the scale free graph models proposed by literature at each temporal step M are added new nodes to the graph, with M defined beforehand, in the system that we have developed, after the analysis of a new document the links related to an unknown number of nodes of the LTM network are updated on the basis of the content of the WM. This is the number of the words that have not been filtered by the stoplist.

Another important difference with other scale free models presented in literature (Dorogovtsev & Mendes, 2001) is the particular fitness function that is used. This function does not depend on a single node but on the considered pair of nodes, i.e,, the fitness value of a word is not constant but depends on the other word that is linked to it. For example the noun "car" should present for the link with "engine" a fitness value greater than the one presented for the link with "economy".

Evaluation of the Scale-Free Properties

To test the validity of the scale free graph model adopted for the WM, we gave 100 files of the Reuters Corpus[2] as input to the system disabling the retrieval of information from the LTM. Two versions of the model have been tested, one with bidirectional links and the other with directed links (in this case we considered $k_i = k_{i(IN)} + k_{i(OUT)}$).

In Figure 2 it is represented an example of a network with bidirectional links. The economic bias of the articles justifies the presence of hubs as "interest rate", "economy", etc., while other frequent words as "child", "restaurant", etc. establish less link with the others.

The following graphs report the average path length between each pairs of nodes (Figure 3) and the clustering coefficients (Figure 4). Some networks with different sizes have been considered filtering less or more words in the analyzed texts.

---

2. Reuters Corpus, Volume 1, English language, 1996-08-20 to 1997-08-19, http://about.reuters.com/ researchandstandards/corpus.

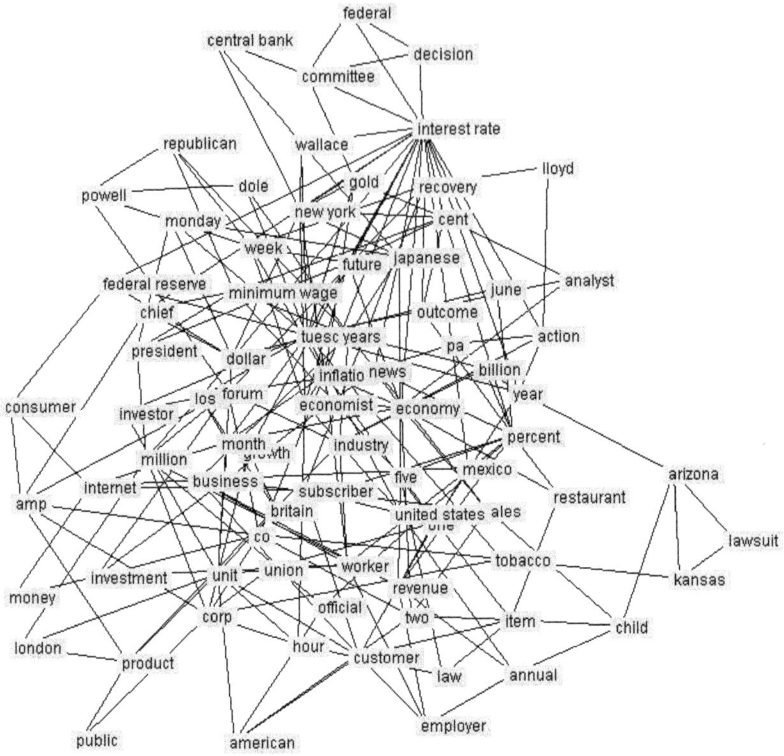

Figure 2 *A network with bidirectional links obtained by the analysis of 100 files of the Reuters Corpus.*

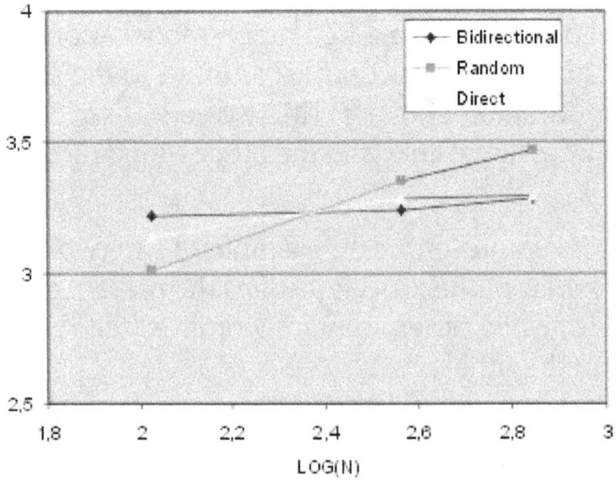

Figure 3 *Comparison of average path lengths of different types of networks.*

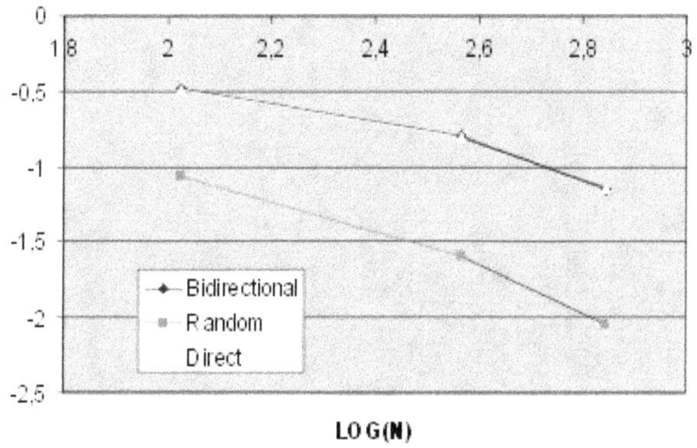

Figure 4 *Comparison of clustering coefficients of different types of networks.*

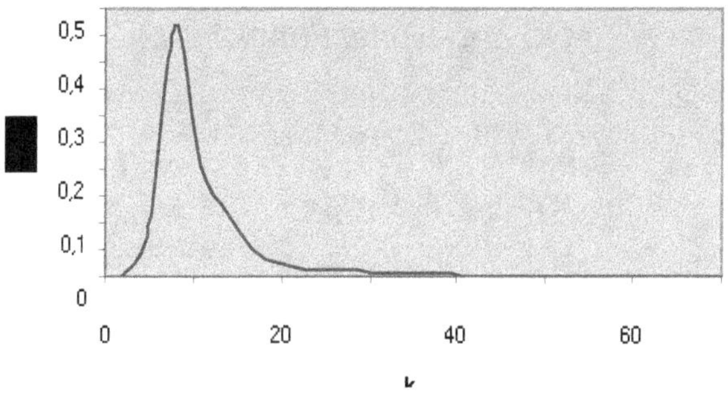

Figure 5 *Degree distribution of a graph with M=5 and bidirectional links.*

All the results seem to confirm those reached by Bianconi and Barabási. The trend of the average path related to random graphs having the same dimensions of the considered scale free graphs has an higher slope. The clustering coefficient of the scale free graph model has an higher order of magnitude in comparison with the one computed for random networks.

Figure 5 and Figure 6 show the degree distribution of a graph with bidirectional links and directed links respectively. In the first case the degrees distribution decays as $P(k) \approx k^{-G}$ with $G = 3.2657$. In the second case the power law trend has a coefficient $G = 2.3897$.

Evaluation of the LTM Representation

We also analyzed the structure of LTM associative network that, as expected, kept the features of a scale-free graph (Table 1). The system was tested enabling the retrieval of information from LTM and the analysis was repeated 30 times computing the coherence rate of the final knowledge representations.

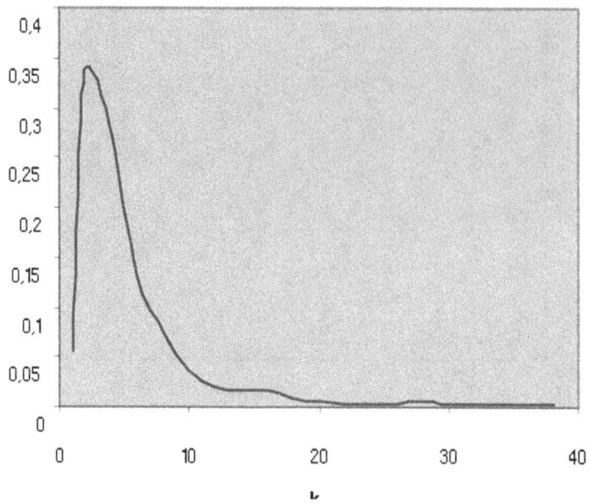

Figure 6 *Degree distribution of a graph with M=5 and bidirectional links.*

M	Average path length	Average degree	Clustering coefficient
1	2.56	5.95	0.32
2	2.49	6.50	0.34
3	2.27	8.30	0.45
4	2.25	9.50	0.43
5	2.23	9.85	0.43

Table 1 *LTM with 40 nodes.*

The coherence rate is obtained by correlating the LTM ratings given for each item in a pair with all of the other concepts[3].

The average coherence rate (0.45) has confirmed that the conceptualization, i.e., the evolution of the associative network, was made by the system on the basis of a precise inner schema.

To evaluate the correctness of this schema we compared the LTM associative network with representations obtained from a group of human subjects. All the participants had college and university degrees and they were aged between 25 and 32.

The subjects were asked to read the same medical article examined by the system, assigning a rate of relatedness to each pair of words that were considered by the system.

A Pathfinder analysis (Schvaneveldt *et al.*, 1985) was performed on the relatedness matrices provided by human subjects and the LTM matrix, in order to extract the so called "latent semantics", i.e., other implicit relations between

3. This rate was computed by the software PCKNOT 4.3, a product of Interlink Inc.

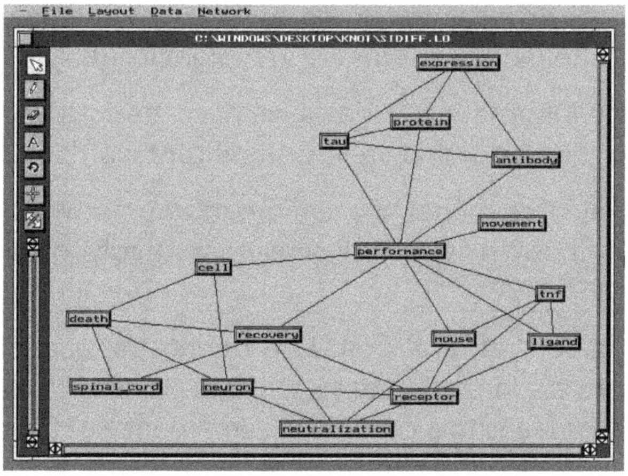

Figure 7 *LTM representation with the diffusion of the activation signal.*

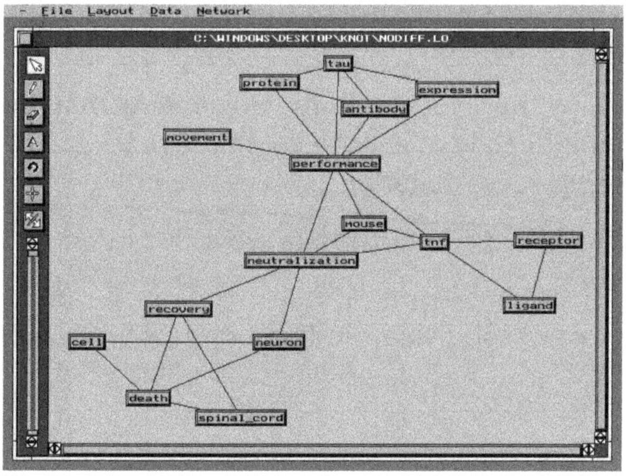

Figure 8 *LTM representation without the diffusion of the activation signal.*

words. The obtained matrices were compared using a similarity rate determined by the correspondence of links in each pair of networks.

The first two layouts (Figures 7 and 8) refer to the representation obtained by the system with and without the information retrieval from the LTM respectively.

The coherence rate of the network that was obtained with the diffusion of the activation signal is lower than the other one due to the fact that more links are created (31 links vs. 27 links). The two networks have 23 common links and the similarity rate of the two representations is very high (0.657). But the first representation presents more significant hubs.

With the information retrieval from the LTM we have the following hubs: performance (degree 9), receptor (degree 6), recovery (degree 5). We must consider that the article deals with the functional recovery by the ligand/receptor systems. So in the first network the hubs are very significant.

Without the information retrieval we have: performance (degree 8), tnf and neutralization (degree 5) that are less significant words.

If only a scale free model or a self organizing network is used to represent the content of a text, its structure converges rapidly towards a state which has an high coherence rate.

The information retrieval from LTM during the analysis of a new text breaks this coherence and the entire system can be considered to be an information amplifier that leads to a continuous emergence of new cognitive structures. It has to be underlined that the expanding of the semiotic domain regards the user-network couple considered as a whole system.

The intrinsic emergence of the system shows itself with the appearance of new codes represented by the structure of the links of the networks after the analysis of each new document.

So the user, by the selection of the documents that must be processed by the system, can indirectly affect the cognitive process which characterizes the system knowledge representation.

Three of the 21 experimental data were discarded for their low coherence rate (below 0.2).

The others were subdivided in three groups whose characteristics are reported in Table 2.

Group no.	Average coherence rate	Average similarity rate	Similarity rate
1	0.780	0.271	0.243
2	0.698	0.218	0.211
3	0.380	0.162	0.122

Table 2 *Characteristics of the group representations.*

The first two reported parameters are the average coherence rate and the average similarity rate of the representations belonging to each group. The last parameter is the similarity rate between the representation of our system and the three group representations that were obtained combining and averaging the relatedness rates provided by the members of each group.

Figures 9-11 show the resulting networks of groups 1-3. It can be noticed that the first group representation presents two fundamental hubs (spinal

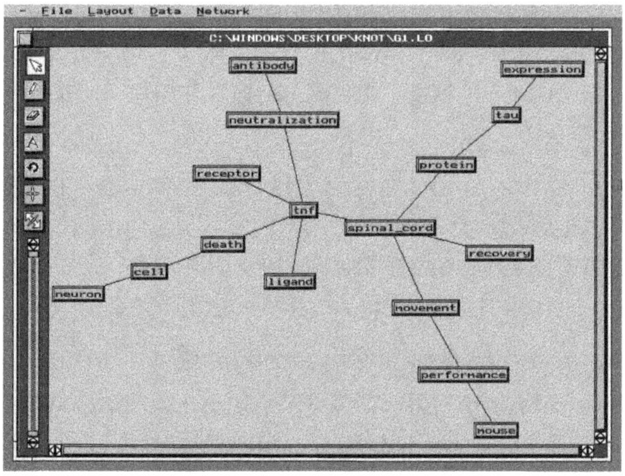

Figure 9 *Group 1 representation.*

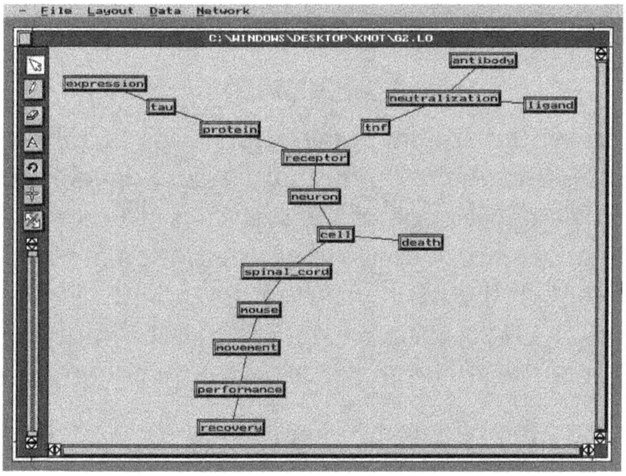

Figure 10 *Group 2 representation.*

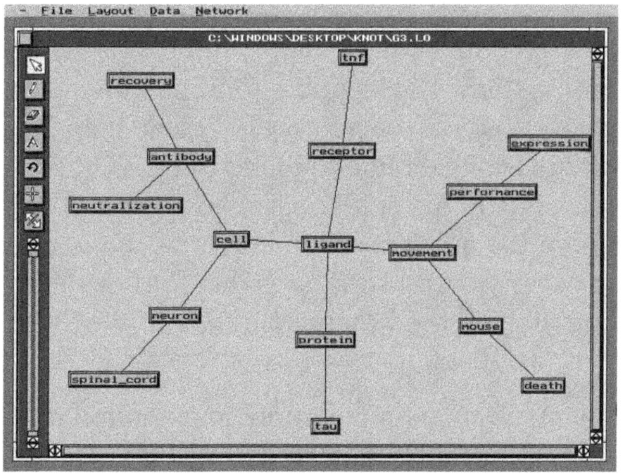

Figure 11 *Group 3 representation.*

chord, tnf) and it has the highest coherence rate. The second group representation has a lower coherence rate and different hubs (cell, receptor, neutralization). The third group representation is the most chaotic one with the lowest coherence rate. The hubs (ligand, movement, cell and antibody) are less representative than the others.

The representation of our system is more similar to the representations of the first two groups which have 9 and 8 common links (out of a total of 15) with the network of the system and the probability to have exactly this number of links in common by chance being very low (0.248%).

Now we are going to repeat the same test on a group of experts in order to evaluate the similarity rate between the system representation and their representation. Furthermore the knowledge base of the system will be enriched by the analysis of other medical articles about spinal chord injuries. In this way we think that, having more a-priori knowledge about the article's topic, the system will be able to conceptualize better the article.

Conclusion

We have presented an innovative knowledge acquisition system based on the long term working memory model developed by Kintsch and Ericsson. The knowledge of the system is structured as an associative network that is dynamically updated by the integration of scale-free graphs that represent the content of the new analyzed documents. By the diffusion of an activation signal in the LTM associative network, all the information necessary to identify the context of the analyzed concepts (terms) is retrieved.

The analysis of the WM and LTM networks have confirmed that they are both examples of scale-free graphs. The computation of the coherence rate of LTM networks revealed that the system acquires knowledge on the basis of precise inner schema whose correctness was evaluated by the comparison with the other associative networks obtained from human subjects.

Certainly our system is susceptible to improvements. Maybe the presence of the external feedback of the human user could help the system to model correctly its knowledge. For example the links in the LTM could be strengthen only when the knowledge representation is used to filter or retrieve documents correctly. Furthermore the association of an age to the links of the LTM could guarantee more plasticity to its structure. This further information could be used in the computation of the fitness values as in the Dorogovtzev models (Dorogovtsev & Mendes, 2000).

The structure of the associative network can be extended using the content of an ontology based on the definitions of WordNet in order to obtain a network of propositions representable by the languages of the Semantic Web.

This ontology is created dynamically considering the context, i.e., the content of the analyzed text and the nodes representing the goals and the requests of the user (query).

This is an innovative technique that could represent the first concrete ontogenetic approach to the problem of the knowledge acquisition.

References

Albert, R. and Barabái, A.L. (2002). "Statistical mechanics of complex networks," *Reviews of Modern Physics*, ISSN 0034-6861, 74: 47-97.

Bianconi G. and Barabasi A.L. (2001). "Bose-Einstein condensation in complex networks," *Physical Review Letters*, ISSN 0031-9007, 86(24): 5632-5635.

Collins, A.M. and Quillian, M.R. (1969). "Retrieval from semantic memory," *Journal of Verbal Learning and Verbal Behavior*, ISSN 0022-5371, 8: 240-247.

Dorogovtsev, S.N. and Mendes, J.F.F. (2000). "Evolution of reference networks with aging of Sites," *Physical Review E - Statistical, Nonlinear and Soft Matter Physics*, ISSN 1539-3755, E62: 1842-1845.

Dorogovtsev, S.N. and Mendes, J.F.F. (2003). *Evolution of Networks: From Biological Nets to the Internet and WWW*, ISBN 9780198515906.

Giordano, W., Lella, L., Licata, I., Montesanto, A. and Tascini, G. (2006). "Scale free graphs in dynamic knowledge acquisition," in G. Minati, E. Pessa and M. Abram (eds.), *Systemics of Emergence*, ISBN 9780387288994.

Ide, N. and Veronis, J. (1998). "Word sense disambiguation: The state of the art," *Computational Linguistics*, ISSN 0891-2017, 24(1): 1-41.

Kintsch, W. (1998). *Comprehension: A Paradigm for Cognition*, ISBN 9780521629867.

Kintsch, W., Patel, V.L. and Ericsson, K.A. (1999). "The role of long-term working memory in text comprehension," *Psychologia*, ISSN 0033-2852, 42: 186-198.

Landauer, T.K., Foltz, P.W. and Laham, D. (1998). "An introduction to latent semantic analysis," *Discourse Processes*, ISSN 0163-853X, 25: 259-284.

McClelland, J.L. and Rumelhart, D.E. (1986). *Parallel Distributed Processing*, ISBN 9780262680530 (1987).

Meyer, D.E. and Schvaneveldt, R.W. (1971). "Facilitation in recognizing pairs of words: Evidence of a dependence between retrieval operations," *Journal of Experimental Psychology*, ISSN 0096-3445, 90(2): 227-234.

Miller, G.A., Beckwith, R., Fellbaum, C,D., Gross, D. and Miller, K. (1993). *Five papers on WordNet*, Cognitive Science Laboratory Technical Report 43, Princeton.

Minsky, M. (1975) "A framework for representing knowledge," in P.H. Winston (ed.), *The Psychology of Computer Vision*, ISBN 9780070710481.

Navigli, R., Velardi, P. and Gangemi, A. (2003). "Ontology learning and its application to automated terminology translation," *IEEE Intelligent Systems*, ISSN 1541-1672, January/February: 22-31.

Schank, R.C. and Abelson, R.P. (1977). *Scripts, Plans, Goals, and Understanding*, ISBN 9780898591385.

Schvaneveldt, R.W., Durso, F.T. and Dearholt, D.W. (1985). "Pathfinder: Scaling with network structures," CRL Technical Report No. MCCS-85-9.

Steyvers, M. and Tenenbaum, J. (2005). "The large-scale structure of semantic networks," *Cognitive Science*, ISSN 1551-6709, 29: 41-78.

van Dijk. T.A. and Kintsch. W. (1983). *Strategies of Discourse Comprehension*, ISBN 9780127120508.

Ignazio Licata is a theoretical physicist and the scientific director of ISEM, Institute for Scientific Methodology for Interdisciplinary Studies in Palermo. His main research interests are Quantum Physics Foundations, mathematical models of cognitive processes and Computation Theory in physical and biological systems.

4. The Application of the Dimensionality Perspective in Organization Study

Zhi Tang
Rochester Institute of Technology, E. Philip Saunders College of Business, USA

Previous research suggests that organizations may apply two opposite complexity mechanisms to cope with environmental uncertainty: absorption and reduction. However, except for some anecdotal evidence, there is no theoretical skeleton established to integrate these two opposite mechanisms in one framework and to prescribe the contingent conditions for employing them. This paper deconstructs organizational complexity at the organizational elemental level and establishes framework that incorporates three dimensions—organizational complexity, organizational dynamism, and organizational variability. This paper also discusses the environmental conditions for applying absorption and reduction mechanisms as well as the benefits and costs of applying these mechanisms. This dimensionality perspective provides a new avenue for researchers and practitioners to understand and handle organizational structuration issues.

Introduction

Organizations are often viewed as open and complex adaptive systems that are dependent on their interactions with environmental resources (Wholey & Brittain, 1989). Previously, two complexity mechanisms have been recommended for organizations to produce a better adaptive capability: reduction and absorption (Cohen, 1999). The reduction mechanism suggests that a firm should standardize its internal processes and simplify organizational systems so as to decrease the number of external agents that it has to face (Anderson *et al.*, 1999). The absorption mechanism, in contrast, suggests that when facing external uncertainty, firms should complicate their systems in order to create a variety of compound options and risk-hedging strategies (Boisot & Child, 1999). To date, no theoretical framework has been built to reconcile these two apparently conflicting logics.

We therefore propose a dimensionality perspective to address this issue. The dimensionality perspective focuses on the structuration (i.e., complexity, dynamism, variability) of an organizational system. Applying this approach, the two mechanisms will be deconstructed and analyzed at the organizational elemental level and incorporated in one framework.

We plan, by centering our discussion on the organizational complexity dimension, to illustrate how to apply the new dimensionality perspective in organization studies. We aim to make two contributions to complexity research. First, this is the first project that looks at organizational complexity phenomena from a dimensionality perspective, and this new perspective helps to reconcile some contradictory findings in previous studies. Second, a thorough benefit-and-cost analysis of organizational complexity will be presented. The antecedents (i.e., environmental complexity and environmental dynamism), the interactions of three organizational dimensions (i.e., complexity, dynamism, variability), and the outcome of these interactions will be systematically discussed to create a theoretical framework to integrate two opposite mechanisms—absorption and reduction.

Organizational Complexity

Ashby (1956) proposed the Law of Variety to set the foundation for studying organizational complexity. In his book *Introduction to Cybernetics*, Ashby stated that in order to cope with environmental variety, firms needed to complicate their organizational structure because only when organizations were more complicated, could they understand, predict, and even "destroy" less-complicated external variety. Following this logic, researchers began to investigate how well a complex organization could cope with environmental uncertainty. Burton and Forsyth (1986) used 14 factors to measure the complexity of organizational structure and processes, including the number of products, the number of product categories, and the number of countries in which operations were conducted, and identified a positive relationship between organizational complexity and firm performance. Damanpour (1996) also found that in general, organizational complexity contributed to organizational innovative capability. Ashmos and his colleagues conducted two studies in the hospital industry in 1996 and 2000 and found that in a turbulent environment, goal complexity, information technology complexity, decision-making complexity, strategy complexity, and network complexity facilitated firms' adaptation process and yielded better performance. Meanwhile, a few studies using computer simulation also confirmed a positive relationship between organizational complexity and firm performance (Moldoveanu & Bauer, 2004; Morel & Ramanujam, 1999; Siggelkow & Rivkin, 2005). Absorption mechanism, which recognizes the value of a complex organizational structure, has been established.

However, when exploring the benefits of organizational complexity, some researchers also found that organizational complexity did not influence firm performance or, even worse, it produced negative influences. Bourgeois (1985) could not detect any significant relationship between goal complexity and firm performance. Organizational structure complexity was found to impede organizational innovativeness and then jeopardize firm performance (Blau & McKinley, 1979). These findings question the effectiveness of organizational complexity (Anderson, 1999; Simon, 1996); therefore, the reduction mechanism, which recognizes the difficulties in leveraging a complex organizational structure and appreciates the logics of simplicity, has been proposed.

In prior studies, organizational complexity was often conceptualized as the number of factors or elements that were involved in a system, such as the number of hierarchical levels in an organization, the number of tasks, or the number of operations and activities an organization employed (Blau & McKinley, 1979). The same numerical concept was also used to describe the complexity of organizational structures, decision-making processes, goal systems, or relation networks (Ashmos *et al.*, 1996). Since organizational complexity is manifested by the number of different elements within a system, the numerical quantity becomes the essence of the complexity definition (Anderson *et al.*, 1999; Ashmos *et al.*, 1996; Simon, 1996). Along with these studies, we deem *organization complexity* as reflected by the number of elements within an organizational system (Boisot & Child, 1999; Daft, 1992; Morel & Ramanujam, 1999). It is worth emphasizing that, in alignment with prior empirical research, we strictly and narrowly define organizational complexity by the quantity of elements. However, the manner in which these elements connect with each other, i.e., the structure of elements, also affects the complicatedness of an organization and thus can be included in a broader definition of organizational complexity (Achrol, 1991; Achrol, Reve, & Stern, 1983). In this paper, we use organizational variability to describe the structure among elements. In other words, in order to better explore complexity phenomena, in this paper, we divide the broad definition of complexity into two dimensions: organizational complexity—the number of elements, and organizational variability—the structure of these elements.

Environmental Uncertainty

Environmental uncertainty, viewed as a defining factor for organizations (Atherton, 2003; Pennings, 1975), requires the presence of both environmental complexity and dynamism (Bourgeois, 1985; Boyd & Fulk, 1996; Damanpour, 1996; Gibbs, 1994; Jauch & Kraft, 1986; Prakken, 2004; Siggelkow & Rivkin, 2005; Tung, 1979). *Environmental complexity* is reflected by the "level of complex knowledge that understands what the environment requires" (Sharfman & Dean, 1991). *Environmental dynamism*, as defined by Miller and Friesen (1983), refers to the rate of change and innovation in an industry as well as the uncertainty or predictability of the actions of competitors and customers.

Environmental complexity describes the structure of an environment, including the number of relevant factors and the interactions of these factors. Of these two sub-dimensions, the interaction generates more complexity than the number of environmental factors. A large number of factors will not create much complexity if all factors are isolated from one another. Environmental complexity is mainly produced by the multi-directional interactions among these factors. In addition, as the number of factors increases, the number of interactions among these factors increases exponentially. For instance, when the number of factors increases from 3 to 6, even regardless of the direction, the number of interactions will increase from 4 to 57. Therefore, environmental complexity is reflected more by the interactions of external factors than by the number of these factors.

Environmental dynamism describes the changes in an environment, including the frequency of changes and the magnitude of each change (Tung, 1979). Among the two sub-dimensions, the magnitude of a change refers to the scale or the significance of a marginal change. Usually radical changes are rarer than incremental changes and we assume that in a regular market, incremental changes are common and can be expected and prepared for by firms. Therefore, environmental dynamism is mainly reflected by the frequency of changes (high frequency means quickly-changing environment, and low frequency means slowly-changing environment).

Organizational Complexity Mechanisms

The Benefit of Organizational Complexity Mechanisms

With an understanding of environmental complexity and dynamism, we can now explore how to apply organizational complexity mechanisms to cope with environmental uncertainty. As discussed, the advantages of absorption and reduction mechanisms are apparently contradictory. By developing networks or multiplying portfolios, the absorption mechanism complicates organizational structure and expands the organization's boundary. It internalizes environmental factors and incorporates them into the span of an organization's control (Oviatt, 1988; Williamson, 1981). Therefore, the number of uncontrollable external factors is decreased. In addition, a complicated organizational system with a strong network will grant firms more channels to collect information and thus a better picture of the environment will be formed (Cohen, 1999; Siggelkow & Rivkin, 2005). The reduction mechanism, in contrast, codifies and abstracts environmental factors into a few categories, which facilitates the diffusion of information within the organizational system (Boisot & Child, 1999). It aids information transfer by simplifying internal structures and decreasing the number of nodes or hierarchies that a piece of information has to go through. As such, these two mechanisms can be used differently in managing environmental complexity and dynamism.

When only environmental complexity is concerned, absorption will outperform reduction in reducing the impacts from environmental complexity. Absorption expands organizational boundaries and incorporates part of the environment into the system and therefore, some of the uncontrollable and unpredictable external factors will be turned into manageable and predictable internal factors within the organizational control. By absorbing the externality, both the number of environmental factors and the number of interactions are eliminated. When the number of environmental factors decreases, the number of interactions decreases even faster; environmental complexity is thus effectively reduced by absorption mechanism. Reduction, on the contrary, codifies or abstracts environmental factors into a few categories for a better generalizeable understanding of the environment. Therefore, the perception of environmental complexity is decreased while the actual environmental complexity has not been affected. In addition, the reduction mechanism may oversimplify or even misinterpret the environment by abstracting everything into merely a few groups. Therefore, where only environmental complexity is concerned, a complicated organizational system suggested by the absorption mechanism will perform better than a simplified organizational system suggested by the reduction mechanism. Such a proposition can also be verified by prior empirical studies.

It has been found that multiple goals push firms to consider more possibilities for future development and to prepare for a variety of uncertainties in the future market (Bourgeois, 1985); multiple strategies give firms various hedging options in dealing with external changes (Ashmos *et al.*, 1996); and complicated structures provide firms enough heterogeneity to collect information about different aspects of an environment (Van Maanen & Schein, 1979). All of these make firms more open to various possibilities, and lead them to collect more information and hedge external complexity by creating more potential options. Therefore, we believe that a complex structure helps a firm to cope with a complex environment.

Proposition 1: As environmental complexity increases, organizational complexity should increase.

On the contrary, where only environmental dynamism is concerned, the reduction mechanism will outperform the absorption mechanism. Absorption complicates organizational structures and processes and creates more hierarchies and formality within the organization, which typically slows down the dynamics of the system (Siggelkow & Rivkin, 2005). Reduction, however, specifies the roles of different departments according to the requirements of the environment (Boisot & Child, 1999). By defining the function and role of each department, the reduction mechanism and the corresponding simpler organizational structure accelerate information transfer within the organization (Garg, Walters, & Priem, 2003). Previous studies have substantially verified that a complex organizational structure helps to collect more information but slows down information transfer, while simpler structure facilitates information transfer but decreases information richness (Siggelkow & Rivkin, 2005).

The advantage in facilitating information transfer has given simpler organizations a greater chance than complex organizations to survive in a turbulent environment. Luo, Tan, & Shenkar (1998) found that an unstable environment required firms to respond effectively and quickly to market competition. Burns and Stalker (1961) also agreed that a turbulent environment required the ability to process alternatives rapidly. Jaszkiewicz (2004) and Rosenthal (1985) found that a clearly hierarchical goal system facilitated strategy execution and implementation processes. Mintzberg (1973), as well, contended that in order to make decisions quickly, managers must obtain inputs from fewer resources, consider fewer alternatives, and limit themselves to fewer analyses. Summarizing these empirical findings, we conclude that:

Proposition 2: As environmental dynamism increases, organizational complexity should decrease.

The Cost of Organizational Complexity Mechanisms

The reduction mechanism implies that the costs of creating and maintaining a certain level of organizational complexity cannot be neglected. Although a complicated organizational form enables firms to collect more information and provide them with better flexibility and more options, keeping such a complex form requires substantial efforts. For example, multiple goals are established at the cost of scanning different environmental segments, searching for right opportunities, and negotiating possible alternatives (Sawyer, 1993). Multiple strategies and complicated structures cause difficulties in coordinating functional departments within the organization as well as in balancing different interest groups outside the organization (Morel & Ramanujam, 1999). The cost concern of organizational complexity suggests that compared with absorption, reduction, with fewer representations of environmental uncertainty and a singular adaptive response mechanism, may reduce frictions and conflicts, and thus lead to a better chance of improved performance (Miller, 1993). For instance, an organization's operating costs decrease as the level of complexity/redundancy within an organization decreases (Lin & Carley, 1997). The cost aspect should thus be taken into consideration whenever the discussion of organizational complexity arises (Tushman & Naddler, 1978). Hereafter, we will elaborate on the interactions between organizational complexity and the other two organizational dimensions, variability and dynamism, to illustrate the cost consideration of maintaining a certain level of organizational complexity.

Organizational variability refers to the compatibilities among the elements within an organizational system. This concept is created to show the difference between the number of elements (i.e., complexity) and the interactions among these elements (i.e., variability). Variability is not only based on the number of elements (i.e., complexity), but also is affected by the compatibility among these elements, which is reflected by the variance of the importance or priorities of the elements that is imposed by an organization (Achrol, 1991; Achrol, *et al.*, 1983). The greater the organizational variability is, the more difficult for a firm to coordinate and manage multiple elements at the same time (Lin & Carley, 1997).

The concept of organizational variability recognizes the potential difficulty created by the elements' compatibilities. However, different from the complexity reduction view, which believes that the complexity cost is unavoidable, organizational variability shows that the costs of initiating and maintaining organizational complexity do not necessarily increase to an unsustainable level as the number of distinguishable elements increases. When elements are compatible with one another, that is, when a clear priority is established, different functionalities can be coordinated and a low organizational variability can be attained. Theoretically, if multiple elements are clearly prioritized, organizations are able to allocate resources and perform strategies with an unambiguous precedence. Efficiency is thus kept or even improved when organizational complexity increases. In contrast, if an organization assigns the same weight to each goal, or evaluates different types of strategies or different structural components as equally important, the potential difficulty of running the business may be intensified due to the extra time and energy associated with the increasing communication, coordination, and management work. The benefit of a complex organization may be compromised.

In sum, the cost represented by organizational variability may not always increase as organizational complexity increases. When organizations are clear about the relative importance of elements, the organizational variability cost may be kept at a manageable level; otherwise, the increase of organizational complexity will entrap the organization by trying to accomplish too many things at the same time.

Proposition 3: As organizational complexity increases, organizational variability may or may not increase depending upon the priority arrangement of elements. Specifically, if there is a priority arrangement of elements, organizational variability may not increase; otherwise, organizational variability will grow.

The second type of cost that is caused by organizational complexity is related to organizational dynamism. *Organizational dynamism* refers to the time that a firm needs to formulate an effective process to respond to environmental changes and execute strategies, such as the dynamism of reporting systems or decision-making processes (Baum & Wally, 2003; Lin & Carley, 1997; Teece *et al.*, 1997).

Organizational dynamism is driven by organizational complexity (Morel & Ramanujam, 1999) and it should be taken into consideration whenever issues of organizational capability, boundary, and process arise (Cohen, 1999). In organization theory, "dynamic" reflects the organizational capability to renew competences to achieve congruence with the changing business environment and to generate innovative responses when time-to-market is critical. The capability of staying "dynamic" secures a firm's market position when the nature of future competitions and markets is difficult to predict (Teece *et al.*, 1997). Organizational dynamism thus reflects a firm's ability to integrate, build, configure, and reconfigure internal and external competences to address environmental changes rapidly (Teece *et al.*, 1997). However, as organizational complexity grows and as the organization evolves towards separable subsystems, the whole system loses the ability to carry out its activity sets or to implement possible changes reliably and predictably (Moldoveanu & Bauer, 2004). As observed, organizational complexity typically undermines the dynamics of the information-transmitting system (Repenning, 2002) and the decision-making system (Baum & Wally, 2003).

However, the effect of organizational dynamism in improving firms' competitiveness across environments is well documented (Baum & Wally, 2003). A greater level of organizational dynamism leads to an early adoption of new products, business models (Jones, Lanctot, & Teegen, 2000), or efficiency-gaining technologies (Baum *et al.*, 2000). Therefore, greater organizational dynamism enables firms to catch external opportunities before they disappear (Stevenson & Gumpert, 1985). Nevertheless, as organizational complexity increases, organizational dynamism will be sacrificed. Organizational complexity increases the nodes or levels that a piece of information has to go through and slows down the whole internal process (Hermann, 1963; Lin & Carley, 1997; Staw, Sanderlands & Dutton, 1981). Therefore, we conclude that:

Proposition 4: As organizational complexity increases, organizational dynamism tends to decrease.

As an extension of this argument, we notice that although the increase in the number of elements can slow down the organization, the way that these elements are structured (i.e., organizational variability) may revise the impact of organizational complexity on organizational dynamism. As discussed, when an organization is able to prioritize different elements, it is likely organizational dynamism need not be compromised by an increase in the element number. For instance, Bourgeois (1985) has found that a goal system with different priorities attaching to different goals (low goal variability) will facilitate the decision-making process, while a system with equally-important goals (high goal variability) will impede the decision-making process by requiring more time for balancing different considerations or interests. Therefore, we propose that:

Proposition 5: As organizational variability decreases, i.e., organizational elements are prioritized, organizational dynamism tends to decrease.

The Trade-Off between Benefits and Costs

We integrate the above benefit-and-cost discussion of organizational complexity in a contingent context, as Figure 1.

In Figure 1, the X-axis represents an increasing environmental complexity, the Y-axis represents an increasing environmental dynamism, and the Z-axis represents organizational complexity. Both front and back planes present an

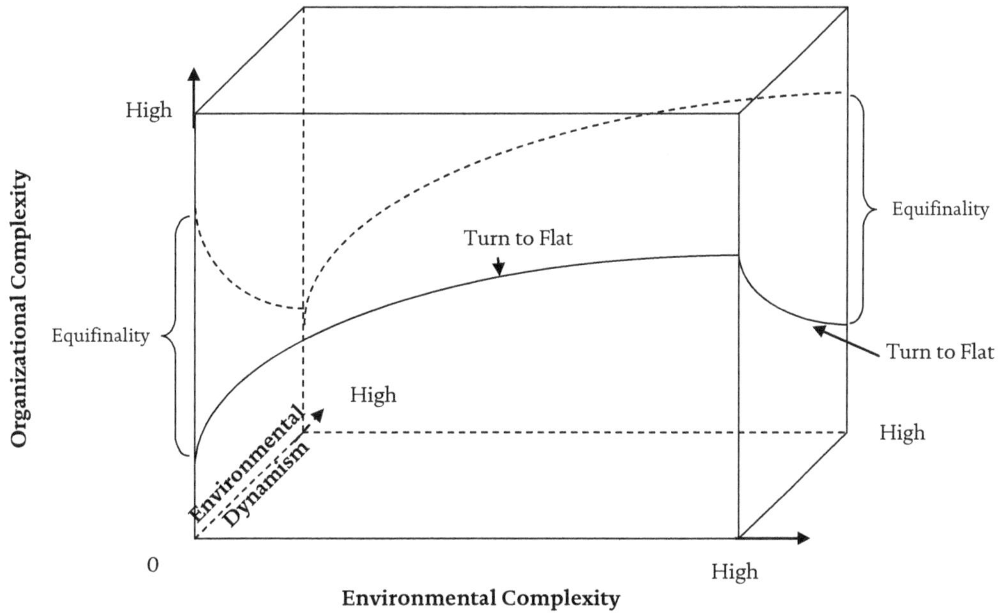

Figure 1 *Organizational Complexity, Environmental Complexity, and Environmental Dynamism.*

increasing level of environmental complexity from left to right; therefore, the pattern of organizational complexity will be the same in these two planes. Also, both right and left planes present an increasing level of environmental dynamism from inside to outside; therefore, the pattern of organizational complexity will be the same in these two planes.

First, in the front plane of Figure 1, when only environmental complexity is concerned, an organization should increase organizational complexity to decrease the number of external factors that are out of the firm's control, as suggested by the absorption mechanism. However, when organizational structure becomes too complicated, the relevant costs (i.e., the increased organizational variability and the decreased organizational dynamism) will offset the benefits attained by organizational complexity. Therefore, the growth of organizational complexity will be decelerated and the positive growth rate will turn into zero at the point when the marginal benefit provided by organizational complexity is equal to the marginal cost. The shape of the organizational complexity curve, therefore, turns into an inverse L-shape.

When only environmental dynamism is concerned, as the right plane of Figure 1, organizations should pursue a reduction mechanism, which employs comparatively simpler organizational structures in order to facilitate information transfer and to produce quicker responses. However, a lower level of organizational complexity hurts the richness of the information collected (Siggelkow & Rivkiv, 2005). As organizational complexity decreases, the change in marginal costs—the decrease of organizational variability and the increase of organizational dynamism—is also decreasing and therefore, the quest for further simplifying the organizational system is undermined. In addition, no matter how simple an organization is, a certain level of organizational complexity is always needed in order to maintain daily routines and to support basic business operations. Therefore, the organizational complexity curve will eventually be formulated as an L-shape.

The interface of the right plane and the back plane of Figure 1 shows a point where equifinality may occur, which indicates that no one-best-fit organizational complexity level can be found at this point and different levels of complexity may all achieve satisficing performance. This interface signifies a situation where the highest environmental complexity and the highest environmental dynamism co-exist. High environmental complexity requires a great level of organizational complexity, while high environmental dynamism requires a low level of organizational complexity. At this point, firms may form a range of organizational complexity levels, depending on managerial perceptions of the environment and the organizational strategic orientations. Firms that are threatened more by external complexity than external dynamism will develop a comparatively complicated structure, while firms that view external dynamism as a bigger menace may develop a comparatively simple structure. Firms who can achieve a delicate balance between collecting more information and facilitating quicker information transfer will perform well. In the back plane, as environmental complexity decreases, the requirements for a complex organizational structure were also mitigated. Therefore, the curve extends to the left side plane with a lower level of organizational complexity. The organizational complexity curve in the back plane presents the same inverse L-shape as in the front plane.

The left side plane copies the pattern in the right plane and the organizational complexity curve presents an L-shape. At the interface between the left plane and the front plane, which signifies the lowest levels of both environmental complexity and dynamism, a range of organizational complexity levels will be observed again. A low level of environmental complexity requires a corresponding low level of organizational complexity, while a low level of environmental dynamism allows a high level of organizational complexity. In other words, a range of organizational complexity may be formed once again and equifinalty may be observed at this point once more.

Firms always need a certain level of organizational complexity to support their basic daily routines. As organizational complexity increases, more potential options and alternatives are offered, which help firms hedge more risks and capture more opportunities whenever they come up (Ashmos *et al.*, 1996; Ashmos *et al.*, 2000). However, as organizational complexity increases, costs offset the benefits provided by organizational complexity and may outweigh benefits at some point. Specifically, when organizational complexity increases, organizational variability may increase and organizational dynamism decreases.

The increased organizational variability is associated with greater coordination and negotiation costs, and the decreased organizational dynamism leads to slower response system. The benefit of organizational complexity is limited by these constraints. The relationship between organizational complexity and its contribution to firm performance is therefore expected to be in an inverted U-shape: a basic level of organizational complexity is always needed to support firm basic operations; a sufficient level of organizational complexity will benefit organizational performance, while too much complexity will hurt performance.

Proposition 6: Organizational complexity has an inverse U-shape relationship with firm performance.

Conclusion and Implications

Scholars have observed that a new perspective is emerging from organization theories but no systematic framework or logic has been established yet (Cohen, 1999). As a consequence, contradictory findings are found in previous studies regarding the application of organizational complexity mechanisms in coping with environmental uncertainty and regarding their benefits to firm performance. The opposite effects of complexity reduction and absorption mechanisms and the resultant difficulty in identifying the main effects of complexity constructs keep researchers suspecting complexity is merely the mask of simplicity (Anderson, 1999; Simon, 1996). This suspicion is also reflected by recent discussions in exploration and exploitation literatures (Benner & Tushman, 2003; Levinthal & March, 1993; March, 1991; Rothaermel & Deeds, 2004). On one hand, complicated organizational forms help firms to collect more information and to explore new opportunities, which serve for the "experiment with new alternatives" (March, 1991: 85). On the other hand, the specialization and standardization implied by simpler organizational structures optimizes the efficiency of current operations and exploits present opportunities by refining and extending the "existing competencies, technologies, and paradigms" (March, 1991: 85). Different levels of organizational complexity reflect different organizational capabilities and how to achieve a delicate balance among them becomes essential in attaining a marketable competitiveness.

As such, we propose this organizational dimensionality perspective to reconcile previous contradictory findings and investigate the conventional management themes through a different angle. By formulating three dimensions that highly abstracts organizational systems, a dimensionality perspective analyzes the structuration of the elements that compose an organizational system and examines the contingent conditions for employing different levels of complexity in this system. By doing so, a dimensionality approach complements the traditional elemental content approach, in which the content of organizational elements, such as different types of goals or strategies, is used to explain the fitness between organizational systems and environments. The integration of the dimensionality approach into the traditional content-analysis approach should yield better results for organizational research. When the content approach predicts what type of organizational structure should be used to adapt to a specific environment, the dimensionality approach can be used to prescribe how to arrange relevant organizational elements in this structure in order to function as designed. We hope that by combining the dimensionality perspective into other theories, the effectiveness of this new approach should be fully demonstrated.

References

Achrol, R.S. (1991). "Evolution of the marketing organization: New forms for turbulent environments," *Journal of Marketing*, ISSN 0022-2429, 55: 77-93.

Achrol, R.S., Reve, T. and Stern, L.W. (1983). "The environment of marketing channel dyads: A framework for comparative analysis," *Journal of Marketing*, ISSN 0022-2429, 47: 55-67.

Anderson, P. (1999). "Complexity theory and organizational science," *Organization Science*, ISSN 1526-5455, 10(3): 216-232.

Anderson, P., Meyer, A., Eisenhardt, K., Carley, K. and Pettigrew, A. (1999). "Introduction to the Special Issue: Applications of complexity theory to Organization Science," *Organization Science*, ISSN 1526-5455, 10(3): 233-236.

Ashby, W. (1956). *Introduction to Cybernetics*, ISBN 9780416683004.

Ashmos, D.P., Duchon, D., Hauge, F.E. and McDaniel, R.R. Jr. (1996). "Internal complexity and environmental sensitivity in hospitals," *Hospital & Health Services Administration*, ISSN 8750-3735, 41(4): 535-555.

Ashmos, D.P., Duchon, D., and McDaniel, R.R. Jr. (2000). "Organizational responses to complexity: The effect on organizational performance," *Journal of Organizational Change Management*, ISSN 0953-4814, 13(6): 577-594.

Atherton, A. (2003). "The uncertainty of knowing: An analysis of the nature of knowledge in a small business context," *Human Relations*, ISSN 0018-7267, 56(1): 1379-1398.

Baum, J.A., Calabrese, T., and Silverman, B.S. (2000). "Don't go it alone: Alliance network composition and startups' performance in Canadian biotechnology," *Strategic Management Journal*, ISSN 0143-2095, 21(3): 267-294.

Baum, J.R. and Wally, S. (2003). "Strategic decision speed and firm performance," *Strategic Management Journal*, ISSN 0143-2095, 24: 1107-1129.

Benner, M.J. and Tushman, M.L. (2003). "Exploitation, exploration, and process management: the productivity dilemma revisited," *Academy of Management Review*, ISSN 0363-7425, 28(2): 238-256.

Blau, J.R. and McKinley, W. (1979). "Ideas, complexity, and innovation," *Administrative Science Quarterly*, ISSN 0001-8392, 24: 200-219.

Boisot, M. and Child, J. (1999). "Organizations as adaptive systems in complex environments: The case of China," *Organization Science*, ISSN 1526-5455, 10(3): 237-253.

Bourgeois III, L.J. (1985). "Strategic goals, perceived uncertainty, and economic performance in volatile environments," *Academy of Management Journal*, ISSN 0001-4273, 28(3): 548-573.

Boyd, B.K. and Fulk, J. (1996). "Executive scanning and perceived uncertainty: A multidimensional model," *Journal of Management*, ISSN 1557-1211, 22: 1-21.

Burns, T. and Stalker, G.M. (1961). *The Management of Innovation*, ISBN 9780198288787.

Burton, R.M. and Forsyth, J.D. (1986). "Variety and the firm's performance: An empirical investigation," *Technovation*, ISSN 0166-4972, 5: 9-21.

Cohen, M. (1999). "Commentary on the organization science special issue on complexity," *Organization Science*, ISSN 1526-5455, 10(3): 373-376.

Daft, R.L. (1992). *Organization Theory and Design*, ISBN 9780314933652.

Damanpour, F. (1996). "Organizational complexity and innovation: Developing and testing multiple contingency models," *Management Science*, ISSN 0025-1909, 42(5): 693-716.

Garg, V.K., Walters, B.A. and Priem, R.L. (2003). "Chief executive scanning emphases, environmental dynamism, and manufacturing firm performance," *Strategic Management Journal*, ISSN 0143-2095, 24: 725-744.

Gibbs, B. (1994). "The effects of environment and technology on managerial roles," *Journal of Management*, ISSN 1557-1211, 20(3): 581-604.

Hermann, C.F. (1963). "Some consequences of crisis which limit the viability of organizations," *Administrative Science Quarterly*, ISSN 0001-8392, 8: 343-358.

Jaszkiewicz, A. (2004). "On the computational efficiency of multiple objective metaheuristics. The knapsack problem case study," *European Journal of Operational Research*, ISSN 0377-2217, 158(2): 418.

Jauch, L.R., and Kraft, K.L. (1986). "Strategic management of uncertainty," *Academy of Management Review*, ISSN 0363-7425, (4): 777-790.

Jones, G.K., Lanctot, A. Jr. and Teegen, H. (2000). "Determinants and performance impacts of external technology acquisition," *Journal of Business Venturing*, ISSN 0883-9026, 16: 255-283.

Levinthal, D.A., and March, J.G. (1993). "The myopia of learning," *Strategic Management Journal*, ISSN 0143-2095, 14(Winter Special Issue): 95-112.

Lin, Z., and Carley, K.M. (1997). "Organizational response: The cost performance tradeoff," *Management Science*, ISSN 0025-1909, 43(2): 217-234.

Luo, Y., Tan, J.J. and Shenkar, O. (1998). "Strategic responses to competitive pressure: The case of township and village enterprises in China," *Asia Pacific Journal of Management*, ISSN 0217-4561, 15(1): 33.

March, J.G. (1991). "Exploration and exploitation in organizational learning," *Organization Science*, ISSN 1526-5455, 2: 71-87.

Miller, D. (1993). "The architecture of simplicity," *Academy of Management Review*, ISSN 0363-7425, 18(1): 1165-138.

Miller, D. and Friesen, P.H. (1983). "Strategy-making and environment: The third link," *Strategic Management Journal*, ISSN 0143-2095, 4: 221-235.

Mintzberg, H. (1973). "Strategy making in three modes," *California Management Review*, ISSN 0008-1256, 16: 44-58.

Moldoveanu, M.C. and Bauer, R.M. (2004). "On the relationship between organizational complexity and organizational structuration," *Organization Science*, ISSN 1526-5455, 15(1): 98-118.

Morel, B. and Ramanujam, R. (1999). "Through the looking glass of complexity: The dynamics of organizations as adaptive and evolving systems," *Organization Science*, ISSN 1526-5455, 10(3): 178-293.

Oviatt, Benjamin M. (1988). "Agency and transaction cost perspectives on the manage-shareholder relationship: Incentives for congruent interests," *Academy of Management Review*, ISSN 0363-7425, 13(2): 214-225.

Pennings, J.M. (1975). "The relevance of the structural-contingency model for organizational effectiveness," *Administrative Science Quarterly*, ISSN 0001-8392, 20(3): 393-410.

Prakken, B. (2004). "Uncertainty, information and reorganization," *Information Society*, ISSN 0197-2243, 20(1): 53-57.

Repenning, N.P. (2002). "A simulation-based approach to understanding the dynamics of innovation implementation," *Organization Science*, ISSN 1526-5455, 13(2): 109-127.

Rosenthal, R.E. (1985). "Principles of multiobjective optimization," *Decision Science Journal*, ISSN 0011-7315, 16(2): 133-52.

Rothaermel, F.T., and Deeds, D.L. (2004). "Exploration and exploitation alliances in biotechnology: A system of new product development," *Strategic Management Journal*, ISSN 0143-2095, 25: 201-221.

Sawyer, O.O. (1993). "Environmental uncertainty and environmental scanning activities of Nigerian manufacturing executives: A comparative analysis," *Strategic Management Journal*, ISSN 0143-2095, 14: 287-299.

Sharfman, M.P., and Dean, J.W., Jr. (1991). "Conceptualizing and measuring the organizational environment: A multidimensional approach," *Journal of Management*, ISSN

1557-1211, 17(4): 681-700.

Siggelkow, N., and Rivkin, J.W. (2005). "Speed and search: Designing organizations for turbulence and complexity," *Organization Science*, ISSN 1526-5455, 16(2): 101-122.

Simon, H.A. (1996). *The Architecture of Complexity*, ISBN 9780262193740.

Staw, B.M., Sanderlands, L.E. and Dutton, J.E. (1981). "Threat-rigidity effects in organizational behavior: A multilevel analysis," *Administrative Science Quarterly*, ISSN 0001-8392, 26: 501-524.

Stevenson, H. and Gumpert, D. (1985). "The heart of entrepreneurship," *Harvard Business Review*, ISSN 0017-8012, 63(2): 85-94.

Teece, D.J., Pisano, G. and Shuen, A. (1997). "Dynamic capabilities and strategic management," *Strategic Management Journal*, ISSN 0143-2095, 18(7): 509.

Tung, R. (1979). "Dimensions of organizational environments: An exploratory study of their impact in Organizational structure," *Academy of Management Journal*, ISSN 0001-4273, 672-693.

Tushman, M.L. and Nadler, D.A. (1978). "Information processing as an integrating concept in organizational design," *Academy of Management Review*, ISSN 0363-7425, July: 613-624.

Van Maanen, J. and Schein, E. (1979). "Toward a theory of organizational socialization," in B.M. Staw (ed.), *Research in Organizational Behavior*, ISBN 9780892320455, pp. 209-264.

Williamson, O. (1981). "The economics of organization: The transaction cost approach," American Journal of Sociology, ISSN 0002-9602, 87: 548-77.

Wholey, D.R. and Brittain, J. (1989). "Characterizing environmental variation," *Academy of Management Journal*, ISSN 0001-4273, 22(4): 867-882.

Zhi Tang is an assistant professor at Rochester Institute of Technology. He graduated from the University of Alabama with a Ph.D. degree in Management. He also has a master degree in Finance from Fudan University, China, and a bachelor degree in International Economics from Shandong University, China. Dr. Tang is especially interested in Organizational Complexity and his dissertation is titled "Organizational Complexity: Assumption, Utility, and Cost". His primary research interests are thus in the areas of organizational complexity, configuration theory, organizational dynamism, and organizational variability. He also conducts research in environmental uncertainty, entrepreneurial orientation, and entrepreneurship in emerging markets. His applications appear in *Entrepreneurship Theory & Practice*, *Journal of Developmental Entrepreneurship*, etc.

5. Temporality in Link Prediction: Understanding Social Complexity

Anet Potgieter[1], *Kurt A. April*[1,3], *Richard J.E. Cooke*[2] & *Isaac O. Osunmakinde*[1]
1 Department of Computer Science, University of Cape Town, ZAF
2 Graduate School of Business, University of Cape Town, ZAF
3 Ashridge, UK

This article describes research into the discovery and modelling of emergent temporal phenomena in social networks. It summarizes experimental results that bring together two views in contemporary science: Bayesian analysis and link prediction, to enhance the current understanding of emergent temporal patterns in social network analysis (SNA), particularly in value creation through social connectedness—an important, and growing, discipline within management science. Traditional link prediction methods use the values of metrics in a graph to determine where new links are likely to arise, and little work has been done on analyzing long-term graph trends. We have found that existing graph generation models are unrealistic in their prediction, and can be complemented through the use of temporal metrics, in the study of some networks. To date, no temporal information has been used in link prediction research, thereby excluding valuable temporal trends that emerge in sociogram sequences and also lowering the accuracy of the link prediction. We extracted information from the Pussokram online dating network dataset, and 9,939 cases of each class were formed. Logistic regression in the Weka data mining system was used to perform link prediction. Our results show that temporal metrics are an extremely valuable new contribution to link prediction, and should be used in future applications. In addition to using metrics to measure the local behaviors of participants in social networks, we used Bayesian networks to model the interrelationships between the metrics as local behaviors and links forming between individuals as emergent behaviors (social complexity). We also explored how the metrics evolve over time using Dynamic Bayesian Networks (DBN).

Introduction

Social networks are complex systems that are characterized by high numbers of interconnected component entities, and a high degree of interaction between these entities. The interrelationships in such a network are dynamic and evolve over time. Temporal changes in social networks are difficult to understand and anticipate. The interrelationships between the component entities in a social network and its global behavior can be so numerous and mostly hidden, and can affect so many different entities throughout the social network that it becomes extremely difficult to comprehend.

Complexity theory is ideally suited to study social networks. Complex adaptive systems theory is a branch of complexity theory that studies systems that consist of agents that are collectively able to evolve in response to environmental changes. The agents in such a system constantly act and react to the actions of other agents and events in the environment. A social network is a complex adaptive system, in which people are agents interacting with each other.

In order to understand social complexity, the local behaviors of the participants must be understood, as well as how they act together and interact with the environment to form the whole. To model this, we use Bayesian network techniques to mine and model emergent relationships between local behaviors and global behaviors in social networks over time.

Social Complexity

The complex structure of any social organization can be thought of as a network of individuals/agents (Nohria & Eccles, 1992: 288; Lincoln, 1982), sometimes termed network actors, that operates and is operated on in an environment which itself is an environment of other distributed organizations (Van Wijk *et al.*, 2003; Potgieter *et al.*, 2006), and the actions of agents within the network are shaped and constrained because of their position and embeddedness in the network (Nohria, 1992).

Complex adaptive systems theory, a branch of complexity theory that is well suited to study social networks, investigates systems that consist of agents that are collectively able to evolve in response to environmental changes (i.e., people who interact with each other in socially complex ways). Social networks are complex systems that are characterized by high numbers of dynamically interconnected component entities, and a high degree of time-based evolutionary interactions between these entities. Temporal changes in social networks are difficult to understand and anticipate, because of the structures of constraint and opportunity negotiated and reinforced between interacting individuals. The interrelationships between the component entities in a social network and its global behavior can be so numerous and mostly hidden, and can affect so many different entities throughout the social network that it becomes extremely difficult to comprehend. Such changes, for modern organizations, can come in the form of globalization, deregulation, competition, technology and political transformation. The agents in such a constantly changing system tend to step out of the traditional boundaries of their network (within-group identification), act and react to, and evolve in relation to, the actions of other agents and events in the environment to achieve a more desirable end, e.g., to learn faster in the changing environment, or to collaborate better, or to craft more relevant identities, or to seek help in making major decisions, or to negotiate and renegotiate relationships of power, etc.

In social network studies, the traditional archetype—which acts as an interpretive schema—of an organization is that it is made up of a number of static departments, which themselves have specialised individuals (agents), and its strategy for competing, or winning, or succeeding, or sustaining itself is fixed for a time period, usually 1-3 years, and is clearly understood and acted upon by its agents. After the specified period, usually the organizational strategy is reviewed, altered, or rewritten or totally revoked in favour of a new strategy, and its agents then enact new behaviors against such a plan. We contend, however, that organizations are not static, atomistic agents, instead they are recurring in dynamic agent linkages and embedded in networks of increasing importance (Brock, 2006) that progressively influence competitive actions (Granovetter, 1985, 1992; Burt, 1995) and may themselves challenge the interpretive schema and supportive structures, and ultimately even delegitimize the existing archetype.

Social Network Analysis and Link Prediction

Social Network Analysis (SNA) is a research area aimed at understanding social complexity by representing and analyzing social networks using mathematical graphs. This was first done by researchers, such as Cartwright and Harary (1956) in the USA, who interpreted Kurt Lewin's social interaction theory into graph theory—thereby helping to transform the study of social networks from description to analysis. In this theory, a graph ***G*** is a structure consisting of a set of nodes ***V*** (also called vertices), and a set of links ***E*** (also called arcs or edges). In our discussion the terms graph, network, social network and sociogram are used interchangeably. A graph can be bi-directed[1], meaning that we are not concerned about the order of the nodes in a link. Alternatively, a graph is directed, meaning that the nodes in a link form an ordered pair. Nodes represent people and links represent some type of relationship between people. For the purposes of this discussion a link exists between two people if they have exchanged a message in the past. Thus links are never removed once they have been formed.

There appears to be one standard textbook on SNA (Wassermann & Faust, 1994), cited by the large majority of researchers in this field. Although SNA has existed for over fifty years, most analysis techniques have been designed for static data, or at least a static archetype of social organization. For example, Wassermann and Faust (1994) contains no mention of temporal metrics, even though it was written in 1994 when electronic networks were well established.

1. Also called "undirected", but bi-directed is used in this paper to emphasise that links are directed and go in both directions.

With the increase in the use of computers, collecting enough data to create numerous graphs over fixed time intervals becomes possible. An example is creating a graph per week from email data, using a server's email log of 'to', 'from', and 'date' fields (Campbell *et al.*, 2003). This series of graphs can be used to study the evolution of the network and the change over time in various metrics. Predicting certain changes to a social network is called the link prediction problem. Liben-Nowell and Kleinberg (2003) explain it as:

> *Given a snapshot of a social network at time* t, *we seek to accurately predict the edges that will be added to the network during the interval from time* t *to a given future time* t'.

Link prediction has many real world applications, especially in the fields of marketing and crime prevention. Examples include:

- Identifying the structure of a criminal network (i.e., predicting missing links in a criminal network using incomplete data).
- Overcoming the data-sparsity problem in recommender systems using collaborative filtering (Huang *et al.*, 2005).
- Accelerating a mutually beneficial professional- or academic connection that would have taken longer to form serendipitously (Farrell *et al.*, 2005).
- Improving hypertext analysis for information retrieval and search engines (Henziger, 2000).
- Monitoring and controlling computer viruses that use email as a vector (Lim *et al.*, 2005).
- Predicting when webpage users will next visit, in order to improve the efficiency and effectiveness of a site's navigation (Zhu, 2003).
- Link prediction might also be useful in ecology, though interdisciplinary sharing between these two fields is still new (McMahon *et al.*, 2001).

Metrics

Link prediction through topological analysis is performed by computing various metrics. A metric is a value calculated from a graph that describes the graph in some way. For instance, it is more likely that two nodes that both have a high degree (number of neighbours) are more likely to form a new link, than two nodes with a low degree (Liben-Nowell, 2005). Most traditional SNA metrics are described and defined in Wassermann and Faust (1994), and summarised in an online book by Hanneman (2001). We defined metrics to quantify local behaviors of agent resources in social resource combinations (Potgieter *et al.*, 2006). The appearance of new links between individual agent resources is emergent behavior of the social network. Table 1 lists the metrics that were used for link prediction in this research. They were chosen since prior research has found

Name	Definition	Description
Degree	$\#\{u_{j,i} : u_{i,j} \in U_n\}$ or $\#\Gamma(v_i)$	The number of links from v_i to any node at time step *n*.
Recency	N/A	One plus the number of time steps elapsed since the node last communicated. Has a value of one if the node sent or received a message in the current time step.
Betweenness (Wassermann & Faust, 1994)	$\sum_{v_j \in V} \sum_{v_k \in V, v_k \neq v_j} \frac{\#P(v_j, v_k, v_i)}{\#P(v_j, v_k)}$	The sum of all shortest paths between all nodes that contain v_i as a percentage of all shortest paths between all nodes. It ranges from 0 to (#*V*-1) (#*V*-2)/2.
Common neighbors (Huang *et al.*, 2005)	$\#\{v_k : u_{i,k} \in U_n, u_{k,j} \in U_n\}$ or $\#\{\Gamma(v_i) \cap \Gamma(v_j)\}$	The number of nodes linked to both focus nodes (i.e., mutual friends).
Adamic\ Adar similarity (Adamic & Adar, 2003)	General case: $\sum_{z: a shared feature} \frac{1}{\log(frequency(z))}$ Common neighbors case: $\sum_{v_z \in \Gamma(v_i) \cap \Gamma(v_j)} \frac{1}{\log(\#\Gamma(v_z))}$	The number of features shared by the nodes, divided by the log of the frequency of the features. This metric rates rarer features more heavily.
Preferential attachment (Huang *et al.*, 2005)	$\#\{v_k : u_{i,k} \in U_n\} \cdot \#\{v_k : u_{j,k} \in U_n\}$ or $\#\Gamma(v_i) \cdot \#\Gamma(v_j)$	The product of the number of edges incident to the two nodes.
Katz measure (Liben-Nowell, 2005)	$\sum_{l=1}^{\infty} \beta^l \cdot \#(paths_{v_i,v_j}^{<l>})$, where $paths_{v_i,v_j}^{<l>}$ is the set of all paths of length *l* from v_i to v_j	The sum of all paths between the nodes exponentially damped by length to weight short paths more heavily. $0 < \beta < 1$.

Table 1 *Metrics used in this link prediction research*

them to be useful. Huang *et al.* (2005) found that the most useful dyadic metrics for link prediction, in descending order, were the Katz measure, preferential attachment, common neighbours and the Adamic\Adar number. Recency is a metric that, to the authors' knowledge, has not been used before. The table uses certain symbols that are now defined:

# denotes the number of elements in a set. $\boldsymbol{U_n}$ is the set of bi-directed links of the social network at time step n. A bi-directed link between nodes v_i and v_j is notated as $u_{i,j}$ and $u_{j,i}$. $\Gamma(v_i)$ is the set of neighbours of node v_i, i.e., the set $\{v_j: u_{i,j} \in U\}$.

$P(v_i, v_j)$ is the set of all shortest paths from v_i to v_j. $P(v_i, v_j, v_x)$ is the set of all shortest paths from v_i to v_j that pass through node v_x.

Existing link prediction techniques use the values of metrics in a graph to determine where new links are likely to arise. Important contributions to the field include Popescul and Ungar (2003), Taskar *et al.* (2004), Popescul and Ungar (2004), and Zhou and Scholkopf (2004). The classic paper on link prediction is by Liben-Nowell and Kleinberg (2003). They tested the predictive power of only proximity metrics, including common neighbors, the Katz measure and variants of PageRank. They found some of these measures had a predictive accuracy of up to 16% (compared to a random prediction's accuracy of less than a percent). A third of Liben-Nowell's doctoral thesis (Liben-Nowell, 2005) was a chapter on link prediction in social networks. A few other link prediction papers are summarized in Getoor and Diehl (2005).

Past Contributions to Temporal Analysis

Leskovec *et al.* (2005) state that little work has been done on analyzing long-term graph trends:

> *Many studies have discovered patterns in static graphs, identifying properties in a single snapshot of a large network, or in a very small number of snapshots; these include heavy tails for in- and out-degree distributions, communities, small-world phenomena, and others. However, given the lack of information about network evolution over long periods, it has been hard to convert these findings into statements about trends over time.*

Their study of trends found that, over time, graphs densify and the average distance between nodes decreases. This was contrary to the existing beliefs that the average nodal degree remains constant and average distance slowly increases. They claimed that existing graph generation models are not realistic and proposed a new "forest-fire" generation model. Desikan and Srivastava (2004) studied the change in metrics of a set of webpages over time for the graph as a whole, and for single nodes (subgraphs are their current research). They found that temporal metrics, such as their Page Usage Popularity, can be effectively used to boost ranks of recently popular pages to those that are more obsolete.

This seems to indicate that temporal metrics are a useful addition to traditional static metrics in the study of some networks.

Using Temporal Analysis to Aid Link Prediction

To date no temporal information has been used in link prediction research (excluding our earlier proposition of the idea in April *et al.* (2005) and Potgieter *et al.* (2006). It is the authors' opinion that ignoring such information excludes valuable temporal trends that emerge in sociogram sequences that may greatly increase the accuracy of link prediction. We have three types of temporal metrics that have been defined, the first two of which are borrowed from finance:

- Return is the percentage increase or decrease of a value over a period of time (Ross *et al.*, 2001). It shows the rate of change of a given metric. For instance, if we are considering the degree of node v_i from time step one to time step fifty, the degree return would be:

$$\frac{\text{degree}_{50}(v_i) - \text{degree}_1(v_i)}{\text{degree}_1(v_i)} \quad ;$$

- Moving averages are used to extract long-term trends from short-term noise. They do not show trends or "movement", like return, but rather serve only to blur the values of the metrics around a point;
- Recency (Table 1), simply shows how much time has elapsed since a node has communicated.

Temporal Analysis Methodology

All the metrics listed in the Table 1 were calculated for a series of hundred networks constructed daily from information extracted from the *Pussokram* online dating network dataset. In addition to the computation standard metrics, their moving averages over twenty, ten and two days were computed, as well as their returns over the same period and the recency of each node. Up to one hundred sample dyads where new links formed on the day, and an equivalent number of unconnected dyads, were chosen for each network. There were 9,939 cases of each class in total. Logistic regression (Hosmer & Lemeshow, 1989) in the Weka data mining system (Witten & Frank, 2005) was used to perform link prediction. The dataset was split into a 70%/30% train/test division, and after the system was trained it predicted whether a given set of test instance metrics characterized a forming link or an unconnected link. The overall classification accuracy, the true positive rates for each class, and the kappa statistic (a measure of a system's accuracy improvement over a random guess) were calculated (Witten & Frank, 2005). The true positive rate of a class is the percentage of all the instances of that class that were predicted correctly. The kappa statistic is defined as:

$$\frac{P(A) - P(E)}{1 - P(A)}$$

where P(A) is the total percentage accuracy of instances predicted by the learning system and P(E) is the total percentage accuracy of instances predicted by random guessing. Since in this research an equal number of positive and negative instances were used we know that P(E) = 0.5 and 1-P(E) = 0.5, which we applied to the kappa statistic. A kappa value over 0.4 has been said to indicate 'good agreement beyond chance' (Fleiss, 1981).

Temporal Analysis Results and Conclusions

Table 2 shows the mean and standard deviation for each class, as well as the significance level of the mean difference using a Normal statistical difference of means test. The last column is the most important and shows the kappa value of each metric when used alone in a logistic regression against the class. The suffix 'F' stands for 'From' (the first node in a dyad instance), the suffix 'T' stands for 'To' (the second node in a dyad instance). 'A' stands for 'Average', 'R' stands for 'return' and the numbers '20', '10' and '2' stand for the number of time steps over which the average and return were calculated. The metric in the middle row of each group is the traditional static metric. The averages are shown above it and the returns are shown below. 'Forming' and 'Unconnected' are emergent properties of the social network. Forming means that a link does not exist in the current time step, but it will appear in the next time step. Unconnected means that link between two nodes does not exist in the current time step, and these nodes will remain unconnected in the next time step.

We can see that almost all metrics had significant differences between each class. Thus the significance levels become meaningless and we consider only the kappa statistic as a measure of worth. A kappa statistic of 40% corresponds roughly to an overall accuracy of 70%, and a kappa of 20% corresponds to an accuracy of 60%. We note that a metric's moving average is a better link predictor than the static metric (except in the case of recency, which is already a temporal metric). Furthermore, the increase in predictive accuracy of moving averages seems to level off somewhere between ten and twenty time steps prior to the current time step. Thus, it is recommended that taking averages further back than twenty time steps is unnecessary and extravagant. Finally, metric returns appear to be completely useless for link prediction and should not be used in the future.

Table 3 shows how the metrics perform in combination to predict links. This is the ultimate test of prediction accuracy—seeing how well we can classify links as forming or not (global behaviors) using any combination of metrics (local behaviors) at our disposal.

The first row shows the accuracy of static metrics alone. The second row shows the accuracy of the same metrics using moving averages. It can be seen that the overall accuracy increases by 6% from the first row to the second. If we include the recency metrics as well the accuracy increases to 82%, with true positive rates above 80% for both classes. This shows that temporal metrics are an extremely valuable new contribution to link prediction and should be used in future applications. Further research to be performed includes suggesting new temporal variations of static metrics and determining exactly the optimum number of time steps over which to compute temporal metrics.

Using Bayesian Networks to Understand Social Complexity

In order to understand social complexity, the local behaviors of the participants/network actors must be understood, as well as how they act together and interact with the environment to form the whole. To model this, we used Bayesian network techniques.

According to Baas and Emmeche (1997), understanding is related to the notion of explanation. A complex adaptive system uses the hyperstructures in its internal model for explanation and understanding. It uses observation mechanisms to create and maintain these hyperstructures. The process of adaptation relies heavily on the observation mechanisms and involves a progressive modification of the hyperstructures (Holland, 1995).

We measured local behaviors using metrics and we used Bayesian networks to model the interrelationships between the metrics as local behaviors and links forming between individuals as emergent behaviors. We also explored how the metrics evolve over time using Dynamic Bayesian Networks.

Link Prediction using a Bayesian Network

Bayesian Networks provide the ideal technology to reason about social resource combinations (Potgieter *et al.*, 2006). A Bayesian network is a directed acyclic graph (DAG) that consists of a set of nodes that are linked together by directional links. Each node represents a random variable or uncertain quantity. Each variable has a finite set of mutually exclusive propositions, called *states*. The links represent informational or causal dependencies among the variables, where a parent node is the *cause* and a child node the *effect*. The dependencies are given in terms of conditional probabilities of states that a node can have, given the values of the parent nodes (Pearl, 1988). Each node has a conditional probability matrix to store these conditional probabilities, accumulated over time.

Bayesian learning can be described as 'mining' the structure of the network and calculating the conditional probability matrices from history data. The

Metric	Unconnected mean	Forming mean	Unconnected standard deviation	Forming standard deviation	Mean Difference	Test statistic	Significance level	Kappa
DegreeFA20	1.9438	8.6818	3.5317	21.5584	-6.7379	-30.75	0.0001	35.48%
DegreeFA10	1.9152	8.3771	3.5181	22.4411	-6.4618	-28.36	0.0001	35.12%
DegreeFA2	1.8927	7.9833	3.5052	22.1150	-6.0906	-27.12	0.0001	29.85%
DegreeF	1.8876	7.8111	3.5006	21.9536	-5.9235	-26.56	0.0001	27.13%
DegreeFR20	0.1361	0.7550	0.5419	2.2471	-0.6189	-26.69	0.0001	10.26%
DegreeFR10	0.0626	0.4599	0.3011	1.8352	-0.3973	-21.3	0.0001	5.86%
DegreeFR2	0.0089	0.0505	0.0952	0.2291	-0.0417	-16.74	0.0001	-8.33%
DegreeTA20	1.9810	8.0030	3.4860	19.1849	-6.0220	-30.79	0.0001	42.46%
DegreeTA10	1.9606	8.0086	3.5178	20.3095	-6.0480	-29.25	0.0001	41.54%
DegreeTA2	1.9256	7.5180	3.5184	20.7846	-5.5923	-26.45	0.0001	32.30%
DegreeT	1.9223	7.0698	3.5223	20.2884	-5.1475	-24.92	0.0001	26.47%
DegreeTR20	0.1426	0.6271	0.5724	1.6312	-0.4844	-27.94	0.0001	15.68%
DegreeTR10	0.0658	0.4169	0.2848	1.5329	-0.3510	-22.45	0.0001	12.97%
DegreeTR2	0.0066	0.0795	0.0734	0.3285	-0.0729	-21.59	0.0001	-2.55%
KatzA20	0.0111	0.0427	0.0225	0.0473	-0.0316	-60.08	0.0001	41.78%
KatzA10	0.0110	0.0415	0.0229	0.0475	-0.0305	-57.7	0.0001	44.92%
KatzA2	0.0107	0.0373	0.0227	0.0471	-0.0267	-50.8	0.0001	37.32%
Katz	0.0106	0.0348	0.0227	0.0467	-0.0241	-46.31	0.0001	28.17%
KatzR20	0.2177	0.5243	0.6967	1.4813	-0.3065	-18.67	0.0001	-2%
KatzR10	0.1344	0.3634	0.6071	1.2972	-0.2290	-15.94	0.0001	-7%
KatzR2	0.0094	0.0846	0.1194	0.6620	-0.0752	-11.15	0.0001	-23.1%
PAA20	3.8428	38.6423	17.7252	97.9615	-34.7995	-34.85	0.0001	45.46%
PAA10	3.7283	37.9504	17.0146	106.8425	-34.2221	-31.54	0.0001	48.36%
PAA2	3.6188	35.7246	16.5470	117.9152	-32.1059	-26.88	0.0001	41.61%
PA	3.6043	33.2047	16.4837	114.0258	-29.6005	-25.61	0.0001	33.27%
PAR20	0.2813	1.6457	0.8481	4.4977	-1.3645	-29.72	0.0001	5.31%
PAR10	0.1404	1.0239	0.5073	3.2094	-0.8835	-27.11	0.0001	1.36%
PAR2	0.0151	0.1359	0.1213	0.4344	-0.1209	-26.72	0.0001	-10.6%

Metric	Unconnected mean	Forming mean	Unconnected standard deviation	Forming standard deviation	Mean Difference	Test statistic	Significance level	Kappa
CNA20	0.0015	0.0306	0.0477	0.2007	-0.0290	-14.02	0.0001	32.2%
CNA10	0.0019	0.0302	0.0507	0.2012	-0.0283	-13.58	0.0001	29.17%
CNA2	0.0019	0.0253	0.0497	0.1865	-0.0234	-12.1	0.0001	12.52%
CN	0.0019	0.0231	0.0501	0.1791	-0.0212	-11.38	0.0001	1.92%
CNR20	0.0000	0.0920	0.0000	0.3282	-0.0920	-27.93	0.0001	-0.87%
CNR10	0.0000	0.0794	0.0000	0.2639	-0.0794	-29.98	0.0001	-1.18%
CNR2	0.0000	0.0112	0.0000	0.1054	-0.0112	-10.57	0.0001	-0.00%
AAA20	0.0015	0.0442	0.0656	0.3151	-0.0427	-13.22	0.0001	32.2%
AAA10	0.0019	0.0417	0.0656	0.2989	-0.0399	-12.99	0.0001	29.17%
AAA2	0.0017	0.0357	0.0619	0.2785	-0.0340	-11.87	0.0001	12.52%
AA	0.0017	0.0329	0.0615	0.2701	-0.0311	-11.21	0.0001	1.92%
AAR20	-0.0393	-0.0091	0.0481	0.2333	-0.0301	-12.62	0.0001	-0.40%
AAR10	-0.0163	0.0539	0.0376	0.2920	-0.0702	-23.78	0.0001	-1.18%
AAR2	-0.0092	0.0066	0.0346	0.1109	-0.0158	-13.59	0.0001	0.07%
RecFA20	38.1636	12.5866	30.5582	16.1022	25.5771	73.82	0.0001	28.38%
RecFA10	38.2149	11.9455	32.0390	16.3050	26.2694	72.85	0.0001	35.75%
RecFA2	38.7039	11.0723	33.3012	16.5778	27.6316	74.05	0.0001	44.52%
RecF	38.7928	10.8518	33.4656	16.6218	27.9410	74.55	0.0001	46.34%
RecFR20	1.5342	1.4653	3.1617	3.5274	0.0688	1.45	NS	-2.92%
RecFR10	0.7021	0.8002	1.5502	1.9719	-0.0981	-3.9	0.0001	1.82%
RecFR2	0.0814	0.1413	0.2480	0.4357	-0.0599	-11.91	0.0001	8.46%
RecTA20	37.3356	15.1974	30.3941	18.7683	22.1382	61.78	0.0001	14.80%
RecTA10	37.5869	14.0277	31.8041	18.7829	23.5592	63.59	0.0001	22.61%
RecTA2	37.7987	11.8050	33.1000	18.2863	25.9937	68.53	0.0001	37.67%
RecT	37.8514	11.1834	33.2870	18.0862	26.6681	70.18	0.0001	42.41%
RecTR20	1.5430	1.5128	3.1033	3.5408	0.0302	0.64	NS	0.00%
RecTR10	0.6878	0.8434	1.5252	2.0300	-0.1556	-6.11	0.0001	8.01%
RecTR2	0.0797	0.1669	0.2564	0.4539	-0.0872	-16.67	0.0001	12.87%

Table 2 *Metric statistics, grouped by category*

Metric subset	Kappa	Unconnected TP rate	Forming TP rate	Overall accuracy
Static metrics: DegreeF, DegreeT, Katz, PA, CN, AA	39.83%	79.8%	59.9%	69.9698%
Static metrics with average 10: DegreeF, DegreeT, Katz, PA, CN, AA, DegreeFA10, DegreeTA10, KatzA10, PAA10, CNA10, AAA10	51.13%	80.4%	70.1%	75.5869%
Static metrics with average 10 and recency: DegreeF, DegreeT, Katz, PA, CN, AA, RecF, RecT, DegreeFA10, DegreeTA10, KatzA10, PAA10, CNA10, AAA10	63.62%	81.3%	82.4%	81.8075%

Table 3 *Metric sets prediction*

data may be incomplete and the structure of the Bayesian network can be unknown.

Figure 1 illustrates the Bayesian network that was mined from the same dataset used in the temporal analysis. It clearly illustrates the cause-effect relationships between the local behaviors (metrics 'PA', 'DegreeT', 'DegreeF', 'Katz', 'AA' and 'CN') and the emergent behavior 'Forming'.

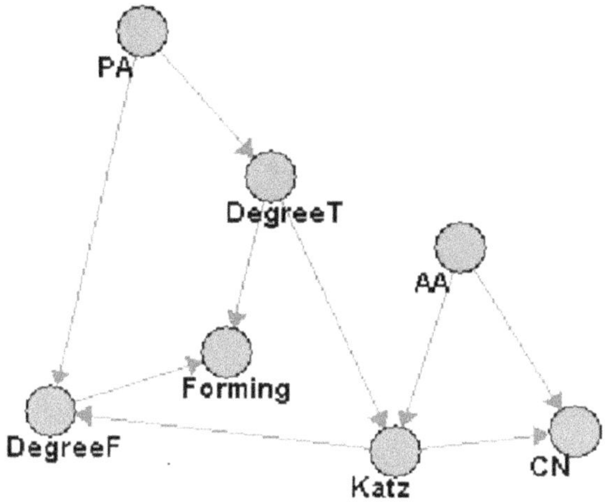

Figure 1 *Link Prediction Bayesian Network*

With a cross validation method, we extracted 30% records at random from the entire dataset of over 19,000 instances as used in the temporal analysis. The test set contained over 8, 000 records. The remaining 70% (over 11, 000) records served as the training set. As one of the pre-processing steps, we discretized the training and the test sets using the equal width algorithm (Osunmakinde, 2006). The pre-processed training set was used by the Hybrid Genetic Algorithm (Osunmakinde, 2006) to mine a Bayesian Network model, shown in Figure 1. This algorithm is described later in this paper.

After training, we used *Bayesian Inference* in the model in Figure 1 to classify the 30% test set. Bayesian inference is the process of calculating the posterior probability of a hypothesis ***H*** (involving a set of query variables) given some observed event (assignments of values to a set of evidence variables ***e***), P(***H***|***e***).

For the link prediction, the hypothesis will be that *'Forming'* is true or false, given the values of the evidence variables 'DegreeT', 'DegreeF', 'Katz', 'PA', 'AA', 'CN':

P(Forming = true | DegreeT, DegreeF, Katz, PA, AA, CN) ?

Or,

P(Forming = false | DegreeT, DegreeF, Katz, PA, AA, CN) ?

The confusion matrix (Ron & Foster, 1998) in Table 4 gives the link prediction and evaluation of the results. The implementation results were computed and generated from Table 4 and presented in Table 5.

Modelling Emergent Social Network Behavior over time using a DBN

A Dynamic Bayesian network (DBN) is ideally suited to model changes in metrics and emergent behavior over time. In Dynamic Bayesian networks, multiple copies of the variables are represented, one for each time step (Pearl & Russell, 2000). A DBN provides a compact representation of temporal probabilistic processes such as Hidden Markov Models (HMM). It contains slices of sub-models which facilitate its probabilistic reasoning over time. That is, we can interpret the present, understand the past and predict the future. We want to compute: $P(X_i(t) \mid E(t_1{:}t_n))$ from the complete joint distribution of variables over one or more time slices as defined as:

$$P(X_0,X_1,\ldots,X_t,E_1,\ldots,E_t) = P(X_0)\prod P(X_i|X_{i-1})P(E_i|X_i)$$

where $(X_i(t))$ represents the i'^{th} hidden or unobservable variable at time t and $E(t_1{:}t_n))$ refers to the observable evidence from time step t_1 to t_n. Observe that this follows a Markov assumption where the current state depends on only finite history of previous states (Russell & Norvig, 2002).

The construction of the DBN requires the following information components:

	Predicted values	
Actual values	True	False
Forming = false (Total 4390)	4390	0
Forming = true (Total 4390)	39	4351

Table 4 *Confusion Matrix*

True positive rate of forming	99.1116173120729%
True positive rate of not forming (unconnected)	100.0%
Kappa	99.1116173120729%
The overall accuracy	99.5558086560364 5%

Table 5 *Implementation results*

1. The prior distribution over the state variables, $P(X_0)$;
2. The transition model, $P(X_t | X_{t-1})$, and;
3. The sensor model, $P(E_t|X_t)$.

The learning of these components depends on the topological connections between the successive slices, the hidden states and the evidence variables. There are many algorithms for learning parameters in DBNs. Where sufficient training data is available, a maximum likelihood estimate (MLE) can be used. Many areas of science and engineering use approximate inference algorithms in DBNs because of the shortcomings of exact inference (Pavlovic *et al.*, 2001).

In this paper, we incorporate a Dynamic Bayesian Network illustrated in Figure 2 which models possible relationships between metrics and forming links. The observable nodes are 'Forming', 'DegreeT', 'DegreeF', 'Katz', 'PA', 'CN' and 'AA' which serve as the evidence variables 'E', to the unobservable variable called 'HiddenNode'. We tested this DBN using the 150 time slices of the dataset used in the temporal analysis, using the metrics defined in the previous sections. The primary objective of this DBN is to predict relationships between metrics (local behaviors) and links forming (global behaviors) in future time steps. Using this, possible links can be predicted between two or more people using the predicted relationships from observed nodes.

Dynamic Bayesian Network Methodology

We used the same dataset as in the temporal metrics analysis, consisting of 150 time steps and used 70% as a training set, while 30% was used as a test set by a hybrid genetic algorithm (see next section). The average number of records in each time step was 201.

In our implementation, we trained the three components of DBN using the MLE (Russell & Norvig, 2002) and Mutual Information in information theory (Cheng *et al.*, 1997). These 3 components were used during DBN inference using the Viterbi algorithm (see next section). We evaluated the relationships predicted by the DBN using the hybrid genetic algorithm. The DBN did not have access

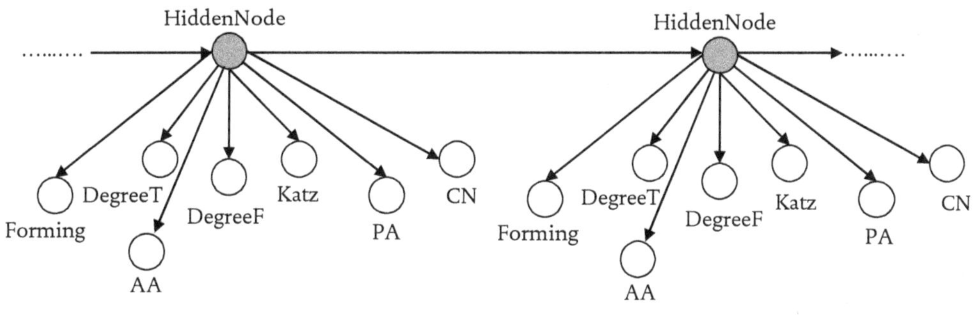

Figure 2 *Dynamic Bayesian Network*

to the 30% dataset, but it was used by the genetic algorithm. That is, the genetic algorithm mined the *actual relationships* between the metrics and forming links at each time step, while the DBN *predicted relationships* between the metrics and forming links at each time step.

Dynamic Bayesian Network Prediction using the Viterbi Approximation

By definition (Pavlovic *et al.*, 2001) the probability of the partial optimal path to a hidden state *i* at time *t* when evidence z_t is observed implies:

$$\delta(i) = \max_j(\delta_{t-1}(j)a_{ji}b_{iz_t})$$

where 'delta' is the probability of reaching a particular intermediate hidden state on the DBN time slices, '*a*' is the transition probability while '*b*' is the corresponding observation probability. Recall the Markov assumption that says that the probability of a hidden state occurring, given the previous history, depends on the previous *i*-states. We wanted to compute the probability of the most probable path to any hidden node X. That is:

$$P(X_t) = \mathrm{Max}_i\, P(i_{t-1})P(X \mid a)P(\mathrm{observation}_t \mid X)$$

We used the knowledge of the previous states; the transition probabilities multiplied by the corresponding observation probabilities. This calculated the hidden node.

The HGA is a variant of classical genetic algorithm operators because it integrates information theoretic measures as learning components and mathematical concepts as population construction. Specifically, some of the information theoretic measures we used were: Mutual Information (MI) model, Shannon's information content, and Minimum Description Length (MDL) model. Also, the mathematical component part of the HGA was a PowerSet lattice which implemented crossover operator. Moreover, PowerSet lattice generated population space from which the best networks could emerge.

One of the interesting features of this methodology is that the components of the HGA were implemented as decomposable systems. The decomposability means the functionalities of every component such as MI perceives its inputs and produces its outputs differently from other HGA components. However, such decomposability of every HGA component is to level complexity into simplicity. More information can be found in Osunmakinde (2006).

Examples of the *predicted relationships* by the Dynamic Bayesian Network for time step 1 were as shown in Table 6.

We compared the *relationships* predicted by the DBN with the *actual relationships* mined using the hybrid genetic algorithm. We calculated the accuracy of each prediction by counting the number of predicted relationships and actual relationships that correspond in structure predicted by the DBN and the actual structure mined by the HGA. We obtained the results detailed in Table 7.

Probability of Relationship between: Forming and CN, at time step: 1 = 0.032570422535211266
Probability of Relationship between: Katz and PA, at time step: 1 = 7.435374363700479E-6
Probability of Relationship between: Katz and Forming, at time step: 1 = 0.002758431904558663
Probability of Relationship between: Katz and AA, at time step: 1 = 0.0016477018677722897
Probability of Relationship between: Forming and AA, at time step: 1 = 0.013710911301140175
Probability of Relationship between: DegreeF and Forming, at time step: 1 = 7.958115165232352E-4
Probability of Relationship between: Katz and CN, at time step: 1 = 0.002924745442351074

Table 6 *Relationships predicted by DBN*

In future research, we will use these predicted relationships to study to see if time graphs densify over time and at what rate the average distance between nodes decreases.

Conclusion

In this paper, we recognise the role of complex adaptive systems theory in shedding light on an increasingly important aspect of organizational life (and survival), namely, social networks, and emergent temporal phenomena in these networks. It is our contention that the fluidity and temporality of the interrelationships, making up such networks, render it dependent on changing contexts—and therefore accurate prediction, going forward, becomes difficult (near impossible) using traditionally static methodologies.

The use of link prediction to improve collaborative filtering in recommender systems was investigated (Huang *et al.*, 2005), and the Katz measure was found to be the most useful, followed by preferential attachment, common neighbors and the Adamic\Adar measure. These path-based and neighbor-based measures outperformed simpler metrics, and so we used these in our experiments. We were able to show that when emergent temporal trends are ignored in sociogram sequences, the link prediction accuracy is significantly lower than when included.

An important contribution of this paper is that we have shown that temporal metrics are an extremely valuable new contribution to link prediction, and should be used in future research and applications. This is significant because, to date, no temporal information has been used in link prediction research, excluding our earlier proposition of the idea in April *et al.* (2005) and Potgieter *et al.* (2006). Further research to be performed includes suggesting new temporal variations of static metrics and determining exactly the optimum number of time steps over which to compute temporal metrics.

Time Step 0: Nr of relationships *predicted* by DBN: 21 Nr of *actual* relationships mined by HGA: 19.0 Accuracy: 90.47619047619048 %
Time Step 1: Nr of relationships *predicted* by DBN: 21 Nr of *actual* relationships mined by HGA: 10.0 Accuracy: 47.61904761904761 %
Time Step 2: Nr of relationships *predicted* by DBN: 21 Nr of *actual* relationships mined by HGA: 21.0 , Accuracy: 100.0 %
Time Step 3: Nr of relationships *predicted* by DBN: 21 Nr of *actual* relationships mined by HGA: 13.0 Accuracy: 61.904761904761905 %
Time Step 4: Nr of relationships *predicted* by DBN: 21 Nr of *actual* relationships mined by HGA: 21.0 Accuracy: 100.0 %
Time Step 5: Nr of relationships *predicted* by DBN: 21 Nr of *actual* relationships mined by HGA: 21.0 Accuracy: 100.0 %
Time Step 6: Nr of relationships *predicted* by DBN: 21 Nr of *actual* relationships mined by HGA: 20.0 Accuracy: 95.23809523809523 %
Time Step 7: Nr of relationships *predicted* by DBN: 21 Nr of *actual* relationships mined by HGA: 19.0 Accuracy: 90.47619047619048 %
Overall DBN Prediction Accuracy: 79.52380952380953 %

Table 7 *DBN Prediction Accuracy*

We have used Bayesian networks for link prediction, and achieved extremely high prediction accuracies. Additionally, we trained the components of the Dynamic Bayesian Network in order to mine the *actual relationships* between the metrics (local behaviors) and forming links (emergent global behaviors) at each time step—of significance is the fact that the DBN was able to *predict relationships* between the metrics and forming links at each time step, as it evolved over time. In future research, we will use these predicted relationships to study more emergent temporal phenomena such as if time graphs densify over time and at what rate the average distance between nodes decreases.

References

Adamic, L.A. and Adar, E. (2003). "Friends and neighbors on the web," *Social Networks*, ISSN 0378-8733, 25(3): 211-230.

April, K.A. (2002). "Guidelines for developing a K-strategy," *Journal of Knowledge Management*, ISSN 1367-3270, 6(5): 445-456.

Baas, N.A. and Emmeche, C. (1997). "On emergence and explanation," *Intellectica*, ISSN 0769-4113, 25: 67-83.

Brock, D. (2006). "The changing professional organization: A review of competing archetypes," *International Journal of Management Reviews*, ISSN 1468-2370, 8(3): 157-174.

Burt, R.S. (1995). *Structural Holes: The Social Structure of Competition*, ISBN 9780674843714.

Campbell, C.S., Maglio, P.P., Cozzi, A.C. and Dom, B. (2003). "Expertise identification using email communications," *Proceedings of the Twelfth International Conference on Information and Knowledge Management*, ISBN 9781581137231, pp. 528-531.

Cartwright, D. and Harary, F. (1956). "Structural balance: A generalization of Heider's theory," *Psychological Review*, ISSN 0033-295X, 63: 277-292.

Cheng, J., Bell, D.A. and Liu, W. (1997). "Learning belief networks from data: An information theory-based approach," *Proceedings of the Sixth International Conference on Information and Knowledge Management*, ISBN 9780897919708, pp. 325-331.

Desikan, P. and Srivastava, J. (2004). "Mining temporally evolving graphs," http://maya.cs.depaul.edu/webkdd04/final/desikan.pdf.

Farrell, S., Campbell, C. and Myagmar, S. (2005). "Relescope: An experiment in accelerating relationships," *Conference on Human Factors in Computing Systems*, 2-7 April 2005, Portland.

Fleiss, J.L. (1981). *Statistical Methods for Rates and Proportions*, ISBN 9780471064282.

Galaskiewicz, J. (1979). *Exchange Networks and Community Politics*, ISBN 9780803911376.

Getoor, L. and Diehl, C. (2005). "Link mining: A survey," *ACM SIGKDD Explorations Newsletter*, ISSN 1931-0145, 7(2): 3-12.

Granovetter, M. S. (1985). "Economic action and social structure: The problem of embeddedness," *American Journal of Sociology*, ISSN 0002-9602, 91(3): 481-510.

Granovetter, M. S. (1992). "Problems of explanation in economic sociology," in N. Nohria and R.G. Eccles (eds.), *Networks and Organizations: Structure, Form and Action*, ISBN 9780875845784, pp. 25-56.

Hanneman, R. (2001). "Introduction to social network methods," http://www.faculty.ucr.edu/~hanneman/nettext/.

Henziger, M. (2000). "Link analysis in web information retrieval," *IEEE Data Engineering Bulletin*, 23(3): 3-8.

Holland, J.H. (1995). *Hidden Order: How Adaptation Builds Complexity*, ISBN 9780201407938.

Hosmer, D.W. and Lemeshow, S. (1989). *Applied Logistic Regression*, ISBN 9780471615538.

Huang, Z., Li, X. and Chen, H. (2005). "Link prediction approach to collaborative filtering," *5th ACM/IEEE-CS Joint Conference on Digital Libraries*, ISBN 9781581138764.

Kautz, H., Selman, B. and Shah, M. (1997). "Referral web: Combining social networks and collaborative filtering," *Communications of the ACM*, ISSN 0001-0782, 40(3): 63-65.

Leskovec, J., Kleinberg, J. and Faloutsos, C. (2005). "Graphs over time: Densification laws, shrinking diameters and possible explanations," Proceedings of the 11th ACM SIGKDD International Conference on Knowledge Discovery in Data Mining, ISBN 9781595931351.

Liben-Nowell, D. (2005). An Algorithmic Approach to Social Networks, PhD thesis at MIT Computer Science and Artificial Intelligence Laboratory.

Liben-Nowell, D. and Kleinberg, J. (2003). "The link prediction problem for social networks," *Proceedings of the 12th International Conference on Information and Knowledge Management*, pp. 556-559.

Lincoln, J.R. (1982). "Intra- (and inter-) organizational networks," in S. B. Bacharach (ed.), *Research in the Sociology of Organizations*, ISBN 9780892321704, pp. 1-38.

Lim, M., Negnevitsky, M. and Hartnett, .J. (2005). "Artificial intelligence applications for analysis of email communication activities," *Proceedings International Conference on Artificial Intelligence in Science and Technology*, pp. 109-113.

Loasby, B.J. (1999). "Capabilities," in B.J. Loasby (ed.), *Knowledge, Institutions and Evolution in Economics*, ISBN 9780415205375, pp. 49-68.

McMahon, S.M., Miller, K.H. and Drake, J. (2001). "Networking tips for social scientists and ecologists," *Science*, ISSN 0036-8075, 293: 1604-1605.

Nohria, N. (1992). "Is a network perspective a useful way of studying organizations?" in N. Nohria and R.G. Eccles (eds.), *Networks and Organizations: Structure, Form and Action*, ISBN 9780875845784, pp. 1-22.

Nohria, N. and Eccles, R.G. (1992). "Face-to-face: Making network organizations work," in N. Nohria and R.G. Eccles (eds.), *Networks and Organizations: Structure, Form and Action*, ISBN 9780875845784. pp. 288-308.

Osunmakinde, I.O. and Potgieter, A. (2006). "Agent-based behavioral modelling for anomaly detection in call data from telecommunication networks," Proceedings of

Southern African Telecommunications Networks and Applications Conference, ISBN 9780620370431, pp. 8-9.

Pavlovic, V., Rehg, J.M., Cham, T. and Murphy, K.P. (2001). "A dynamic Bayesian network approach to figure tracking using learned dynamic models," *Proceedings of the Seventh IEEE International Conference on Computer Vision*, pp. 94-101.

Pearl, J. (1988). *Probabilistic Reasoning in Intelligent Systems: Networks of Plausible Inference*, ISBN 9780934613736.

Pearl, J. and Russell, S. (2000). "Bayesian networks," http://ftp.cs.ucla.edu/pub/stat_ser/R277.pdf.

Pepper, S. (2003). "ISO 13250:2002-Topic maps: An international standard knowledge representation for humans and agents," http://www.csc.liv.ac.uk/~valli/WG-Ontology/Helsinki/Id3-Ontopia.ppt#649,1,ISO%2013250:2002-Topic%20Maps.

Popescul, A and Ungar, L. H. (2004). "Cluster-based concept invention for statistical relational learning," Proceedings of the Tenth ACM SIGKDD International Conference on Knowledge Discovery and Data Mining, ISBN 9781581138887, pp. 665-670.

Popescul, A. and Ungar, L. H. (2003). "Structural logistic regression for link analysis," http://www.cis.upenn.edu/~popescul/Publications/popescul03mrdm.pdf.

Potgieter, A., April, K. and Cooke, R. (2005). "Using Bayesian agents to enable distributed network knowledge: A critique," http://www.mngt.waikato.ac.nz/ejrot/cmsconference/2005/abstracts/socialnetworks/Potgieter.pdf.

Potgieter, A., April, K., Cooke, R. and Lockett, M. (2006). "Adaptive Bayesian agents: Enabling distributed social networks," *South African Journal of Business Management*, ISSN 0378-9098, 37(1): 41-55.

Potgieter, A., April, K. and Bishop, J. (2005). "Complex adaptive enterprises," in M. Khosrow-Pour (ed.), *Encyclopedia of Information Science and Technology*, ISBN 9781591405535, pp. 475-480.

Pujol, J., Sanguesa, R. and Delgado, J. (2002). "Extracting reputation in multi-agent systems by means of social network topology," *Proceedings of the 1st International Joint Conference on Autonomous Agents and Multi-agent Systems*, ISBN 9781581134803, pp. 467-474.

Ron, K. and Foster, P. (1998). "Special issue on applications of machine learning and the knowledge discovery process," *Journal of Machine Learning Research*, ISSN 1532-4435, 30: 271-274.

Ross, S., Westerfield, R., Jordan, B. and Firer, C. (2001). *Fundamentals of Corporate Finance*, ISBN 9780074707999.

Russell, S.J. and Norvig, P. (2002). *Artificial Intelligence: A Modern Approach*, ISBN 9780137903955.

Ryle, G. (1949). *The Concept of Mind*, ISBN 9780226732954.

Taskar, B., Wong, M-F., Abbeel, P. and Koller, D. (2004). "Link prediction in relational data," *Proceedings of Neural Information Processing Systems.*

Van Wijk, R., Van Den Bosch, A.J. and Volberda, H.W. (2003). "Knowledge and networks," in M. Easterby-Smith. and M.A. Lyles (eds.), *The Blackwell Handbook of Organizational Learning and Knowledge Management*, ISBN 9781405133043, pp. 428-453.

Wassermann, S. and Faust, S. (1994). *Social Network Analysis: Methods and Applications*, ISBN 9780521382694.

Witten, I. and Frank, E. (2005). *Data Mining: Practical Machine Learning Tools and Techniques*, ISBN 9780120884070.

Zhou, D. and Scholkopf, B. (2004). "A regularization framework for learning from graph data," http://www.cs.umd.edu/projects/srl2004/Papers/zhou.pdf.

Zhu, J. (2003). "Mining web site link structures for adaptive web site navigation and search," http://kmi.open.ac.uk/people/jianhan/thesis.pdf.

Anet Potgieter received her PhD from the University of Pretoria, with her research titled: *The Engineering of Emergence in Complex adaptive Systems*, combining software engineering and complex adaptive systems theory. In her PhD research, Anet developed a novel adaptive agent architecture that can learn from and adapt to uncertain and dynamically changing environments. Anet is a Director of Complex Adaptive Systems (Pty) Ltd (South Africa), also a Founder Member of the Robotics and Agents Research Laboratory at the Mechanical Engineering Department, University of Cape Town (South Africa). Her research interests include complex adaptive systems, distributed artificial intelligence, sensor networks, social networks and gaming.

Kurt April lectures in Leadership, Diversity & Inclusion, and is a tenured Professor at the Graduate School of Business (University of Cape Town, SA), Research Fellow of Ashridge (UK), Associate Fellow of Saïd Business School (University of Oxford, UK), Adjunct Faculty of Duke University Executive Education (USA), and Visiting Professor of Erasmus Rotterdam School of Management (Erasmus University, Netherlands), while previously having been the Academic Director of the Center for Leadership & Public Values (CLPV) in conjunction with Duke University (USA), and ran a Leadership Workshop for Harvard Business School, as part of their Making Markets Work programme. Kurt is a Fellow of the Royal Society of Arts, Manufactures and Commerce (UK), holds a PhD, Diploma (Japanese Production), MBA, MSc (Electronic Engineering), HDE, BSc (Electrical Engineering), National Diploma (Electrical Engineering), and National Diploma (Logic Systems), which he obtained through the University of Cape Town (South Africa), University of Oxford (UK), AOTS Nagoya (Japan) and Wingfield College (South Africa). Outside of academia, Kurt is Managing Partner of LICM Consulting, is an Advisory Council Member at Novartis International (Switzerland), is Head of an

International D&I Faculty to the NHS (UK), and is a Non-Executive Board Member of Achievement Awards Group (South Africa). Kurt has coauthored seven books and over 80 academic articles in various academic journals around the globe.

Richard Cooke has a BBus. Science (Hons), BSc(Hons), and an MSc from the University of Cape Town (South Africa) in Computer Science, with a dissertation on temporal social networks. He currently programs in .NET for Saratoga Software.

Isaac Olusegun Osunmakinde is completing his PhD in the Computer Science Department of the University of Cape Town (South Africa). He has a MSc in Computer Science (University of Cape Town, South AfricaCT), a PgDS in Mathematical Science from Stellenbosch University (South Africa), a BSc Hons in Computer Science, and a HND in Computer Technology. Previously, Isaac was a Senior Researcher for Complex Adaptive Systems (Pty) Ltd. (South Africa). He is a member of the IEEE Computational Intelligence Society. His current research interests in AI are advanced probabilistic and mathematical modelling. He has authored and coauthored over 10 refereed academic papers.

6. The Emergence of the Communicative Value of Silence

Maria Pietronilla Penna, Sandro Mocci & Cristina Sechi
The University of Cagliari, ITA

Within traditional theories of communication the silence is often devoid of any communicative value. When the latter is taken into consideration, it is viewed as depending on the intentionality of the agent producing the communicative act. Unfortunately there are diverging opinions about the role to be attributed to intentionality. Moreover, its detection by the receiver is often difficult or impossible, a circumstance which prevents from building a theory of a number of interesting communication phenomena. We hold that the previous problems can be dealt with by resorting to a systemic view in which communication is nothing but a macroscopic phenomenon, emergent from the interactions between elements of a communicative system. This perspective allows to introduce the methodological tools of Systemics to better describe all kinds of communication, grasping their emergent meanings. Only in this way the emergent communicative value of silence can be detected. Such an approach is endowed with a strong potential usefulness when dealing with the communicative interactions within both small and large organizations.

Introduction

Human communication is characterized by its *multidimensionality*, stemming from its cognitive, social, cultural, economic, political implications, as well as from its close interconnection with human actions. This implies a plurality of different approaches to communication, each one emphasizing specific features, as a function of particular goals and disciplinary competencies. For this reason we still lack an universally accepted definition of communication and this fact gives rise to a confusion about this subject, which makes difficult any attempt to transform theoretical statements into practical applications.

Historically, all attempts to build a general theory of communication have been dominated by two main influences: the one of Shannon and Weaver's traditional *Information Theory* (1949) and the other based on the psychological perspective introduced by Watzlawick and his coworkers (Watzlawick *et al.*, 1967). However, they were pushing towards opposite directions. As regards Shannon and Weaver's approach, it undoubtedly offered a firm theoretical basis to the study of communication, being deeply grounded on classical statistical thermodynamics. Its shortcomings derive from the fact that the latter theory entails a strictly unidirectional view of communication process, described as a signal flow from an emitter to a receiver through a medium. Within this context most aspects of human communication are lost. For instance, silence, being coincident with absence of signals, does not communicate.

Watzlawick's psychological perspective tries to remedy the flaws of previous approach by introducing the concepts of *intentionality* and of *behavioral interaction*, so as to include within the category of communication processes all behaviors occurring within a given interactional context. However, while this approach allows to account for a number of important communication processes, impossible to describe within Shannon and Weaver's theory (this time silence is a form of communication), it fails to give a precise definition of communication. And such a circumstance, in turn, prevents from building a useful theory of communication, as even the subject of this theory cannot be defined. We thus arrived to a sort of *impasse*: while, relying on principles of classical physics, almost nothing is communication, on the contrary, relying on psychology, all is communication. In such a situation, all theorizing about communication risks to be useless, being poorly grounded even the starting point of all arguments.

In this paper we hold that a way for sorting out from such an unpleasant state of affairs is to adopt a truly systemic view. Within it, communication is identified with a macroscopically emergent process, coming out from the interactions occurring between the elements of a *communicative system*. Thus, differently from Watzlawick's perspective, it is not true that all is communication. Communication is viewed as occurring only when there is an emergent meaning, associated to behavioral acts, shared by the elements of the communicative system. The study of communication processes thus becomes a particular chapter of a more general theory of emergence, when applied to systems of interacting agents. And this fact allows to take advantage of methodological tools of Systemics and of theory of emergence. In this way it would be possible, for instance, to detect the cases in which silence communicates and the ones in which, instead, the latter is a form of miscommunication. While aware of the intrinsic conceptual limitations of most tools of this kind, we feel they could be useful in a number of contexts, including therapeutic, educational, and even organizational ones.

This paper will be devoted to a discussion of arguments supporting this view. We will choose as a privileged context the one of silence, because within it the shortcomings and advantages of different views are easier to detect.

The Role of Intentionality

At a first sight the simplest way for overcoming the *impasse* described in the previous section would seem the one based on a suitable generalization of Shannon and Weaver's theory. To be more precise, the latter would consist in endowing the emitter with a cognitive system, including, among the other features, the *intentionality*, that is the presence of goals, not only driving behavioral acts, but also shaping its communicative behaviors (see also Searle, 1983; Dennett, 1987). If we suppose the receiver equipped, in turn, with a device able to detect the intentionality of the emitter, it is easy to understand that in this way it would be possible to grasp the communicative meaning of all behavioral acts of the latter. Of course, the correct operation and the reliability of this device would be critical in the cases of non-verbal communications, and mostly in the case of silence, as verbal communication offers a number of greater number of cues for detecting intentionality. In this regard, it is very easy to make examples of situations in which detection of intentionality allows to grasp the communicative meaning of a silence. The cases of the restaurant and of the airplane introduced by Watzlawick and coworkers are now famous:

> *The man at a crowded lunch counter who looks straight ahead, or the airplane passenger who sits with his eyes closed, are both communicating that they do not want to speak to anybody, or be spoken to, and they neighbors usually "get the message" and respond appropriately by leaving them alone* (Watzlawick *et al.*, 1967: 49).

"I do not want you disturb me" is the message the silences of the man at the table and of the airplane passenger seem *to shout*, without any possibility of misunderstanding for "their neighbors [who] usually 'get the message'...".

It is to be remarked that the role of intentionality as a distinctive feature of communicative acts has been emphasized by a number of authors. For instance, it grounds Berne's transactional analysis of psychological processes (see, among the others, Berne, 1964; more recent accounts are contained in Stewart and Jones, 1987, and in Thaler Singer and Lalich, 1996). In turn, Miller and Steinberg (1975) hold that every human communication, whatever be its nature, is *per se* intentional, being finalized to elicit an answer or, in general, to influence the potential receiver. Whence, in absence of intentionality, the communication itself would be impossible. The introduction of intentionality, however, raises a number of problems, which can be shortly listed as follows:

1. What is intentionality? How to find a reliable psychological definition of this construct?
2. How to detect intentionality? What are its cues? What cognitive-emotional processes allow us to detect intentionality? What is the neurobiological basis for these processes?
3. Is intentionality a unitary construct? Could we distinguish between different kinds of intentionality, as already proposed by Grice (1975)?

So far, almost all previous questions appear as unanswered and it seems highly probable that answering them will require a lot of theoretical and experimental work. In order to cope, at least in a partial way, with these difficulties Buller and Burgoon (1994) introduced a theory of communication including, as a crucial element, the *perception*, by the receiver, of the intentionality of the emitter. In other words, a communication would occur only if two conditions are simultaneously meet: the emitter having the *intention* of communicating (and concretely revealing such an intention), and the receiver *detecting* such an intention. This perspective entails that, in absence of explicit cues of emitter intention of communicating, the receiver can *attribute* communicative processes to the emitter, but cannot properly define them. In other terms, we should not have a true communication. Likewise, when the receiver does not detect the emitter intention of communicating, even if the latter can be clearly detected because declared in an explicit way (or observed by other persons), we still lack a communication, but we are only in presence of an *attempt* to communicate. Finally, when both the intention and the detection of it are absent, we are not in presence of communication but, in the best case, only of simple *behaviors*.

Though interesting, the proposal of Buller and Burgoon is unable to account for a number of situations in which communication actually occurs, even if, according to the previous criteria, it would be absent. Let us consider, for instance, the case in which two old friends quarrelled. One of them, feeling himself offended by the other, remains silent when the other, trying to make it up with him, talks to him again and again. The communicative meaning of the silence is, in this case, evident and well understood by both persons. It means: "You offended me. Notwithstanding I would like a reconciliation. However, to reach it, you must make a strong effort and try many times to apologize". However, according to Buller and Burgoon criteria, in this situation we should have complete absence of communication.

A seemingly opposite case is the one of a firm in which there is a previously designed communication channel allowing an interchange of information between the management and the staff. If technical information flows in both directions along this channel, Buller and Burgoon would conclude that we are in presence of a communication between management and staff, as their conditions for the occurrence of a communication are trivially meet. However, if the employees are unsatisfied, their performance is low and firm productivity is scarce, we should conclude that there has been no true communication between the management and the staff. Namely each one of them is unaware of the needs and feelings of the other. And surely this is a situation in which the firm is running the risk of a sudden collapse, probably in absence of evident premonitory cues.

To conclude this section, we remark that Buller and Burgoon proposal, while suited to describe a number of highly simplified communicational processes, appears as inadequate when we deal with complex interactional contexts such as the ones sketched in the examples made above. The difficulty seems to derive from the fact that, even within their perspective, the communication still continue to be viewed as a unidirectional process, in the Shannon-Weaver sense. In the next section we will try to show that, by abandoning the postulate of unidirectionality, we could build a more realistic theory of communication.

A Systemic View of Communication

A systemic perspective of communication entails, as a first step, that all elements (emitters, receivers, but even external observers) involved in communication processes be considered as belonging to a unique *communicative system*. This step, however, is not enough. It must be followed by a second step in which we take into account that every element is endowed with a cognitive system, allowing him/her to interact with other elements. These interactions, in the specific case of communication, are mediated by the exchange of *signs* (words, letters, gestures, behavioral acts, and even poses and attitudes, such as being silent) which, in turn, can trigger in each element of the system a process of *inference* about the possible meaning of each sign or of a constellation of signs. Within a communicative system the inferences made by the single elements constitute a sort of net, as each inference can influence (and modify) another inference made by the same or by other elements owing to the behavioral acts ensuing from inferences themselves. In short, this net allows a *circulation of inferences* associated to a change with time of net global and local state.

The global equilibrium states of this net, emergent from the interactions between elements, allow the existence of a *meaning* of signs, shared by all elements. This is the situation in which we can speak of a true communication, vehiculated by signs themselves. Therefore communication appears as a macroscopic, global process, which can be defined only in relation to a whole system, rather than a unidirectional, single-channel, relationship between emitter and receiver. In this regard we remark that just the phenomenon of silence can help to understand this new perspective. We already dealt elsewhere (Penna & Mocci, 2005) with silence as a *communicative phenomenon of systemic nature*, showing that it becomes meaningful only if occurring within a system in which the circulation of inferences gives rise to the emergence of a global meaning.

We remark that in this way the problem of defining and detecting the intentionality of a single agent is entirely avoided, Namely the whole communicative system includes not only the actual and potential communicating agents, but even their possible relationships, behaviors, attributes, expectancies. The intentionality, the sharing of meanings, the awareness thus become emergent systemic properties, *distributed* within the system itself, and not always detectable when observing a single element or a single behavioral act.

It is to be stressed that our perspective is based on a pragmatic view, which takes into account essentially the observable interactions and behaviors taking place within a given communicative system. This allows, in principle, to introduce qualitative and even quantitative tools of systemics and of theory of emergence to describe, and sometimes forecast the time evolution of a communicative system. Owing to such a circumstance the systemic view of communication sketched above appears as endowed with potential usefulness in a number of applications. An example is given by medium or large organizations, whose operation and performance is crucially based on the nature of processes of communication acting inside the organization itself and/or relating the latter to the external social/physical environment. In these cases the existence of well defined structures as well as of a large number of agents and of interactions allows the introduction of specific system-theoretical models, which could be very useful to describe the state of communication system taken into consideration, assess its evolution and its role within the organization.

The Methodological Tools Useful to Study the Emergence of Communication

Before discussing the methodological tools available within emergence theory we remark that two main views of emergence are possible: the one based on an *external* observer, and the other based on an *internal* observer. The former, borrowed from physics, assumes the existence of an observer which is external to the system itself, and can interact with the latter through both physical and mental operations. While physical operations can be synthetically qualified as *measures*, mental operations consists mainly in selecting (for instance measurement outcomes, relevant variables, and so on) and performing symbolic processing (typically through models describing the system under study). According to this view, interactions can induce changes both in the observed system and in the external observer (even modifying its mental operations). In the case of systems following the laws of classical physics these changes are very small and can be neglected while, in the case of quantum systems, can be dramatic. However these changes cannot produce a breaking of the barrier separating the system from the external observer: the latter will never become a part of the system under study. On the contrary, the view based on an internal observer (the most prominent supporter of it is Ralph Stacey; see, for instance, Stacey, 2001; 2003; 2005; 2007) assumes from the starting that the observer is a part of the system itself. Within it, therefore, we cannot speak of cognitive systems endowed with autonomous existence independent from the interactions taking place within the system which individuals and observers belong to. Nor traditional models can capture the essence of self-organization processes occurring within the system. Namely none of members of the system can go outside the system to have a general overview of its evolution, eventually in order to control it. All a member of a system can do is only to interact (in a local way) with other members, and this same interaction underlies the emergence of 'mind', 'self', 'intention', 'meaning'. However, while recognizing the intrinsic inadequacy of models based on external observer view, the approach grounded on internal observer concedes that some of these models can offer a useful metaphor of what can occur within systems. Thus, in order to take into account the important suggestions made by holders of internal observer view, we will shortly discuss the methodological tools of Systemics as if they were designed to give rise to new and better metaphors of emergence and, in our case, of emergent phenomena associated to communication. Among the latter tools (for formalized descriptions see Minati & Pessa, 2006) we will quote the distinction between *microscopic* and *macroscopic* description levels. They can be synthetically characterized as follows:

1. The microscopic level individuates the system through its borders, its elements, the attributes and the relationships between the elements themselves. This individualization phase may be considered as a typical *micro* description, since it describes single punctual actions, elementary features, relationships caught in their stillness. Examples of micro-descriptions are the social role or a single behavioral act of a given agent, the detailed structure of a particular social system as, for instance, a firm, the script of a drama. They describe a single relational element, each element of a group, each component of an organization, each possible behavior.
2. The macroscopic level deals with the average effects of the system. It includes a phenomenological description: what we may observe at a global level, that is the average effects of many variables. An example may be the buzz of a crowd as the average effect of countless overlapped conversations.

Theory of emergence deals with the interactions between these two levels, which are both of bottom-up and top-down nature. Bottom-up interactions allow the occurrence of macroscopic entities and properties, while top-down ones are produced by processes occurring to keep the global coherence of macroscopic entities and properties against the destabilizing effect of microscopic disturbances. At a modelling level these interactions are described by resorting to suitable macroscopic *order parameters*, driving system dynamics and measuring the degree of emergence of new entities and properties.

This methodology of description can be used in different ways, for instance by embedding it within arguments making use of natural language, or resorting to formalized description to be used within mathematical theories. However, owing to the usually high complexity of most communication systems, it seems unavoidable to encounter situations where both conceptual reasoning and mathematical arguments are impossible or unreliable. In these cases the resort to computer simulations of suitable models of communication systems can help the modelers to explore the consequences of some hypotheses, characterize possible scenarios, and, mostly, individuate the weaknesses of modelling schemata, addressing the research towards more productive directions. In this regard, a number of computer simulations of models of *emergence of communication within systems of agents* appeared in the last years (see, for instance, Werner & Dyer, 1992; Oliphant, 1996; Cangelosi, 2001; a useful review is contained in Perfors, 2002). These models are mostly based on artificial neural networks and on genetic algorithms. Their present status appears as too rudimentary to account for real phenomena occurring in human communication systems. They, however, seem to be promising, particularly if coupled with the tools of the so-called network dynamics (see, for a review, Albert & Barabási 2002; Barabási, 2002), widely used, for instance, in Econophysics (see Mantegna & Stanley, 2000; McCauley, 2004).

Conclusions

The main advantage of systemic perspective in the study of communication consists in a greater generalization ability: it allows to account for a larger number of communication phenomena, with respect to traditional theories. Silence, or pauses within a conversation, can be considered as meaningful communicative acts just because endowed with a pragmatic valence and producing the emergence of meanings. Moreover, only a systemic theory of communication can resort to methods of theory of emergence to deal with difficult and still unsolved problems such as the emergence and development of language. Finally, in presence of a microscopic description detailed enough, this approach appear as potentially useful to deal with description, forecasting, and eventually control, of processes of communication—and miscommunication—occurring within large and medium organizations. The study of communicative value of silence led us to recognize the need for introducing the systemic methodology and the theory of emergence to build more general models of communication processes. It is worth exploring the wide world of these models both through conceptual and mathematical reasoning as well as computer simulations, thus opening a new field of application for theory of emergence itself.

References

Albert, R. and Barabási, A. (2002). "Statistical mechanics of complex networks," *Reviews of Modern Physics*, ISSN 0034-6861, 74: 47-97.

Barabási, A. (2002). *Linked: The New Science of Networks*, ISBN 9780738206677.

Berne, E. (1964). *Games People Play: The Basic Handbook of Transactional Analysis*, ISBN 9780345410030 (1996).

Buller, D.B. and Burgoon, J.K. (1994). "Deception: Strategic and non-strategic communication," in J.A. Daly and J.M. Wiemann (eds.), *Strategic Interpersonal Communication*, ISBN 9780898599572, pp. 191-223.

Cangelosi, A. (2001). "Evolution of communication and language using signals, symbols, and words," *IEEE Transactions on Evolutionary Computation*, ISSN 1089-778X, 5: 93-101.

Dennett, D.C. (1987). *The Intentional Stance*, ISBN 9780262040938.

Grice, H.P. (1975). "Logic and conversation," in P. Cole and J.L. Morgan (eds.), *Syntax and Semantics: Vol. 3, Speech Acts*, ISBN 9780127854236, pp. 41-58.

Miller, G. and Steinberg, M. (1975). *Between People: A New Analysis of Interpersonal Communication*, ISBN 9780574175014.

Minati, G. and Pessa, E. (2006). *Collective Beings*, ISBN 9780387355412.

Oliphant, M. (1996). "The development of Saussurean communication," *BioSystems*, ISSN 0303-2647, 37: 31-38.

Penna, M.P. and Mocci, S. (2005). "The systemic approach to the communicative silence," *Proceedings of the 6th Systems Science European Congress*, Paris, September 19-22, 2005, Paris, CD-ROM.

Perfors, A. (2002). "Simulated evolution of language: A review of the field," *Journal of Artificial Societies and Social Simulations*, ISSN 1460-7425, 5(2).

Searle, J.R., (1983). *Intentionality: An Essay in the Philosophy of Mind*, ISBN 9780521273022.

Shannon, C. and Weaver, W. (1949). *The Mathematical Theory of Communication*, ISBN 9780252725463 (1998).

Stacey, R.D. (2001). *Complex Processes in Organizations: Learning and Knowledge Creation*, ISBN 9780415249195.

Stacey, R.D. (2003). *Complex and Group Processes: A Radically Social Understanding of the Individual*, ISBN 9781583919200.

Stewart, I. and Jones, V. (1987). *TA Today: A New Introduction to Transactional Analysis*, ISBN 9781870244008.

Thaler Singer, M. and Lalich, J. (1996). *Crazy Therapies: What Are They? Do They Work?* ISBN 9780787902780.

Watzlawick, P., Beavin, J.H. and Jackson, D.D. (1967). *Pragmatics of Human Communication: A Study of Interactional Patterns, Pathologies, and Paradoxes*, ISBN 9780393010091.

Werner, G. and Dyer, M. (1992). "Evolution of communication in artificial organisms," in C. Langton, C. Taylor, J.D. Farmer and S. Rasmussen (eds.), *Artificial Life II*, ISBN 9780201525700, pp. 659-687.

Maria Pietronilla Penna is Full Professor in General Psychology at "Facoltà di Scienze della Formazione", University of Cagliari, ITA. His research involves the study of the Artificial intelligence, neural network models, systems theory, cognitive psychology, psychophysiology, psychology of communication, man-machine interfaces, usability assessment, e-learning design. From 1993, Specialist in "Cognitive Psychology and Neural Networks". Besides, she worked on perception, memory, knowledge representation, both through laboratory experiments and theoretical formulation of new cognitive processes models.

Sandro Mocci is Philosophy and Psychology graduate, and Ph. D. Student in History, Philosophy and Teaching of Science at the University of Cagliari, ITA. His research involves the study of the Artificial intelligence, systems theory, cognitive psychology, psychology of communication, usability assessment, e-learning. He produced about twenty papers accepted for publication in scientific journals, book chapters, as well as communications in scientific meetings. His currently research concerns the Text Comprehension, the Silence and communication processes in the organizations and memory, perception and mental representations.

Cristina Sechi is Psychology graduate, and Research Assistant at "Department of Psychology", University of Cagliari, ITA. His research involves the study of the Artificial intelligence, systems theory, cognitive psychology, psychology of communication, usability assessment, e-learning design. She produced about twenty papers accepted for publication in scientific journals, book chapters, as well as communications in scientific meetings. His currently research concerns the design of e-learning courses usability based, educational technology, memory, perception and mental processes.

7. Power Law Distributions in Pattern Dynamics of Attacker-Defender Dyads in the Team Sport of Rugby Union: Phenomena in a Region of Self-Organized Criticality?

Pedro Passos[1], Duarte Araújo[2], Keith Davids[3], João Milho[1] & Luis Gouveia[4]
1 Universidade Lusófona de Humanidades e Tecnologias, POR
2 Universidade Técnica de Lisboa, POR
3 Queensland University of Technology, AUS
4 Universidade de Lisboa, POR

In the region of self-organized criticality (SOC) interdependency between multi-agent system components exists and slight changes in near-neighbor interactions can break the balance of equally poised options leading to transitions in system order. In this region, frequency of events of differing magnitudes exhibits a power law distribution. The aim of this paper was to investigate whether a power law distribution characterized attacker-defender interactions in team sports. For this purpose we observed attacker and defender in a dyadic sub-phase of rugby union near the try line. Videogrammetry was used to capture players' motion over time as player locations were digitized. Power laws were calculated for the rate of change of players' relative position. Data revealed that three emergent patterns from dyadic system interactions (i.e., try; unsuccessful tackle; effective tackle) displayed a power law distribution. Results suggested that pattern forming dynamics dyads in rugby union exhibited SOC. It was concluded that rugby union dyads evolve in SOC regions suggesting that players' decisions and actions are governed by local interactions rules.

Introduction

Modeled as dynamical systems, multi-agent systems like team sports display important characteristics of complexity due to the potential for interactions that emerges between system components (i.e., performers) over time (e.g., an attacker and a defender are two components of a dyadic system in 1 vs. 1 subphases) (Schmidt, O'Brien & Sysko, 1999; Guerin & Kunkle, 2004; McGarry & Franks, 2007; Passos *et al.*, in press). Previous work has shown how, despite differences in individual constraints (e.g., technical, tactical knowledge, emotional and cognitive skills), both agents in a dyadic sub-system of team sports such as rugby union and basketball explore the space available in front of them to maintain or de-stabilize sub-system symmetry (e.g., Araújo *et al.*, 2004; Passos *et al.*, 2006). For example, previous research found that, in the initial stable state of the sub-system in the team sport of rugby union, the defender started closest to the try line and if an attacker passed the defender with the ball, system organization was destabilized (e.g., Passos *et al.*, in press). After a change in the dyad's structural organization (i.e., the attacker passed the defender and became the closest player to the try line), a try occurred (i.e., when the attacker reached the try area, touching the ball on the ground to score). Also when the connection between the system agents changed (i.e., from non-physical to physical) this outcome was consistent with an effective tackle or a tackle when the attacker passed the defender.

These observations were consistent with Juarrero's (1999) insights suggesting that self-organization under constraints in complex systems is characterized by system agents becoming systematically re-organized in qualitatively novel ways with changes in connection type or structural re-organization occurring between them. Initially it was proposed that each player's behavior was regulated by first order contextual constraints such as the rules of the game, the performance area dimensions and boundary markings, and by each player's role (i.e., the attacker aims to score a try and the defender attempts to successfully tackle the adversary). In accordance with Juarrero's (1999) conceptualization we defined these constraints as a *first order* kind because they initially define the perceptual-motor performance space where action takes place. This type of constraint enhances the probability that specific actions will emerge, such as a defender committing to a tackle at a specific position when defending the score area or an attacker selecting a particular running line trajectory to score a try instead of running randomly across the playing field. Throughout the approach phase in 1 vs. 1 sub-phases of rugby union, there exists a relative independence of both players' decisions and actions. In this phase, one player's actions will not immediately affect the actions of another in a dyad. But as captured by Passos and colleagues in previous research, in order to achieve their personal tactical performance goals, the flow of running line trajectories pulls the players towards the same basin of attraction (i.e., players' trajectories create a region on the field that attracts the dyadic system and where a sudden change in system structural organization could emerge) (Passos, Araújo, Davids, Diniz, Gouveia and Serpa, 2007). It was observed that the players' interactions attracted each other (due to decreasing interpersonal distance) to a critical region of the field where the decisions and actions of each player no longer remained independent. The decrease in interpersonal distance between an attacker and a defender in a dyad was suggested to be inversely correlated with each player's relative dependence and characterized the 1vs1 performance sub-phase of team sports like rugby union. Over time, players' actions in a dyad turned out to be systematically interrelated and each player's intentions did not make sense if separated from each other's decisions and actions (see ideas on complexity of Kauffmann, 1993).

This context dependency led to the emergence of another category of constraints, termed *second order constraints* which exposed the self-organization, emergent tendencies of behavior in complex social systems (Juarrero, 1999). Due to the influence of second order constraints (Juarrero, 1999), attacker-defender behavioral dependence is an emergent property of dyadic systems in team sports, signifying that new behavioral repertoires become available to the dyad as a coevolving system. In spite of the many diverse running lines trajectories available to the players, second order contextual constraints that emerge throughout this stage of dyadic system interactions typically box the sub-system into three possible outcome states: i) *physical contact takes place but the attacker does not pass the defender* and initial system organization is conserved. Nevertheless, the type of connection between the dyad components changes (from non-physical to physical) resulting in the system undergoing a new phase in the self-organizing, emergent process; ii) *physical contact takes place and the attacker passes the defender*. Due to physical contact, the type of connection between the dyad components changes but the main difference between this new emergent state and the previous one is that a change in within-system organization occurs, and the attacker is now the player closest the try line; or iii) *the attacker passes the defender without physical contact* and the connection between the two players remains non-physical. However, the dyad undergoes a phase transition since the players' within-system structural organization changes with the attacker now nearer than the defender to the try line.

Self-Organizing Criticality in Team Sports

In the past decades Per Bak and colleagues developed the sandpile model to describe the existence of self-organizing criticality in nature (Bak *et al.*, 1988). This model highlights that in nature every open system evolves throughout several dynamical states due to catastrophic events. Moreover these abrupt (i.e., catastrophic) changes in system structural organization are due to self-organizing behaviors that evolve to critical states through dynamical processes that occur between system components and are not led by an external agent.

According to Bak's (1996) insights, it could be construed that, in the performance context of team sports, most of the changes in the attacker-defender symmetry can occur through catastrophic events. For example, despite the many small fluctuations that may occur in a pattern of play, the attacker-defender balance can be abruptly broken rather than undergoing smooth gradual transitions. One moment the defender can be counterbalancing the attacker's actions, the next moment the attacker can suddenly break this symmetrical organization, passing the defender and running free to the try line. System evolution to this critical state occurs without the direct design of an outside agent (e.g., the coach). Rather it emerges due to the influence of inherent pattern forming dynamical processes within the system. This conceptualization of system organization is in line with Kauffman's (1993, 1995) modeling of coevolving agent adaptation in complex evolutionary systems. In the coadaptation model of evolutionary processes, when one ecological system component undergoes a change it may be provided with a fitness advantage over other complex system components (e.g., in a predator-prey complex). These components are forced to respond with related changes to enhance their fitness. This cycle perpetuates as the system undergoes continual transitions as each component seeks to gain an evolutionary advantage. Processes of coevolution aim to attain the edge of chaos for a particular system, which according to Kauffman (1995) is a close cousin of self-organized criticality.

These ideas provide a sound theoretical foundation to describe emergent decision making processes (along different timescales) in the dynamics of interpersonal interactions of multi-agent systems in team sports. Processes of coadaptation also exist in multi-agent systems in team sports. These processes set in play a web of compromises where each player progresses towards behavioral goals, but none can be sure about the consequences of his/her own optimal next step due to the constraining influence of other players. As in any biological complex system, at this poised state between order and chaos, agents cannot predict the unfolding consequences of their actions (Kauffman, 1995). This critical state is shaped by the constraint imposed by local dynamical interactions rules among individual system components (i.e., the players). Bak (1996) suggested that in nature the critical states in system order that emerge due catastrophic events are self-organized.

To conceptualize team sports as complex dynamical systems it is important to note that the unpredictable nature of this performance context is due to the intrinsic variability that is available to the players as system agents. During team sport competitions the decisions and actions of each player are constrained by multiple causes that generate multiple effects, and according to Bar-Yam (2004), this is a crucial feature in considering complexity in team games. The potential for interaction between players in a rugby union match, viewed as a complex system, signifies that it is not possible to accurately describe a specific outcome that occurs in a game sustained by a single cause-effect relationship (Passos, Araújo, Davids & Shuttleworth, 2008).

Game Evolution Through Stasis and Quiescence Intermittency

The apparent equilibrium displayed at certain moments of a rugby union match (e.g., a ruck; or a line-out) exemplify periods of stasis that exist between intermittent bursts of activity and volatility in which specific dynamical patterns are formed, such as an attacker-defender balance (e.g., when a defender can counterbalance an attacker's actions). These system patterns can be annihilated as new patterns emerge, such as when the attacker passes the defender in a dyad and runs free until matched by another defender and a new dyadic sub-system is formed. This process in team sports exemplifies the phenomenon of "punctuated equilibrium" which is at the heart of the pattern forming dynamics of complex systems (Bak, 1996). Figure 1 displays data exemplifying the "punctuated equilibrium" that characterizes dyadic interactions in team sports like rugby union. In this instance a clean try is the outcome of attacker-defender interpersonal dynamics in a 1 v 1 dyad.

The phenomenon of punctuated equilibrium in team sports can be characterized as a continuous change in the attacker-defender balance (i.e., at one moment the defender has the advantage, the next it is the attacker who can gain an advantage). This phenomenon manifests itself in changes in critical control parameter values (denoted as 'nested' control parameters by Passos *et al.*, 2008), resulting in rich variations of behavior in the dyadic complex system. Near the critical state, interactions between players and nearest neighbors (i.e., team mates and opponents) can become correlated, in a type of domino effect, capturing global system dynamics and leading to a sudden reduction from multiple options to one. In the critical state a slight change in circumstances (e.g., small changes in players' relative velocity) characterizing near neighbor interactions will break the balance of equally poised options leading to an abrupt transition in system order. In other words the coadapting moves of the players shift the local system towards a critical region of the landscape in which it is poised for a transition. Criticality provides the platform for a functional fusion of creativity and constraint in dynamic performance settings. It affords new opportunities for behavior which can fit newly arising circumstances. In this region of criticality it was hypothesized that the frequency of events (i.e., the attacker attempts to pass the defender and the defender counterbalances the attacker's actions) of differing magnitudes exhibits a power law distribution. Such a function is captured when smaller dyadic system fluctuations arise more frequently than larger fluctuations, characterizing system changes such as phase transitions, as agents

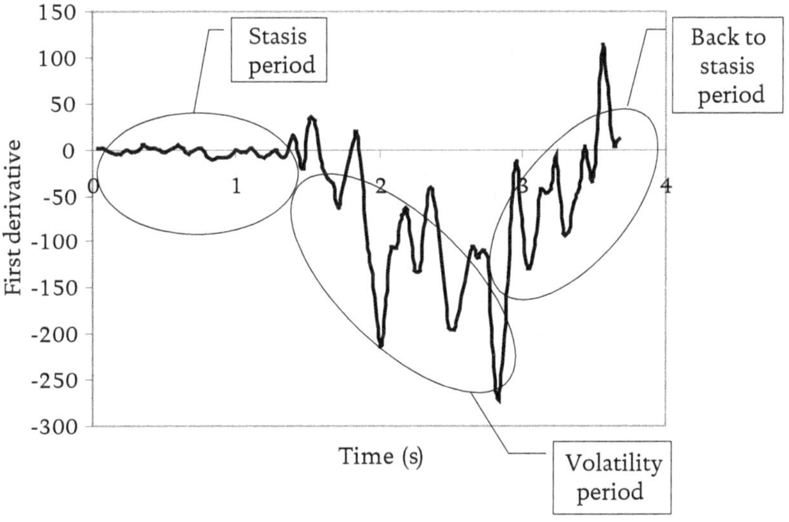

Figure 1 *The Punctuated Equilibrium.*

(players) search for optimal 'fitness' solutions (i.e., breaking or maintaining the equilibrium of the dyad). As each individual in the coadapting dyadic system seeks to enhance his/her success, the whole system inexorably moves (i.e., is attracted) in the cocreated landscape towards a region of self-organizing criticality poised for a transition (Bak & Chialvo, 2001).

To summarize, self-organizing criticality is an underlying principle for complexity and this principle is expressed with power laws (Bak, 1996). In this line of reasoning the aim of this paper is to investigate whether a power law distribution exists for attacker-defender interactions in the team sport of rugby union. If such a distribution in interpersonal interactions were observed, it may be possible to understand when the pattern forming dynamics of a 1 v 1 dyadic system in the team sport of rugby union may enter a region of self-organizing criticality.

Methods

In previous work an experimental task was designed that was representative of a typical sub-phase of rugby union with the minimum number of players involved i.e., the ubiquitous 1 vs. 1 (i.e., attacker vs. defender) situation close the try line. In this research program a methodology was developed, based on videogrammetry, and using two digital video cameras to capture player motion on the performance field. TACTO 7.0 software (Fernandes & Caixinha, 2003) was used to digitized and convert the player images to numbers, and artificial neural networks (ANN) were used to solve the stereo resolution problem (i.e., convert time series data from the two video cameras to three dimensional real world coordinates, x, y, z) (for further details see Passos *et al.*, 2006).

The aim of this study was to identify whether a power law distribution existed for pattern forming dynamics emerging from the attacker-defender dyadic interactions in rugby union (i.e., clean try; unsuccessful tackle; and effective tackle). Bak (1996) proposed that every phenomenon that displays a straight line on a double logarithmic plot may be called a "power law". In this study, the power law relationship referred to log-log plots that were calculated using the 'magnitude of adjustments' on the x axis, and 'frequency (Log of N adjustments)' on the y axis. This power law relationship displayed a specific quantity N (i.e., frequency of changes in attacker-defender relative positioning, called 'frequency (Log of N adjustments') which was expressed as a power of the magnitude of frequency of changes in attacker-defender relative positioning, termed 'magnitude of adjustments', in the following equation:

$$\log N(s)=-\tau \log s$$

To capture the pattern forming dynamics in attacker-defender dyads in rugby union, we identified a collective variable that was suitable for describing dyadic system behavior: that is the angle between a vector from defender to attacker with an imaginary line parallel to the try line (Passos *et al.*, 2008). Data from the collective variable analysis allowed us to identify three different outcomes (i.e., coordination patterns) of dyadic pattern forming dynamics: i) effective tackle; ii) tackle with the attacker passing the defender; and iii), clean try (Passos *et al.*, 2008).

Supported by the collective variable data we calculated the first derivative data for each performance situation which allowed us to characterize the rate of change of relative position between attacker and defender which differentiated the three outcomes of interpersonal interactions of attackers and defenders functioning in dyads (i.e., clean try; unsuccessful tackle; and effective tackle). This procedure allowed us to analyze how rapidly the players adjusted their relative positions over time. If the values remained at 0 ms^{-1}, this result signified that there were no adjustments in the players' relative positions. Conversely, any adjustment in the players' relative positions led to fluctuations in the first derivative values (Passos *et al.*, 2007). An increase in the magnitude of first derivative fluctuations may be interpreted to suggest that the system was entering a self-organized state of criticality, and was poised for a transition (Figure 2).

Results

Figure 3 displays three exemplar data curves each representing a coordination pattern displayed in the pattern forming dynamics of the dyadic system: clean try situation, unsuccessful tackle, and effective tackle. Our analysis showed that these three emergent patterns from dyadic system interactions revealed a power law distribution with exponents between 1.4 and 1.8, defined as the slopes of the curve. The magnitude of adjustments for a clean try situation acquired values between 1.5 and 2.4; for an unsuccessful tackle values reached between 1.8 and 2.7; and descriptive curves for effective tackles acquired values between 1.5 and 2.5. The *R* square values for the three coordination patterns revealed a good fit with the model (Figure 3).

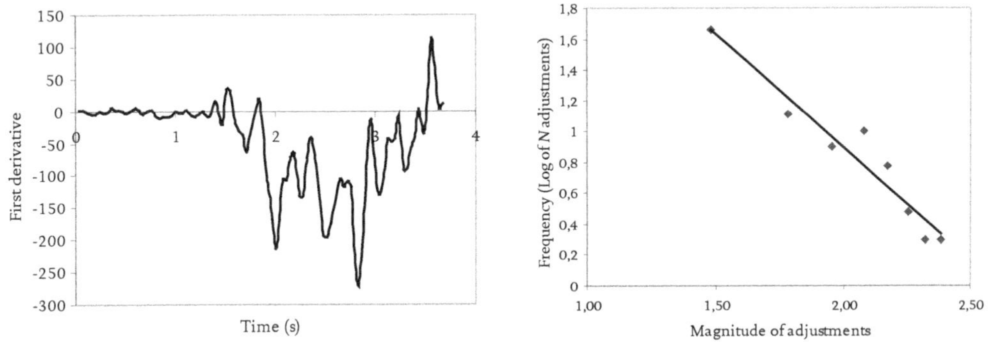

Figure 2 *The rate of change of players relative position and the power law distribution.*

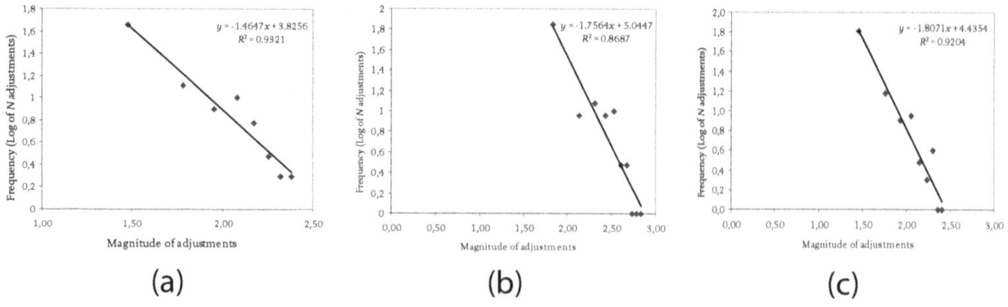

Figure 3 *Power law distribution for the three coordination patterns. (a) Clean try, (b) Unsuccessful tackle, and (c) Effective tackle.*

Discussion

During each attacker-defender dyadic performance trial it was observed that, adjustments in players' relative positioning with higher magnitudes occurred less frequently than smaller magnitude adjustments. This observation signified that, due to a decrease in interpersonal distance, as well as a number of other task constraints (e.g., players' individual goals), the players' behavior was continuously probing the stability of the dyadic system. This testing of system stability seemed to occur until a specific region of state space where an abrupt and unpredictable change in the dyadic system structural organization occurred. These findings suggested that pattern forming dynamics observed in attacker-defender dyads in the team sport of rugby union exemplified the natural phenomenon of SOC. In this region the distribution of adjustments in players' relative positioning approximated a straight line, exemplifying a power law in accordance with Bak's (1996) underlying principle for system complexity. Supported by local information rules, players spontaneously adjusted their relative positions with information available on each specific performance setting, such as an opponent's relative position and speed. Required adjustments were "scale free" since, due to context uncertainty, there were no typical sizes of variations.

However in geophysical and biological systems evolution from stable states to phase transitions does not involve the need to tune a control parameter. Various studies in our program of work on team sports have shown how the interactions of the individual agents of the complex system seemed to follow their own simple local rules and created unique, poised, global system dynamics in which the motion of one agent might affect any other agent in the system (see also Bak, 1996). In previous work on rugby union, we proposed that the local rules might be assumed to act as potential, nested control parameters (Passos *et al.*, 2007). In that research investigation we identified a specific relation between interpersonal distance and the relative velocity of attacker and defenders in dyads which underwent a phase transition to one of the three possible coordination patterns (i.e., clean try; unsuccessful tackle; and effective tackle). Taken together these findings confirm Bak's (1996) observation that the behavior of complex systems cannot be dependent on the tuning of a single parameter (e.g., temperature in a weather system). Furthermore, in accordance with the sandpile model (Bak *et al.*, 1988), the magnitude of an event is unrelated to the parameter that triggers it. In other words large and small events can be triggered by the same kind of cause (Kauffman, 1995). This insight seems to be consistent with the notion of 'multiple causes for multiple effects', a primary feature in considering complexity of team games (Bar-Yam, 2004). Moreover the term "control parameter", which emanated from synergetics (Haken & Wunderlin, 1990), implies that there needs to be an external agent tuning the parameter. The use of this term may not be appropriate here because according to Bak (1996) self-organized critical systems can evolve to a complex critical state without any interference from an external agent. However, building on our previous work, our data suggest the need to maintain usage of the term "nested system parameters", without the qualifying descriptor "control" to identify the important role of local interaction rules as a mechanism that can move a complex system to phase transitions. In our research the nested system parameters that we identified (i.e., interpersonal distance and attacker-defender relative velocity) are emergent system constraints because they became spontaneously coupled without interference from an external agent. At critical values (i.e., corresponding to a bifurcation point, see e.g., Prigogine, 1996), their interaction provoked an abrupt change in structural organization of the dyadic system that led to multiple effects (i.e., the coordination patterns previously presented). These findings correspond to Bak's (1996) notion of criticality and (therefore complexity) emerging "for free"... without any watchmaker tuning the world".

The data imply that in dyadic systems in team sports fluctuations are unavoidable. The large fluctuations observed in 1 vs. 1 attacker defender dyadic systems in rugby union was indicative of coadaptive agent behavior operating in a self-organizing criticality state. In this region minor changes in system parameters (i.e., interpersonal distance and attacker-defender relative velocity) might lead to abrupt changes that characterize the stasis and quiescence intermittency observed in the evolution of competitive team games. Attacker-defender dyads evolve in SOC regions poised at the edge of chaos, where players' decisions and actions are governed by local emergent interactions rules rather than the *a priori* instructions provided by external agents (such as coaches, trainers, parents, significant others). These findings provide some important consequences for training in team sports like rugby union, considered as complex systems. Too much prescriptive advice on decision making and action should be avoided in practice contexts for team sports. Rather, training methods should be built on a 'constraints based approach' with players provided with extensive opportunities to explore the perceptual-motor performance workspace in order to harness emergence and maintain goal directed behavior when entering SOC regions (Araújo *et al.*, 2004). Submitting the players to this kind of training philosophy increases their attunement to the relevant information constraints that can lead them to functional decisions and actions.

References

Bak, P., Tang, C. and Wiesenfeld, K. (1988). "Self-organized criticality," *Physical Review A*, ISSN 1050-2947, 38: 346-376.

Bak, P. (1996). *How Nature Works: The Science of Self-Organizing Criticality*, ISBN 9780387987385.

Bak, P. and Chialvo, D.R. (2001). "Adaptive learning by extremal dynamics and negative feedback," *Physical Reviews E*, ISSN 1539-3755, 63: 031912.

Bar-Yam, Y. (2004). *Making Things Work: Solving Complex Problems in a Complex World*, ISBN 9780965632829.

Guerin, S. and Kunkle, D. (2004). "Emergence of constraint in self-organizing systems," *Nonlinear Dynamics, Psychology and the Life Sciences*, ISSN 1090-0578, 8: 131-146.

Haken, H. and Wunderlin, A. (1990). "Synergetics and its paradigm of self-organization in biological systems," in H.T.A. Whiting, O.G. Meijer and P.C.W. van Wieringen (eds.), *The Natural-Physical Approach to Movement Control*, ISBN 9789062568932.

Juarrero, A. (1999). *Dynamics in Action: Intentional Behavior as a Complex System*, ISBN 9780262100816.

McGarry, T. and Franks, I. (2007). "System approach to games and competitive playing: Reply to Lebed (2006)," *European Journal of Sport Science*, ISSN 1746-1391, 7(1): 47-53.

Passos, P., Araújo, D., Davids, K., Gouveia, L. and Serpa, S. (2006). "Interpersonal dynamics in sport: The role of artificial neural networks and three-dimensional analysis," Behavior and Research Methods, ISSN 1554-351X, 38(4): 683-691.

Passos, P., Araújo, D., Davids, K., Diniz, A., Gouveia, L., Serpa, S. and Milho, J. (2007). "Interpersonal pattern dynamics and decision making in Rugby," Symposium Presentation at XII European Congress of Sport Psychology, Halkidiki, Greece, 3rd to 9th September 2007.

Passos, P., Araújo, D., Davids, K. and Shuttleworth, R. (2008). "Manipulating constraints to train decision making in Rugby Union," International Journal of Sports Sciences and Coaching, ISSN 1747-9541, 3(1):125-140.

Passos, P., Araújo, D., Davids, K., Diniz, A., Gouveia, L., Serpa, S. and Milho, J.(2008). "Decision making in team sports: Data, model and applied perspective," 116th Annual Convention of American Psychological Association, 14-17 August 2008, Boston, MA.

Prigogine, I. (1996). *The End of Certainty: Time, Chaos, and the New Laws of Nature*, ISBN 9780684837055.

Schmidt, R.C., O'Brien, B. and Sysko, R. (1999). "Self-organization of between-persons cooperative tasks and possible applications to sport," *International Journal of Sport Psychology*, ISSN 0047-0767, 30: 558-579.

Pedro Passos is Assistant Professor at the Faculty of Physical Education and Sports at Lusófona University of Humanities and Technologies in Portugal. He has recently gained his PhD in Sport Sciences, at Faculty of Human Kinetics, Technical University of Lisbon. His research involves the study of the dynamics of interpersonal coordination in team sports, which led him to produce several papers accepted for publication in scientific journals, book chapters, as well as communications in scientific meetings. He currently maintains is research work regarding interpersonal coordination in team sports, searching for new methods of analysis and extending is collaborations with researchers in Portugal, Australia and United States. He supervises master students from Portugal. Parallel to research activity he is also the head coach of an Under-18 Rugby team.

Duarte Araújo is Assistant Professor at the Faculty of Human Kinetics at Technical University of Lisbon in Portugal, where he gained his PhD in Sport Sciences. His research involves the study of the ecological dynamics of expert decision-making in sport, which led to him receiving a young researcher awards at the European College of Sport Science in 2001 and at the Association des Chercheurs en Activités Physiques et Sportives (ACAPS) in 2002. He was elected as a member of the managing council of the International Society of Sport Psychology in 2001. He holds editorial board positions for th International Journal of Sport Psychology, Revista Iberoamericana de Psicologia del Deporte, and Open Sport Sciences Journal. He is a founder of the European Network of Young Specialists in Sport Psychology, being now its past coordinator. He currently supervises doctoral students from Brazil, Italy, and Portugal.

Keith Davids is currently Head of School, Human Movement Studies, Queensland University of Technology. He was a first-class honours graduate of the University of London and gained a PhD at the University of Leeds in 1986. Between 1993 and 1999, he led the Motor Control group at the Department of Exercise and Sport Science at Manchester Metropolitan University, UK. Currently he supervises doctoral students from Portugal, UK, Australia and New Zealand. He is co-editor in chief of the International Journal of Sport Psychology and holds editorial board positions for the Journal of Sports Sciences, Infant Behavior and Development and the Psychology of Sport and Exercise. He is interested in the coordination and control of human movement, and their acquisition, considering human movement systems as belonging to a class of nonlinear dynamical systems. A major focus concerns how ideas from ecological psychology and nonlinear dynamics can be integrated into a Nonlinear Pedagogy in sport.

João Milho is Assistant Professor in the area of Sports Biomechanics at the Faculty of Physical Education and Sports at Lusófona University of Humanities and Technologies in Portugal. He is also a Researcher in the area of Mechanical Systems Dynamics at Institute of Mechanical Engineering in Lisbon. He is currently concluding his PhD in Mechanical Engineering at Instituto Superior Técnico,

Technical University of Lisbon. Currently, his research work within a collaborative group of researchers from Portugal, Australia and United States, is focused on human movement capture methods and analysis methods applied to interpersonal coordination in team sports.

Luís Gouveia is an Assistant Professor at the Faculty of Pharmacy, University of Lisbon, in Portugal where he has gained his PhD in Pharmaceutical Technology. His current research involves mainly modeling and simulation of pharmaco-kinetic problems. He has authored or coauthored several papers, as well as communications in scientific meetings. He currently maintains his research work within the iMed.UL—Research Institute for Medicines and Pharmaceutical Sciences supervising PhD students from Portugal.

8. Understood Complexity: Ibsen's 'An Enemy of the People'— On Complexity, Sense-Making, Understanding, and Exit/Voice/Loyalty

Tom Eide
Diakonhjemmet University College, NOR

'Understood complexity' is a term of Albert Hirschman (1976) whose economic-political theory of 'exit' ('vote with your feet') versus 'voice' (feedback or use your influence for change) (1970), has often been used to (try to) understand whistleblowing (Alford, 2001; Maclagen, 1998). Real complexity is not linear and cannot be adequately studied an model of 'A causes B'. Complexity entails 'A causes B' in a situation wherein 'B causes A'. Bateson in his 'ecology of the mind' understood the circularity of the hermeneutic of complexity; while Weick did not in his theory of sense-making. I argue in this article, via an examination of a play of Ibsen, that circular thinking spiraling towards new insight(s) is much more a possibility of literature (studies) than of social science. Social complexity theory needs (at least partially) I believe to methodologically merge with literary studies.

Introduction: An Enemy of the People: A Drama on Whistleblowing

Literature is an indirect phenomenon. On the one hand, it is a product of the author's artistic imagination; on the other, it represents aspects of our lives and the world in which we live. I will explore Ibsen's representation of complexity in his realistic drama *An enemy of the people* (1882).

The plot takes its point of departure in the discovery that the water at a Spa is polluted and in a concerned employee's unsuccessful attempt to make the management take action to stop the pollution. The plot develops as a case of whistleblowing. Ibsen exposes the organizational response to the whistleblower, resulting in the whistleblower's persecution and in retaliation against him. To my knowledge, the drama is the first literary work to make the pollution of the environment, the struggle for environmental protection, and whistleblowing, its central issues. My focus is on how Ibsen dealt with the complex (organizational) phenomenon of whistleblowing and making moral sense of it.

Whistleblowing is discussed in major research literature, starting about 100 years after Ibsen wrote his play (Bok, 1981; Elliston *et al.*, 1985; Petersen & Farrall, 1986; Alford, 2001; Johnson, 2002; Bowers *et al.*, 2007; Micely, Near & Dworkin, 2008). There is no consensus either on the term 'whistleblowing' or on the role of morality involved. I understand 'whistleblowing' as an act of an employee to make information on illegitimate practices within an organization known to the responsible management and if necessary going to the public with that information. The whistleblower grapples with 'exit' (to leave the organization and/or avoid conflict) and 'voice' (make the problematic situation known), wherein 'loyalty' (to the organization, to the society at large and/or to one's own morality) plays a major role in what gets decided and/or happens (Hirschman, 1970). Ibsen's drama mirrors in many ways the chaos and complexity of whistleblowing. As a work of dramatic art, the play has a clear plot and a tight structure composed in five acts, which suggests order, balance and meaning. I believe that the harmony of the form helps the reader make sense of the complexity and chaos of the content. The play, I will argue, offers 'understood complexity'.

The drama takes place in a small town at the Norwegian coast. The prosperity of the town is based on the running of a spa—the Baths, which every summer attracts a lot of visitors and patients. The doctor at the Baths suspects that the drinking water is polluted. He sends samples to a university laboratory and his suspicions are verified. He then sends a report to his superior, the chairman of the board of the Baths, who also is the Mayor of the town. In this report he makes it clear that the water is poisonous and may cause sickness amongst the visitors. The water pipes need to be renewed. He also breaks the bad news to his family and friends, among them Hovstad the editor of the local newspaper, and Aslaksen the representative of the local small businessmen. Both of them declare their support, and assure him that he also will have the majority of the citizens behind him. The Mayor refuses to present the report to the Board and commands the doctor to keep the information to himself. The Mayor approaches Aslaksen and Hovstad, and makes them change their mind about supporting the doctor. The Doctor calls a public meeting to communicate the danger(s). The assembly decides to forbid him to talk about the pollution and declares him an enemy of the people. After the meeting, a mob attacks his home and throws stones through his windows. He is fired; his daughter loses her job; his two young sons are sent home from school. Then the Mayor pays a visit, telling him that he might get his job back if he would only recant and admit his mistake. The Doctor refuses and throws the Mayor out. Having considered immigration to America with his family, he decides to stay and to take care of his son's education himself.

The drama is in five acts and is perfectly symmetrically constructed. In the first act the characters, social environment, and roots of the conflict are introduced. We learn that the water is polluted. The first climax happens in the second act in the discussion between the doctor and the Mayor about the content of the report. The turning point and second climax comes in the middle of the play in the third act, when the doctor in a dramatic and comic confrontation with the Mayor, is deprived of his illusion that he has the community behind him and he realizes that Mayor and his former allies are all now against him. The third and final climax comes in the fourth act, when the doctor at the public meeting is declared an enemy of the people. In the fifth and final act, the thematic threads are collected and the Mayor and the doctor are put to final tests of moral integrity.

An enemy of the people is a one of Ibsen's so-called problem plays, which by some are termed critical realism and by others modern cotemporary drama (Hemmer, 1994). It is commonly read as an exploration into the individual's struggle for truth and justice in opposition to the suppressing forces of society (Bull, 1924; Garton, 1994; Kittang, 2002; Hemmer, 2003). In the historical-biographical context, it is considered to be Ibsen's response to the negative reception of his former play Ghosts (1881), and especially as a reaction to the criticism that play received in the liberal press, which Ibsen felt betrayed basic liberal ideas and values (Koht, 1954; Meyer, 1971; de Figueiredo, 2007).

Representing Complexity

Hirshman (1970) has assumed that 'exit' and 'voice' each have their own costs and benefits. If there is too much 'exit', there is no control or feedback and little chance of reform or improvement. If there is too much 'voice', there is constant opposition, conflict and an excess of chaos. If persons are 'loyal'—i.e., ethically committed to their situation, society and/or organization, they will make use of 'voice'. Whistleblowing, thus, can be understood as a product of voice and loyalty. But mis-placed loyalty, and voice out-of-control, can endanger society. Whistleblowing as a calculated strategy of self-enrichment and/or of power grabbing is not explored here. What the play (but not so much the 'plot') has on offer is 'complexity understood'—i.e., the irresolvable complexity of good causing bad, and of bad causing good; or of trust (social involvement) causing conformity, and of social cohesion (belonging) destroying honesty. The play allows the audience to see and (vicariously) experience this complexity; a complexity that (seemingly, from the plot) no one in the drama can consciously grasp. The level of knowing of the 'audience' is principally different from the one that Ibsen attributes to the 'normal citizenry' represented in the play. This paradox of (im-)possible knowing—i.e., of the tension between what (supposedly) is commonly thinkable and what the artist tries to make thinkable in his audience, is crucial to the complexity of artistic investigation. Art (the play) makes the socially unthinkable, thinkable. And this is exactly what the revealing of complexity requires.

In the following, I do close reading of some key dialogues in the play with the aim of exploring Ibsen's representation of the complex dynamics emerging from moral dilemma(s). Such literary criticism is far more contextual, specific and experiential, than is (normal) social science. Methodologically, I am convinced that this entry into literary criticism is a good way to come close to the study of complexity, which cannot be achieved via linear procedures or abstract methodology. The method has to match the content; the form of writing has to be consistent with what one wants to say. Abstract, generalizing argumentation can hardly represent complexity—it can point to it from somewhere else—i.e., from a linear non-emergent episteme, but it cannot actually be one with it.

The Opening Lines

The opening sequence of a narrative or of a dramatic work, like a first meeting with another person, is often crucial. The opening dialogue—as the first images of a film, or the first stanza of a poem, or the first sentence of fiction—often has the function of cuing the reader to what is to come. Note that I say cuing and not coding; it is a matter of relationship and not of unidirectional information. The first dialogue of Ibsen's play takes place in the doctor's living room, between Mrs. Stockmann the doctor's wife and Mr. Billing a journalist of the *People's Herald* who has come (too late) to dinner. The opening lines:

Evening. Dr. Stockmann's living room, simply but tastefully furnished. (...)

Mrs. Stockmann. Well, if you will arrive an hour late, Mr. Billing, you'll have to put up with everything being cold.

Billing [*eating*]. It's absolutely delicious, really excellent.

Mrs. Stockmann. You know how strict my husband is about keeping punctually to his mealtimes...

Billing. It doesn't matter to me in the least. In fact I almost believe it tastes better, sitting down like this to it, alone and undisturbed.

Mrs. Stockmann. Ah well, as long as you enjoy it... (p. 22).

Nothing much seems to be happening, except small talk about coming too late and the pleasures of eating. What is this journalist doing here? Is he a friend of the house or is he invited as a journalist? He arrives late and does obviously not care very much about etiquette or his hostess' feelings. On the contrary, he seems to take pleasure in coming late and in eating leftovers. Is not cold meat the favorite food of a scavenger—of vultures and hyenas? What kind of person is this journalist? What is his role in the play going to be? We don't know. But Ibsen is certainly cuing the reader' (consciously or unconsciously) to the scavenger scenario, i.e., a possible ecological drama of the survival of the fittest.

The First Act: The Exposition

Ibsen starts in the periphery and little by little exposes the main characters and the core of the conflict. The reader or spectator is given time to mentally tune in and open up to the drama. While talking to Mr. Billing, Mrs. Stockmann suddenly turns towards the hall, and listens. A new visitor is arriving, the Mayor who is also the doctor's brother. In the list of characters he is presented as: 'Peter Stockmann the doctor's elder brother, Mayor, Chief of Police, Chairman of the Board of the Baths, etcetera.' Ibsen describes him in the following way:

[PETER STOCKMANN, the Mayor, enters; he is wearing an overcoat and his mayor's hat, and he carries a stick.] (p. 23).

His attributes—coat and hat, and walking stick—signal protection, position and personal power. The Mayor is the main authority in the small community, a stiff representative of formal position; the Doctor is (seemingly) the opposite. When he receives the test results, the doctor responds spontaneously:

DR. STOCKMANN [waving the letter]. Well! Here's a bit of news that will set a few tongues wagging about the town!

BILLING. News?

Mrs. Stockmann. What news?

DR. STOCKMANN. A great discovery, Kathrine!

Hovstad. Really?

MRS. STOCKMANN. Which you've made?

DR. STOCKMANN. Which I've made, yes. [*Walks up and down.*] Now let them come as they always do, and say it's some madman's crazy idea! Ah, but they'll watch their steps this time! They'll watch out this time, I'll bet. (pp. 36-37)

The doctor's enthusiasm seems to ignore the fact that the water is polluted and that this represents a serious threat both to the health of patients and to the prosperity of the small community. He obviously bears a grudge against somebody and his discovery will be a weapon in his hands: 'they'll watch their steps this time!' Ibsen is unveiling the conflict and its agents. The doctor uses metaphors from his own profession: the Baths are the 'artery', 'nerve centre' and 'throbbing heart' of the town (p. 37). But the discovery of the pollution may turn the imagery around: 'The Baths are nothing but a cesspool', and further:

DR. STOCKMANN. The whole establishment is a whitened poisoned sepulcher, I tell you. A most serious danger to health! All that filth up at Mölledal, where there's such an awful stench—it's all seeping into the pipes that lead into the pump-room. And that same damned, poisonous muck is seeping out on the beach as well. (p. 38)

Act II: The Confrontations

Ibsen delays the confrontation between the doctor and the Mayor, signaling again and again that the Mayor and his associates may not welcome the findings. The doctor however is confident that the Mayor will respond (favorably) to his discovery: 'He can't help but be pleased that an important matter like this has been brought to light, surely' (p. 41). The doctor is a man of science, he knows he is right and believes firmly that everybody will view the situation like he does. But the doctor's daughter is skeptical. The same goes for his wife and his father-in-law: 'Oh, you should never trust anybody. You can be taken in almost before you know where you are', his father in law says (p. 43); and soon after: 'you'll never get the Mayor to believe a thing like this' (p. 44). The complexity of the matter is also pointed to by the journalist Hovstad who introduces the perspective of politics and power:

HOVSTAD. You said yesterday that the water was contaminated by impurities in the soil.

DR. STOCKMANN. Yes, there's no doubt it all comes from that poisonous swamp up at Mölledal.

HOVSTAD. You'll forgive me, Doctor, but I think it comes from a very different swamp.

Dr. Stockmann. What swamp?

HOVSTAD. The swamp that our whole community is standing rotting in.

DR. STOCKMANN. What kind of damned nonsense is this you're talking, Mr. Hovstad?

HOVSTAD. Everything in this town has gradually found its way into the hands of a certain group of officials . . .

DR. STOCKMANN. Come now, not every one of them is an official.

HOVSTAD. No, but those that aren't officials are friends and hangers-on of those that are—the wealthy ones of the town, and the well-connected. These are the people in control. (p. 47)

Like Hovstad, Aslaksen gives his full support to the doctor, but not on political, but on commercial grounds—i.e., on behalf of the small businessmen and the 'majority' of towns people. Ibsen is continuously feeding the reader with new information, challenging the reader to imagine what the Mayor's reaction might be:

ASLAKSEN. Well then, I've just called to say that I am ready to give every support to a thing like that. (...)

DR. STOCKMANN. That's extremely kind of you, thank you very much; but...

ASLAKSEN. Because you might easily find you need some middle-class support to back you up. We now form what you might call a compact majority here in town—when we really want to, that is. And it's always a good thing to have the majority on your side, Dr. Stockmann.

DR. STOCKMANN. That is undoubtedly true. It's just that I don't quite understand why it should be necessary to take any special measures of that kind here. When it's such an ordinary straightforward thing, it seems to me...

ASLAKSEN. Ah, you never know but what it mightn't be a good thing anyway. I know well enough what the local authorities are like. Those in charge are never very keen on any kind of proposal that *other* people put forward. And that's why I think it wouldn't be a bad thing if we made a bit of a demonstration. (p. 49)

What to the doctor seems like a practical problem related to a dysfunctional water system emerges gradually to the reader (audience) as a possible matter of politics, power and personal pride. But the doctor does not see any danger, insisting 'it'll be all plain sailing', 'nothing but plain sailing!' (p. 52). In other words, Ibsen has created his main character as the archetypical whistleblower who—according to research more than a century later—'reports malpractice in the belief that senior management will be grateful for the information' (Rothschild & Miethe 1999). Not until the second part of Act II does Ibsen reveal the Mayor's response. By postponing the confrontation and cuing the reader to imagine what the Mayor's response will be, Ibsen creates dramatic and moral tension. The first confrontation is a test of morality:

Dr. Stockmann. Yes. Surely you are also convinced yourself by now!

Mayor. Is it your intention to present this document to the Board as an official report?

Dr. Stockmann. Of course. Something will have to be done about this thing. And quick. (p. 53)

The Mayor fails this first test. He is indifferent to the heart of the matter—the pollution and its threat to the patients' health; and instead focuses on procedure and form:

Mayor. I received from you yesterday, after office hours, a report concerning the state of the water at the Baths.

Dr. Stockmann. Yes. Have you read it?

Mayor. Yes, I have.

Dr. Stockmann. And what have you got to say about it?

Mayor. [*with a sidelong glance*]. Hm . . . [*after a pause*]. Was it necessary to make all these investigations behind my back?

Dr. Stockmann. Yes, until I know with absolute certainty . . .

Mayor. And now you do, you mean?

Dr. Stockmann. Yes. Surely you are also convinced yourself by now! (. . .)

Mayor. As usual, you use some rather emphatic expressions in your report. Among other things, you say that what we offer our summer visitors is sheer poison.

Dr. Stockmann. Well, Peter, what else can you call it? Just think! That water's poison whether you drink it or bathe in it! And this is what we offer those poor invalids who come to us in good faith and pay good money hoping to get their health back!

Mayor. And then you conclude by stating we must build a sewer to deal with these alleged impurities from Mölledal, and that the present water pipes must be re-laid.

Dr. Stockmann. Well, can you suggest any other solution? I can't. (pp. 53-54)

By reducing the pollution to 'alleged impurities,' the Mayor contests the scientific evidence, trivializing the matter, and downplaying the possible implications. While the doctor's project is ethical—he wants to do something about the problem and to protect the patients, the Mayor appears to have something quite different in mind. The Mayor's primary concern seems to be financial. According to him, repairs will be expensive and the work will take at least two years:

Dr. Stockmann. Two years, eh? Two whole years?

Mayor. At least. And what's to be done with the Baths in the meantime? Shall we shut them? We'll have to. You don't think people are going to come all the way here if the rumor got around that the water was polluted?

Dr. Stockmann. But Peter, that's what it is.

Mayor. And all this has to come out just when the Baths were beginning to pay their way. A lot of other places in the district could equally well develop into health resorts. Can't you see they would set to work at once to divert all our tourist traffic to themselves. Of course they would; no doubt whatever. And we'd be left sitting there with all that expensive plant on our hands; we'd probably have to abandon the entire project. The whole town would be ruined, thanks to you!

Dr. Stockmann. Me...? Ruined...? (p. 54-55)

Again Ibsen puts the Mayor to a moral test. Does he want to do something about the problem or not? The Mayor is afraid of the negative consequences for the Bath's image and accordingly for the profits and the prosperity of the town. In contemporary terms, he practices organizational ecology (Hannan & Freeman, 1989) and fears firm mortality. When the doctor insists on sticking to the truth, the Mayor responds by blaming the messenger. Again he fails the moral test. The Mayor denies the facts, downplays the seriousness of the matter, and tries to shift the moral responsibility. And when this does not do the trick, the Mayor questions the doctor's professional competence and integrity:

Mayor. As I said before, I think you exaggerate considerably. Any competent doctor would surely be able to meet this situation—take some suitable precautionary measures and treat any noticeable injurious effects, if there actually turned out to be any.

Dr. Stockmann. Well? And what then?

Mayor. The existing water-supply of the Baths is now an established fact, and must be treated as such. (p. 55)

The Mayor's defense of the status quo trumps both truth and morality. The Mayor is a representative of moral neglect, who challenges the reader to moral reflection. Concern for organization and prosperity is represented by the Mayor and concern for the health of the patients and the citizens is represented by the doctor. The reader is left in the middle, trying to understand why the Mayor is turning a deaf ear to the moral concerns and insists on doing (almost) nothing:

MAYOR. (...) But it is reasonable to suppose that in time the Directors might not be disinclined to consider how far, in the light of the prevailing financial situation, it would be possible to initiate certain improvements.

DR. STOCKMANN. Do you honestly think I would lend myself to that sort of sharp practice?

MAYOR. Sharp practice?

DR. STOCKMANN. Sharp practice, yes! That's what it would be. A swindle, a fraud, an absolute crime against the public and against society!

MAYOR. As I remarked earlier, I have not been able to persuade myself that there is any actual imminent danger.

DR. STOCKMANN. Oh yes, you have! You couldn't help it. My report is absolutely correct and clear, I know that! And you know it too, Peter, but you won't admit it. You were the one responsible for having the Baths and the water-supply sited where they are now. And it's *this*—this damned blunder of yours—that you won't admit. Puh! Do you think I can't see right through you?

MAYOR. And even if that were so? Even if I may seem to guard my reputation somewhat jealously, it's all for the good of the town. (pp. 56-57)

This is not just a question of organizational ecology or the ability to compete. Defending the status quo by blaming the Doctor and questioning his integrity, are evidently the Mayor's way of protecting himself from a (past) serious mistake. But the doctor—at least to a certain degree—seems to enjoy being able to point to the Major's 'damned blunder' and 'tremendous piece of stupidity.' Corporate social responsibility involves responding adequately to an ethical dilemma, which entails organizational ecology as well as the psychology of leadership. In the play's case, moral neglect or deficiency operates as a—conscious or unconscious—strategy of self-protection, defending the self from shame and the risk of social condemnation or degradation. The whistleblower's moral flaw is in his disposition for arrogance, hubris, and perhaps also vindictiveness. Both the Mayor's leadership and the whistleblower's morality are problematic. Having unveiled the Mayor's personal interest in keeping the information secret, Ibsen continues by exposing the Mayor's rationale for obstructing and counteracting transparency:

MAYOR. (...) Without some measure of moral authority, I should not be able to guide and direct public affairs in the way I consider best serves the common weal. Therefore—and for various other reasons—I consider it imperative that your report should not be presented to the Board. In the public interest, it must be withheld. Then I shall bring the matter up later, and we'll do all we can privately. But nothing, not a single world, of this disastrous business must be made public. (p. 56)

In exposing the Mayor's hypocrisy, Ibsen makes use of dramatic irony, i.e., the technique of letting the audience perceive underlying significances that are not apparent to the characters themselves. The Mayor pretends—or believes—to act in the best interest of the town, while the reader knows that he is acting out of self-interest. The conflict continues with the doctor refusing to give in to the Mayor's demands. This represents yet another moral test to the Mayor, which he fails:

DR. STOCKMANN. My dear Peter, I doubt if we can prevent that now.

MAYOR. It must and shall be prevented.

DR. STOCKMANN. It's no use, I tell you. Too many people know about it already.

MAYOR. Know about it already? Who? I only hope it's not those people on the *Herald*...?

DR. STOCKMANN. Oh yes, they know already. The progressive ad independent press will see to it that you do your duty.

MAYOR [*after a short pause*]. You are an astonishingly indiscreet man, Thomas! Did you never think what consequences this might have for you personally?

DR. STOCKMANN. Consequences? For me?

MAYOR. For you and your family (p. 56)

The Mayor responds with subtle threats to the Doctor and his family. When the Doctor shows no sign of changing his mind, the Mayor appeals to his loyalty: 'You seem to forget that it's me you have to thank for your appointment here as medical officer to the Baths...' (p. 56). And when the Doctor does not respond, the Mayor becomes explicit:

MAYOR. Since you have been so indiscreet as to discuss this delicate matter with certain unauthorized persons—despite the fact that it should have been treated as a matter confidential to the Board—things can of course no longer be hushed up. All sorts of rumors will spread, and the more spiteful ones among us can be relied on to embellish them with all sorts of extras. It will therefore be necessary for you to make a public denial of these rumors.

DR. STOCKMANN. For me! How? I don't understand you.

MAYOR. We shall expect you, after making further investigations, to come to the conclusion that the matter is not by any means as dangerous or as serious as you in the first instance imagined it to be.

Dr. Stockmann. Aha! So that's what you expect, is it?

Mayor. Furthermore we shall expect you to make a public declaration of your confidence in the Board, in its efficiency and its integrity, and in its readiness to take all necessary stems to remedy such defects as may arise. (p. 58)

There are echoes here of Galileo and the abjuration enforced upon him. The Mayor's threats, force, and denial of facts are meant to make the Board seem trustworthy. By letting the Mayor put pressure on the doctor, Ibsen exposes the Doctor's moral integrity to testing. Unlike the Mayor, the doctor passes the test. He refuses to give in:

Dr. Stockmann. Yes, but don't you see, you'll never do anything just by fiddling with the problem, hoping to patch things up. I'm telling you straight Peter, and I', absolutely and utterly convinced...

Mayor. As an employee you have no right to any private opinion.

Dr. Stockmann [falters]. No right...?

Mayor. As an employee, I mean. As a private individual—good Lord, yes, that's quite different. But as a subordinate member of the staff of the Baths, you have no right to express any opinions that conflicts with that of your superiors. (…)

Dr. Stockmann. And supposing I don't ... obey?

Mayor. Then we shall ourselves issue a statement to reassure the public.

Dr. Stockmann. Indeed. Well, then I shall contradict you in the newspapers. I shall stand up for myself. I shall prove that I'm right and you're wrong. And then what will you do?

Mayor. Then I shall not be able to prevent you from being dismissed. (pp. 58-60)

Ibsen's Mayor increasingly represents the corruption of mind by threatening the doctor with dismissal and soon after appealing to the Doctor's wife to try to make the Doctor change his mind—implicitly threatening her with dire consequences for the whole family. Whistleblowing results from the Mayor's refusal to listen and effort to suppress the truth:

Dr. Stockmann. I'm the one with the real best welfare of the town at heart! All I want to doe is expose certain things that are bound to come out sooner or later anyway. Oh, I'll show them whether I love this town or not.

Mayor. All you are really doing, by your sheer blind obstinacy, is cutting off the main source of the town's prosperity.

Dr. Stockmann. That source is poisoned, man! Are you mad! We live by peddling filth and corruption! The whole of town's prosperity is rooted in a lie!

Mayor. Fantastic nonsense—or worse! Any man who can cast such aspersions against his own birthplace is nothing but a public enemy. (p. 60-61)

Ibsen dramatizes two conflicting rationales; on the one hand there is the organizational concern for profit and organizational survival; and on the other, the professional concern for the health of the patients and citizens. But at the same time, Ibsen also explores underlying conflicting psychological and existential motives interfering with and sabotaging these (simple) moral concerns. The Mayor needs to protect himself from losing face and the doctor takes pleasure in being right. With this complexity, Ibsen exposes whistleblowing as an emergent phenomenon that cannot be simply modeled.

Act III: Organizational Sense-Changing

During the first two acts, Ibsen exposed the conflict central to the play. On the one hand, we find the doctor and his allies, the free press and the representatives of the citizenry. And on the other hand, we find the Mayor and the representatives of the leading citizens, and their organizational power. The third act is staged in the editorial office of the local newspaper. Ibsen starts out by exposing the Doctor's fighting spirit and the enthusiasm of his supporters when they learn that he has decided to blow the whistle:

HOVSTAD [*crosses to him*]. Ah, it's you, Doctor. Well?

DR. STOCKMANN. Print it, Mr. Hovstad!

HOVSTAD. Has it come to that?

BILLING. Hurrah!

DR. STOCKMANN. Print away, I tell you. Yes, it *has* come to that. Now they're going to get what's coming to them. This is war, Mr. Billing! (p. 65)

Billing, the radical journalist, finds the report 'absolutely devastating', like a 'blow from a sledge-hammer'; he can almost hear 'the revolution coming' (p. 64). Likewise, editor Hovstad finds the report an 'absolute masterpiece' and will print it immediately (p. 65). Aslaksen, the representative of the small businessmen of the town declares the Doctor 'a true benefactor to the town, a real benefactor to society' (p. 68). The Doctor may be an enemy of society to the Mayor, but he is a hero to Aslaksen and the journalist. But the Mayor has no intention of giving in. He decides to pay a visit to the newspaper's editorial office and approaches the Doctor's allies. He starts with Hovstad, the editor:

MAYOR [*looking about him*]. You've settled yourself in here nice and comfortably. Very nice.

HOVSTAD. Oh...

MAYOR. And here I come without any appointment, and proceed to take up all your precious time.

HOVSTAD. *Please*, Mr. Mayor, I'm only too delighted to be of service. Let me take your things. [*He puts the* MAYOR'S *hat and stick on a chair.*] Now, won't you sit down?

MAYOR [*sits at the table*]. Thank you.

[HOVSTAD also sits down at the table]

MAYOR. I have an extremely disagreeable matter to deal with today, Mr. Hovstad.

HOVSTAD. Really? Of course, with so many things to see to...

MAYOR. This particular matter has been raised by the Medical Officer of the Baths.

HOVSTAD. By the Doctor?

MAYOR. He's written a kind of report about a number of alleged shortcomings at the Baths, and sent it to the Board.

HOVSTAD. Has he? (p. 75)

The Mayor excuses himself for taking the editor's time and offers to share a personal unpleasant experience of his. *Self-disclosure* or *self-sharing* is an efficient strategy for building trust and creating personal alliances and loyalty. The mayor's seemingly humble and almost pleasing attitude, downplays his authority, making him look much less of an antagonist than he really is. By letting the editor put away the Mayor's *attributes of power*—his hat and stick, Ibsen signals to the reader that the Mayor's strategy is working. The conflicts of political interest are being put aside and a new alliance is born. With the symbols, Ibsen reminds the reader who the Mayor really is—a man of power with means of retaliation and punishment. Again dramatic irony is used to expose the Mayor's slyness or falsehood; he refers to the doctor's report as 'a kind of report' and talks about 'alleged shortcomings'. The reader (audience) knows that the report is based on conscientious scientific testing. The Mayor is a strategic and manipulative communicator. Having got the editor on his side, he tackles Mr. Aslaksen approaching him in a friendly, flattering, almost intimate way to get his message through, namely that he has everything to lose by staying loyal to the Doctor:

MAYOR. If all these extensive alterations are considered desirable, the town itself must pay for them. (...) The most ruinous thing is that we'll be forced to close the Baths for a couple of years.

ASLAKSEN. Yes, but Heavens! We could never last out that long, Mr. Mayor. What would people like us live on in the meantime? (pp. 77-78)

After having actualized Aslaksen's self-interest as a businessman, the Mayor can undermine his loyalty to the Doctor.

MAYOR. I regret to say that is an extremely difficult question to answer, Mr. Aslaksen. But what do you expect us to do? Do you think anybody is going to come here if you get people going around making up these stories about the water being polluted, and about the place being a cesspool, and the whole town...

ASLAKSEN. Do you think the whole thing might just be imagination?

MAYOR. With the best will in the world, I cannot come to any other conclusion.

ASLAKSEN. Then I must say Dr. Stockmann is being a most irresponsible in all this. You must forgive me, Mr. Mayor, but...

MAYOR. I regret what you say is quite true, Mr. Aslaksen. My brother has always been rather impetuous, unfortunately. (p. 78)

The Baths may be polluted, but self-interest is powerful. The Mayor is a powerful schemer, capable of undermining the doctor's allies' commitment by lies and slander. Having established the conception of the doctor as untrustworthy, the Mayor takes communicative control of the situation, offering the editor a written statement.

ASLAKSEN. Are you still prepared to support him after this, Mr. Hovstad?

HOVSTAD. But who would have thought...?

MAYOR. I have drawn up a short statement of the facts, putting a rather more sober interpretation on them; and in it I have suggested some ways in which such defects as may come to light could reasonably be dealt with without going beyond the present resources of the Baths.

HOVSTAD. Have you this statement with you, Mr. Mayor?

MAYOR [*fumbling in his pocket*]. Yes, I brought it with me on the off-chance that... (p. 78)

With a series of small strategic moves, the Mayor succeeds in turning both morality and reality upside down, and in making the doctor seem completely irresponsible. Having prevented the publication of the doctor's report, the Mayor makes his own statement. The process is displayed of making a lie into an official truth. In the climax, which is the turning point of the play, the doctor realizes what the audience already knows, namely that the citizenry are now all against him. No one will print his article. The doctor is furious:

DR. STOCKMANN. You think you can gag me and silence the truth! You'll not get away with this so easily. Mr. Aslaksen, will you please take my manuscript and print it for me at once as a pamphlet—at my own expense, and on my authority. I want four hundred copies—no, five ... six hundred, I want.

ASLAKSEN. Not if you offered me its weight in gold could I let my printing press be used for a thing like that. I daren't offend public opinion. You'll not get anybody in town to print this, I shouldn't think. (p. 84)

The representative of leadership and power is on the winning hand. The whistleblower is discredited, his actions are considered irresponsible, people are afraid of being associated with him, and he is denied every possibility of making his views known. Sense has (been) changed—true seems false, responsible action seems irresponsible, responsibility seems a threat to society. Not only the representatives of power are involved, but also the average citizens silently consent or actively contribute to what is going on. Ibsen lets the Doctor call a mass meeting: 'All my fellow citizens shall hear the voice of truth!' (p. 85). In other words, the Mayor's task of silencing the Doctor is not quite completed yet.

Act IV: Taking Control Of Public Opinion

In Act IV Ibsen stages the confrontation between the doctor and the people of the town. A big crowd of townspeople has turned out. The Mayor arrives late, 'eases his way through the crowd, bowing politely' (p. 89). The Mayor insists that the meeting needs a chairman and proposes Aslaksen who he knows will be his loyal representative. Having secured his power position, the Mayor is ready for his following step, preventing the Doctor from speaking:

MAYOR. Mr. Chairman!

ASLAKSEN. Yes, Mr. Mayor.

MAYOR. In view of the close relationship which, as is doubtless well known, exists between me and the present Medical Officer of the Baths, I should have much preferred not to speak this evening. But my connections with the Baths, to say nothing of my concern for the vital interests of the town, compel me to put forward some sort of proposal. I think I may safely assume that not a single one of us present here today wants to see irresponsible and exaggerated accounts put about concerning the sanitary conditions at the Baths and in the town generally.

MANY VOICES. No, no! Certainly not! We protest!

MAYOR. I should like to propose, therefore, that the Medical Officer be not permitted by this meeting to present his account of the matter.

DR. STOCKMANN [*flaring up*]. Not permitted! What is this...?

MRS. STOCKMANN [*coughing*]. Hm! hm!

DR. STOCKMANN [*composing himself*]. Ah! Not permitted, eh!

MAYOR. In my communication to the *People's Herald*, I acquainted the public with some relevant facts, and every right-thinking person can quite well form his own opinion. It clearly shows that the Doctor's proposal—apart from being a vote of censure on the leading citizens of the town—simply means saddling the ratepayers with an unnecessary expenditure of at least several thousand crowns.

[Cries of disapproval, and whistles.] (p. 91)

The Mayor has put a person he trusts in the chair, obstructed the Doctor from giving his lecture, and deprived the Doctor of credibility by branding his accounts as 'irresponsible and exaggerated' and calling the reparation costs an 'unnecessary expenditure'. The Mayor drives a wedge of self-interest between the crowd and the Doctor with the image of the Doctor, as a threat to society's 'vital interests'. And the crowd responds with a spontaneous condemnation of the Doctor as an enemy of the people:

A MAN [*in the crowd*]. That's the talk of an enemy of the people!

BILLING. That, God damn me, was the voice of the people!

THE WHOLE CROWD [*shouting*]. Yes! Yes! He's an enemy of the people. He hates his country. He hates his people. (p. 102)

The Doctor is expelled from the community and is unable to make his voice heard. The Mayor has achieved his aim. Truth is defeated. The concept of 'pollution' no longer refers to a quality of the drinking water, but now to the society as a whole:

DR. STOCKMANN. I am going to make a great exposure, gentlemen! And the revelation I am going to make to you is incomparably bigger than this petty business about the water-supply being polluted and the Baths standing over a cesspool.

SEVERAL VOICES [shouting]. Don't talk about the Baths! We don't want to hear it! None of that!

DR. STOCKMANN. I have said I am going to speak about the tremendous discovery I have made in the last few days ... the discovery that all our spiritual sources are polluted and that our whole civic community is built over a cesspool of lies. (p. 93)

The concrete (water) has become the symbolic (truthfulness and moral responsibility) and the interrelationship of the two is asserted.

Act V: Final Steps To Corruption Of Mind

The decor for the final act is in the doctor's home. The crowd has smashed the windows—stones and glass are spread on the floor and the landlord has given notice. The Doctor's daughter Petra has been fired from her job as a teacher, and the two young boys have been thrown out of school, all the results of public opinion. Then, the Mayor shows up:

MAYOR. If I may give you some advice, it's this: go away for some while...

DR. STOCKMANN. Yes, I had actually been thinking of going away.

MAYOR. Good. And after you'd had six months or so to think things over, and if after mature consideration you then felt you were ready to write a few words of apology, admitting your mistake...

DR. STOCKMANN. Then I might perhaps get my job back again, you mean?

MAYOR. Perhaps. It's not altogether impossible. (p. 113)

Ibsen puts both the Mayor and the Doctor to a final moral test. So far the Mayor has won both the struggle for power and of sense-making. With a few exceptions, everyone perceives the doctor as a public enemy. He has lost his job, so he is not a threat to anyone anymore. Nevertheless, the Mayor proposes a piece of horse-trading—trying one more time to bribe the Doctor into abjuration. As before, the doctor refuses to give-in. Appealing to his economic responsibility for his family, does not do the trick. The Mayor makes a final attempt to force the Doctor to play by his rules. But the Doctor refuses. The only thing he has left is his moral integrity. And he does not compromise, despite pressure, retaliation, persecution, vandalism, dismissals and repeated attempts at bribery and corruption. Instead, the Doctor throws the Mayor out of the house and concludes that 'the strongest man in the world is the man who stands most alone'.

Understanding Complexity

Bateson understood the ecology of the mind (1972) in terms of how reality produces text, just as much as how text (concepts / assumptions) produces reality, and how circumstances (for instance, human evolution) produce consciousness just as much as how consciousness (perception) produces circumstances. In the play, 'pollution' is a quality of nature and of mind, where the two are ironic, complementary, causal, and in conflict. This is an aspect of the complexity of the play. To use complexity terminology, the play's logic is a 'strange attractor'. It is very uncertain what the social and moral system leads to. There are many ecologies here: there is the environmental ecology of the pollution of the water, there is the organizational ecology of the Baths trying to survive within their competitive environment, there is the social ecology of wealth, profit, and power, and there is the moral ecology of responsibility and 'truth-telling'. The pollution of the drinking water develops into a metaphor for the moral corruption of society and of its representatives. In his article on Ibsen and organization, Sørhaug concludes:

> *Science has a lot to learn from art regarding the handling of complexities and contradictions (Engelstad, 2006). In its depiction of social processes, aesthetic judgment is more able than scientific calculation to perceive and retain the combinations of emotionality and rationality and to generalize without losing specificity. Science has a tendency to erase contradictions, 'solve' paradoxes and simplify complexity. Art has an adverse tendency to develop contradictions and to explore and exploit complexity. The art of science is committed to a theoretical and methodological rigor, but it is in need of an aesthetical supplement preserving the sense of complexity and keeping an eye on 'the sources of novelty'. Without an active aesthetic supplement, science may lose out on both reality and the emergent* (Sørhaug, 2007: 1294).

Weick's sense-making rejects Bateson's complex hermeneutic, and stresses the pragmatic value of 'knowing where you are', even if the map is 'not the right one' (Weick, Sutcliffe & Obstfeld, 2005). Weick assumes that the subject has to 'know'—in sense-making one has to 'know' where one is. If 'reality' seems ambiguous, emergent or paradoxical, 'sense-making' will be retrospectively needed (applied) to stabilize and make it knowable. Ibsen has no trust in any such sense-making, or in socially desirable or sanctioned 'truth'. What complexity (theory) has to offer is the realization that socially motivated retrospective labeling may not do justice to motives, circumstances or evolving change. Weick in effect prescribes some sort of socially anchored nominalism wherein the subject(s) decide(s) what 'is', based on their (pragmatic) need(s) to know. Ibsen clearly tries to debunk such ethical opportunism.

Hirschman's theory of 'exit, voice and loyalty' (1970) is here illuminative. The Doctor had indeed to face such forces: does he stay in the community or leave, does he have and/or can he achieve 'voice', to what is he (ultimately) loyal? But these terms do not coalesce into a simple system—they remain dynamic, emergent and operate as strange attractors. The basic themes of moral responsibility and truth, and the resulting actions, thoughts and results, remain ethically unresolved. The Mayor retains his social authority however dishonest his actions may be and the Doctor stays on the moral high ground however compromised he is by vanity and/or revenge. But the ambivalence of good and evil, morality and effectiveness, self-interest and common good, remains.

In Ibsen's play the representation of whistleblowing and the ethical processes involved are ambiguous and complex. As Sørhaug puts it, art has a tendency to develop contradictions and to explore and exploit complexity. However creating a work of art implies more than exploring and exploiting complexity. There are formal aspects to art as well. Complexity needs a form actualizes and represents difference and recognition, unity and chaos, familiarity and strangeness. The work of literary art understood as a strange attractor, offers a form or structure for the representing of complexity as a series of 'points' in language. To some extent, these 'points' challenge the reader (audience) to co-create and discover moral phenomena not easily otherwise seen or reflected upon. I have touched upon four such formal and/or structural elements in the play: composition, dramatic irony, the use of metaphor ('pollution') and the ethical method of putting the main character to moral tests leaving moral reflection and judgment to the reader (Eide, 2001, 2004).

So what might (social) science learn from this? Ibsen does not give any answers. On the contrary, he is continuously raising questions, exploring complexities, (re)presenting ambiguities and challenging moral intuitions and assumptions. On the one hand, a researcher might learn a lot from the author's attitude (tolerating complexity, chaos) and methods (expressing contradiction, paradox, emergence). On the other hand, the work of art (play) might be considered to be a mimetic (Aristotle) investigation into a complex reality, including representation of how things are (or are not) linked together. The work of art may be considered a source of theory; for instance, what Ibsen's *An enemy of the people* tells us about the phenomenon of whistleblowing.

Seven Steps to Corruption of Mind

Taking Ibsen's method of putting his main characters to moral tests as a point of departure, it seems that the text is built around a succession of test cases (practical situations) on the one hand, and moral values (truth and moral responsibility) on the other. The question is whether one lives up to the moral standards of truthfulness and moral responsibility, or not. The text does not give any simple answers. The reader (audience) is the moral judge. The text appeals indirectly to the reader's sense of truth and morality. In the following I will briefly summarize the Mayor's moral tests and give a brief 'translation' of the 'answers' given by his actions. These are so many steps to the corruption of the mind. I see seven such steps:

1. This test entails how to respond to a message of organizational malpractice? One starts with turning a deaf ear to the problem. Keywords are moral inertia and indifference. One overlooks moral cues, shows no interest in information and keeps a safe distance from the matter in question;
2. This test involves how to respond if the strategy of indifference does not make the problem go away and the informant insists on having voice? You reject the information and refuse to listen. If necessary, one contradicts the facts and plays down any possible negative implications. A possible follow-up strategy is to blame the messenger;
3. This test centers on how to respond if simple information rejection does not work? One argues all the stronger against taking action and against change. All arguments that work are used, whether true or false. Try to convince others that everything is going to be OK; appeal to their sense of community and loyalty. Try 'you are one of us' and 'we are here to help each other,' and the 'what I expect of you', strategies. If argumentation and appeals to loyalty do not work, appeal to self-interest;
4. This test examines how to respond if appeals to rationality and loyalty do not work? One can try threats and intimidation. Even if the messenger may seem ethically confident, he (she) probably has some weak spots. Try to threaten her (him) with the loss of respect and of reputation. That will put his (her) self esteem to the test. You can threaten to fire her (him). Make it clear that he (she) will lose position, income and the ability to taking care of family. Threaten personal freedom and put close relations at risk. A possible follow-up strategy is to try to bribery;
5. This test focuses on how to respond if threats and intimidation do not work? One can organize campaigns and turn her (his) colleagues against her (him). See to it that lateral communication canals are closed. Deny voice; make it impossible for her (him) to speak. Isolate him (her) in the workplace, and make him (her) into 'not one of us'. If this does not work, try dismissal;

6. This test aims at how to act if not even dismissal makes the problem disappear? Retaliation and persecution may be necessary. Create rumors, slander and make her (his) motives look fishy and self-interested. Talk to everybody and see to it that she (he) does not get a new job. If sanctions against her (his) family are within reach, fine. Attacks on private property might make him (her) change his (her) mind and attitude, and;
7. This final test asks how to act if nothing works, nothing at all? Stick to the status quo; what 'is' (or has been) is 'True'. Do whatever is needed to avoid change, whatever that might be. When step 1 to 6 are tried without success, wait some time and try again. Do as if you wanted the best for the whistle-blower and give her (him) a new chance on the condition that she (he) will change her (his) mind.

Does this imply that ethical processes are linear after all? As dramatized by Ibsen, each of these steps towards the corruption of the mind is based on conscious or unconscious decision—(im)moral choices guided by self-interest and/or a need for the protection of self. Such a psychological interpretation may be valid. This is a question of perspective and of the reader's ability to embrace the complexity of the text, which surely also has other elements (like those of organizational ecology). Turning to the other main character, the Doctor or the moral 'hero' of the play, the question remains to whom or to what does he stay loyal. Why does he choose voice above exit, and why is he not successful in causing change when he has the scientific 'truth' of 'pollution' on his side. Neither protagonist accepts complexity; they are both locked into their own (linear) logic, and proceed blindly (like a tragic hero) from a basic (fixed) assumption to their fate.

The argument for complexity is grounded in the ambiguity of the play's form and language, not in its characters or plot. Ibsen's constant use of dramatic irony is complex, like the moral tests and his use of metaphor. The playwright and the audience share awareness complexity. Here we find a dynamic aggregate of interacting parts or components triggered by difference. But we need motivation, moral imagination, determination and intellectual hard work to see, understand and position ourselves vis-à-vis the complex chains of moral challenge, choice and action.

Acknowledgements

The author would like to thank Associate Professor Einar Aadland and the Ethics Research Group at the Diakonhjemmet University College for comments on an early draft, and the anonymous reviewers and Professor Hugo Letiche for suggesting important improvements in the final phase.

References

Alford, C.F. (2001). *Whistle-Blowers: Broken Lives and Organizational Power*, ISBN 9780801487804 (2002).

Alford, C.F. (2007). "Whistle-blower narratives: The experience of choiceless choice," *Social Research*, ISSN 0037-783X, 74(1): 223-248.

Aristotle (1965). "On the art of poetry," in T.S. Dorsch (ed.), *Aristotle/Horace/Longinus*, ISBN 9780140441550.

Aristotle (1989). "Poetics," in D.H. Richter (ed.), *The Critical Tradition: Classic Texts and Contemporary Trends*, ISBN 9780312101060.

Bateson, G. (1972). *Steps to an Ecology of Mind: Collected Essays in Anthropology, Psychiatry, Evolution, and Epistemology*, ISBN 9780226039053.

Bateson, G. (1980). *Mind and Nature: A Necessary Unity*, ISBN 9780553242836.

Bok, S. (1981). "Blowing the whistle," in J.L. Fleishman, L. Liebman, and M.H. Moore (eds.), *Public Duties: The Moral Obligations of Government Officials*, ISBN 9780674722316.

Bowers, J., Fodder, M., Lewis, J. and Mitchell, J. (2007). *Whistleblowing: Law and Practice*, ISBN 9780199299584.

Calland, R. and Dehn, G. (eds.) (2005) .*Whistleblowing Around the World: Law, Culture and Practice*, ISBN 9781919798561.

Dawson, S. (2000). "Whistleblowing: A broad definition and some issues for Australia," http://www.uow.edu.au/~bmartin/dissent/documents/Dawson.html.

Eide, T. (2001). *Ibsens Dialogkunst* [Ibsen's Art of Dialogue], ISBN 9788215001463.

Eide, T. (2004). "Ibsen's ethical method," http://www.ibsensociety.liu.edu/conferencepapers/ethical.pdf.

Hemmer, B. (2003). *Ibsen: Kunstnerens Vei* [Ibsen: The Artist's Path], ISBN 9788241902796.

Hannan, M.T. and Freeman, J. (1989). *Organizational Ecology*, ISBN 9780674643499 (1993).

Hirschman, A.O. (1970). *Exit, Voice and Loyalty: Responses to Decline in Firms, Organizations, and States*, ISBN 9780674276604.

Hirshman, A.O. (1978). "Exit voice and the state," *World Politics*, ISSN 1086-3338, 31(1): 90-107.

Ibsen, H. (1882). *An Enemy of the People* [En Folkefiende], The Oxford Ibsen, Volume VI, J.W. McFarlane (ed. & trans.), London: Oxford University Press, 1960.

Ibsen, H. (1882). En Folkefiende [An Enemy of the People], København: Gyldendal.

Johnson, R.A. (2002). *Whistleblowing. When it Works, and Why*, ISBN 9781588261397.

Maclagen, P.W. (1998). *Management and Morality: A Developmental Perspective*, ISBN 9780803976795.

Miceli, M.P., Near, J.P. and Dworkin, T.M. (2008). *Whistle Blowing in Organizations*, ISBN 9780805859881.

Sørhaug, T. (2007). "The drama of society: Capitalism, labor and love," Organization Studies, ISSN *1741-3044, 28(8): 1285-1296.*

Weick, K.E., Sutcliffe, K.M. and Obstfeld, D. (2005). "Organizing and the process of sense-making," *Organization Studies*, ISSN 1741-3044, 16(4): 409-421.

Tom Eide (PhD) is Associate Professor of literature & leadership at the Diakonhjemmet University College, and head of Peer Gynt Academy, Norway. He has published extensively on literature, especially on Ibsen, and on communication and ethics in healthcare and social work. His research interests include: (i) leadership and ethics as represented in literature, (ii) moral reflection and learning, and (iii) value-based interventions in organizations. He has been the director and coordinator of interdisciplinary programmes for ethics research at the Oslo University and the Norwegian Research Council, and has taught literature at Groningen University (Netherlands).

9. A Fractal Approach to Sustainable Networks

Kathryn Pavlovich
University of Waikato Management School, NZL

This paper qualitatively illustrates how and why interdependence becomes significant in building coherent and sustainable network systems based upon human flourishing. Ethnographic case data of an icon tourism destination is provided to examine the structure, process and patterns that are essential for understanding network organization. The notion of fractals has been applied to more deeply understand the multi-dimensionality of networks. Through the fractal characteristic self-similarity, the data revealed aspects of volume-filling, reciprocity and enfoldment that were central to the transforming power of network organization.

Behind the divisible there is always something indivisible.
Behind the disputable there is always something indisputable.

Chuang-Tzu

Introduction

In the world of measurement and deduction, objects are separated into small detail in order to be understood through a quantifiable process of analysis, known as Cartesian and Newtonian scientific approaches. While the outcome of these mainstream approaches has been an industrial revolution of scientific innovation, the disconnect that results from this separation remains consequential (Bohm, 1980). For example, our earth is organized by a string of social, ecological and environmental connections between people, place, space and other life forces. Production of our own food, clothing, warmth and shelter connect us intimately with the cycles of nature. However, our modern industrial system has disconnected us from our environment and we have lost this intimate awareness. Conventionally we see nature as something to be controlled, as a malevolent intrusion on our world, with potentially disastrous consequences that may emerge from this absence of intimate awareness (Snyder, 1990).

Thus, we need additional lenses to understand this interconnected world of relationships, processes, patterns and context that Capra (2005) claims are central to the coherence and integration of network systems. Yet our knowledge of such complementary approaches is emergent and raw. Drawing on the new science of complexity, this paper examines fractals as one possible perspective that may assist us in understanding how the intimacy of interdependent processes are not only connected but are also self-organizing and mutually constructed. Fractals demonstrate patterns of interconnection, and while they are primarily mathematical models, this paper applies the fractal concept to an organizational context to *qualitatively illustrate how and why interdependence becomes significant in building coherent network systems.* This paper identifies themes from a review of fractal literature and illustrates how they apply to an organizational network system. The contribution from this paper is its translation of hard science into an 'observable reality'.

Network Interdependence

Networks are claimed to be a defining contributor in reshaping our social communities in the 21st century (Parkhe *et al.*, 2006). Yet their complex multi-dimensionality has signalled the inadequacy of our current linear and predictive theories (Galaskiewicz, 2007). In most network analysis, there is a preoccupation with structure (the nodes and connections) through the relations amongst actors and their individual positional location. These 'measurement' orientations designate a group of relationships that are fixed and controllable which are contrary to the distinctive characteristics of networks. Rather, their plural pathways whereby information is transmitted through any possible connecting point (as opposed to more vertical and horizontal forms in hierarchies) (Powell *et al.*, 2005), their nodal complexity whereby networks become interdependent based upon complementary resource transfers obligating one organization to another (Uzzi, 1997; Gulati, 1999) and the fluidity that emerges from this structural complexity when partners 'switch on the dance floor' (Powell *et al.*, 2006), each indicate complex, flexible and non-linear modes of self-organization.

Conventional network analysis then, displays little acknowledgement of dynamics, of the fluidity emerging from the connecting points and of the 'empty space' within which the network resides. Process is assumed to be determined by structure "and there is no sense in which the mutuality of pattern, process and structure is accommodated" (Lowe, 2010: 54). The current state of network analysis therefore generally avoids examining these interrelationships that Capra (2002) claims are central to the practice of organizing, primarily because of the difficulty in capturing them for analysis. Lawler (1982) adds further support as to why this difficulty exists by highlighting that our European languages have no linguistic form related to processes that do not have material carriers, such as the words pure, quality and even God. Our words require that verbs always be associated with nouns. This helps to explain why our western science too struggles to acknowledge and clarify such integrating and holistic processes as the immaterial.

The purpose of this research then is to more deeply understand the interdependent multi-dimensionality of networks. What is the relationship between structure, process and pattern? How do they impact on the organizing mechanism of the network? And how does self-organization within the network allow us to understand the quality of the system? Understanding this more complex and multidimensional form of organization therefore requires qualitative approaches that embrace this non-linearity and complexity.

Process, Pattern and Organization

Within this inquiry, we observe the metaphysical assumption that network nodes "become more like markers around which dense networks of relationships are acted out instead of fixed objects which predetermine our mental and physical movement" (Cooper, 2005: 1691). The focus of examination therefore, is on the realization of the network through process, pattern and structure. Following Capra's (2002) claim that process is the 'bringing forth' of a reality, this inquiry considers how this engagement is enacted to create networks of sustenance that include economic, social, environmental and spiritual codependency.

While sociologists have long examined the role of reciprocity and obligations as part of the organizing norms in society (Granovetter, 1985), in management literature on networks these features are noted but not universally prioritized (Ibarra, 1992). Yet norms of reciprocity are essential if we are to engage in ecological networks for sustainable futures. Turner (1995: 45) notes that we must take our "residence in a biological order; to become whole we must live as part of a larger system of plant and animal communities governed by reciprocity". This more system-based approach means that the norms and rules of collaboration, reciprocity and interdependence become inherent in their self-organization. The whole is emphasized more than the parts, with recognition of the intrinsic values of all that inhabit the system as an interdependent web of life (Naess, 1973).

This gives rise to the importance of the network system as a unit of analysis, with it enabling us to more closely observe the patterns of interconnection that emerge from process within the context. While not intending to give primacy to the structure as a fixed and objective reality, it does provide a physical configuration for conveying the movement of energy within the system. The patterning, Capra (1997: 167) defines is "the configuration of relationships that gives a system its essential characteristics. Thus, the patterns illustrate the qualitative configuration and organization of relationships.

Fractal Organization

The focus, then, is on the patterns that emerge from the network configuration, and fractals provide a language for understanding this qualitative nature of self-organization. Fractals are geometric shapes that provide a lens to study spatial order through the relationships that form from patterns of interconnection, indicating how fluidity and movement can be viewed multi-dimensionally. The pattern then is illuminated through the fractal shape and becomes significant in creating form to the relational system through its spatial multi-dimensionality. As Lawler (1982: 5) suggests, "the content of our experience results from an immaterial, abstract, geometric architecture which is composed of harmonic waves of energy, nodes of relationality". This reinforces the multi-dimensionality of the interconnected system necessary when thinking in ecological terms, and the bonding of the material world and the spaces in between. Capra (1991: xi) too brings to our consideration that "all forms are associated with processes, all interrelations with interactions, and opposites are unified through oscillations". These oscillations highlight the movement and change that becomes of primacy of network organization.

Conventionally, fractal is the measure of the irregularity of the shape of an object: it is neither a straight line nor a smooth curve. Based upon mathematical equations, fractals allow examination of irregular forms, thus critiquing the 'measurement' paradigm of fixed and stable points. The common analogy of fractals is the measurement of a coastline, which becomes greater with increasing precision as you take into account ever smaller irregularities along the length. Popularized by Mandelbrot (1983) who worked hard to demonstrate that fractals have application in the real world and were not just a mathematical curiosity, they are also seen in nature as ferns, twigs, cauliflowers, broccoli, clouds and ocean waves. Their central characteristics are that they are a) irregular all over, b) have the same degree of irregularity on all scales, and c) look the same whether examined close up or far away. Thus, fractals are shapes that are systematic even when their components are uncertain in outcome and form. Their rules governing growth ensure that small-scale features become translated into large-scale ones which indicate the forms of stability that inherently reside within the system. By successively magnifying a part of the whole reveals a further structure that is nearly a copy of the original.

The fractal concept therefore enables us to explore the notion of quality as it represents the dynamics of the whole, whereas measurement examines the parts. The following case discussion examines these ideas in depth through empirical data. The paper then offers insights into how fractal patterning can assist in further understanding of network interconnection for creating systems of co-creative sustenance.

Method

This paper examines these premises of structure, pattern and process as organizing mechanisms in a case study of an icon tourism destination in New Zealand, the Waitomo Caves, famous for its underground caving activities and attractions. While tourism began in 1887, the focus for 100 years was mostly on the Glowworm Cave with its imposing limestone formations, and a float through the cave-stream viewing its famous fauna, the glowworms, which twinkle like a universe of stars in the darkness. Yet a history of exploitation by the State owned agent, the Tourism Hotel Corporation, stymied the development of the destination. By the late 1980s, however, a growing number of younger people were coming to New Zealand as free and independent travellers, and this preempted the development of a complementary production system based around adventure caving. These six hour caving trips necessitated additional symbiotic interdependencies among the primary and support activities, food and overnight accommodation providers to cater for this market. The development of this small but comprehensive network system brought significant changes to the destination, and for the first time enabled complementary knowledge flows to exist. Only the core tourism product, the Glowworm Cave, has independent corporate management and the other organizations are locally owned. They vary in size from the Glowworm Cave with 70 employees at the height of the summer season, to mid-range adventure operations with 30 employees, to micro-owner/operated organizations.

This context is significant for analysis in that its geographic isolation provides some limitations for a study on a network system. Further, it relies upon the sustainable management of its fragile environment, and thus, economic, social, environmental and also spiritual imperatives that emerge from the indigenous cultural traditions embedded in the landscape are present in its organization. This makes the network itself a multi-dimensional and complex integration of imperatives rich for this form of study.

The paper builds on earlier ethnographic work that identified a series of collaborative groupings that emerged from the thematic analysis of case data collected over a five-year period, 1996-2000. The data collection process included formal interviews with each of the twelve owner/managers of each firm within the destination each year, plus an additional eight associated external institutions. In accordance with ethnographic research, a significant amount of archival documentation, informal conversations and personal observations also comprised the data collection. A further follow up study in 2007 was undertaken to give a 10 year history of collaboration in the destination. This included a small questionnaire to each organization examining their six most strategic partnerships, both within or external to the destination, and a second set of questions regarding the economic status of their business. Currently, 15 of the 24 organizations responded, and the following preliminary data are based on these responses. Further qualitative interviews were also conducted with key organizations, both within and external to Waitomo.

The analysis in this paper conforms with an 'abductive' process whereby one begins with fragments of understanding, and then one builds more holistic pictures of an observable reality. This form of 'invention' includes acts of faith, as abduction forms the "rules that are postulated to explain the observed facts" (Weick, 2006: 1731). As Shotter (2005: 4) notes, "it is not a matter of acquiring some new information or data, but of redirecting or reorienting our attention, to noticing things which no-one has doubted, but which have escaped remark only because they are always before our eyes".

The Network System

With this research inquiry focusing on the patterns that emerge to more deeply understand network organization, mapping of the data were significant to illustrate these configurations. As noted earlier, this context has a history of exploitation. However, Figure 1 illustrates significant growth in new business development over the last 20 years, with the network now sustaining a rich diversity of core, supporting and external firm interconnectivity. This map is intentionally portrayed from a holistic perspective to illustrate the collective integration of the system, the purpose of which will emerge in the discussion section. The full acronyms are included in Appendix A.

Alongside this structural change in the context over the last 20 years is also substantial increase in the diversity of external connectivity. This diversity has been identified as significant in integrated networks, as they not only allow plural pathways of information transfer but are also strategically important in assisting complex specialised knowledge to be accumulated (Powell *et al.*, 2005). The density of relationships indicate the fluidity and movement within the network through the informal 'comings and goings' among suppliers (economic

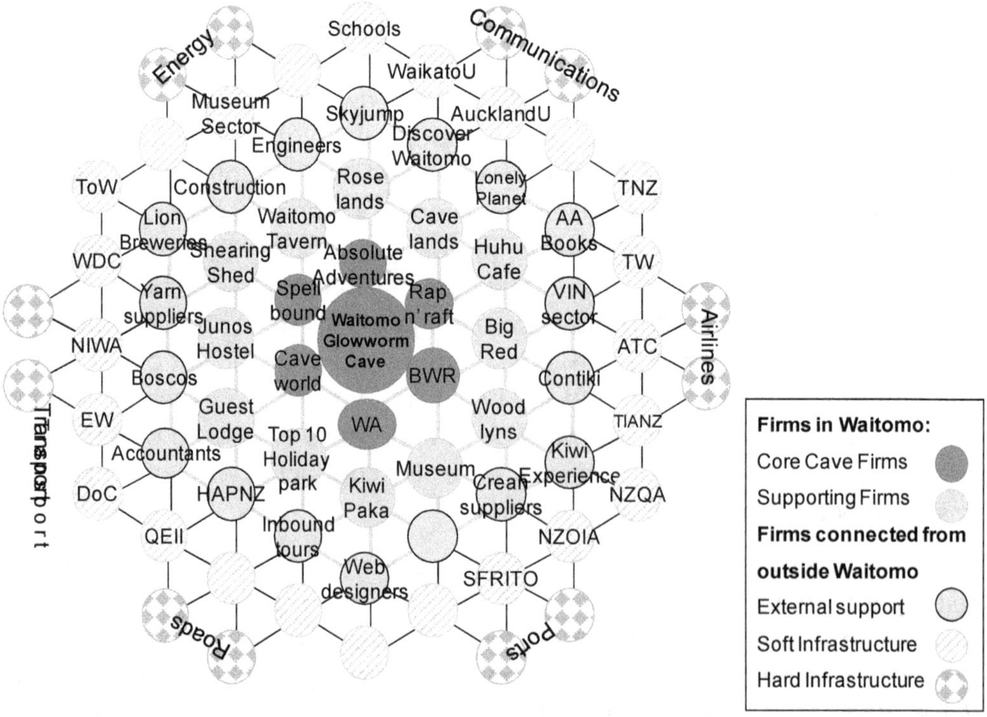

Figure 1 *Structural Arrangement in Waitomo Caves Tourism Destination, 2007.*

and social relationships), employees (economic and social relationships), the broader community (social, environmental and spiritual relationships), external institutions (economic, regulatory and environmental relationships), and visitors (economic and social relationships). It is the density of these diverse types of activities and relationships within the network that form the multi-dimensional patterns that underlie its organization. Further, this growth in business development is also represented in the increase in economic contributions: from $3.5 million in 1985, to $9 million in 1995, to $23 million in 2005.

Collaborative groups

It is this multi-dimensionality that now becomes the focus of this paper as it relates to non-linear organization and system-related sustainable outcomes. Predicating this discussion on fractal organization, a brief description of the collaborative groups that emerged from the earlier data analyses is given to set the context to explore network interdependence and ecological sustainability. While other collective groups emerged from the data, in this paper the three examples illustrated as they relate to sustainability are: Landcare, adventure risk management, and underground karst management.

Above-Ground Landcare

The first and most robust of these collaborative groups was the above-ground Waitomo Landcare Group which was formed in 1992 after mounting community concerns regarding the environmental degradation of the underground caves and the inadequate management procedures. These karst environments are recognized for their 'twin landscapes', as activities occurring above the ground can impact on the underground system. For example, roading improvements can result in flow-offs into the river streams, and this silting can cause cave erosion, fauna damage, and glowworm disruption in areas some distance from the initial activity. These negative impacts have the potential to destroy the limestone cave formations, the glowworm populations in the caves, and the economic viability of the tourism destination.

Facilitated by the Regional Council, Environment Waikato (EW), the group included the local farming community (donating land to be fenced and replanted), tourism operators (contributing financial inputs), community members (providing voluntary labour) and other outside knowledge-based institutions (Waitomo District Council, Department of Conservation, and universities). The purpose of the Landcare Group was to protect the Waitomo Caves system from sedimentation through appropriate and sustainable land management practices. Collaboration among these internal and external stakeholders included fencing off the waterways and forest replanting programmes, primarily to improve water quality for sustaining glowworm populations and the cave environment. The group is promoted nationally as a 'best practice' example of a Landcare group with a 60% reduction in sediment loading in the Waitomo stream, less soil erosion, and over 500 hectares of bush now protected. Its initial purpose of improving the longevity of the cave environment has been achieved, and the participation of the community has resulted in widespread consensus and acceptance of the land management practices to sustain the destination environment. The pattern of connection is illustrated in Figure 2.

Adventure Risk Management

The second collaborative group involved managing the visitor risk attached to the underground caves. Given that they are dark, cold, wet and often confined spaces, the ability to manage these potentially risky and dangerous situations requires an advanced standard of guiding to ensure client safety. Some of these adventure activities involve abseiling, floating in the underground river streams, and clambering around limestone rocks. Organizational collaboration in Waitomo around this issue has lead to several outcomes. First, the informal information exchanges among operators and employees (i.e., after-work drinks at the local tavern) resulted in better safety practices being developed for managing operational risk. From these exchanges and concerns, the Waitomo Caves Rescue Team was formed, with specialist guides from the adventure caving organizations working together in emergency situations—most commonly arising from situations involving recreational cavers. Finally, the adventure caving organizations in Waitomo have been instrumental in building national caving standards throughout New Zealand, as their Waitomo-based collaborative systems and standards (Waitomo Rope Standard Technique - WRST) have been incorporated into institutional frameworks (New Zealand Qualifications Authority, Sport Fitness Recreation Industry Training Organization and the New Zealand Outdoor Instructor Authority). Thus, these organizations are seen as adventure sector leaders in New Zealand with local knowledge spilling into the larger institutional education framework and in the broader tourism industry (Adventure

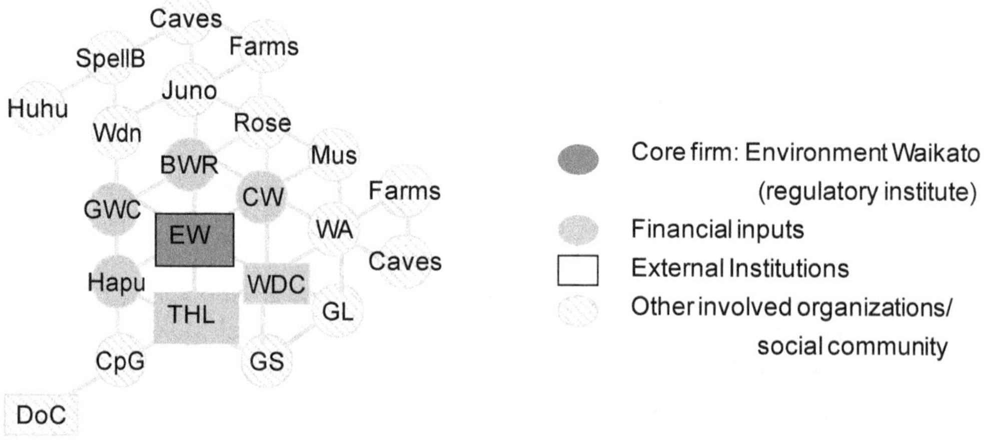

Figure 2 *Waitomo Landcare Group*

Tourism Council, Tourism New Zealand). This pattern of interconnection is illustrated in Figure 3.

Underground Karst Management

Finally, a third area of emerging collaboration that has not yet attained an integrated focus, is that of underground environmental management (illustrated in Figure 4). While there has been access to 'weak-tied' external knowledge for decades, the restrictive administration practices of the previous central actor (Tourism Hotel Corporation) stymied any development of knowledge-building in this area. Committed investment into processes, practices and analysis of underground cave management has only occurred over the last ten years in the Glowworm Cave (GWC). Currently, other members of the community, such as tourism operators using other caves, are becoming involved. Together these organizations are endeavouring to form a system of managing the interdependent karst context—humidity, glowworms, carbon dioxide levels—building a data-

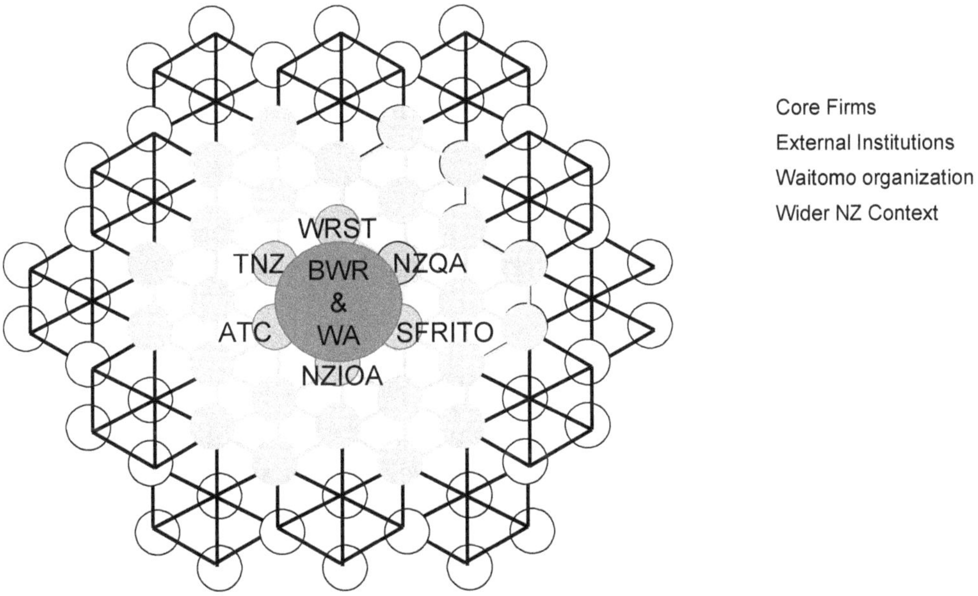

Figure 3 *Adventure Risk Management*

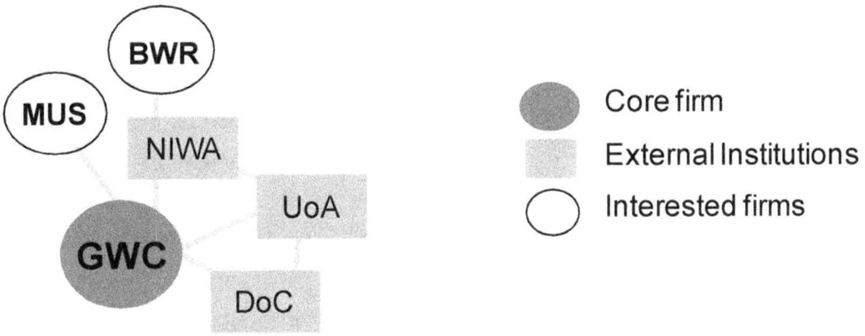

Figure 4 *Underground Karst Monitoring*

base of systems, practices and experiences that can restore the environmental habitats. These linkages are primarily internal ties, and again the limited nature of these ties demonstrates the absence of flow in building strategic capabilities.

In the original analysis of these data, these were the most significant of the collective groups and they relate to knowledge specialization that does not emerge when only focusing on commercial-related questioning. Thus, like the landscape, they were beneath the surface, and yet are fundamental to providing sustenance to this destination. Both the Landcare group and underground monitoring include both environmental and social imperatives. Interdependence is evident in the adventure risk management group as it has the potential to tarnish the reputation of Waitomo through any serious accident. For instance, if heavy rain falls, the river level can rise by over one metre within an hour—such conditions need to be carefully monitored. Spiritual imperatives too are evident in this landscape through the 'waahi tapu'—places that are locally known as sacred and have to be honoured as such. This has meant finding alternative avenues for commercial activities in order to not disturb these sacred sites.

Thus, the implications of identifying these groups are that they include imperatives other than solely economic, and provide more holistic ways of management than lineal models conventionally applied.

Fractal Organization

This section discusses a number of insights on fractals that emerged from the data regarding the searching for a deeper understanding of network self-organization.

Patterns as structure

First, the figures above illustrate the density of nodal connection within each of the collective groups, illustrating the complexity and quality of relational connectivity. Figure 4 depicts limited connectivity, both within Waitomo and with external institutions, and the consequence of this has been restricted knowledge creation around underground monitoring processes. In comparison, Figure 2 illustrates a relative complexity of agents, resource complementarity, and reciprocal information sharing. Waitomo Landcare is New Zealand's most acclaimed landcare group with significant specialised knowledge and routines now residing within the community. The third pattern, Figure 3, illustrates interdependence, and demonstrates the development of specialised knowledge on risk management stemming from within the destination and flowing into a broader community. Mutual interdependence in the destination is evident whereby a single action by one firm can implicate the safety of visitors underground with disastrous consequences on the reputation of the whole network. Fortunately, in this context, there has been significant responsibility taken to ensure competent and professional delivery of services, lead by the core organizations in the destination.

These patterns, therefore, give us insights into the relational connectivity within the context and from this we can induce the extent of multi-dimensional connection within the network. This is significant as Lawler (1982) claims it is the intensity of these patterns that determine the quality of the system. This suggests that reciprocal information flows around multiple issues have the potential to add intensity into the network; that is, the process of collaboration facilitates complex structural patterns within a context. Deleuze and Guattari (1987) define these multi-dimensional configurations as 'multiplicities', in that they emerge in any form, in any direction and from any point. Thus, they are multi-dimensional and this heightens our understanding of non-linear organization.

Fractals and Enfoldment

In the second insight, Figure 1 illustrates a holistic pattern of interconnected organizations, and the above multiplicities become the fractal sub-sets *enfolded* within this (Figures 2-4). They take on a fractal form because unlike quantitative measures where the concept of (intrinsic) length makes little sense, fractals identify notions of (relative) length measured at any given resolution qualitatively (Sornette, 2004). Thus, fractals embody the qualitative expressions of the network through the relative intensity of their connections. With the characteristics of fractals stemming from self-similarity, these are clearly evident in the patterning of land care, adventure risk and underground monitoring: the same core organizations belong to each of these multiplicities; there are overlaps in the specialised knowledge activities regarding the identity of the tourism destination (e.g., caves, management of people, tourists, short stays, interdependence on primary and secondary activities); and the expression and identity of the iconic Glowworm Cave that acts to pull visitors to Waitomo is embedded throughout the network.

These patterns then, demonstrate fractal characteristics that are self-similar (whereby an infinite number of smaller shapes can emerge, each still resembling the whole), multi-dimensional, and formed through iteration whereby although the copying of the pattern may become more complex, each still retains core attributes that are observable throughout the network structure. They are self-energizing and replicating. Each situation and event always contains an essence of the past and it is this fractal patterning that holds the memory acting as the coordinating mechanism of the network. This implicitly references that the patterns underlying organization are a unifying force.

Fractal patterns, then, demonstrate the nature of non-linear self-organization with the same overall structure being maintained despite an ongoing flow and change of components (Capra, 2005). No matter how many multiplicities are developed, there is the same core information inherent in each. Bohm (1980: 224) clearly expresses this in that "each part contains content information about the whole object... that is, the form and structure of the entire object may be said to be *enfolded* within each region".

This contribution, then, illustrates a cohesion that is demonstrated through organizations with the same resource configurations replicating different activities within each of these collaborative specializations (fractals). That is, they become *enfolded* within each other and within the overall system. Thus, as Bohm (1980: 222) insightfully continues, "the interactions between different entities constitute a single structure of indivisible links, so that the entire universe has to be thought of as an unbroken whole".

Volume-Filling through Process

The third insight focuses on the 'space' within which the network resides, and demonstrates that the fractal patterns are created and maintained through reciprocal relations. As noted earlier, conventional network research focuses on the nodes and connections, yet we now understand that what was regarded as 'empty' space is actually full of excitable energy, and our physical world is merely a small manifestation that emerges from this background of total energy (Bohm, 1980). As Merleau-Ponty (1962: 243) recalls, we were "imaging (space) as some sort of ether in which all things float... [rather] we must think of it as a universal power enabling them to be connected". Lawler (1982: 22) too describes this space as a magnetic field of energy filled with "whirling, fiery bodies from the heavens".

The focus turns to this background energy and its role in organization through the fractals. All of the illustrated fractals are multi-dimensional. That is, they comprise multiple actors from multiple sectors—with each fractal focused on a specialised activity. They are multi-dimensional in that they are complex. If enfolded into each other there would be some overlap while also considerable extension of diversity beyond the simple pattern. Thus, they become 'volume-filling'. They occupy more 'space'. Network connectivity, then, acts to actualize that latent space, through creating possibility from that universal power of whirling energy into physical form. Turner (2002: 13) claims that such forms "have the odd property of filling up the space available to them with more and more detail, a line densely kinking up to fill a place, or a surface folding up a volume...". This clearly references how the fractal patterns emerging in Waitomo over the last twenty years can be described as 'volume-filling' through their multi-dimensionality, and how knowledge specialization developed in this context.

Yet how and why did this occur at this time? As noted earlier, the absence of connectivity over the 100 year life of the destination was characterized by limitation and exploitation. With the new network forming in the 1990s, the plural pathways assisted information transfer (Powell *et al.*, 2005), and the *process,* or intention underlying this sharing was central to the transformation of the network. The intention of the previous central actor (1910-1986) was to stymie new business development in the destination, and to extract short term profit, with little investment in the Glowworm Cave over those years. However, from 1987 onwards, more collective intentions emerged from the new actors through a process of inter-sector collaboration that has benefited the network as a whole.

Process, then, becomes central to the organizing phenomena of networks, as it facilitates 'volume-filling' from the immaterial through creating coherent fractal patterns that add richness and texture. In this way, when it is argued that fractals represent the quality of the system, this refers to how the quality of reciprocity is reflected through the intensity of the multi-dimensional patterns.

Thus, organization occurs to space and time within the local context by the connectivity based upon "an invisible and missing wholeness" Cooper (2005: 1700). This transformation is dependent on the extent and intensity of local interactions forming durable connections. As this connectivity deepens, multi-dimensional multiplicities form based upon areas of collaborative specialization. Lawler (1982: 21) reminds us that, "All elements flow out from a central unity in accordance with the law of inversion or reciprocity".

Ecology and Multi-Dimensionality

The previous sections have presented an argument that network organization occurs through its fractal patterns of self-similarity. The self-replicating aspect is able to be actualized into physical form through reciprocal relations. This final section argues that more ecologically sustainable networks can be developed through this multi-dimensional approach, through understanding how and why interconnection takes place.

While complexity theory discusses the self-similarity that characterizes fractals, Bohm (1980) uses the word *enfoldment* to describe a similar effect. Enfoldment, however, explicitly references that through reciprocity, an essence is retained in the process of connection. This essence connects all living nature. Sheldrake (1981) too notes that systems develop and self-regulate through a cumulative memory that 'knows' how to look (form) and how to act (function). This 'morphic resonance' is realized in a multi-dimensionality not evident in physical form, with this memory becoming enfolded through reciprocal inter-connection.

The multi-dimensional nature of the fractals and their 'volume-filling' action begins to bring that integral coherency, that universal power noted by Merleau-Ponty (1962), to the surface. The impact of enhancing this implicit 'know-how' is reflected in the case data through the fractal patterns. The *Landcare group* plays a significant role in the environmental management of the above and below landscapes. As noted earlier, considerable community support was given to fence waterways and to replant native bush. The landcare group has also assisted in the protection of the indigenous sacred sites, known as a 'waahi tapu'. It has been the power of connection that has supported the 'enfolding' of increased knowledge specialization regarding best environmental practices within the landscape, additional economic benefits from improved management of this resource, greater community spirit through their volunteer work with fencing and replanting, and a deeper understanding of the significance of sacred sites. Thus, economic, environmental, social and spiritual imperatives all contribute to the intensity of this multi-dimensional fractal.

Protecting people has been central to the *risk management* multiplicity, and on many occasions the Waitomo search team have been called upon to rescue people trapped in other parts of the country. In 2007, a well known doctor was trapped in a recreational cave for over 48 hours, and more than half of the rescue team were from Waitomo (including a father and son team). As noted earlier, significant development of caving-related standards have stemmed from this context and they are now embedded in many of the national institutional frameworks. Again, this demonstrates the enfoldment that emerges with interdependence stemming from the caving community in terms of specialised knowledge, love of caving and awareness of risk factors.

Finally, the longevity of the tourism destination depends upon the health of the limestone caves and the glowworms. Previously there have been incidents where the glowworm habitat has been disturbed (through altering airflows in the cave), resulting in closure of the cave for four months. Today, this would have significant economic impact on the destination, not to mention on the cave ecology. The scientific *monitoring and research* of this environment now contributes to improved economic and environmental outcomes of the destination regarding relationships between cave hydrology, carbon dioxide levels and climate changes. Yet the embedded memory of exploitation continues to flow through this fractal, and there is currently significant effort being put to improve the reciprocal exchanges needed to build a more multi-dimensional aspect to this fractal.

These small illustrations highlight how the fractal patterns add intensity and texture to the network to assist in its organizing function. Central to this process of organization has been the multi-dimensionality of connections. The data illustrated that transformation occurred through connectivity, through the creation of multi-dimensional fractals patternings of specialization within the network. The 'volume-filling' aspect of these were formed by a process of reciprocity, thereby allowing the immaterial to flourish in new forms.

Original Ref—E:CO Vol. 11 No. 3 2009 pp. 49-60

Conclusion

The concept of network organization through interdependence has been examined in this paper through a qualitative focus on structure, pattern and process that Capra (2002) claims is central to our understanding. A reflective process of abducting from an ethnographic study, allowed the case data to illustrate how fractals offers a complementary perspective in understanding the organization of complex interdependent networks. More specifically, it qualitatively revealed how and why interdependence becomes significant in building coherent network systems.

The data revealed further essential characteristics of fractals beyond the self-similarity reported in the literature. Characteristics of volume-filling, reciprocity, multi-dimensionality and enfoldment play an important role in our understanding of network organization. This is significant in that collaboration involving these features create more holistic and ecologically sustaining interdependent network contexts that may include economic, environmental, social and spiritual values to assist in transforming lineal imperatives to more integrated ecologically sound notions of organization. Central to this is that all interaction is enfolded within each other; that there is no 'out there'. When addressing the notion of organization, we need to address ecological imprints with a more conscious and deep awareness: its depth, its richness and its presence. As Capra (2002) noted, meaning is formed from patterns of relationships. As we have seen, reciprocity allows richness and intensity to develop, so that we can see the object's depth and its presence. To understand our relationship with the world as sacred, we need to focus on the unseen and accept that our earth is organized by a string of social, ecological and environmental connections between people, place, space and other life forces. We live in an interdependent web of life where the invisible connections hold a universal power. Our journey is to connect. To conclude, in the words of Lawler (1982: 20):

> *We exist in groups, determined by various levels of energetic affinities, repelling, exchanging and absorbing through interconnected subtle energetic communications. We have to learn that there is nowhere that we can dispose of things we have finishing using… everything remains here with us; the cycles of growth, utilisation and decay are unbroken. There is no through away bottle.*

References

Bohm, D. (1980). *Wholeness and the Implicate Order*, ISBN 9780710009715.

Capra, F. (1991). "Foreward," in J.-E. Berendt (ed.), *The World Is Sound: Nada Brahma: Music and the Landscape of Consciousness*, ISBN 9780892813186, pp. xi-xiii.

Capra, F. (2002). *The Hidden Connections: A Science for Sustainable Living*, ISBN 9780385494724.

Capra, F. (2005). "Complexity and life," *Theory, Culture and Society*, 22(5): 33-44.

Cooper, R. (2005). "Relationality," *Organization Studies*, ISSN 0170-8406, 26(11), 1689-1710.

Deleuze, G. and Guattari, F. (1987). *A Thousand Plateaus: Capitalism and Schizophrenia*, ISBN 9780816614028.

Galaskiewicz, J. (2007). "Has a network theory of organizational behavior lived up to its promises?" *Management and Organization Review*, ISSN 1740-8776, 3(1): 1-18.

Granovetter, M. (1985). "Economic action and social structure: The problem of embeddedness," *American Journal of Sociology*, ISSN 0002-9602, 91: 481-510.

Gulati, R. (1999). "Network location and learning: The influence of network resources and firm capabilities on alliance formation," *Strategic Management Journal*, ISSN 0143-2095, 20: 397-420.

Lawler, R. (1982). *Sacred Geometry: Philosophy and Practice*, ISBN 9780500810309.

Lowe, S. (2010). *Managing in Changing Times: A Guide for the Perplexed Manager*, ISBN 9788132102335.

Mandelbrot, B. (1983). *The Fractal Geometry of Nature*, ISBN 9780716711865.

Merleau-Ponty, M. (1962). *The Phenomenology of Perception*, C. Smith (trans.), ISBN 9780415045568.

Naess, A. (1973). "The shallow and the deep, long range ecology movement: A summary," *Inquiry*, ISSN 0020-174X, 16: 95-100.

Parkhe, A. Wasserman, S. and Ralson, D. (2006). "New frontiers in network theory development," *Academy of Management Review*, ISSN 0363-7425, 31(3): 560-568.

Powell, W., White, D., Koput, K. and Owen-Smith, J. (2005). "Network dynamics and field evolution: The growth of interorganizational collaboration in the life sciences," *American Journal of Sociology*, ISSN 0002-9602, 110(4): 1132-1205.

Sheldrake, R. (1981). *A New Science of Life: The Hypothesis of Formative Causation*, ISBN 9780856341151.

Shotter, J. (2005). "The role of 'withness'-thinking in 'going on' inside chiasmically-structured processes," Keynote address at the first European Group of Organization Studies Summer Workshop, June 12-13, Santorini. Greece.

Snyder, G. (1990). *The Practice of the Wild*, ISBN 9780865474543.

Sornette, D. (2004). *Critical Phenomena in Natural Sciences: Chaos, Fractals, Self-Organization and Disorder: Concepts and Tools*, ISBN 9783540308829.

Turner, F. (2002). "Creating a culture of gift," The Philanthropic Enterprise: Working Paper 3. Dallas: Univeristy of Texas.

Turner, J. (1995). "Gary Snyder and the practice of the wild," in G. Sessions (ed.), *Deep*

Ecology for the 21st Century, ISBN 9781570620492, pp. 41-46.

Uzzi, B. (1997). "Social structure and competition in interfirm networks: The paradox of embeddedness," *Administrative Science Quarterly*, ISSN 0001-8392,42: 35-67.

Weick, K. (2006). "Faith, evidence and action: Better guesses in an unknowable world," *Organization Studies*, ISSN 0170-8406, 27(11): 1723-1736.

Appendix A: Key to Waitomo Organizations

Central Cave

GWC—Glowworm Cave

Hapu—Indigenous Maori family

THL—Tourism Holdings

Adventure Caves

BWR—Blackwater Rafting

WA—Waitomo Adventures

SB—Spellbound

Support Activities

Jun—Junos

Rose—Roselands

Mus—Museum

Wdn—Woodlyns

T10—Top 10 holiday park

CL—Cavelands general store

GL—Waitomo guest lodge

External Organizations

ATC—Adventure Tourism Council

DOC—Department of Conservation

EW—Environment Waikato (regulatory body)

NIWA—National Institute of Water and Air

NZOIA—New Zealand Outdoor Instructors Association

NZQA—New Zealand Qualifications Authority

SFRITO—Sport Fitness Recreation Industry Training Organization

TNZ—Tourism New Zealand

UoA—University of Auckland

WDC—Waitomo District Council

Kathryn Pavlovich is an Associate Professor at the University of Waikato Management School in Hamilton, New Zealand. She has more than 20 international referred journal publications that include her two research and teaching areas. The first area focuses on "relational competencies" (strategic alliances, networks, clusters and partnerships), and are theorized from a metaphysical perspective. The second area of "inner technologies" includes the development of intuitive knowledge, reflection, and inner leadership for conscious living.

10. Moving Far From Far-From-Equilibrium: Opportunity Tension as the Catalyst of Emergence*

Benyamin B. Lichtenstein
College of Management, University of Massachusetts, USA

Complexity scholars have identified two distinct catalysts of emergence: (1) Far-from-equilibrium dynamics that trigger order creation, and (2) adaptive tension (McKelvey, 2004) which can push a system toward instability, leading to the emergence of new order. Each of these provides a necessary but incomplete explanation of the catalyst for emergent order. In particular, the far-from-equilibrium framework, when taken to its logical ends, would conclude that most dynamic and fluid organizations are the ones farthest-from-thermodynamic equilibrium—like Exxon or GM, for example. Adaptive tension on the other hand identifies an exogenous force of market change, but doesn't explain how emergence is actually triggered. As a solution I propose "Opportunity Tension," which integrates the endogenous intention of an entrepreneur to create a new venture to the exogenous changes that open up an entrepreneurial opportunity—a market that will exchange money for the value being created. Opportunity tension occurs in "pulses," each cycle leading to a new dynamic state of the system. This model, which is consonant with the notion of "dynamic creation" (Chiles *et al.*, 2010), contributes to a complexity science that is moves us beyond a far-from-equilibrium framework.

Introduction

In our search for the driver of order creation, management scholars have developed two contrasting catalysts of emergent order: far-from-equilibrium dynamics (e.g., Meyer *et al.*, 2005), and adaptive tension (e.g., McKelvey, 2004). Although on the surface these two approaches seem similar, technically the constructs are different in significant ways, which have important implications for an organization science of complexity.

The more common approach for describing the origin of new order is through the onset of "far-from-equilibrium" dynamics. Far-from-equilibrium approaches "elucidate the non-linear mechanisms that actually drive [discontinuous] change forward" (Meyer *et al.*, 2005: 470a). In this theoretical framework, organizing far-from-equilibrium is what leads to "...emergence and ongoing, perpetual novelty" (Meyer *et al.*, 2005: 450b). Choi, Dooley and Rungtusanatham (2001: 356) also use this framework to explain the origin of systemic state change, arguing that such change is triggered "...when the system is far from equilibrium." In a similar way, Browning, Beyer & Shetler (1995) show that the emergence of an alliance was sparked by "...a period of disequilibrium [which required the semiconductor industry] to operate qualitatively differently than in the past." In sum, a range of authors focus on how far-from-equilibrium processes catalyze the emergence process.

On the other hand, McKelvey has offered a different explanation for the driver of emergence, namely *adaptive tension*. In McKelvey's understanding of Prigogine's dissipative structures theory (Nicolis & Prigogine, 1989), order creation is caused by and initiated through "energy differentials" which are imposed onto the system. New order is created when one of these energy differentials crosses a threshold (McKelvey, 2004: 319): "...[when] an imposing energy differential, what I term *adaptive tension*, exceeds...the lower bound of the region of emergent complexity." Plowman and her colleagues build on McKelvey's formulation, suggesting that periods of organizational instability are often "... full of adaptive tension and tension gradients; it is in this state that emergent self-organization and creative destruction occur" (Plowman *et al.*, 2007: 520). In these models, adaptive tension is the push, the catalyst, the driver that initiates a dynamic state that leads to emergence and order creation.

In sum, we have a bit of a conflict around causality: What actually initiates the emergence of new order? In the far-from-equilibrium approach, the entire system moves into a regime that is away from equilibrium; this "far-from-equilibrium" organizing leads to non-linearities and perturbations or experiments that generate novelty. Under continuing far-from-equilibrium conditions, new order will emerge. This is represented in Figure 1 below.

In contrast, the Adaptive Tension model suggests that the onset and increase of adaptive tension will push the system far away from equilibrium. At a threshold this process may generate a new state of emergent order. This process is represented in Figure 2 below.

On the surface one could say that this apparent conflict is not a real problem, only an issue of semantics. Furthermore, *both* of these constructs can be traced to the original applications of dissipative structures theory into management, sociology, and social evolution, starting with Jantsch (1980) and including important editions by Ulrich & Probst (1984) and Weber, Depew, & Smith (1990). Further, according to one study (Lichtenstein & Plowman, 2009) these two constructs can mean essentially the same thing. So, in what ways is this a conflict, and how is that a problem?

For those of us seeking a general theory of emergence, the problem is in identifying the catalyst of order creation—what is the driver that initiates an emergence process? Although far-from-equilibrium conditions are central to

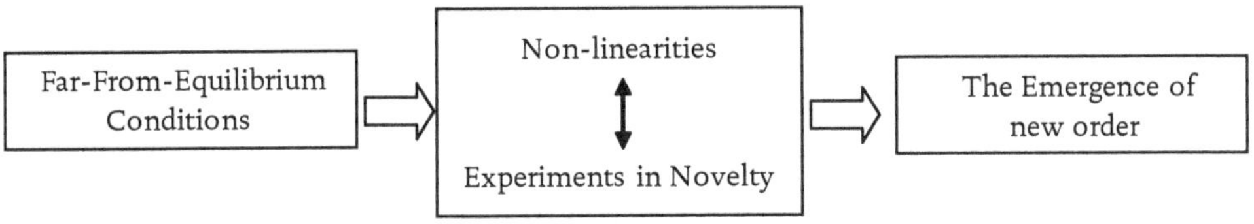

Figure 1 *Far-From-Equilibrium as the Driver of Order Emergence*

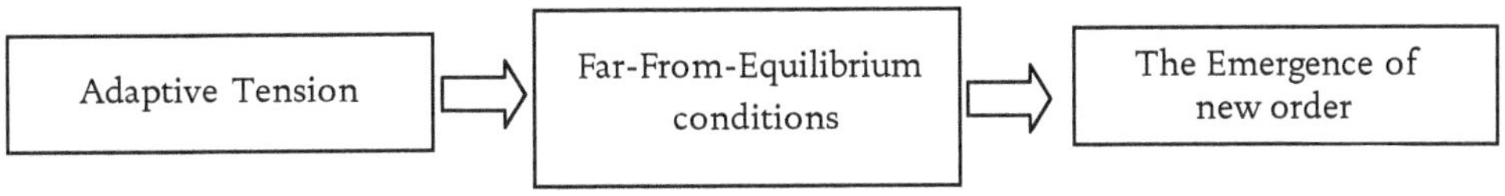

Figure 2 *Adaptive Tension as the Driver of Order Emergence*

this problem, it's unclear whether those conditions initiate the process or mediate the process leading to emergence. As a means of unraveling this distinction, I take a deeper look into the functional meaning of far-from-equilibrium conditions in economic organizations. This exploration will set the stage for a third choice, a catalyst that integrates both of the other views.

The Far-From-Equilibrium View of Organization

What is Life?—A Thermodynamic Explanation

One of the earliest explications of the far-from-equilibrium view of organization was developed by Erwin Schrödinger (1944) in his remarkable book, *What is Life?* Following his significant contributions to quantum mechanics in the 1920s and 1930s, this book presents an integrated theory about how the biological world operates—from the bottom all the way up. Essentially his task is to explain how atoms organize themselves into human beings; in so doing he presents a thermodynamic explanation for the way that all living matter is maintained, i.e., through the importation of what he called "negative entropy." The book explains how this order-creating process operates at each level of analysis, i.e., in atoms, molecules, macro-molecules, and so on. Finally he reaches the level of an organism, where he posits an explanation for how order gets maintained in larger organisms (Schrödinger, 1944: 73; *sic*):

> *Thus the device by which an organism maintains itself stationary at a fairly high level of orderliness (= fairly low level of entropy) really consists in continually sucking orderliness from its environment.*

What does it mean to "suck orderliness" from its environment? Essentially he's suggesting that "order" is a compact form of energy; for an organism to maintain itself, it needs to import high degrees of energy into itself. This insight became the core of open systems thinking (Katz & Kahn, 1966; Thompson, 1968) and general systems thinking (von Bertallanffy, 1956; Miller, 1978); it also provides a theoretical framework for autopoiesis and similar models (Maturana

& Varela, 1980; Drazin & Sandelands, 1992). Moreover, the empirical details of Schrödinger's idea—that systemic order is only maintained through a constant, far-from-equilibrium influx of energy from the environment—were worked out by Prigogine and his associates as the theory of dissipative structures (Prigogine & Glansdorff, 1971; Nicolis & Prigogine, 1989). Their approach was nicely summarized by Maguire and McKelvey (1999: 29), and it forms the starting point for McKelvey's long-time exploration of complexity in management:

> *The key question becomes: What keeps emergent structures in states of equilibrium far above entropy, that is, in states that violate, locally, the 2nd law? Prigogine & Stengers (1984) observe that energy importing, self-organizing, open systems create structures that in the first instance increase neg-entropy... [These structures] are labeled "dissipative structures," because they are the sites where imported energy is dissipated.*

By this formulation, the more entropy that is dissipated, the more order is created, through a web of nested and coevolving ecologies (Weber *et al.*, 1990). Thus, as further levels of order are created, the system as a whole moves farther and farther away from equilibrium. Or does it?

Far-, Farther-, and Farthest-Away-From-Equilibrium

To draw out the metaphor in economic terms, consider the analogy of the firm as an "energy conversion system" (Slevin & Covin, 1997) of inputs, transformations, and outputs (Scott, 1981). In the most simple example these inputs are essentially its revenues, cash flows and core resources; internal transformations represent the business functions and entrepreneurial activities that produce value for customers (Afuah, 2004), and the outputs are the goods or services being offered by the organization. In exchange for the value customers receive through these goods or services they purchase these offerings using money, which in this analogy represents a highly compact form of energy—a generalizable medium of exchange. In sum firms "convert" potential pools of resources, i.e., a potential market, into revenue, by creating products or services which serve this market; the money which customers exchange for the value they receive is the energy that literally sustains the firm.

In its simplicity (by removing the many complexities that would be required to explain most interesting organizations) this example offers a thermodynamic response to the question of how organizations operate in "far-from-equilibrium" conditions. We begin with the definition of equilibrium in thermodynamic terms: equilibrium represents the absence of potential, the eradication of difference, the complete cessation of all activity (Adams, 1988; Dyke, 1988; Odum, 1988; Swenson, 1989, 1992). "Economic equilibrium, therefore, constitutes a point of rest from which there is no endogenous tendency for any individual, form, or market to change" (Meyer *et al.*, 2005: 459). In contrast, as soon as an entrepreneur engages in value-creation activities which are enjoyed by customers who exchange the value they receive for money, the 'system' moves away from equilibrium and toward a non-equilibrium dynamic. If the market need and the venture offerings are sustainable, the entrepreneurial firm can exist in a dis-equilibrium dynamic state for quite some time. From this logic, the higher the exchanges between the firm and the environment—in terms of resources, including cash flow or revenue—the farther away the firm is from its zero-activity equilibrium point.

The implications for this thermodynamic framework are significant. At one end of the spectrum consider the classic mom and pop grocery store, or small manufacturing firm. These organizations exemplify a simple business model design (Zott & Amit, 2007) that generates a dynamically stable organization, in which net revenues from sales, i.e., operating revenues with Cost of Goods Sold subtracted, are sufficient to pay for Fixed Costs including employee and executive salaries. "Lifestyle ventures" like these are organized primarily to create a stable lifestyle for the lead entrepreneur(s) and those they employ. From this view, *all* companies operate in some degree of non-equilibrium conditions (Dyke, 1988; Robb, 1990). We can, for example, define the "distance from equilibrium" of a firm as its overall cash flow, revenues (or some other financial ratio), such that the more of these revenues flowing through the firm, the farther-away-from-equilibrium the firm would be operating.

Pushing the metaphor further, some ventures identify larger pools of economic potential, i.e., larger markets (Bhide, 2000) and more valuable technologies or business models; clever entrepreneurs find new ways to capitalize on these potentials by creating and accessing the resources necessary to produce value in a sustainable way (Stevenson & Gumpert, 1985). The more resources and the more value that can be produced—assuming there is a sufficiently large market that will purchase the offering—the farther away from equilibrium the firm will be operating. Thus, a small manufacturing firm will be a moderate distance away from thermodynamic equilibrium, a large company will be much farther away, and huge multi-national corporations are the farthest-away-from equilibrium. By this reasoning, the farthest-from-equilibrium firms in the world—measured by net revenues—would be Walmart, GM and Exxon/Mobile—the latter earned $40.6 Billion dollars in net profit in 2007, with more than $400,000,000,000 in revenues.

However, this scientific analysis runs counter to the common management metaphors associated with "far-from-equilibrium" conditions; including "...the constructive role of disorder and even crisis in organizations, as well as the merits of management relinquishing and distributing control to a more autonomous work-force" (Maguire *et al.*, 2006: 173). Far-from-equilibrium conditions that managers face are neither predictable nor linear, instead they lead to "ongoing, perpetual novelty" (Meyer *et al.*, 2005: 470).

In contrast, Exxon and GM seem to exemplify the opposite: Firms that abide by a model of General Linear Reality (Abbott, 1988), with managers who lead by control, under the assumption Gaussian averages rather than Paretian extremes (Andriani & McKelvey, 2009). The recent "Great Recession" has shown, if nothing else, the inaccuracy of these assumptions, highlighting the dis-equilibrium nature of our social world.

Creating Emergence?

To review: The goal of this essay is to highlight a contradiction in how we as complexity researchers explain what catalyzes emergence. I started by identifying the two versions of those explanations: far-from-equilibrium dynamics, versus adaptive tension (Figures 1 and 2, above). Then I reviewed the underlying assumptions of far-from-equilibrium dynamics, by presenting a simplified version of the firm as an "energy conversion system." In this economic analogy to a thermodynamic system, distance from equilibrium technically refers to the degree of energy flowing through the system, measured by the amount of revenues (or cash flow) in the firm. Extending this metaphor to its logical ends shows that the farthest-from-equilibrium companies are those with the highest revenue streams.

However—and this is the core of my argument—these firms don't exemplify the novelty, flexibility, and adaptability that could catalyze emergent order. In other words, huge companies like GM, Exxon/Mobile, and WalMart—far-from-equilibrium though they may be—do not appear to embody the conditions that initiate the creation of new order. Rather, when we say "far-from-equilibrium" we seem to imply something quite different from its technical meaning, something much more dynamic, creative, innovative, and non-linear. But these meanings are imputed as metaphors, they are not embedded within a rigorous analogy to the thermodynamic science.

On the other side, McKelvey's "adaptive tension" presents a useful but also incomplete view of how emergence is catalyzed. Adaptive tension is caused by *external* situations, whether a market-based performance gap or an external directive to change course. Specifically, McKelvey (2004: 319) defines adaptive tension as "an imposing energy differential" that can spark a phase transition; this differential reflects an "externally influenced" change to the "control parameters" of the system.

However as any entrepreneur knows, the creation of new business venture is almost never sparked by external influences alone. Organizational emergence requires significant initiative and entrepreneurial action (Gartner, 1993). Even the most basic view of opportunity recognition requires an entrepreneur who is alert to external changes (Kirzner, 1997), and who takes risks that reflect his/her experience, expertise, and access to resources (Ardichvili, Cardozo & Ray, 2003).

To be clear: adaptive tension represents a crucial aspect of emergence, namely *exogenous* market changes that lead to new (and newly perceived) pools of potential resources. Far-from-equilibrium conditions, on the other hand, represent an *endogenous* source of variation and change, which also plays a crucial role in catalyzing emergence[1]. Both elements are important, but neither one alone is sufficient to initiate the process of order creation. Formally, neither far-from-equilibrium conditions nor adaptive tension on their own may be sufficient to catalyze emergence. Can these two constructs be integrated in some way, offering a more dynamic view that incorporates the endogenous *and* exogenous elements of emergence?

"Opportunity Tension" as a Catalyst of Emergent Order

As an alternative catalyst I propose the notion of "Opportunity Tension." Opportunity tension represents an internal drive—the entrepreneur's intention—that arises in conjunction with his/her perception of a business opportunity—a market need that can be filled through entrepreneurial action. Opportunity tension starts when an entrepreneur perceives/identifies a pool of potential resources, creating an opportunity and simultaneously constructing a way to capitalize on that economic potential through a unique and sustainable business model (Zott & Amit, 2007). In opportunity tension, the endogenous drive to create a new venture is bound up with the exogenous changes that are revealed through the creation of a market opportunity.

Creating and enacting an opportunity takes a huge amount of time and effort, a great deal of intention and action. Empirical evidence shows that the greater this drive to action, the more likely that a business will actually emerge as an independent start-up venture (Lichtenstein *et al.*, 2007). An entrepreneurial pursuit of an opportunity requires a high degree of personal commitment as well as a degree of personal passion (Adler & Obstfeld, 2007); these reflect the "creative tension" within the entrepreneur (Fritz, 1984; Senge, 2006) that motivates the process of capitalizing on the opportunity. At the same time, this intention—a commitment to pursue the venture—is based on the perception of a market potential, an entrepreneurial opportunity to create new value for customers who want it. If I feel I can really do this, the opportunity becomes alive, initiating a creative tension within. When there is opportunity *and* (in)tension, the organizing process begins (Lichtenstein *et al.*, 2007).

In this way, the core of opportunity tension integrates the combination of internal and external motivation. "Energy differentials need to have a motivational valance attached before they can be expected to be felt as tension by agents" (McKelvey, 2001: 195). This motivational valence is opportunity tension: the *felt* belief that the emerging opportunity is worthy of pursuing. More formally opportunity tension describes the combination of *externally* imposed

1. This wise distinction was presented by one of my reviewers.

changes leading to exogenous energy differentials and pools of potential resources, with *internally* motivated action leading to the entrepreneurial activities that organize a business (Gartner & Brush, 2007). Thus, the intrinsic motivation to pursue the idea is supported by the extrinsic motivation of the potential market for the idea. At the same time, this extrinsic recognition of a market opportunity feeds back to increase intrinsic motivation, since the size and viability of the perceived market provides a measure of confidence that supports the intrinsic drive for organizing. As a whole, opportunity tension provides the catalyst for order creation.

Opportunity Tension is Rare

Although novelty and improvements are happening all the time, the actual initiation of opportunity tension is not a common activity. Only a unique or unusual event (Andriani & McKelvey, 2007) would have the power to push an entrepreneur or an organization beyond its structural inertia, to start a process of novel organizing and emergent order creation. McKelvey (2001) shows this in his exemplar of adaptive tension: Jack Welch, as the CEO of GE, catalyzed adaptive tension in every business there by giving them only *one* success metric: "Be #1 or #2 in your industry, or be reorganized, split up, or sold." This charge is not an "average" or "normal" event—it is extreme, perhaps one of the most extreme challenges ever given to a group of companies. And the result, conditioned by Welch's remarkable leadership, was also extreme, as GE became one of the most financially successful corporations in the past 100 years.

This kind of extreme event is driven by opportunity tension: Welch as entrepreneur creates an *internal* tension in each of his divisional executives by using an *external* 'bar' that sparks a new kind of thinking—an entrepreneurial, opportunity-driven mindset (Lumpkin & Lichtenstein, 2005) through which these executive managers identify, create and act on new business opportunities. As Gartner (1993; Gartner *et al.*, 1992) shows, organizational emergence often starts in the same way: in rare moments of tension an individual sees an opportunity for change, and at the same time feels the internal drive to act on that opportunity. Such unique experiences, along with the commitment and follow-up they catalyze, are often at the origin of opportunity tension.

Cycles of Opportunity Tension

This urgency, this push, does not last forever; the internal tension that drives action does not exist indefinitely at that initial very high rate. Instead, a flush of creative energy and hard work that can last for a period of weeks or months ultimately settles down into a more sustainable degree of intensity. Thus, opportunity tension occurs in ebbs and flows, just exactly like creative tension (Fritz, 1989). Specifically the energy of creative tension, which is at the heart of opportunity tension, is generated through a polarization—the 'distance' between one's envisioned goal and one's present state: The farther from the goal, the higher the creative tension to achieve it (Senge, 2006). Once the goal has been achieved, creative tension is reduced.

Like the source of creative tension in artists, opportunity tension motivates action for an intense period of time, sometimes generating a kind of "flow" state (Csikszentmihalyi, 1990). Within this flow state, organizing continues until the initial goal is achieved (Lichtenstein *et al.*, 2007); sometimes the goal itself changes through the organizing process (Lichtenstein, Dooley, & Lumpkin, 2006; Sarasvathy, 2001). Yet once reached, even for short period of time, the degree of intensity is relieved.

Opportunity tension is thus a *pulse of activity*—a committed intention to pursue this project and generate an emergent result. The project can be the creation of a new business, the re-creation of an existing company, or a much more contained initiative that extends some quality of the organization. Given the right set of resources and conditions new order may come into being; this emergent order literally converts the environmental (market) potential into real value for customers who pay for that value using resources (e.g., money) that keeps the organizational operating.

As others have pointed out, this process of emergence is often expressed as a "state change" within an organization (Cheng & Van de Ven, 1996; Dooley, 1997) as the firm or company move from one "dynamic state" to another (Levie & Lichtenstein, in press). For example, Plowman and her colleagues show how Mission Church transformed from a "silk-and-stockings" congregation in one era, to a social activist community in a new era that worked tirelessly to improve conditions in its inner-city neighborhood. Chiles and his colleagues show how the emergence of Branson, MO occurred in a series of "epochs," each era of change representing a transformative shift from one set of disequilibrium conditions to another. In a similar fashion Lichtenstein, Dooley & Lumpkin's (2006) analysis of one start-up venture identified an "emergence event"—a punctuated shift in strategy and tactics that reflected a new goal and vision for the company, and the entrepreneurial pursuit of a new opportunity.

In each of these cases, entrepreneurial action forms around one opportunity; this may be followed by another opportunity, and so on; each of these initiates a new cycle of opportunity tension. Each of these cycles represents a distinct phase of activity, initiated by the combination of exogenous forces of change and endogenous forces of creative expression, resulting in new products and offerings that yields more novelty in the market, attracting more customers (additional revenue streams) and maintaining the organization at this next dynamic state (Levie & Lichtenstein, in press). Entrepreneurial leaders are always on the lookout for new opportunities, each of which may catapult the organization to yet another new dynamic state.

Chiles and his colleagues (Chiles *et al.*, 2010a, 2010b) have explored this cyclic process deeply, providing a powerful framework for understanding the dynamics of markets in continual disequilibrium. In their view, any given "round" of competition—each cycle of opportunity tension—may be followed by another round or cycle of emergent change:

> *Craig (1996: 305) characterizes these phases as 'rounds' of hypercompetition... [each of which is] "followed...by a subsequent hypercompetitive 'round,' equally intense but based on different competitive dimensions." ...Such phases of hypercompetition may be understood in the context of continual disequilibrium as examples of "punctuated disequilibrium," in which kaleidic shifts further disrupt markets already in disequilibrium, driving them...from one disequilibrium phase to another* (Chiles *et al.*, 2010b: 21).

Moving Far From Far-From-Equilibrium: Dynamic Creation

Taking this further allows us to envision a truly dynamic systems science that is built around disequilibrium processes as the norm rather than the exception (Meyer *et al.*, 2005; McKelvey, 2006; Andriani & McKelvey, 2009; Chiles *et al.*, 2010b). In the words of Meyer and his colleagues (2005: 471) such an approach would "...embrace notions like co-evolution, CAS, field configuration, network formation, autocatalytic feedback, niche evolution, and emergence." An excellent term for this science is "dynamic creation," through which "...plans, resources, and market forces are constantly in flux. This ferment spawns new plans, new resources, and new forces in an unending cycle of reiterative creation" (Chiles *et al.*, 2010a: 7).

Chiles and his colleagues have been artfully developing this dynamic model through their explication of a Radical Subjectivist Austrian economics. Their analysis focuses on the self-organized creation of economic markets; opportunity tension is the entrepreneurial drive that underlies the novelty in those markets. In a broad sense Chiles's approach and mine are not distinct but rather are complementary and interdependent. In particular, they argue (Chiles *et al.*, 2010a: 26) that an entrepreneur's expectations about the needs of potential customers "...generates opportunity tension," (a term that I suggested to Todd Chiles upon reading their working paper). This leads to the creation of new products and services, i.e., a "source of novelty...that moves markets away from equilibrium and initiates order creation." They see the novelty of new offerings as the basis of systemic heterogeneity, which is itself the driver of economic order: "This heterogeneity in turn begets still more novelty, and so on in an ongoing rivalrous process that generates increasing product heterogeneity as the market moves farther from equilibrium."

They suggest that these processes build on themselves, generating a momentum that increases the likelihood of further organizing. Although that may be true at a macro level, in entrepreneurial terms these order creation process are less common; they represent an "extreme event" that can push the organization/ecology system beyond its norm and into a new level of order (McKelvey & Andriani, 2005).

Implications and Conclusion

Sustainability and Value Creation

Further—and this is a very important distinction—unique opportunity tension events are viable *only if* there is a potential pool of resources that the entrepreneur believes s/he can capture through this new round of organizing. That's because in order to maintain a newly expanded system, a higher amount of "negative" entropy is needed, i.e., a system farther-from-equilibrium requires more net resources to maintain itself in its new expanded state (Levie & Lichtenstein, in press). In ecological terms, *dynamic creation is only viable to the degree that it is sustainable,* for new order can only be created and maintained through the transformation of energy that exists in some form. For a time entrepreneurs can draw on the energy of personal passion, social connection and shared vision, but as any leader knows even these have some limit. These limits are far more visible when the resources needed to create value are more physical, as in any company that makes and sells tangible objects. The current levels of environmental degradation around the world show the severe limitations to a belief in unlimited material growth. For these reasons, dynamic creation must be bound up with an economics and a systems science of sustainability.

A second implication, less severe but more broad, is that all organizing must generate value of some kind if it is to survive for any length of time. In formal terms, emergent order is always based on value creation, because continuing to produce that value—whether social or economic—*requires* a consistent input of energy that maintains the system in a sustainable way. We've already seen the business examples: A firm that produces value through the 'widget' it creates is maintained by the revenues it receives from the customers who pay for the widget in exchange for the value it provides them.

This implication extends much further than organization creation. This paper emerged because of the value that I perceived it would create for my readers and for me; that perceived value has sustained me through the writing and revision process. By extension, the emergence of *Emergence: Complexity & Organization* is based on the value that it provides for scholars and practitioners; that value is paid for through subscriptions which sustain the activities required to produce and publish the journal (albeit to a lesser degree than the publisher would like, I am sure…!). In this way, I posit that the sustained emergence of any entity can be traced to the value which it creates; when the perceived value of the offering is less than what it 'costs' (in energy, time, money, etc.) to produce it, the venture or initiative will cease to exist (unless it is maintained by some other source).

Summary

From an entrepreneurial perspective, value creation is the goal of opportunity tension: When an entrepreneur or any leader perceives an opportunity, what they see/create is a new means for generating value for others; the creative tension inherent in this vision leads to a cycle of entrepreneurial organizing that is aimed at capitalizing on the opportunity. Emergence—whether of a project, an initiative, or a company—is catalyzed by opportunity tension. Opportunity tension thus integrates the endogenous commitment of an entrepreneur with the exogenous market changes that lead to her/his perception and creation of a business opportunity. The result is the emergence of a product/service that adds heterogeneity to the market, pushing the economic region further and further from equilibrium (Chiles *et al.*, 2010a).

As a whole, the {opportunity tension → emergence} process exemplifies the core of dynamic creation. When grounded in social and environmental sustainability a dynamic creation approach can explain the emergence and development of human societies (Carniero, 1970, 1987), of cities (Dyke, 1988), and of expanding order in society (Adams, 1988; Coren, 1998). Dynamic creation, driven by cycles of opportunity tension, offers a view of emergence that begins to move us far from far-from-equilibrium, and toward a more powerful science of complexity for management.

References

Abbott, A. (1988). "Transcending general linear reality," *Sociological Theory*, ISSN 0735-2751, 6: 169-186.

Adams, R.N. (1988). *The Eighth Day: Social Evolution as the Self-Organization of Energy*, ISBN 9780292720619.

Adler, P. and Obstfeld, D. (2006). "The role of affect in creative projects and exploratory search," *Industrial and Corporate Change*, ISSN 0960-6491, 16: 19-50.

Afuah, A. (2004). *Business Models: A Strategic Management Approach*, ISBN 9780072883640.

Andriani, P. and McKelvey, B. (2009). "From Gaussian to Paretian thinking: Causes and implications of power laws in organizations," *Organization Science*, ISSN 1047-7039, 20: 1053-1071.

Ardichvili, A., Cardozo, R. and Ray, S. (2003). "A theory of entrepreneurial opportunity identification and development," *Journal of Business Venturing*, ISSN 0883-9026, 18: 105-124.

Bhide, A. (2000). *The Origin and Evolution of New Businesses*, ISBN 9780195170313.

Browning, L., Beyer, J. and Shetler, J. (1995). "Building cooperation in a competitive industry: Sematech and the semiconductor industry," *Academy of Management Journal*, ISSN 0001-4273, 38: 113-151.

Carniero, R. (1970). "A theory of the origin of the state," *Science*, ISSN 0036-8075, 169: 733-738.

Carniero, R. (1987). "The evolution of complexity in human societies and it mathematical expression," *International Journal of Comparative Sociology*, ISSN 0020-7152, 28: 111-128.

Cheng, Y. and Van de Ven, A. (1996). "The innovation journey: Order out of chaos?" *Organization Science*, ISSN 1047-7039, 6: 593-614.

Chiles, T., Tuggle, C. S., McMullen, J., Bierman, L. and Greening, D. (2010a). "Dynamic creation: Elaborating a radical Austrian approach to entrepreneurship," *Organization Studies*, ISSN 0170-8406, in press.

Chiles, T., Vultee, D., Gupta, V., Greening, D. and Tuggle, C., (2010b). "The philosophical foundations of a radical Austrian approach to entrepreneurship," *Journal of Management I*nquiry, ISSN 1056-4926, in press.

Choi, T., Dooley, K. and Rungtusanatham, M. (2001). "Supply networks and complex adaptive systems: Control vs. emergence," *Journal of Operations Mana*gement, ISSN 0272-6963, 19: 351-366.

Coren, R. (1998). *The Evolutionary Trajectory: The Growth of Information in the History and Future of the Earth*, ISBN 9789056996017.

Csikszentmihalyi, M. (1990). *Flow: The Psychology of Optimal Experience*, ISBN 9780060162535.

Dooley, K. (1997). "A complex adaptive systems model of organization change," *Nonlinear Dynamics, Psychology, and the Life Sciences*, ISSN 1090-0578, 1: 69-97.

Drazin, R. and L. Sandelands (1992). "Autogenesis: A perspective on the process of organizing," *Organization Science*, ISSN 1047-7039, 3: 230-249.

Dyke, C. (1988). "Cities as dissipative structures," in B. Weber, D. Depew and J. Smith (eds.), *Entropy, Information and Evolution: New Perspectives On Physical and Biological Evolution*, ISBN 9780262231329.

Fritz, R. (1989). *The Path of Least Resistance*, ISBN 9780449903377.

Gartner, W. (1993). "Words lead to deeds: Towards an organizational emergence vocabulary," *Journal of Business Venturing*, ISSN 0883-9026, 8: 231-239.

Gartner, W. and Brush, C. (2007). "Entrepreneurship as organizing: Emergence, newness and transformation," in T.H.M. Rice (ed.), *Entrepreneurship: The Engine of Growth*, ISBN 9780275989897, pp. 1-20.

Gartner, W., Bird, B. and Starr, J. (1992). "Acting as if: Differentiating entrepreneurial from organizational behavior," *Entrepreneurship Theory and Practice*, ISSN 1042-2587, 16(3): 13-30.

Jantsch, E. (1980). *The Self-Organizing Universe*, ISBN 9780080243122.

Katz, D. and Kahn, R. (1966). *The Social Psychology of Organizations*, ISBN 9780471023555, (1978).

Kirzner, I.M. (1997). "Entrepreneurial discovery and the competitive market process: An Austrian approach," *Journal of Economic Literature*, ISSN 0022-0515, 35: 60-85.

Levie, J. and Lichtenstein, B. (in press). "Terminal assessment of stages theory: Toward a dynamic states model of entrepreneurship," *Entrepreneurship: Theory and Practice*, ISSN 1042-2587,

Lichtenstein, B. and Plowman, D.A. (2009). "The leadership of emergence: A complex systems leadership theory of emergence at successive organizational levels," *The Leadership Quarterly*, ISSN 1048-9843, 20: 617-630.

Lichtenstein, B., Carter, N., Dooley, K. and Gartner, W. (2007). "Complexity dynamics of nascent entrepreneurship," *Journal of Business Venturing*, ISSN 0883-9026, 22: 236-261.

Lichtenstein, B., Dooley, K. and Lumpkin, T. (2006). "Measuring emergence in the dynamics of new venture creation," *Journal of Business Venturing*, ISSN 0883-9026, 21: 153-175.

Lumpkin, T. and Lichtenstein, B. (2005). "The role of organizational learning in the opportunity recognition process," *Entrepreneurship: Theory and Practice*, ISSN 1042-2587, 29(4): 451-472.

Maguire, S. and McKelvey, B. (1999). "Complexity and management: Moving from fad to firm foundations," *Emergence*, ISSN 1521-3250, 1(2): 19-61.

Maguire, S., McKelvey, B., Mirabeau, L. and Oztas, N. (2006). "Complexity science and organization studies," in S. Clegg, C. Hardy, W. Nord and T. Lawrence (eds.), *Handbook of Organization Studies*, ISBN 9780761951322, pp. 165-214.

Maturana, H. and F. Varela, (1980). *Autopoiesis and Cognition*, ISBN 9789027710154.

McKelvey, B. (2001). "What is complexity science? It is really order creation science," *Emergence*, ISSN 1521-3250, 3(1): 137-157.

McKelvey, B. (2004). "Toward a complexity science of entrepreneurship," *Journal of Business Venturing*, ISSN 0883-9026, 19: 313-342.

McKelvey, B. (2006). "Van de Ven & Johnson's 'Engaged Scholarship': Nice try, but...," *Academy of Management Review*, ISSN 0363-7425, 31: 822-829.

McKelvey, B. and Andriani, P. (2005). "Why Gaussian statistics are mostly wrong for strategic organization," *Strategic Organization*, ISSN 1476-1270, 3: 219-228.

Meyer, A., Gaba, V. and Colwell, K. (2005). "Organizing far from equilibrium: Nonlinear change in organizational fields," *Organization Science*, ISSN 1047-7039, 16:456-473.

Miller, J.G. (1978). *Living Systems*, ISBN 9780070420151.

Nicolis, G. and Prigogine, I. (1989). *Exploring Complexity*, ISBN 9780716718598.

Odum, H. (1988). "Self-organization, transformity, and information," *Science*, ISSN 0036-8075, 242: 1132-1139.

Plowman, D.A., Baker, L., Beck, T., Kulkarni, M., Solanksy, S. and Travis, D. (2007). "Radical change accidentally: The emergence and amplification of small change," *Academy of Management Journal*, ISSN 0001-4273, 50: 515-543.

Prigogine, I. and Glansdorff, P. (1971). *Thermodynamic Theory of Structure, Stability, and Fluctuations*, ISBN 9780471302803.

Prigogine, I. and Stengers, I. (1984). *Order out of Chaos*, ISBN 9780553340822.

Robb, F. (1990). "On the application of the theory of emergence and of the law of maximum entropy production to social processes," *Systems Practice*, ISSN 1573-9295, 3: 389-399.

Sarasvathy, S. (2001). "Causation and effectuation: Toward a theoretical shift from economic inevitability to entrepreneurial contingency," *Academy of Management Review*, ISSN 0363-7425, 26: 243-263.

Schrödinger, E. (1944). *What is Life?* ISBN 9780521427081 (1992).

Scott, R. (1981). *Organizations: Rational, Natural, and Open Systems*, ISBN 9780136419778.

Senge, P.M. (2006). *The Fifth Discipline*, ISBN 9780385517256.

Slevin, D. and Covin, J. (1997). "Time, growth, complexity, and transitions: Entrepreneurial challenges for the future," *Entrepreneurship: Theory and Practice*, ISSN 1042-2587, 22(2-Winter): 53-68.

Stevenson, H. and Gumpert, D. (1985). "The heart of entrepreneurship," *Harvard Business Review*, ISSN 0017-8012, 63(2-Mar/Apr): 85-94.

Swenson, R. (1989). "Emergent attractor and the law of maximum entropy production: Foundations to a theory of general evolution," *Systems Research*, 6(3): 187-197.

Swenson, R. (1992). "Order, evolution, and natural law: Fundamental relations in complex system theory," in C. Negoita (ed.), *Cybernetics and Applied Systems*, ISBN 9780824786779, pp. 125-147.

Thompson, J. (1967). *Organizations in Action*, ISBN 9780765809919 (2003).

Ulrich, H. and Probst, J.B. (eds.) (1984). *Self-Organization and Management of Social Systems*, ISBN 9780387134598.

von Bertalanffy, L. (1956). "General system theory," *General Systems*, 1: 15-36.

Weber, B.H., Depew, D.J. and Smith, J.D. (eds.) (1990). *Entropy, Information, and Evolution*, ISBN 9780262231329.

Zott, C. and Amit, R. (2007). "Business model design and the performance of entrepreneurial firms," *Organization Science*, ISSN 1047-7039, 18: 181-199.

Benyamin B. Lichtenstein is Assistant Professor of Management and Entrepreneurship at the University of Massachusetts, Boston. His research and consulting focuses on applying the insights from complexity science to leadership and transformation, in order to support more sustainability and learning in entrepreneurial companies, large corporations, and non-profit organizations. 'Professor Benyamin,' as his students call him, has published over 40 papers and chapters including articles in internationally recognized journals such as *Organization Science, MIT Sloan Management Review, Journal of Business Venturing, Entrepreneurship Theory and Practice*, and the *Academy of Management Executive*, where he received the "Article of the Year" award in 2000. In addition to his professional work, he finds great joy connecting with his wife Sasha and their two children, Simeon and Moriah.

11. The Complexity of Complexity Theory: An Innovative Analysis

Steven E. Wallis
Independent Organizational Consultant, USA

As more scholars join the conversation around complexity theory (CT), it seems a useful time to ask ourselves if we are talking about the "same thing?" This concern is highlighted by the present survey, which finds more conflict than agreement between definitions. In contrast to the conflict, a path toward common ground may be found by applying the idea of a "robust" theory. A robust theory is expected to be more effective in application and more reasonably falsifiable. In this paper, Reflexive Dimensional Analysis (RDA) is used to analyze existing definitions of CT. These definitions are deconstructed, redefined as scalar dimensions, combined, and investigated to identify co-causal relationships. The robustness of CT is identified as 0.56 on a scale of zero to one. Paths for advancing the theory are suggested, with important implications for complexity science.

Introduction: Seeking the Core of Complexity Theory

Given the breadth, depth, and growth of the current conversation, it seems reasonable to ask—exactly what is this thing called "complexity theory?" For although there are many definitions of CT, it has been suggested, that there is no unified description (Axelrod & Cohen, 2000: 15; Lissack, 1999: 112). While this plurality may reflect the many voices engaged in the conversation, it also calls into question the validity of the theory because there is no common sense as to what the theory "is." Indeed, the general assumption seems to be that we are all talking about the "same thing." Like blind men discussing an elephant, such assumptions may lead to false conclusions and unnecessary conflict.

While the academic process thrives on the differences between points of view, the extent of those differences calls into question whether scholars are, indeed, talking about the same thing. After all, if one author states that CT may be understood through concepts "A, B, and C" while another author states that the relevant concepts are "C, E, and F," there is some conceptual overlap, but there are also inherent contradictions. Although according to their authors, these descriptions fit under the general rubric of CT, these differences may be seen as representing a conflict in the common understanding of CT, and so reflect differences in our understanding of systems from atoms to institutions.

The issue of understanding of a body of theory has been of concern for decades. In one attempt to make sense of the issue, theories are described as having of a "hard core" of unchanging assumptions, surrounded by a more changeable "protective belt" (Lakatos, 1970). When a theory is challenged, a theorist may rise to defend it with a new concept that changes the belt, but presumably leaves the core intact. In the present paper, I seek to identify the core of CT. This effort will provide general and specific support for the continued development of CT.

If the core is defined as "that which is generally accepted," it might be easy to define the core of CT. Unfortunately; no such commonality seems to exist (as will be explored in greater depth below). Some other indicator is then needed for the core.

Where the social sciences might be generally said to have highly variable protective belts of theory, it should be noted that Ohm's I=E/R is a robust theory. I use the term robust in the same way that it is used in physics and mathematics, to describe a theory where each dimension of the theory may be determined by the other dimensions (this will be discussed in greater detail below). In the present article, I will identify how an understanding of CT might be shifted from the shifting obfuscation of Lakatos's outer belt, toward an enduring and useful law. When our theories attain this level of advancement, we may anticipate meaningful changes in the way we study institutions.

Leaving that lack of effective theory unquestioned is like ignoring our fundamental assumptions. And, as Lichtenstein (2000a: 539) suggests, "...since these assumptions are rarely discussed, many of the potential insights from complexity science have not been fully developed."

Although a complete study is beyond the scope of the present paper, it may be suggested that a more robust version of CT may prove more effective in application. This claim is based on the idea that robust theories of physics are more effective than the ever-flexible theories commonly found in the social sciences.

"A proposition is a declarative sentence expressing a relationship among some terms." Van de Ven (2007: 117). Further, a proposition may be of three types. Atomistic propositions are very simple, essentially claiming, "A is valid." Linear propositions are more complex noting a causal linkage between two concepts such as, "Change in A causes change in B." More complex relationships are found in concatenated proposition. In this, I use the sense expressed by Van de Ven (2007) where two aspects of a theory are shown to influence the third aspect. For an abstract example, a concatenated proposition might state that changes in A and B will cause changes in C. Similarly, a theory may be generally understood as a collection of interrelated concepts—an idea that will be explored in depth below.

In this process, I will adopt a Knowledge Management (KM) approach. KM includes the study of creating, transferring, and sharing knowledge (Kak-

Original Ref—E:CO Vol. 11 No. 4 2009 pp. 26-38

abadse, Kouzim & Kakabadse, 2001). Codification is seen as a critical step in the process of social interaction (Leydesdorff, 2002). Codified knowledge is considered more easily transmittable, although it may seem less related to a given context. The following analysis might also be seen as a form of scientometrics and bibliometrics. The concise descriptions of CT as found in the study of organizational theory (OT) will be taken as codified knowledge and this paper may be seen as another step in the process of codification.

Although the present studies will avoid a historical approach to the analysis of theory, it may be useful to note here that Van Dijkum (1997) suggests that CT is closely related to cybernetics as both include questions of subjectivity, self-organization, and self-steering founded primarily on the work of Prigogine, Haken, and Casti. In the present paper, the focus will be on eight publications.

Although each author studied in the present paper is writing within the organizational field, they draw their influences from a variety of sources. Pascale (1999) traces his influences to the interdisciplinary work at the Santa Fe Institute. Frederick (1998) draws on mathematics, chemistry, biology, and Darwinian evolution. Axelrod and Cohen (2000) pull from Darwin, Smith, Simon, and "many fields of study," while Brown & Eisenhardt (1998) mention physics, biology, economics, and strategy. Stacey, *et al.* (2000) note their preference for the works of Prigogine, Kaufman, and Goodwin, and against the works of Gell-Mann and Holland. Kernick (2006) suggests that the study of complexity began with computer-based models and spread to other disciplines, while Hurtado (2006) draws primarily on more recent works from organizational theory. Finally, Dagnino (2004) suggests the origins were in biology and physics and that these origins represent the Santa Fe approach and the European approach.

It should be noted here that the diversity of sources, including fields of study and specific authors, does not seem to suggest a shared sense of what CT is.

Drawing on Davidson and Layder, Romm (2001) suggests that researchers use "triangulation" (where multiple research methods are used to reduce subjectivity in research). In the present paper, I provide two studies of CT, where both studies use the same source of data found in those eight publications.

In the first study, I draw on techniques of content analysis to identify the range of concepts found in the theory. Content analysis (e.g., Hjørland, 2002; Hood & Wilson, 2002) essentially involves looking at the words used by the authors as reasonable representations of the concepts that they convey. Grasping the range of concepts helps to identify the similarities and differences between the various versions of theory. The first study identifies a high level of disagreement between authors.

The second study is informed by narrative analysis (e.g., Pentland, 1999) to focus on the propositions found in CT. In this study, I use an innovative, though easily replicable, method of analysis (inspired by insights from CT) to identify relationships between propositions and calculate an objective measure of the robustness of CT. Importantly, an objective path for developing a more robust CT is suggested. A robust version of CT may be expected to provide meaningful benefits to academics and practitioners alike.

Study #1—The Atomistic Concepts of Complexity Theory

This study identifies the range of concepts of CT within organizational theory (OT) and analyzes that collection of concepts from two perspectives. As developed by Wallis (2008), these diverse views, in some sense, may be seen as reflecting the existing diversity in the field of CT. With each perspective, a different view of CT is provided. The new versions of CT created here may be seen as schools of thought, or as newly evolved versions of CT. In addition to identifying the range of concepts present in CT, this study also suggests that more differences than similarities exist between versions of the theory. And, these differences represent an inherent conflict.

Based on a word-search of the ProQuest™ database, I found 683 matches for "complexity theory" with "definitions." By focusing on those papers within the subject area of organizational theory, this number was reduced to 85 matches. Additionally, I followed promising leads and reviewed books from my own shelves. Next, the versions of CT were tested for conciseness to see if they were of a reasonable level of complexity for this kind of study. They should not too long (extending across many pages because such extensive descriptions of CT might suggest a study that would not be of suitable length for this publication.

From this search, I found eight definitions of CT in the field of OT where those definitions were relatively concise (less than one page). For this paper, I will analyze CT based on relatively concise definitions developed by scholars working in the area of OT. The authors drawn on for this study are: (Axelrod *et al.*, 2000; Brown *et al.*, 1998; Dagnino, 2004; Frederick, 1998; Hurtado, 2006; Kernick, 2006; Pascale, 1999; Stacey *et al.*, 2000).

From that collection of writings, I deconstructed each description into the authors' component concepts. For example, Dagnino (2004: 61) suggests (in part), "These subsystems are therefore subject to evolutionary pressures." From such a statement, the concepts of nested systems, and evolutionary pressures may be drawn. While another reader might develop a different list, it is expected that such lists would be substantively similar to the one developed in this study.

From the above eight sources, I found 47 easily differentiable concepts—from the idea that Agents act, to the existence of Unexpected change. The complete set of concepts will not be listed, due to limitations of space. That list of concepts might be seen as representing the whole of CT as found in OT from a conceptual perspective. It should be noted that at this "survey" stage, no concept appears to be closer to the core of CT than any other. The next task is to search for concepts held in common by the authors. It may be assumed that greater commonality represents some shared acceptance of some core concepts, and less conflict, around the question of what CT is. The following is a list of the concepts that seemed most popular among the eight descriptions, each concept being noted by three or more authors:

- Meta systems have midi agents;
- Connections are of varied types;
- Firm behavior changes;
- Evolutionary pressures;
- Agents act;
- Unexpected change occurs, and;
- Agents are of varied types.

These seven concepts stand in stark contrast to the 47 concepts that comprise the complete list of concepts discussed as CT. Also, it is worth noting that a new version of CT has been created based on this focus.

Moving from one form of popularity to another, of the eight publications in this study, two could clearly be seen as the "most cited" according to Google Scholar™. The version of CT developed by Axelrod and Cohen (2000), as well as the version by Brown and Eisenhardt (1998) were each cited well over 500 times suggesting a certain level of authority in the field. The next less frequently cited source received about 280 citations while others were much lower. While this division may be somewhat arbitrary, it serves to provide a useful observation.

Focusing on the two "most cited" versions of CT, there are significant differences and no similarities. The Axelrod and Cohen version takes note of eleven concepts focusing on agents, their variations, interactions, strategies, and success. In contrast, the Brown and Eisenhardt version notes only three concepts, focused tightly on the idea of the "edge of chaos." While it is possible to infer a connection between the two definitions, such an inference would require the interpretation of the authors' works—and so beyond the scope of the present article.

Rather than creating what might be considered a single "authoritative" version of CT, this comparison essentially creates two new description of CT, where each publication may be seen as representing its own school of thought. The lack of overlap (finding zero concepts in common) stands in stark contrast to the 14 concepts that the two do not share. Rather than creating a useful consensus, this study suggests bifurcation and implied conflict between versions. Either version, it should be noted, has less conceptual breadth than the whole body of CT.

There are obvious limitations to this study such as sample size and the amount of interpretation applied to understand exactly what the authors meant to convey with each specific word. However, it should also be noted, that any discussion around investigating some deeper level of understanding must consist of still more concepts. Therefore, a deeper exploration might be expected to produce more concepts and so more levels of conflict.

The present study may stand as an example of an expansion of Lakatos's outer belt. The number of theories was easily increased, yet this expansion does not seem to have increased the understanding of the core.

Of course, many such studies are possible. For example, a new study of this sort might be conducted based on the authors' influences, field of origin, date of publication, and so on. The point to be taken here is that each focus seems to create a new version of CT; and, that new version will have fewer conceptual components than the body of CT as a whole. Further, each new theory will similarly lack a discernible core. In short, it may be concluded that CT is a highly contested field of study, where each new theory adds to the conflict.

This form of study may be understood as useful to the extent that it identifies the range of concepts within a body of theory. This form of study is also useful because it highlights the difficulty of working with disconnected (atomistic) concepts. In the following study, I will use an innovative methodology to investigate the same body of data. Rather than focusing on atomistic concepts, however the focus will be on connected concepts found in the causal propositions of the theories. Before that study, the importance of such a focus is explained in the following section in a discussion on the structure of theory.

The Structure of Theory

Understanding the structure of theory is foundational to understanding the second study, and to identifying the core of CT. Given the diverse versions of CT, the question presents itself: is it possible to ascertain the legitimacy of a theory through its structure? The answer would seem to be in the affirmative because Popper's arguments for falsification include the idea that the structure and composition of a theory will add to the testability of that theory (Popper, 2002: 111-114). The concept is of such importance, however, that some explication seems suggested.

In the validation of theory, Kaplan (1964) describes three norms (correspondence, coherence, and pragmatism). A pragmatic test (one which is tested through application) is beyond the scope of this paper. A norm of coherence suggests that we question whether a theory fits within the existing body of theory. As the present analysis and theory development draws entirely on existing theory, there should be little difficulty in maintaining that norm. Then, there is the norm of correspondence. Kaplan takes pains to note how, for this norm, the truth is plainly useless because every appeal to the facts rests on presuppositions. Therefore, he suggests, what counts in the validity of a theory is the "concatenation" of the evidence. As noted above, concatenation is seen where two aspects of a theory are shown to influence the third aspect In this, I use the sense expressed by (Van de Ven, 2007). For an abstract example, a concatenated proposition might state that changes in concept A and concept B will cause changes in concept C.

In the present study, however, I found few propositions that were concatenated. The second study, therefore, will use Reflexive Dimensional Analysis (RDA) as an analytical tool to "shift" propositions toward increasing levels of concatenation (Wallis, 2006a, b).

Generally, RDA begins with a set of linear propositions. A linear proposition might be seen abstractly in a causal relationship between the aspects of the theory such as, "Changes in A cause changes in B." The propositions found in a body of theory are investigated to determine how they may be combined to create concatenated propositions. For example, one linear proposition (e.g., Changes in A cause changes in B) may be conceptually related with another linear proposition (e.g., Changes in C cause changes in B) to suggest a concatenated proposition (Changes in A and C cause changes in B).

The increasing concatenation may be understood as representing increasing levels of interrelationship, or increasing levels of structure. Dubin (1978) suggests that there are four levels of efficacy in theory; and, these levels seem to reflect the structure of the theory, they are:

1. Presence/Absence—what concepts are contained within a theory;
2. Directionality—what are the causal concepts and what are the emergent concepts within the theory;
3. Covariation—how several concepts might impel change in one another, and;
4. Rate of change—to what quantity does each of the elements within the theory effect one another.

The first study (above) tested for the presence/absence of concepts. In the second study, I will use RDA to combine causal propositions as a means of advancing the structure of CT toward the third level, that of covariation. Achieving the fourth level is beyond the scope of this paper.

In a view that seems similar to Dubin's, Weick (1989) suggests that a theory may be understood as an, "ordered set of assertions." Viewing Weick's statement as a concatenated proposition seems to suggest that more assertions with more order between them, will tend to result in better theory—where better theory may be generally understood as having some preferable aspects of quantity and quality. If a theory is held to be useful in practice, we may infer that a better theory might be more useful in practice. Therefore, the structure of a theory may be said to have some bearing on the efficacy of that theory when it is applied in practice.

If we are looking at the assertions or propositions of a theory as being "interrelated," the propositions of that theory might be seen as, "reciprocally or mutually related" (Dictionary, 1993: 998). With such a view, a body of theory might be seen as a kind of system and, "...any part of the system can only be fully understood in terms of its relationships with the other parts of the whole system" (Harder, Robertson & Woodward, 2004: 83, drawing on Freeman). It seems, therefore, that every concept within a theory would best be understood through other concepts within that body of theory. This view seems to fit some assumptions of CT with regards to importance of the interrelated nature of varied components.

An objective level of relationship is easily calculated by identifying the total number of concepts within a theory as well as the concatenated aspects of the theory, then dividing the number of concatenated aspects by the total number of aspects. This method provides a measure of "robustness" between zero and one Wallis (2008). Briefly, those aspects of a theory that are more closely related to one another may be understood as being closer to the core of the theory.

As an example, Ohm's law (I=E/R) contains three aspects. Each of those aspects is concatenated from the other two. Therefore, Ohm's law has a robustness of one (the result of three divided by three. In contrast, a theory that simply lists its component concepts, without identifying how the concepts are related to one another, would have a robustness of zero. For example, in the above study, the new version of theory is a list of concepts and, as such, has a static robustness of zero because no conceptual aspect is said to have any effect on any other.

Comparing the great usefulness of Ohm's law (fully robust), with the limited usefulness of many social theories (low robustness) suggests important implications for practice.

In the following section, study #2 will use RDA to develop the concatenated set of propositions and to measure the level of interrelationship between propositions in CT.

Study #2—Investigating Relational Propositions

In this section, I analyze CT using RDA. The process of RDA consists of six steps:

1. Define a body of theory;
2. Investigate the literature to identify sources that define it;

3. Code the sources to identify relevant components;
4. Clump the components into mutually exclusive categories;
5. Define each category as a dimension, and;
6. Investigate those dimensions—looking for a robust relationship.

For the first step, the body of theory chosen for this analysis is defined as those concise versions of CT as found in the study of OT. The second step (investigation of definitional literature) was accomplished by a survey of the ProQuest ™ database and additional sources as described above. The third step (coding propositions) proved more difficult than expected. The sheer variety among the propositions meant that few were found to be identical. Also, the complexity of some (long and convoluted) propositions inhibited the coding process.

To facilitate the process, I bypassed coding and went to the fourth step where related propositions were grouped into mutually exclusive categories. By mutually exclusive, for example, we might see that agents cannot be understood as interactions (and visa versa). Many of the categories developed were of that sort of straightforward variety. Others were more challenging. For example, there seems to be considerable congruence between concepts of adaptation, evolution, emergence, and change. Frederick (1998) links concepts of self-organization, emergence, and complexity.

The goal of RDA is to identify the fewest possible dimensions that may be used to relate the greatest range of understanding that may be attributed to a body of theory. I was able to combine a number of terms without loss of conceptual sense-making capacity by first noting the essential similarity between the concepts of evolution and adaptation as both are said to be maximized at the EOC (Brown *et al.*, 1998; Pascale, 1999). Then too, each may be understood as emergence because both are described as causing and resulting from change to a system in the context of its environment. Therefore, the concepts of evolution and adaptation may be seen as dimensions that have been "renamed" or perhaps "derived," rather than as dimensions that are essential to the model. There also appears to be general congruence between the concepts of complexity and emergence, because both may be explained as causing unpredictability at some level.

In this way, I combined the wide variety of concepts. Through the clumping process, at least one concept did not fit neatly with the others. That is the negative result of a fitness test. The propensity of a firm to exit a market is not well understood in CT, perhaps because of a more appreciative approach to firm creation and survival. We seem limited to the idea that firms may disintegrate in chaos (Dagnino, 2004) or fail through excessive stability (Pascale, 1999). Both are possible, however a greater understanding would be beneficial. Again, in the interest of parsimony, it may be said that the destruction of a firm may be described as a lack of emergence based on the idea that a firm that is created, changed and sustained through emergence will fail if the level of emergence drops below a critical level. Or, from another view, if emergence is maximized at the EOC (Brown *et al.*, 1998), it may be suggested that emergence is minimized at the extremes of stability and chaos. If firms fail at those extremes, we may conclude that extinction is the opposite of Emergence. Or, that a certain level of emergence is required to maintain a firm (nourished by a flow of people, ideas, capital, etc.).

Within the causal propositions of these eight definitions, nine mutually exclusive categories are suggested. These may be seen as mutually exclusive in that (for example) systems might engage in action, but no action seems possible without systems; therefore, the two (as concepts) are mutually exclusive.

Although space restrictions preclude a full investigation, this step of the RDA process may be summarized as identifying the mutually exclusive dimensions of CT as:

1. Predictability;
2. Emergence;
3. Fitness;
4. Systems;
5. Action;
6. Information;
7. Goals;
8. Strategies, and;
9. Time.

The fifth step of RDA is to define each categorical cluster as a dimension. That means the category is understood as a dimensional representation of the relevant data and/or observations. So, for example, in observing a firm, a consultant might characterize that firm as having greater or lesser Fitness. This may create perceptual difficulties in the understanding of systems. If we have a dimensional representation of systems, do we say the dimension is said to represent "more" and "fewer" systems or does the dimension represent a single system that is "larger" and "smaller?" The most general answer may prove, in the long run, to be the most useful. That is to say for the present model, what is really being represented is not so much any existing conceptualization of systems, rather a broader sense of systems. That, as yet undefined, sense of "system-ness" is the unexpected result of combining the perspectives of multiple theorists. For the purpose of the present analysis, we may hold such a dimension to represent some sense of systems that is, as yet, undetermined. The explication must be left to a future study.

In the final step of the RDA process, the interrelationships between all dimensions are investigated to identify the relationships between them—what we have been calling "concenated." In this step, the concepts and propositions within each dimensional category may be "translated" to better match the shared terminology of the dimensions. The intent of this process is not to alter the meaning of the original authors, rather it is to identify how dimensions might be related to one another as they exist within a shared body of theory and so suggest paths toward greater testability and falsifiability.

The statements of the eight authors suggest the interrelationships diagrammed in Figure 1. In this figure, each arrow represents a causal direction and each box represents a scalar dimension. For example, more Information results in more Fitness and more Predictability. This creates a richer, and we hope, more accurate view of the aspects of complexity theory than could have been obtained by any analysis of the atomistic concepts, or the linear propositions of any one author.

This model may (or may not) be an accurate representation of the "real world." However, such is not the goal of the present analysis. This model is only intended to be a representation of CT as it is found in the realm of OT. This clarification of the core of CT may serve as a solid foundation for scholarly conversation around the nature of that theory.

From Figure 1, it may be seen that there are nine aspects within CT. Of these nine, there are five that may be understood as concatenated. That is to say, changes in each of those five aspects may be understood as resulting from changes in two or more other aspects. Therefore, the robustness of this model is 0.56—the result of five (concatenated aspects) divided by nine (total aspects). This measure of robustness may serve as a benchmark for the advancement of CT.

As the atomistic version of CT from the present study #1 has a robustness of zero (because it is a list), and the RDA version of CT has a higher robustness than the atomistic versions of CT, the RDA version may be seen as an improvement and a significant step toward robustness, and the present paper serves as an example of advancing theory towards a robust form. Further, the closely related nature of the included aspects suggests that this model may be a good representative for the core of complexity theory.

Conversation

With Figure 1, two challenges become immediately apparent. First, there are three aspects that appear to be strictly causal (Time, Information, and Goals). As such, these aspects are essentially undefined from the view of the model. Where more Strategies may be enabled by increased in Goals and Actions, nothing causes an increase in Goals. They seem to simply exist.

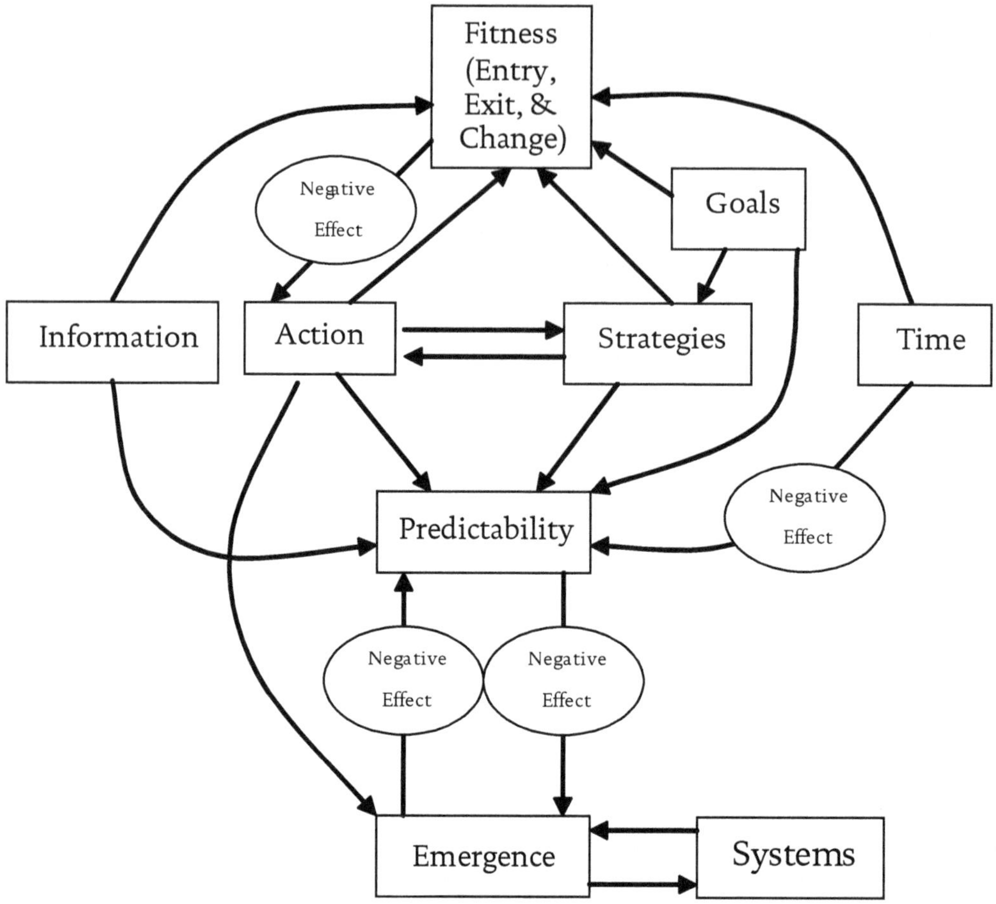

Figure 1 *Causal Relationships Between Aspects of Complexity Theory*

In order to improve the robustness of the CT model, Time, Information, and Goals should be described as concatenated in relation to two or more causal aspects. If those aspects may be legitimately found among the existing aspects of the present model, so much the better. We may retain some level of parsimony.

The second clear challenge to this model is that Systems seems to be both caused by, and be causal to Emergence, and nothing else. This relationship seems to suggest that the concept of Emergence and Systems may be synonymous. If they do indeed represent the same thing, than one aspect may be removed, which will increase the overall robustness of the model. If they are different, that difference must be described in terms of the causal relationships between those aspects and other aspects.

When those two challenges are met, the CT model may achieve full robustness. And, as such, may be more readily falsified, and so more rapidly improved.

Practice

CT is often applied for organizational change efforts as a vague metaphor that is useful for stimulating thought and conversation (Fuller & Moran, 2000). As such, it may be seen that every aspect of the metaphor is indeterminate. The usefulness depends on what aspects the client finds interesting. As a result, the client understands only a portion of the metaphor and does not get a picture of the whole system.

As the concept of non-linear dynamics is an important aspect of CT (e.g., Dent & Holt, 2001; Lichtenstein, 2000b), it may be worth noting that the relationship between the co-causal aspects of this model. For example, the level of Predictability might be completely deterministic if one possesses knowledge of the causal aspects (Information, Action, Strategies, Goals, Time, Emergence). If one of these is unknown, Predictability becomes less determinate. Having a model where a consultant understands which aspects of the model are more determinate and which might be less so, suggests that this kind of model may be developed into a useful tool for practitioners.

Consultants might provide a robust model, and help the client organization to "fill in the blanks." Such a process would provide a roadmap for the client that would suggest different areas for exploration. This way, unknown aspects may be inferred from known aspects. As every client has limited resources, it behooves the consultant to help the client apply those resources in the most effective way possible. If the client wants to gain a CT perspective of its situation using the present model, we might imagine that the client has more knowledge of some aspects than others.

Let us say that the client wants to gain a CT based understanding of their Fitness in an uncertain market. Let us also say, that they are well aware of their Actions, Strategies, and Goals. According to the present model, they might, instead of investigating Fitness directly, choose in investigate Information. This is because Information might be more easily investigated than Fitness, and because by knowing those five aspects, Fitness may be inferred.

Wholeness of Analyses

A theory of low robustness seems to suggest the opportunity for authors to choose which concepts they use as from a menu. This may be thought of as cognitive prejudice, attentional bias, or simply a case of the author not explicating the reasons for choosing those particular concepts. In contrast, a robust model suggests the need to include all aspects of the model in each investigation because each aspect is clearly connected to the other. In other words, a robust theory may be effective in removing the blind spots from analysis. Essentially, when using a theory as a whole model, researchers would be impelled to include descriptions of each part of the model.

The dimensional relationships presented in this model suggest new requirements for the study of CT. For example, if organizational Fitness may be seen as caused by five other dimensions (Goals, Information, Action, Strategy, and Time), a study of Fitness should be reducible to those five dimensions and no others. If it is determined that this is not the case, the theory will be successfully falsified and should be modified appropriately.

Although this form of study would be necessarily more difficult, it is certainly within the realm of expertise of most contributing scholars. Importantly, by the efforts of authors, reviewers and editors, such an effort has the potential to rapidly accelerate the advancement of theory for the benefit of scholars, practitioners, and society at large.

Summary and Conclusion

There are two primary contributions of this paper, one theoretical, the other metatheoretical. The metatheoretical contribution of this paper includes arguments that link the structure of theory with the potential efficacy of the theory. The theoretical contribution of this paper is seen in the advancement of CT from a non-robust model towards a level of greater robustness.

Where previously a reader might have believed that authors conversing around the topic of CT are describing the same thing, the present paper has suggested that is not the case.

In the present paper, I've presented two analyses each drawing on the same eight versions of CT. The first study focused on the 47 conceptual components found in those eight versions. This study found more conflict than commonality between authors. Further, this study found that approaching the whole body of CT from various perspectives resulted in more versions of the theory, but did not seem useful in identifying the core of CT.

In the second study, the focus shifted to the relationship between concepts based on the idea that more complex relationships suggest a higher level of structure—and therefore the opportunity for greater efficacy in theory. The second study used RDA to shift linear propositions of CT into concatenated propositions with higher internal integrity. In this study, I found that the robustness of CT is 0.56 on a scale of zero to one.

In a sense, it may be said that the core must be found from "within" the theory, rather than from any particular point of view from "outside" the theory. The belt, then, may be understood to consist of the 47 concepts that are relatively linear or atomistic, and so less concatenated than those at the core. It is suggested in the second study that the core of CT consists of those concepts that are concatenated. The present study suggests that the core of CT include five aspects (Action, Fitness, Strategies, Predictability, and Emergence), and, importantly, the specific causal relationships between them.

It may be seen by this analysis that a dimensional representation is a more useful tool for analysts than atomistic representations because an atomistic report might say simply whether a firm is fit or not, while a dimensional report could relate a level of fitness, or reflect how one firm were may be more fit than another. Thus, in a sense, a dimension may be said to "contain" (or support the communication of) more information than its atomistic counterpart. Thus, it may be said, that the process of "dimensionalizing" the information supports further investigation in to CT because it allows and encourages theorists to quantify their observations and insights.

In this same way, the dimensional perspective supports observations and analysis for practitioners. For example, a consultant facilitating a conversation among executives might encourage an organizational analysis from a CT perspective by asking atomistically, "Do you see emergence occurring in this firm?" The answer, according to the definitions explored here, should be yes (otherwise, the firm would not exist). In comparison, a more generative conversation would be result if the facilitator were to ask dimensionally, "How *much* emergence do you see in this firm now compared to last year (or compared to our competitors or our clients)?" The atomistic question might lead to a check mark on the flipchart paper, while the dimensional question will more likely lead to conversation, comparative analysis, and action.

While the present study suggests lines of investigation for advancing CT toward robustness, there are also limitations. Most notably, the sample size of eight concise versions of theory might not be considered statistically significant. The validity and usefulness of this line of research may be improved by repeating the study with a larger sample.

Second, it is assumed by the present author that this method of analysis is repeatable—that if performed by others, it would lead to substantially similar results. That assumption should be tested.

Finally, in this paper, I deepen the conversation on structure of theory to suggest that there may be a correlation between the robustness of a theory, and the efficacy of that theory in application. However, using robustness to measure of the potential of a theory for testability and applicability is itself untested. Further investigation will need to occur to test the a priori assertions developed in this paper—perhaps through innovative modeling techniques.

Within academia, it has been generally understood that theory is advancing due to some natural evolution of ideas emerging from the back-and-forth between scholars. Applying a measure of robustness to these theories suggests an objective method for testing that assumption.

Conversations on the structure and construction of theory seem likely to continue and increase as the general understanding of theory increases. It remains an open question, for example, whether the same level of conflict exists in other branches of CT. Measures of robustness, inspired by insights from CT, will provide useful tools for advancing that conversation.

Acknowledgements

I am indebted to two anonymous reviewers whose suggestions helped me to create a stronger paper.

References

Axelrod, R. and Cohen, M.D. (2000). *Harnessing Complexity: Organizational Implications of a Scientific Frontier*, ISBN 9780465005505.

Brown, S.L. and Eisenhardt, K.M. (1998). *Competing on the Edge: Strategy as Structured Chaos*, ISBN 9780875847542.

Dagnino, G.B. (2004). "Complex systems as key drivers for the emergence of a resource- and capability-based interorganizational network," *Emergence: Complexity & Organization*, ISSN 1521-3250, 6(1-2): 61-69.

Dent, E.B. and Holt, C.G. (2001). "CAS in war, bureaucratic machine in peace: The U.S. Air Force example," *Emergence*, ISSN 1521-3250, 3(3): 90-107.

Dictionary. (1993). *Random House Unabridged Dictionary*, ISBN 9780679429173.

Dubin, R. (1978). *Theory Building* (rev. ed.), ISBN 9780029076200.

Frederick, W.C. (1998). "Creatures, corporations, communities, chaos, complexity," *Business and Society Review*, ISSN 0045-3609, 37(4): 358-390.

Fuller, T. and Moran, P. (2000). "Moving beyond metaphor," *Emergence*, ISSN 1521-3250, 2(1): 50-71.

Harder, J., Robertson, P.J. and Woodward, H. (2004). "The spirit of the new workplace: Breathing life into organizations," *Organization Development Journal*, ISSN 0889-6402, 22(2): 79-103.

Hjørland, B. (2002). "Domain analysis in information science: Eleven approaches-traditional as well as innovative," *Journal of Documentation*, ISSN 0022-0418, 58(4): 422-462.

Hood, W.W. and Wilson, C.S. (2002). "Analysis of the fuzzy set literature using phrases," *Scientometrics*, ISSN 0138-9130, 54(1): 103-118.

Hurtado, P. (2006). "Will the real complexity in strategic management please stand up?" *Competition Forum*, ISSN 1545-2581, 4(1): 175-182.

Kakabadse, N.K., Kouzim, A. and Kakabadse, A. (2001). "From tacit knowledge to knowledge management: Leveraging invisible assets," *Knowledge and Process Management*, ISSN 1099-1441, 8(3): 137-154.

Kaplan, A. (1964). *The Conduct of Inquiry: Methodology for Behavioral Science*, ISBN 9780765804488.

Kernick, D. (2006). "Wanted-new methodologies for health service research. Is complexity theory the answer?" *Family Practice*, ISSN 0263-2136, 23: 385-390.

Lakatos, I. (1970). "Falsification and the methodology of scientific research programmes," in I. Lakatos and A. Musgrave (eds.), *Criticism and the Growth of Knowledge*, ISBN 9780521096232, pp. 91-195.

Leydesdorff, L. (2002). "The communication turn in the theory of social systems," *Systems Research and Behavioral Science*, ISSN 1092-7026, 19(2): 129-136.

Lichtenstein, B.B. (2000a). "The matrix of complexity: A multi-disciplinary approach for studying emergence in coevolution," in A. Lewin and H. Voldberda (eds.), *Mobilizing the Self-Renewing Organization: The Coevolution Advantage*, ISBN 9789058920485.

Lichtenstein, B.M.B. (2000b). "Emergence as a process of self-organizing: New assumptions and insights from the study of non-linear dynamic systems," *Journal of Organizational Change Management*, ISSN 0953-4814, 13(6): 526-544.

Lissack, M.R. (1999). "Complexity: The science, its vocabulary, and its relation to organizations," *Emergence*, ISSN 1521-3250, 1(1).

Pascale, R.T. (1999). "Surfing the edge of chaos," *Sloan Management Review*, ISSN 1532-9194, 40(3): 83.

Pentland, B.T. (1999). "Building process theory with narrative: From description to explanation," *Academy of Management Review*, ISSN 0363-7425, 24(4): 711-724.

Popper, K. (2002). *The Logic of Scientific Discovery*, ISBN 9780415278447.

Romm, N.R.A. (2001). *Accountability in Social Research: Issues and Debates*, ISBN 9780306465642.

Stacey, R.D., Griffin, D. and Shaw, P. (2000). *Complexity and Management: Fad or Radical Challenge to Systems Thinking*, ISBN 9780415247610.

Van de Ven, A.H. (2007). *Engaged Scholarship: A Guide for Organizational and Social Research*, ISBN 9780199226306.

van Dijkum, C. (1997). "From cybernetics to the science of complexity," *Kybernetes*, ISSN 0368-492X, 26(6/7): 725.

Wallis, S.E. (2006a). "A sideways look at systems: Identifying sub-systemic dimensions as a technique for avoiding an hierarchical perspective," paper presented at the *International Society for the Systems Sciences*, Rohnert Park, CA.

Wallis, S.E. (2006b). *A Study of Complex Adaptive Systems as Defined by Organizational Scholar-Practitioners*, unpublished Theoretical Dissertation, Fielding Graduate University, Santa Barbara.

Wallis, S.E. (2008). "From reductive to robust: Seeking the core of complex adaptive systems theory," in A. Yang and Y. Shan (eds.), *Intelligent Complex Adaptive Systems*, ISBN 9781599047171, pp. 1-25.

Weick, K.E. (1989). "Theory construction as disciplined imagination," *Academy of Management Review*, ISSN 0363-7425, 14(4): 516-531.

Glossary

Aspect

A broad term relating to the part of theory that represents a concept, idea, or other representation of something that is used in a model. May be represented by a scalar, atomistic concept.

Atomistic

Refers to any reductive proposition representation such as "A is valid."

Concatenated

More complex relationships are found in concatenated proposition. In this, I use the sense expressed by Van de Ven (2007) where two aspects of a theory are shown to influence the third aspect. For an abstract example, a concatenated proposition might state that changes in aspect A and aspect B will cause changes in aspect C.

Dimension or scalar dimension

An aspect of theory that represents quantitative or qualitative variations. For example, a dimension of "size" might be used to represent whether a system is smaller or larger.

Linear

Refers to propositions that describe causal relationship between two concepts. Such as, "Changes in A cause changes in B."

Proposition

"A proposition is a declarative sentence expressing a relationship among some terms." (Van de Ven, 2007: 117).

Reflexive Dimensional Analysis (RDA)

Six step process for analyzing a body of theory and shifting the component propositions of that theory from linear to concatenated.

Static Robustness

Ratio describing the interrelatedness between aspects of a theory on a scale of zero to one. Robustness is calculated by dividing the number of concatenated aspects by the total number of aspects in a theory.

Theory

An ordered set of assertions. Weick (198: 517. Drawing on Southerland).

Steve Wallis is a Fellow of the Institute for Social Innovation, which helps individuals and organizations address societal problems via research, leadership, and organizational development. As an organization development consultant he has a decade of experience in Northern California across a wide range of industries. He earned his Ph.D. in 2006 from Fielding Graduate University. As the founder and director of the Foundation for the Advancement of Social Theory (FAST), a non-profit venture under the auspices of ISI and Fielding, Steve is dedicated to supporting scholars and practitioners as they identify and pursue objective methods for advancing theory across the spectrum of the social sciences. His papers and chapters serve to advance the metatheoretical conversation and identify innovative paths for advancing social theory. He is the editor of "*Cybernetics and Systems Theory in Management: Tools, Views and Advancements*" (in press) and guest editor for a special issue on Emerging Perspectives on Metatheory and Theory, with the "*Integral Review*" (forthcoming).

12. Competitive Scenarios, Community Responses and Organizational Implications

Fabio Boschetti & Markus Brede
CSIRO Marine and Atmospheric Research, AUS

We analyze four scenarios commonly encountered in social processes undergoing competitive pressures: resource depletion by individuals acting greedily ('tragedy of the commons'), wasted opportunity due to over protective players ('tragedy of the anti-commons'), crowd following ('majority wins') and competition for niches ('minority wins'). We show that these scenarios are extremes of a continuous resource exploitation problem and that complex and counter-intuitive behaviors are found at the transitions between 'pure' scenarios. We discuss the likely community behaviors and under what conditions a centralised management intervention may play a role in the resource and community resilience.

Introduction

It is common practise to view many organizational and social processes in the light of competition: human organizations, communities and societies compete for resources, niches, information, supremacy etc. However, the life of individuals (that is the members of such groups) comprises a delicate balance of a wide range of interactions which include competition as much as collaboration and altruistic behavior.

The Darwinian interpretation of such an apparent paradox is that seemingly unselfish behaviors provide for indirect advantages within the context of complex and larger human interactions. This view presupposes the existence of different levels at which the dynamics of human affairs develop and leads quite naturally to complexity theory: interactions at one level (the individual, say), via a chain of feedback effects may result in group competitive advantage whose subtleties may remain obscure to the group members themselves; at the next level, a mix of competitive and collaborative interactions between groups may give similarly unexpected advantages to larger communities, which in turns may feedback to the level of individuals. Similar nesting may progress for many levels, in principle including all forms of life from single cells up to mankind as a whole. The keyword here, albeit a slippery one, is emergence: local interactions (reduced competition and increased collaboration) may result in unexpected global outcomes (improved competitive advantage).

The purpose of this work is to discuss some of these phenomena and to show how counter-intuitive behaviors may arise even under simpler settings. We model only two levels (individuals and a single community), a simple interaction process (exploitation of a common resource) and unintelligent, emotionless, rule-based agents who follow stochastic strategies. By simply changing the way we model the resource availability and distribution we can analyze several competitive scenarios we encounter in our daily business and private life.

The relevance of this work is two-fold; we want to understand, and thus possibly predict, the broad response of communities under different competitive challenges as a function of the strategies they decide to adopt. At a management level, we would also like to anticipate what challenges a community (a business organization for example) may face and what behavior (competitive versus collaborative, centralised versus de-centralised) should be fostered in order to respond to such challenges. One of the main lessons we draw from this work is that not only may the scope for intervention change depending on the details of the challenge, but also that, in certain specific circumstances, global management intervention may be a crucial factor in determining the efficacy of strategies at an individual level.

These insights are the outcome of work we have carried out over a number of years (Boschetti, 2007; Boschetti & Brede, 2008; Brede & Boschetti, 2008; Brede *et al.*, 2008; Brede & De Vries, 2008) using a game-theoretical and evolutionary economics framework (Gintis, 2000). Here we intend to summarize the results and infer some general conclusions from tests carried out under different model scenarios. Our discussion is purposely non technical and results are presented qualitatively while we refer to our previous papers the reader interested in the details of the mathematics, modelling implementation and numerical results of the specific experiments.

The Scenarios

Let's consider a community of individuals who compete for a renewable resource. This could be a natural resource (a farmland or a fishery), a business opportunity (niche in a market) or a non-monetary commodity (access to a natural park or a restaurant). The individuals desire to access as much of a resource as possible, in other words they are naturally selfish (but not necessarily competitive; we will address the difference below). We also assume each individual has a physical, economic or legal limit to how much resource it can access, which may represent a fishing quota, an anti-trust regulation or the maximum space it can occupy in a restaurant. In this section we illustrate a number of scenarios which can arise as a function of resource availability and distribution and the implications these different scenarios have on the individuals' likely exploitation strategies, whose outcomes will be the focus of the rest of the paper.

Original Ref—E:CO Vol. 11 No. 4 2009 pp. 39-47

Homogenous Resource

Here we assume the resource can be accessed at a single location or via a single process or approach; in other words individuals do not need to try to outcompete each other by searching for least exploited niches.

Suppose the resource is plentiful in comparison to the community demands so that each individual can satisfy his wishes. This is a trivial case to analyze: whatever behavior or strategy an individual employs will be successful. In the rest of the paper we call this a 'Promised Land' scenario. Now let's imagine the resource is scarce in comparison to the community demands so that, if shared equally, no one can satisfy his wishes. Obviously competition will occur and most individuals, if not all of them, will be left unsatisfied. If no additional regulation is in place and if the resource is not fully renewable, the resource will be over-exploited and most likely crash leaving no further resource for future utilization. In this situation, despite the inevitable outcome is obvious to everyone, no structure is in place to limit exploitation since it is still convenient to each individual to extract the little resource available rather than seeing it exploited by others: competition is inevitable, collaboration uneconomical and disaster unavoidable. This scenario is known in the literature as 'tragedy of the commons' (Hardin, 1968) and has attracted attention for a several decades with regards to global exploitation of world resources (Ostrom, 1990; Sethi & Somanathan, 1996; Batten, 2007).

Now let's assume some individuals decide to address the doom scenario and impose constraints on the resource exploitation. This may take the form of an exploitation tax or a ban on some exploitation gear. Alternatively, some individuals may develop a very effective exploitation tool but prevent others from using it by the use of patents.

In other words an individual may try to limit resource access to others and, in order to account for some sort of self-regulation, such resource limitation comes at a cost to the individual. This may work initially, but should the use of such constraints be itself uncontrolled (for example every individual may impose some ban on others or tax grow to be excessive) then it may become impossible or not economical for anyone to access the resource. This will prevent resource crash but also any economic benefit from its use. This scenario is known in the literature as 'tragedy of the anti-commons' (Heller, 1998; Heller & Eisenberg, 1998; Benkler, 2004).

To our knowledge the tragedies of the commons and anti-commons have been mostly analyzed independently. In Brede and Boschetti (2008) we discussed them as two extremes of a continuum determined by the amount of regulation individuals (or a system manager) may decide to impose: for zero regulation we have a tragedy of the commons and for extreme regulation an anti-commons. The behavior at the transition between these extremes and the implications for a management team which attempts to maximise resource exploitation are insightful and will be discussed below.

Distributed Resource

So far we assumed that the resource can be accessed at a single location, that is, individuals do not need to choose where to access the resource. The dynamics change considerably if the resource is distributed so that a choice needs to be taken. Here for distribution we intend either a spatial division (fishing in two separate zones, farm in two separate land allocations, choose in which restaurant to dine) or virtual division (deciding which market niches to target).

Let's assume the resource can be accessed at two locations (extension to more locations is trivial) and the individuals have the same exploitation capacity so they need to share equally the amount of resource at the location they chose to target. In this setting, it is convenient for an individual to target the least exploited location, that is the location chosen by the minority of the individuals: the resource will then be shared with fewer people and the return will be higher. This scenario is known in the literature as 'minority game' (Arthur, 1994; Challet & Zhang, 1997; Zhang, 1999) since the 'minority wins'.

As for the commons and anti-commons situation, the 'minority wins' setting has mostly been studied in isolation, though this does need not be so. Suppose we have a resource equally distributed between two locations (each location has 50% of the resource); now suppose we smoothly reduce the amount of resource at one location (location A, say). While the resource at location A diminishes, it becomes less profitable for individuals to target it, even if they are in the 'minority'. When the resource at location A reaches zero, all individuals need to target location B since they have no alternative choice, however the amount of resource (50% of the original amount) is insufficient to satisfy all requests and we end up in a 'tragedy of the commons'.

Summarizing this section, we described 4 scenarios:

1. A 'Promised Land' where nothing can go wrong since there is enough resource for everyone whatever strategy individuals adopt;
2. A 'tragedy of the commons' where there is not enough resource and still individuals have no options but overexploit such resource;
3. An 'anti-commons' in which over-regulation prevents anyone from exploiting a healthy resource, and;
4. A 'minority wins' situation in which individuals try to target the least exploited among several resources.

In the following we study the smooth transition among these 4 scenarios, analyze the likely community behaviors, their likely outcomes and management implications.

Scenario Outcomes

From Commons to Anti-commons—Amount of Regulation

As we have seen, communities face a 'tragedy of the commons' or 'anti-commons' when resources are scarce compared to the potential demand. Two types of regulations naturally arise when a decision to intervene and manage a resource is taken: de-centralised regulations, which are implemented and controlled by the individuals and centralised regulations, imposed and managed by a resource administrator or community representative. Here we discuss and summarize the analytical and numerical work we carried out in (Brede & Boschetti, 2008), to which we refer the reader for further details.

Let's first address a de-centralised regulation and let's suppose it is implemented by an individual's limiting the use of certain strategies or tools (via patents for instance) or by lobbying the resource management body for certain kinds of interventions. The global amount of regulation can thus be quantified by the number of individuals imposing the regulation. Its effect on the community is described with the aid of Figure 1 which gives a sketch of the relation between the amount of regulation (X axis) and the income (Y axis) arising from the resource exploitation (for precise numerical plots, see Brede & Boschetti, 2008).

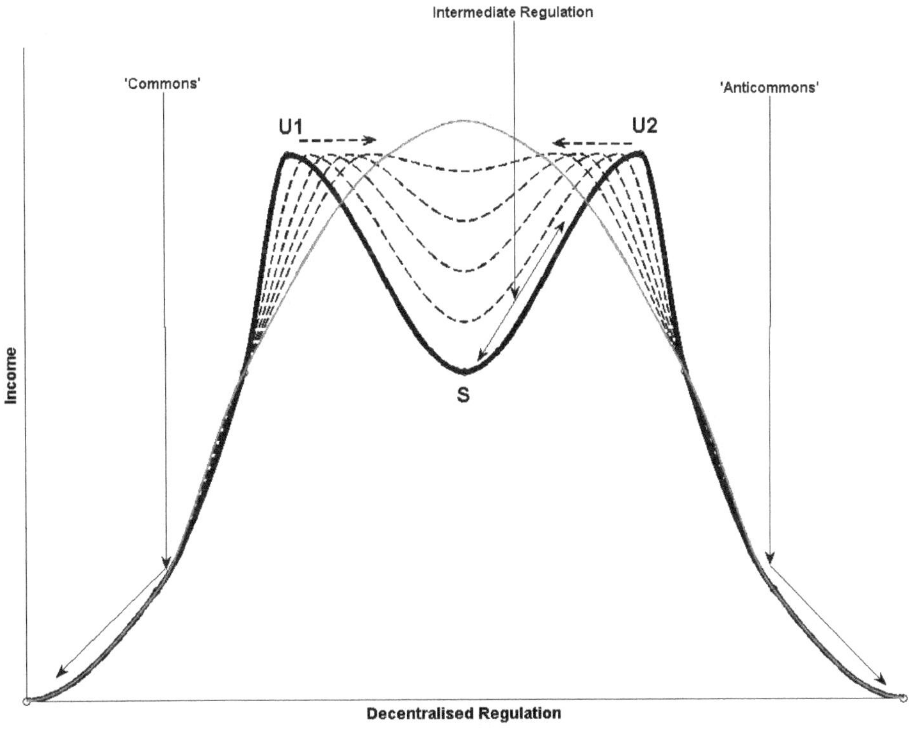

Figure 1 *Transition between 'commons' and anticommons' as a function of amount of regulation.*

In the figure the black thick line represents the solution surface of the dynamical process and the system will naturally tend to reach its stable points. We can envisage a number of options:

1. When regulation is very low (left-hand-side of the unstable point U1) the resource is overexploited, only few people try to impose a regulation, such regulation does not pay off and is soon abandoned. This leads to a crash and no income for the community in the long run (tragedy of the commons);
2. When regulation is very high (right-hand-side of the unstable point U2) many people try to impose regulations, the few who don't lose out and soon will join the regulating majority until the point at which global regulation effectively prevents (on either legal or economic basis) resource exploitation and no income is possible (anti-commons);
3. When the regulation is at some intermediate level (between points U1 and U2) level a number of options are possible:
    a. The community may converge towards the stable point S: here slight variations in regulation amount, resource abundance or economic value impose only local perturbations to the system which will then return to S; the economic outcome to the community is however not optimized.
    b. Under economic pressure the community may strive for better outcomes at points U1 and U2; these points however are unstable and small and unpredictable oscillations in resource abundance or value could result in the system falling into a tragedy of the commons (from point U1) or anti-commons (from point U2).

Let's now analyses centralised regulations. These can be represented by the cost placed on individuals to exploit a resource (a tax) or to apply for a decentralised regulation (the cost of lobbying) or the efficacy of the regulation itself (how extensively it is applied, how compliance is ensured, etc.). Alternatively, the role of this centralised regulation can be interpreted as a management attempt to optimize the overall economic return (a corporation, for example). In our work, we modelled the amount of centralised regulations as an increase in the individuals' cost of imposing regulation. This effectively changes the shape of the solution surface of the dynamical process as depicted in Figure 1 by the dashed lines. As we can see, for an increasing amount of centralised regulation points U1 and U2 get progressively closer (dashed arrows) which in turn makes point S less and less stable. Although the details of such behavior are beyond the scope of this paper (see Brede & Boschetti, 2008) it is eventually possible to reach a solution surface like the red line in Figure 1, for which the point S disappears altogether and the only likely outcomes from resource exploitation are a tragedy of the common or an anti-commons, with no viable alternatives.

The message for these experiments can be summarized as follows:

1. De-centralised regulation can, under certain circumstances, prevent both 'tragedies';
2. Centralised intervention can help the de-centralised regulation adjust towards stable configurations, and;
3. However, the need to optimize for economic return is likely to push the system into a configuration which inevitably leads to either the tragedy of the commons or anti-commons.

As is well known in the field of sustainable and renewable resource management, strict economic optimization and resource sustainability rarely walk hand and hand.

From Abundance to Scarcity in Distributed Resources—The Concept of Impact

The dynamics of the exploitation of a distributed resource is even more counter-intuitive than for a homogenous resource, due to a self-referential feedback loop resulting in a complex behavior. As we discussed above, in this setting the minority wins: to be competitive an individual needs to belong to the minority. However, a competitive strategy tends to be copied in any evolutionary (=competitive) dynamics. As soon as the competitive strategy is copied one too many times (it is adopted by one too many individuals) an individual following it will belong to the majority, and will thus lose its competitive advantage and will become a losing strategy. In other words, if we find a niche market which provides wealth because unexploited, it is better not to advertise the discovery or else others will try to invade the niche making it no longer unexploited. Curiously, other niches, previously over-exploited, may be abandoned by the very competitors who switched niche and may suddenly become under-exploited and thus attractive.

The obvious consequence of this phenomenon is a continuous oscillation with strategies losing and gaining competitive edge and resource locations gaining and losing attractiveness. These oscillations never settle, even in the very long run. Important for our analysis, the amplitude of these oscillations represents lost income (Boschetti, 2007; Boschetti & Brede, 2008). Suppose the resource is spread equally between two locations. The best resource exploitation approach is represented by individuals splitting their effort equally among both resources: any departure from this (=oscillation) results in income loss either because of unexploited resource or because of increased exploitation cost. Ample literature shows that this income loss will not be avoided as long as individuals try to optimize their own personal gain (Arthur, 1994; Challet & Zhang, 1997; Zhang, 1999).

A 'solution' to this dilemma was developed in the field of computer science and is surprisingly simple (Wolpert *et al.*, 2000; Wolpert & Tumer, 2001; Wolpert *et al.*, 2004); its application to renewable resource exploitation and management is discussed in (Boschetti, 2007; Brede & De Vries, 2008). The idea is to ask individuals neither to maximise their own personal gain (= not to be selfish), nor to concern only about the good of the community (=not to be naively unselfish), rather somewhere in between: individuals should try to maximise the impact their action has on the community behavior. This is 'measured' as the difference in income between what the community gains minus what the community *would* have gained *had* that specific individual *not* participated in the community action. In other words an action which results in an individual gain but no community gain is penalized. This is very much in line with the attitude towards rewarding employees based on the demonstrated benefits they provide to a company's performance. The ethical concerns for and against this approach are significant, but beyond the scope of this work.

The 'solution' discussed above goes under the name of Collective Intelligence and relies on a quantity which can not be measured: the consequence of an action (absence of an individual) which did not occur. However, under certain circumstances an approximation can be computed and its numerical performance is surprisingly good (Boschetti, 2007; Brede & De Vries, 2008).

Returning to the distributed resource problem, we want to study what exploitation strategy among the following is most effective for an individual:

1. It may decide not to put any effort in choosing a strategy and simply target the resource at a random location;
2. It can decide to act selfish and always target the location which appears to provide the most profitable return, and;
3. Or it can follow a Collective Intelligence approach and try to maximise his impact on the community income.

Under an evolutionary economics framework, we also assume that the most profitable of the three above mentioned strategies will thrive and spread in the population[1]. The results are summarized in Figure 2 (for more details see Boschetti & Brede, 2008; Brede & De Vries, 2008).

The X axis shows the global amount of resource decreasing rightward and the Y axis shows the average economic return per individual. The leftmost part of the plot represents the 'Promised Land' in which abundant income is available for anyone under any circumstance; in this setting there is no reason to strive to develop the best strategy, even a random selection of a location will deliver as much as an individual can collect ('Everything goes' in the figure). Moving rightward the resource decreases slowly; in this setting the collective intelligence approach prevents resource waste by avoiding overcrowding single

1. Notice that in this context by strategy we do not refer to the selection of a resource, but rather to the criterion used to select the resource.

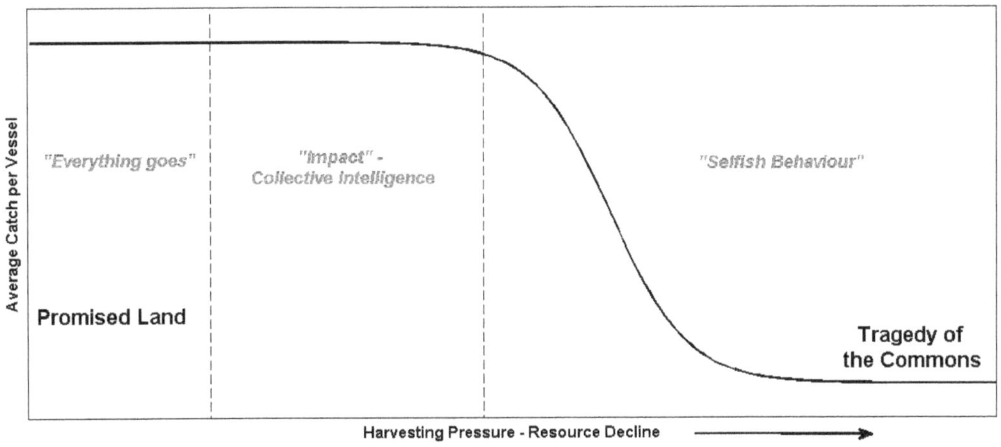

Figure 2 *Transition between a Collective-Intelligence-dominated to a selfish-behaviour-dominated state as a function of resource decline*

locations ('Impact—Collective Intelligence' in the figure). When the resource decreases further, three events occur simultaneously: a) the collective intelligence approach fails to be profitable, b) individuals start to act selfishly ('Selfish behavior in the figure) and c) the resource declines abruptly towards a complete crash past which no income is possible: once again the tragedy of the commons has been reached. Transitions of this kind are typical in the dynamics of distributed renewable resources.

The abrupt resource decline defines a sharp boundary between two different regimes which deserves further analysis: why is accounting for an action's impact rather than acting selfishly profitable before the boundary? And why does this approach fail to succeed past the boundary?

When the resource is relatively abundant, the collective intelligence prevents resource waste, this leads to the community increasing its total income by spreading its effort proportionally to the available resource and as a result the average income of an individual also increases. By accounting for his role within the community an individual indirectly provides a benefit to the overall community as well as to himself: his strategy is not selfish, but nevertheless competitive and optimal.

Things change abruptly when the available resource falls below a certain threshold. When resource is scarce, all available resource is exploited by the community and the individuals have to share the little resource available. Whatever choice an individual takes, whether effective or not, will not change the overall community income, since all remaining resource will be exploited anyway by his competitors; in other words his action can not have any impact on the overall community income. In this setting, following the collective intelligence is equivalent to acting randomly since no information is available on how to influence the community outcome. All an individual can do is to act for his own interest disregarding global outcomes; the result is the tragedy of the commons (Boschetti & Brede, 2008).

The threshold of abrupt decline is also very important for centralised management as we discuss in detail in (Boschetti & Brede, 2008). First, incentives based on an individual's impact on the community can work only when such impact can occur; as we have seen, under considerable hardship this is not possible. Under more fortunate conditions a management body may need to ensure that the proper initiatives are in place to make impact achievable in practise or else this potentially useful approach will not succeed. Second individuals need to be able to evaluate such impact; in very large organizations this may not be possible and extra effort should be placed in providing such information to individuals. Finally, if the management was able to monitor the number of individuals who successfully account for their impact and the ones who merely act selfish, it could in principle obtain some feedback on the resource status: the more the successful individuals who follows a collective intelligence approach, the healthier and more abundant the resource is likely to be. For more details about these conclusions we refer the reader to (Boschetti & Brede, 2008), where the effect of centralised versus decentralised management is also discussed.

Discussion and Conclusions

Resource depletion by selfish individuals (tragedy of the commons), wasted opportunity due to over protective players (tragedy of the anti-commons), crowd following (majority wins) and competition for niches (minority wins) are settings we commonly encounter in our social, professional and business life. The economics and game theoretical community have studied these scenarios carefully but most often in isolation. Our contribution is to show that they all can be seen as extremes of a continuous resource exploitation problem; as so often occurs in complexity science, the most interesting, challenging and counter-intuitive results are found at the transition between 'pure' scenarios.

We summarize our results graphically with the help of Figure 3. To keep the figure simple we limit ourselves to 3 dimensions which are represented by axis X, resource distribution (single or multiple locations), Y, amount of available resource and Z, amount of regulation (here intended both as centralised and de-centralised).

The blue box (low values of the X and Y axis) represents a scarce homogenous resource; in this scenario very little can be done if individuals do not have an external incentive not to exploit the resource, in which case the 'tragedy of the commons' may be unavoidable[2]. For high values of the Y axis, the red box of

2. Several authors suggest that social factors usual come into place to prevent the full extent of a tragedy of the commons to occur in real situations. The historical analysis of whether

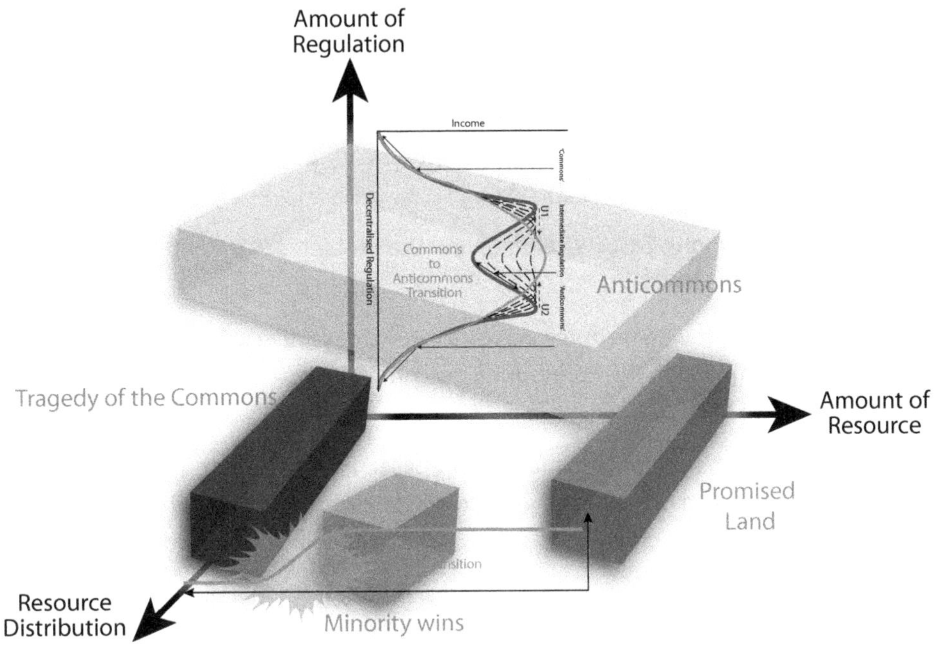

Figure 3 *Summary of the scenarios discussed and their transitions as a function of resource distribution, resource amount and amount of regulation.*

plentiful resource depicts the 'Promised Land' scenario in which nothing can go wrong. The yellow box of maximum regulation represents the 'tragedy on the anti-commons'. Parallel to the Z axis (that is for change of regulation and fixed amount and distribution of resource) we overlaid the plot from Figure 1, which represents the transition from tragedy of the commons to anti-commons. For distributed resources (X axis >1) the green box represents the 'minority wins' scenario and the plot taken from Figure 2 shows the transition for a collective intelligence dominated strategy to a selfish one.

The role of the centralised management (not plotted in the image) lies in recognizing in what scenario the system is. In certain cases, this may be obvious. In others, some monitoring may be devised, for example by measuring the ratio between collective intelligence and selfish agents as suggested in the previous section. Intervention may be possible in the distributed resource settings by favouring impact seeking individuals and providing for the information necessary to detect such impact or in the de-centralised regulation setting, by providing incentive to avoid economic optimization which may decrease system sustainability as discussed in section "From Commons to Anti-commons—amount of regulation". In other scenarios ('tragedy of the commons') intervention may be futile unless it allows to increase the size of the system, for example by providing additional income from external sources which may induce individuals not to overexploit a resource close to exhaustion.

---

these claims apply is beyond the scope of our discussion, which is based on the relatively simple individual behavior discussed above.

References

Arthur, W.B. (1994). "Inductive behavior and bounded rationality," *The American Economic Review*, ISSN 0002-8282, 84: 406-411.

Batten, D. (2007). "Are some human ecosystems self-defeating?" *Environmental Modeling & Software*, ISSN 1364-8152, 22(5): 649-655.

Benkler, Y. (2004). "Commons-based strategies and the problems of patents," *Science*, ISSN 0036-8075, 305: 1110-1.

Boschetti, F. (2007). "Improving resource exploitation via collective intelligence by assessing agents' impact on the community outcome," *Ecological Economics*, ISSN 0921-8009, 63(2-3): 553-562.

Boschetti, F. and Brede, M. (2008). "An information-based adaptive strategy for resource exploitation in competitive scenarios," *Technological Forecasting & Social Change*, ISSN 0040-1625, 76(4): 525-532.

Brede, M. and Boschetti, F. (2008). "Commons and anti-commons in a simple renewable resource harvest model," *Ecological Complexity*, ISSN 1476-945X, 6(1): 56-63.

Brede, M. and De Vries, H. (2008). "Harvesting heterogeneous renewable resources: Uncoordinated, selfish, team-, and community-oriented strategies," submitted to *Environmental Modeling & Software*, ISSN 1364-8152.

Brede, M., Boschetti, F. and McDonald, D. (2008). "Strategies for resource exploitation," *Ecological Complexity*, ISSN 1476-945X, 5(1): 22-29.

Challet, D. and Zhang, Y.C. (1997). "Emergence of cooperation and organization in an evolutionary game," *Physica A*, ISSN 0378-4371, 246(3-4): 407-418.

Gintis, H. (2000). *Game Theory Evolving: A Problem-Centered Introduction to Modeling Strategic Interaction*, ISBN 9780691009438.

Hardin, G. (1968). "The tragedy of the commons," *Science*, ISSN 0036-8075, 162: 1243-1248.

Heller, M.A. (1998). "The tragedy of the Anti-commons: Property in the transition from Marx to markets," *Harvard Law Review*, ISSN 0017-811X, 111: 622-688.

Heller, M.A. and Eisenberg, R.A. (1998). "Can patents deter innovation? The anti-commons in biomedical research," *Science*, ISSN 0036-8075, 280: 698-701.

Ostrom, E. (1990). *Governing the Commons: The Evolution of Institutions for Collective Action*, ISBN 9780521405997.

Sethi, R. and Somanathan, E. (1996). "Evolution of social norms in common property resource use," *American Economic Review*, ISSN 0002-8282, 86(4): 766-788.

Wolpert, D. and Tumer, K. (2001). "Optimal payoff functions for members of collectives," *Advances in Complex Systems*, ISSN 0219-5259, 4(2/3): 265-279.

Wolpert, D., K. Wheeler and Tumer, K. (2000). "Intelligence for control of distributed dynamical systems," *Europhysics Letters*, ISSN 1286-4854, 49(6).

Wolpert, D., Tumer, K. and Bandari, D. (2004). "Improving search algorithms by using intelligent coordinates," *Physical Review E*, ISSN 1539-3755, 69(1): 017701.

Zhang, Y.C. (1999). "Modeling market mechanism with evolutionary games," *Europhysics News*, ISSN 1432-1092, 29: 51-53.

Fabio Boschetti is as an applied mathematician. He has worked on geoscientific applications for several years and more recently has focussed on Complex System Science and ecological modelling with a view to improve our understanding of how ecosystems and their interaction with human activities can be best modelled.

Markus Brede is a physicist. He has worked on a variety of problems of dynamical systems coupled by networks, but has recently also focussed on ecological modelling of the interplay of resource dynamics and human decisions about resource use.

13. Social Entrepreneurship Effects on the Emergence of Cooperation in Networks

Arianna Dal Forno & Ugo Merlone
Department of Statistics and Applied Mathematics, University of Torino, ITA

Even in simple contexts, the dynamical interaction between agents creates complex features. The presence of agents of change affects dramatically the underlying social structure. Some agents seem to be important in shaping the evolution of interactions: traditionally, these agents have been referred to as leaders; nevertheless, recently scholarly interest has been attracted by social entrepreneurs. Do social leaders and social entrepreneurs act differently? Can a social entrepreneurship culture, one that aims for a large number of social entrepreneurs, be welcomed? This paper presents a model of interaction among agents in a community, and sheds light on the catalytic role that some individuals have on the social structure. The results provide some implications about the role of social entrepreneurs and the differences between social entrepreneurship and leadership.

Introduction

The language of social entrepreneurship may be new, but the phenomenon is not. The interest in it has been rising in recent years, and other forms of social entrepreneurship, beyond that occurring within the nonprofit sector, have also grown. The concept of social entrepreneurship is gaining popularity, but, at the same time, the term has undertaken several meanings, and many authors agree on the fact that this can be confusing (Dees, 2001; Austin *et al.*, 2006; Mair & Martí, 2006).

For this reason, social entrepreneurship is still an emerging area for academic inquiry. Theoretical supports have not been sufficiently explored, and contributions to theory and practice are necessary. In addition, boundaries of social entrepreneurship to other fields of study remain fuzzy. Several contributions have been proposed in order to clarify the topic. Among others, Austin (2006) presents studies of collaborations in social entrepreneurship, such as alliances and networks, hoping for interdisciplinary research. Empirical approaches examine the sociological aspects behind the exploitation of social entrepreneurial opportunities, deriving from the existing entrepreneurship theory of opportunity. Robinson (2006) considers the relationship among three factors: the decision to enter a particular market, the social networks in which entrepreneurs are embedded, and the existing types of institutions which can help the development of the initiative. Considering the integration of sustainability and the environment, Clifford *et al.* (2006) suggest that successes related to the mission-driven values and ideals and to creating networks of mutual benefiting stakeholders.

Following Austin *et al.* (2006), the social entrepreneur must focus on building a network of contacts, developing the skills to manage the different relationships in this network effectively. Furthermore, networking across organizational boundaries seems to be essential, because the goals of creating social value do not imply that value can be captured within boundaries. An interesting case which emphasizes this aspect is studied in Rhodes and Donnelly-Cox (2008).

We do not intend to put much emphasis on social entrepreneurs as individuals, focusing on personality traits that may contribute to entrepreneurial success. Rather, we are interested in what social entrepreneurs do; in fact, it has been already observed that the right question to ask is not "who the entrepreneur is" (Gartner, 1988). Furthermore, as Light (2006: 50) underlines, the available evidence suggests that success depends less upon personality than on teachable skills. According to Light's definition, social entrepreneurships "can also come from small groups or groups of individuals, organizations, networks, or even communities that band together to create pattern-breaking change". Moving away from *who* becomes an entrepreneur to *what* they seek, the number of social entrepreneurs expands. The level and the intensity of social entrepreneurship can vary greatly: because of continuous changes in circumstances, this activity might pause, stop and restart. With this definition, social entrepreneurship can be found almost everywhere, aiming to include in the list of names for possible study also characters like "sometimes-entrepreneurs" or "on-hold entrepreneurs".

In previous researches (Dal Forno & Merlone, 2007) social leaders were considered. They were characterized mainly by an assigned role which could not change; here we are interested in a more dynamical role. The aim of this chapter is twofold. First we propose a model of interaction where the role of social entrepreneurs can be studied and assessed. This will allow us to propose answers to some questions: Do social leaders and social entrepreneurs act differently? Can a social entrepreneurship culture, one that aims for a large number of social entrepreneurs, be welcomed? Moreover, we analyze how the role of social entrepreneurs differs from that of social leaders, allowing for the presence of "sometimes-entrepreneurs" or "on-hold entrepreneurs" as in Light (2006). In fact, in the model we present, it is possible to observe how their presence affects the dynamics of different linkages and interactions among agents. In this sense, *social entrepreneurs may be interpreted as facilitators in creating agreement on projects between individuals with different opinions*. Similar roles can be found when considering non-profit organizations, cross sector social partnerships, and NGOs (for a discussion, see Mohan, 2002; Peredo & McLean, 2006). Finally, we shed light on the modifications each of the two roles, leader versus social entrepreneur make on the underlying social structure of the community in terms of emerging work groups.

The structure of the paper is as follows. In the next section we present the model of interaction. We will then explain the computational model and discuss the results. The article concludes by summarizing our implications and presenting further research.

The Model

Among the different scholarly contributions on social entrepreneurship, Waddock and Post (1991) considered the catalytic changes on complex social networks enabled by social entrepreneurs. In order to explore these aspects we consider a model of community where individuals may form groups to commit on projects. We are interested in shedding light on the emergence of groups when individuals do not act directly to the group formation process, rather they act on the social network which shapes the community structure. In particular, while in our model all the agents exhibit the same simple behavior in order to form groups, some of them, acting on the social network, indirectly affect the emergence of groups.

Our purpose is to analyze, through a formal model, how group composition evolves as the result of the catalytic influence of the social entrepreneurs on the social network. While our model is general enough to consider also how the individual behavior, in terms of partner selection, and exerted effort may influence the social network and the group composition, in this paper we keep the first aspects simple and restrict our attention to the role of social entrepreneurs as summarized in Figure 1.

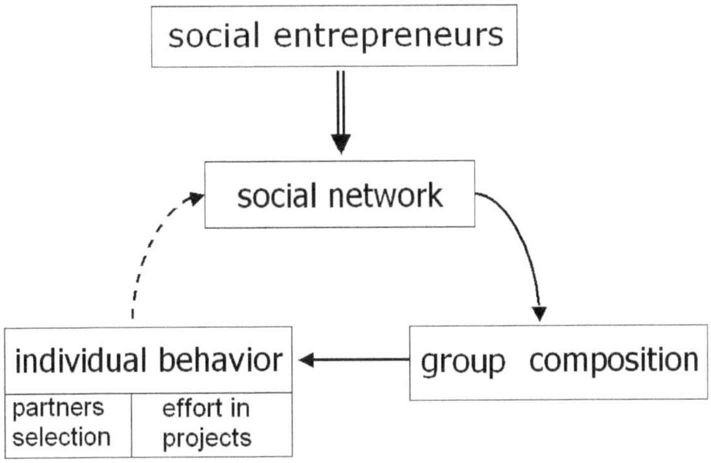

Figure 1 *Group Composition as the Result of Social Network and Social Entrepreneurs*

To identify the role of social entrepreneurs in group formation, we introduce and compare different behavioral rules in the computational model of interaction among the artificial agents we consider. This way we are able to break down the agents' behavior in micro phases. This allows us to assess the relative importance of each of these individual aspects of behavior when leading towards the emergence of some macro behaviors in the artificial society we consider. In particular, the model enables us to study how the percentage of agents taking the role of social entrepreneur influences the emergence of groups.

Interaction Formalization

The community consists of n agents univocally identified by an index $i \in N = \{1,2,\ldots,n\}$. Agents interact forming groups to work on a project in which at most m members can participate. In the artificial simulations, we fixed $m=7$ (for an empirical motivation of this choice see chapter II in Miller and Rice, 1967).

Each agent can choose its partners from a subset $M \subseteq N$ of known people. Knowledge of agents in the community is described using a sociomatrix **K**. Each element k_{ij} of the sociomatrix **K** indicates whether agent i knows agent j: zero indicates that i does not know j; conversely, value one indicates that i knows j. We assume that each agent knows itself; as a consequence, all diagonal entries are set to one. **K** is not necessarily a symmetric $n \times n$ matrix. Agents can participate in at most two projects; in each of them their decision is two-fold:

1. They must specify the designated members of the group, and;
2. They must specify the effort they will exert in each group.

When all participating agents agree on the group composition, this, together with their efforts, constitute an *implemented project* and is univocally determined.

The relation "i works with j in an implemented project" defines a non-dichotomous symmetric matrix **W** where element $w_{ij} \in \{0,1,2\}$ is defined by the number of projects in which agents i and j work together. Matrix **W** defines the project network; when n agents work together on an n-member project, we say they form a *size n clique* since in the graphical representation of matrix **W**, they are depicted as a clique with n nodes.

Within each implemented project, the agents play a public goods game. The efforts of the participants are aggregated and used to produce a collective benefit function f; the output is shared among the members of the group. We denote c_i agent i's cost of effort, and assume that greater effort means greater cost to the agent and increasing marginal cost. The individual benefit of agent i in project p can be formalized as follows:

(1) $$\beta_{i,p} = \frac{f\left(\sum_{j \in T_p(i)} e_j\right)}{n} - c_i(e_i)$$

where e_i is agent *i*'s effort and $T_p(i)$ is the set of partners of agent *i* in project *p*. We assume: 1) There exists a unique level of effort maximizing the agent's individual benefit; 2) when all the agents exert the same effort, the optimal collective benefit increases with the number of members participating in the project. In order to keep the math simple we considered in our experiments the following individual benefit function:

(2) $$\beta_{i,p} = \frac{\left(\sum_{j \in T_p(i)} e_j\right)^2}{n} - e_i^3$$

In this case, it is easy to prove that the socially optimal effort for a *n*-member project is $e_n^s = 2n/3$. With this formulation, when everybody exerts the socially optimal effort, the individual benefit increases with the number of agents in the project.

In Dal Forno and Merlone (2005, 2007) human subjects behavior when interacting in the proposed model is analyzed; there was considerable evidence that free riding was limited in these situations, as the group members exerted optimal effort to the group and did not free ride. This evidence makes the proposed model quite appealing for analyzing social entrepreneurship in artificial societies since, when agents do not free ride, the benefit to the social entrepreneur is the same as to other members in its group. This is consistent with Austin *et al.* (2006), who claim that "social entrepreneurs are seeking to attract resources for the social good, rather than for financial returns".

The Computational Model

As in Dal Forno and Merlone (2007), each turn is divided in three phases. First, in the communication phase, the agents propose and discuss the group composition for projects. Then, some agents—the social entrepreneurs—under certain conditions, try to expand the sociomatrix in order to increase the probability of obtaining a larger consistent project. Finally, agents coordinate on the two best projects that emerged in the discussion and the game is played.

N	size 2 cliques	size 3 cliques	size 4 cliques	size 5 cliques	size 6 cliques	size 7 cliques	Total
7	6	15	20	15	6	1	64
8	7	21	35	35	21	7	127
9	8	28	56	70	56	28	247
10	9	36	84	126	126	84	466
11	10	45	120	210	252	210	848
12	11	55	165	330	462	462	1486
13	12	66	220	495	792	924	2510
14	13	78	286	715	1287	1716	4096
15	14	91	364	1001	2002	3003	6476
16	15	105	455	1365	3003	5005	9949
17	16	120	560	1820	4368	8008	14893
18	17	136	680	2380	6188	12376	21778
19	18	153	816	3060	8568	18564	31180
20	19	171	969	3876	11628	27132	43796

Table 1 *The Number of Possible Projects in a Population Consisting of n Individuals*

Communication and Discussion among Agents

In our model, we implemented the discussion and proposal of projects allowing each agent to propose up to 50 projects and having agents to choose the best feasible common project. This is a sort of brainstorming process (Osborn, 1991) in which agents propose whatever comes into their mind.

This brainstorming process is necessary since, when agents choose their project compositions randomly, the probability of obtaining a seven-member project is very low because they must know each other and the number of possible projects each subject may propose is very high. In fact, assume a population of *n* individuals knowing each other. We denote with *r* the number of group members. If $r \leq n$, then groups of different dimension can be formed, up to *r*-member groups. For the sake of simplicity we call an *r*-member group a *size r clique*. Therefore, the number of projects that can be obtained with exactly *r* individuals (i.e., the number of *r*-cliques) is equal to the binomial coefficient $C_{n-1,r-1}$. In Table 1 we show how many size *r* cliques, with $r = 2, ..., 7$, are possible in a population of *n* individuals.

Comparing the figures in Table 1 allows us to understand how the sociomatrix can play an important role. In fact, in a *n*-agent society, where all the agents know each other, the probability of obtaining size 7 cliques is very low: For example, in the case of a 7-agent society, the probability that the seven-agent project is implemented is $(1/64)^7 = 1/2^{42}$. This makes the coordination of several agents on a single project a complex task, due to the combinatorial explosion of the number of cliques. This factor is related to computational complexity (see Garey & Johnson, 1979). The agents rank their feasible projects assuming that they exert the socially optimal effort.

Individual Diversity in Social Interaction and Social Entrepreneurship

According to the feasible projects that emerged during the discussion among agents, some of them may decide to act on the sociomatrix. While in Dal Forno and Merlone (2007) only some agents were assigned this capability, in this research we assume that, under some conditions, all the agents may act on the sociomatrix. This assumption is consistent to Light (2006), since this author bases some of his arguments on the belief that several social entrepreneurs are either ready to emerge or are already working hard to make a difference.

At each round the density of matrix **W** is computed through an appropriate measure; we recall that the density of a graph is the ratio of the number of lines present, to the maximum possible (Wasserman & Faust, 1999). Each agent has an individual density threshold which is assigned at the beginning of the simulation. The random individual threshold value is uniformly distributed in $[0,b]$ where $b \in [0,1]$. At each turn, when the individual threshold is greater than the current density, the agent assumes the role of a social entrepreneur with probability p. Vice versa, social entrepreneur quit their role whenever matrix **W** density equals at least their individual threshold.

When acting as a social entrepreneur an agent may decide to introduce its acquaintances to others and/or expand the sociomatrix including all the agents known by its acquaintances. The social interactions have effect on the following turn. If we follow Mair and Martí (2006) and other researchers in defining social capital as actual and potential assets embedded in relationships among individuals, we can think of social entrepreneurs as agents who increase the social capital.

At the moment, we do not consider situations in which the entrepreneur may bargain with other agents on the group composition because, even with the simple kinds of social interaction so far implemented, the model is rather interesting.

Game Interaction

According to the results emerged in the communication phase, each agent proposes the two best projects. Here the agents may decide whether to play the socially optimal effort as in the communication phase. Actually, as a consequence of the empirical evidence in human subjects discussed in Dal Forno and Merlone (2005, 2007), we assume agents exerting the social optimal effort. Then benefits are computed and payoffs are given to agents.

Classes of Behavior and Structural Properties of the Sociomatrix

Given the complexity of the model, the behavior of artificial populations depends on four dimensions, namely: the *initial sociomatrix* form, the *group selection behavior*, the *effort determination behavior*, and the *social entrepreneurship behavior*.

The Role of the Initial Sociomatrix

As discussed previously, the number of known agents is crucial in terms of the ability to select groups with several agents. This has been modeled using the dichotomous sociomatrix **K** describing the knowledge relationship between agents. Our agents are assumed to be located on a circle graph where all nodes are interchangeable. At the beginning of each simulation the mutual knowledge between agents may assume different forms. Specifically, the following cases are relevant.

Total Mutual Knowledge

Every agent knows each other. In this case, sociomatrix **K** is unitary. However, due to the combinatorial explosion mentioned above, this initial matrix configuration is not particularly interesting in terms of simulation results. In fact, a complete mutual knowledge affects the efficiency of the discussion phase, not allowing the emergence of clusters in a reasonable number of iterations.

n-Neighbor Knowledge

This graph generalizes the circle graph we discussed above. While in the circle graph each agent knows just one agent on each side, in the *n*-neighbor knowledge graph each agent knows *n* neighbors on each side. For 1-neighbor knowledge, that is the circle graph, only size 2 cliques are possible, while, in order to have size 7 cliques, we need 6-neighbor knowledge. In this case, each agent knows 13 individuals.

Previously Observed Sociomatrix

Initial sociomatrix may be any previously saved one. For example, it is possible to start with one of the final sociomatrices that emerged during other simulations.

Group Selection Behavior

Since, as we saw previously, the combinatorial aspects of the group selection process must be carefully considered in order to have agents converging towards large projects, we developed different approaches in modeling this aspect of behavior. The most interesting were:

1. Considering the project with the largest number of members and expanding it by adding one more new agent;
2. Considering the two best projects and expanding the second one, either adding one more subject or proposing a new project with at least one more agent than the second best project.

The behavioral rules—allowing agents to keep their best projects and to expand them—proved to be extremely important in the simulations. These kinds of rules allowed, for example, the emergence of strong connections between agents.

Effort Selection Behavior

The behavior we implemented in terms of effort selection consists uniquely in playing the socially optimal effort. This behavior can be justified recalling the structure of the game the participants are called to play. In fact, since the game bears the structure of a public goods game, effort selections behavior can be interpreted as the strategy of a multiplayer Prisoner Dilemma, where playing the socially optimal effort stands for "cooperating". In the experiments we observed (Dal Forno & Merlone, 2007) the human subjects almost always exerted the socially optimal effort. This can be explained by assuming that players' behavior is driven by considerations of fairness and equity (e.g., Fehr & Schmidt, 1999).

Social Entrepreneurship Behavior

When modeling social entrepreneurship we are interested in considering quite simple behaviors. Assuming that agents are located on a circle graph, all the agents are interchangeable and know just their closest neighbors; in this case only size 2 clique projects are possible. As mentioned above, *social entrepreneurs* may wish to act on the social network in order to allow the selection of larger projects. Specifically, for these agents we considered several actions in order to expand the sociomatrix. The most interesting were:

- Provided that the social entrepreneur knows less than thirteen agents, when the first best project is not a size 7 clique or, the first best project is a size 7 clique but the second project is not, then it introduces each other to all the agents it knows;
- When the social entrepreneur knows less than seven agents and the best project is not a size 7 clique or, the agent knows less than eight agents and the best project is a size 7 clique but the second best project is not, the social entrepreneur expands the vector of its known agents in order to include all the agents that in the sociomatrix have a geodesic distance smaller than three; in a friendship relation this would simply mean that "the friends of my friends become my friends".

These combined procedures were the most effective for the emergence of connected clusters similar to those we observed in the human subjects experiment. In fact, the social entrepreneurs expand the sociomatrix only when it is necessary to improve the size of projects. While the first procedure acts on the other agents, the second one increases the number of agents the social entrepreneur may contact. Given the combinatorial intractability of the project selection process, it is important to balance carefully the sociomatrix expansion with the number of members of implemented projects. Since in our model we do not consider explicitly communication, what we obtain seems to be a good approximation of these behaviors.

Results

The computational model has been implemented in C++; we obtained an interactive platform where the composition of the community can be chosen together with different parameters. We fixed a community of 63 agents and compared the results when considering different social entrepreneur distributions in the population. We assumed that agents, observing a density value for matrix **W** smaller than their individual threshold, acted as social entrepreneurs with probability $p = 1$. Our interest was directed to analyze how the distribution of social entrepreneurs in the community could affect the cooperation among individuals.

With our approach we measure the effect of social entrepreneurship in terms of effective projects. In other words, social entrepreneurs act as catalysts on the social network in order to enable the formation of larger groups. To keep things simple, for the first set of simulations we considered the 1-neighbor knowledge as the initial sociomatrix; this choice allows at most size 2 cliques in the circle graph; when considering our assumption on the number of project to be implemented, this results in each agent implementing two 2-agent projects as depicted in Figure 2.

We consider different series of simulations in a fixed size 63 agents population where agents play with socially optimal effort. The artificial population data are the mean of five independent replications; we decided not to consider many more replications since the results were quite stable and the time for each replication was quite long (about 5 hours on a Pentium 4 CPU 2.80 GHz, Ram 512 MB). Specifically, we observe evolution at turns 1, 2, 3, 4, 5, 6, 10, 20, 30, 40, 50, 60, 100, 200, 300, 400, 500, 600, 1000, 2000, 3000, 4000, 5000, 6000 and 10000. This time choice allows observation of both the short and long term evolution.

In order to measure the role of social entrepreneurs we considered different values for the upper bound b of the uniform distribution which determines the individual threshold. The following values of b were considered: .25, .50, .75, and 1.00. As it concerns the value .25 some comments are in order. In

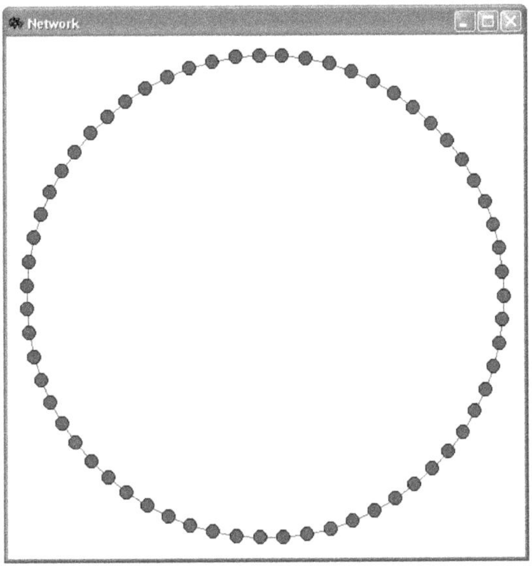

Figure 2 *Project Emergence with 1-neighbor Knowledge Initial Sociomatrix.*

fact, observe that when considering a number of agents which is a multiple of *m*, the density of the initial projects depicted in Figure 2 is greater than .25; this means that having agent with individual threshold level distributed in [0,.25] results in no agent assuming the social entrepreneur role. The consequence is that in a circle graph population, agents keep on proposing and playing size 2 cliques forever, since no one acts on the sociomatrix.

A first series of simulations analyzes the optimal distribution of social entrepreneurs and compare their effect to the role of social leader described in Dal Forno and Merlone (2007). The results of this comparison may help to understand how social entrepreneurs differ from social leaders. In fact, while both social entrepreneurs and social leaders expand the sociomatrix when it is necessary to improve the size of projects, the latter are placed every six agents, and keep their role throughout the simulation since they do not have any internal threshold. The results of these first series are depicted in Figure 3.

We can observe that, among social entrepreneurs the most effective threshold upper bound seems to be .50; with higher values, the number of social entrepreneurs makes sociomatrix **K** expansion too fast, and the combinatorial complexity mentioned above hampers the formation of groups. Finally, when the threshold upper bound value equals .25 no agent acts as a social entrepreneur.

The comparison with social leader provides several insights. In fact, it must be observed that, while in terms of number of links, social leaders are more effective, their role is exogenously assigned and they are strategically positioned

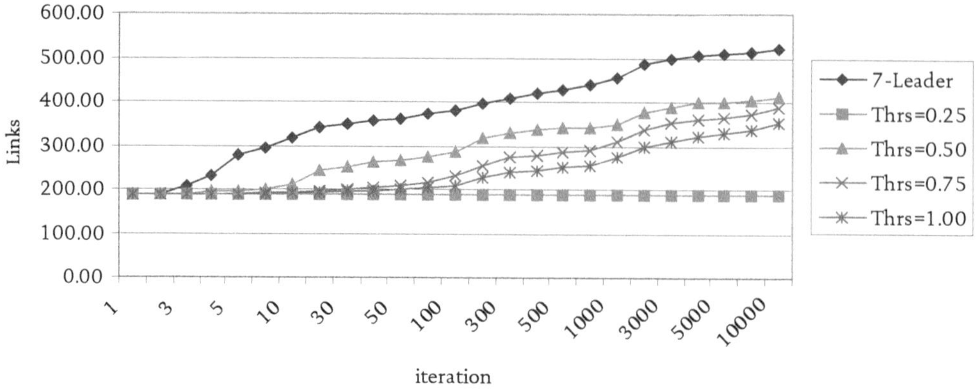

Figure 3 *Number of Links in Different Populations*

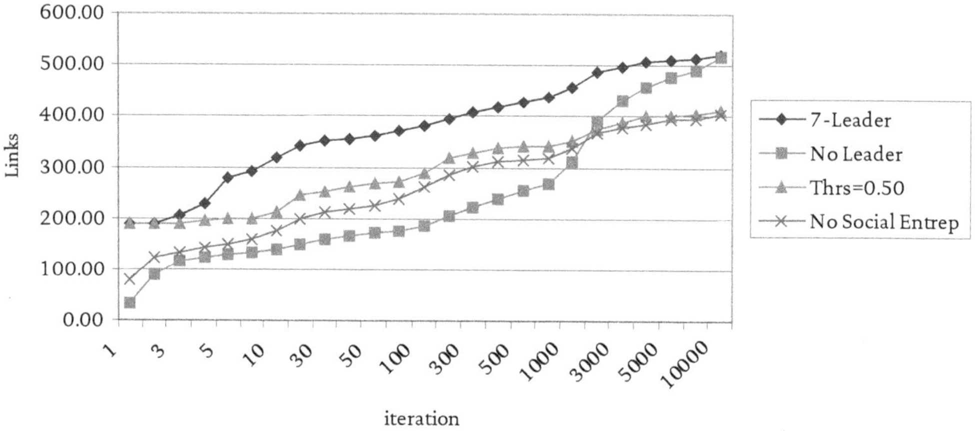

Figure 4 *Number of links when comparing the Differences between Social Entrepreneurs and Social Leaders*

every other 6 agents. This, on one hand would suggest that having agents with assigned role would be more effective, but on the other would steer away from the motivations of social entrepreneurs.

In order to better understand what makes the social entrepreneurs peculiar, we performed a second series of simulations where we took as initial sociomatrix, instead of the circle graph, the final sociomatrix obtained at the end of the previous simulations. These series were run with neither social leaders, nor social entrepreneurs, and considering the final sociomatrices obtained in the previous simulations; the results are presented in Figure 4. In this figure we can observe the evolution of sociomatrix **W** when the initial knowledge matrix **K** is the same as the one obtained at the end of the 7-leader and at the end of the .50 threshold social entrepreneurs simulations depicted in Figure 3

This approach allows for incorporating one of the effects of having social entrepreneurs and leaders in the population: in fact, the sociomatrix is already expanded. While theoretically these sociomatrices are necessary to have the same simulation outcomes we obtained when social entrepreneurs and social leaders were present, the results are quite different. In fact, when considering social entrepreneurs the two kinds of sociomatrices allows for the emergence of a different number of links, what really matters is that, when considering social entrepreneurship, the evolution in terms of links is quite similar. On the contrary, when considering social leaders this difference remains wide for a large number of iterations. This may indicate that social entrepreneurs differ starkly from social leaders on their actions on the sociomatrix. While for social leaders the results are similar only on the long run, as it concerns social entrepreneurs, the results become closer earlier. This may be interpreted in the sense that social entrepreneurs' actions are absorbed more deeply in the sociomatrix when compared to actions performed by social leaders.

The results we have presented so far are in aggregate form; in the following we present the same comparison when considering one single simulation for each kind of the simulations depicted in Figure 4.

Comparing Figures 5 and 6, we can see that, on the first turns, the situation is completely different as the result of the initial sociomatrix. While with a circle initial sociomatrix, the agents start implementing projects with their neighbors; in the other case some agents in the very first iterations remain isolated due to the large number of agents they already know. Furthermore, it can be observed that later on, in both cases, the same group structures emerge even if in the case of no social entrepreneurs the network evolution is slightly slower.

By contrast, when comparing Figures 7 and 8, we can see that differences are not limited to the first iterations. In fact, in case of no social leader the network evolution is much slower: in fact, in Figure 8 the working projects are much sparser and densely connected working groups emerge later on in the evolution.

Finally, comparing Figures 6 and 8, we can observe that the change the social entrepreneurs induced on the sociomatrix allowed for a faster emergence of clusters. This confirms the differences between social entrepreneurs and social leaders, in terms of effects on the sociomatrices; also in this case, we could say that changes inducted by social entrepreneurs on the sociomatrix are more deeply embedded and retained.

Conclusions and Further Research

In this chapter we presented a model of interaction among individuals in a community, where the network of interactions may allow for the emergence of cooperative projects. The model consists of two coupled networks modeling respectively knowledge and cooperation among individuals. We showed how the combinatorial explosion of the number of possibilities on which agents

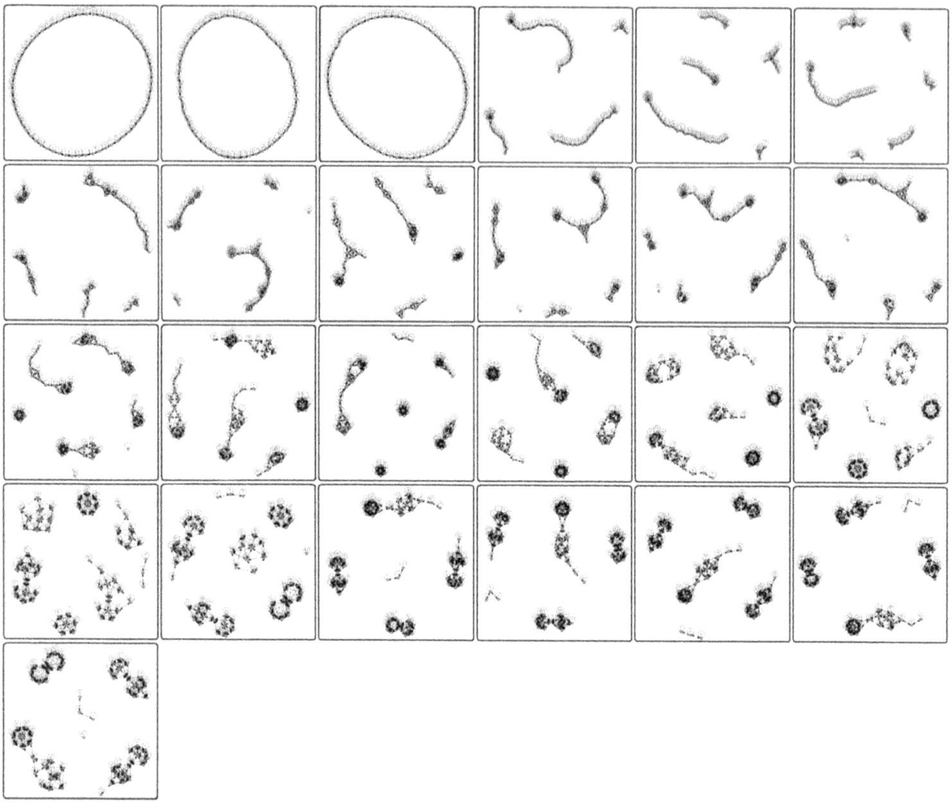

Figure 5 *Project Network Evolution in Populations with Threshold .50 Social Entrepreneurs and Circle Graph Initial Sociomatrix, at Iterations 1-6, 10-60, 100-600, 1000-6000, 10000.*

may cooperate makes the coordination a complex process. Our model sheds light on the role of leadership in complex systems by examining two different roles: the first one -the social entrepreneur—shared and mediating internal drive with community needs and the second one—the social leader (Dal Forno & Merlone, 2007)—more individualistic and intrinsic.

Two aspects of complexity are intrinsic in the model: the large number of group combinations, and the dynamical interactions among agents. In order to have the best outcome—i.e., the largest number of wide groups which, as illustrated in Figures 2 and 3, is measured in terms of number of links in the network—the action of agents of change is needed. These particular agents must be able to help the realization of effective groups enriching the net by exploiting old connections and looking for new ones between members of the community. But, at the same time, they must be able to state the pace of a balanced expansion of the social matrix. In other words, they tame the combinatorial explosion while fostering mutual knowledge among agents, making possible projects otherwise impossible. In fact, as we saw when discussing Table 1, as the number of connections increases the probability of agreeing on 7-agent projects decreases exponentially. Social entrepreneurs, by acting on the knowledge

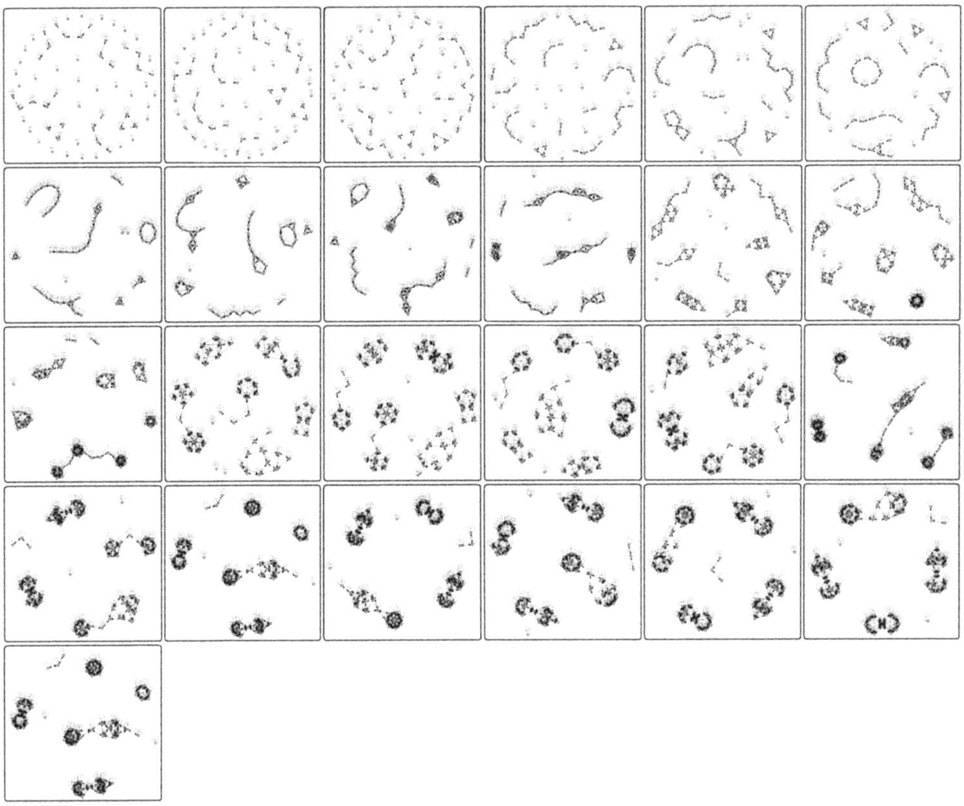

Figure 6 *Project Network Evolution in Populations with no Social Entrepreneurs and Previous Case Final Sociomatrix as the Initial One.*

matrix, catalytically allow the emergence of cooperation projects. In our model, social entrepreneurs do not suggest projects; rather they act on the social network and may help the emergence of projects in the discussion phase. Any member of the community can be a social entrepreneur, as long as they believe it is appropriate.

We compared the effectiveness of social entrepreneurs in terms of number of links in the community. According to our findings the number of social entrepreneurs and their relative location is extremely important. In this sense, the social entrepreneurship role is necessary for a sort of implicit coordination of agents. Different levels of social entrepreneurship diffusion in the community may have different effects on the emergence of projects; when the level is low the coordination remains limited, while a too high level of social entrepreneurship, on the contrary, hampers coordination. Only a moderate level of social entrepreneurship is effective to approach the complexity of coordinating on larger projects. This suggests that a culture for a moderate social entrepreneurship would be effective in promoting projects; deciding who should act as a social entrepreneur can be considered a coordination game.

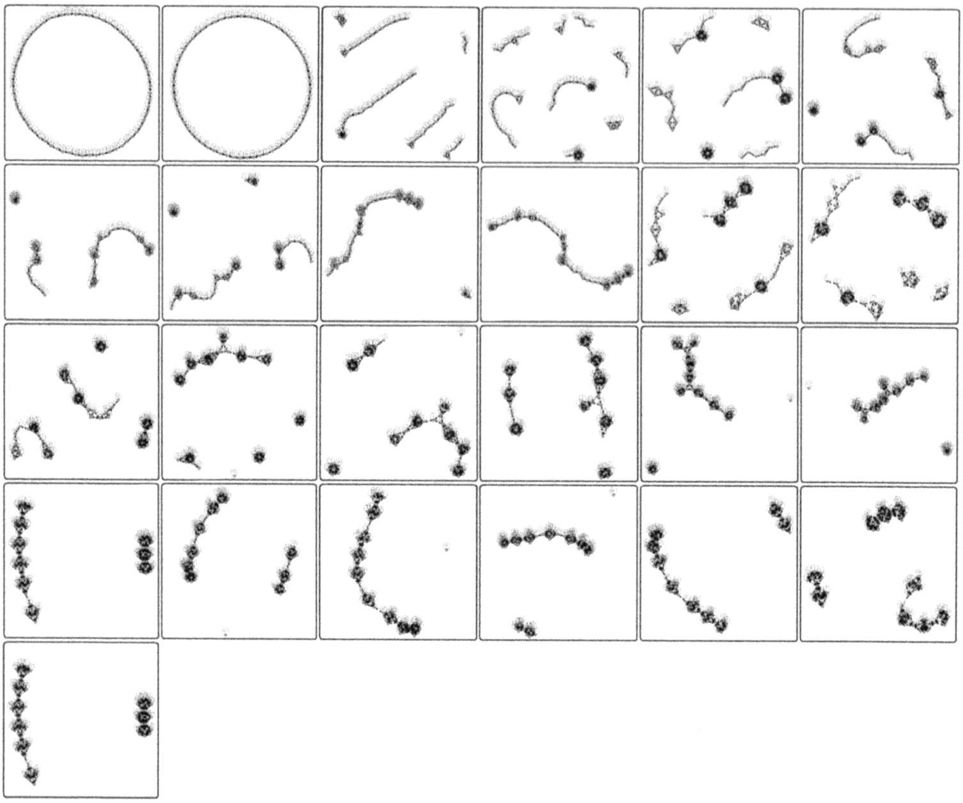

Figure 7 *Project Network Evolution in Populations with 7-interval Social Leader and Circle Graph Initial Sociomatrix.*

Finally, social entrepreneurs were compared to social leaders. As a matter of fact in our model they do not act differently; rather the differences among the two roles are in terms of strategic position, persistence in acting on the knowledge matrix and the fact that, while the level and intensity of social entrepreneurs' action can vary greatly, social leaders are more constant in their actions. This turns to the consequences of Light's (2006) definition of social entrepreneurs, which is more directed to what they do instead of considering their personality traits. While social leaders outperform social entrepreneurs in terms of number of links, their actions on the knowledge matrix are retained at a different level. While the action of social entrepreneurs seems to be more delicate, it seems to be embedded in the knowledge sociomatrix as it can be replicated when active social leaders are removed. On the contrary, while the action of social leaders is more effective, when they are removed the community cannot replicate the evolution in the short run.

In further research it would be interesting to find the most effective threshold in fostering effective cooperation links among agents. Another aspect which would be interesting might be introducing a sort of decay in the knowledge matrix, i.e., to consider the tendency for relationships to weaken and, even-

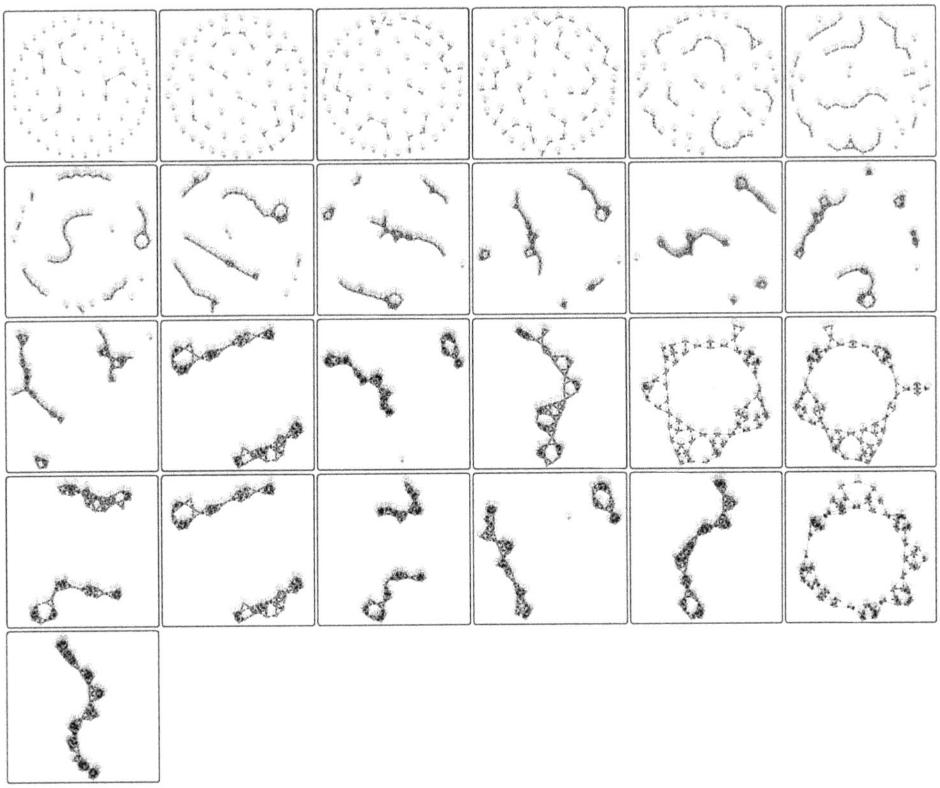

Figure 8 *Project Network Evolution in Populations with no Social Leaders and Previous Case Final Sociomatrix as the Initial One.*

tually, disappear (Burt, 2000). This would help to keep effective the knowledge sociomatrix and would be more realistic in how human networks evolve over time. Finally, since our results showed that too many social entrepreneurs could be counterproductive it would be interesting to model the coordination game implicit in this.

References

Austin, J., Stevenson, H. and Wei-Skillern, J. (2006). "Social and commercial entrepreneurship: Same, different, or both?" *Entrepreneurship: Theory and Practice*, ISSN 1042-2587, 30 (January): 1-22.

Austin, J.E. (2006). "Three avenues for social entrepreneurship research," in J. Mair, J. Robinson and K. Hockerts (eds.), *Social Entrepreneurship*, ISBN 9781403996640, pp. 22-33.

Burt, R.S. (2000). "Decay functions," *Social Networks*, ISSN 0378-8733, 22 (January): 1-28.

Clifford, A. and Dixon, S.E.A. (2006). "Green-Works: A model for combining social and ecological entrepreneurship," in J. Mair, J. Robinson, and K. Hockerts (eds.), *Social Entrepreneurship*, ISBN 9781403996640, pp. 214-234.

Dal Forno, A. and Merlone, U. (2005). " The evolution of coworker networks. A comparison between experimental and computational results," in K.G. Troitzsch (ed.), Representing Social Reality, ISBN 9783934795334, pp. 223-227.

Dal Forno, A. and Merlone, U. (2007). "The evolution of coworkers networks: an experimental and computational approach," in B. Edmonds, C. Hernï¿½ndez Iglesias, and K.G. Troitzsch (eds.), *Social Simulation: Technologies, Advances and New Discoveries*, ISBN 9781599045221, pp. 280-293.

Dees, J.G. (2001). "The meaning of 'social entrepreneurship,'" http://www.caseatduke.org/documents/dees_sedef.pdf.

Fehr, E. and Schmidt, K.M. (1999). "A theory of fairness, competition and cooperation," *Quarterly Journal of Economics*, ISSN 0033-5533, 114 (August): 817-868.

Garey, M.R. and Johnson, D.S. (1979). *Computers and Intractability*, ISBN 9780716710455.

Gartner. W. (1988). "'Who is the entrepreneur?' is the wrong question," *American Journal of Small Business*, ISSN 0363-9428, 12 (October): 11-32.

Light, P.C. (2006). "Reshaping social entrepreneurship," Stanford Social Innovation Review, ISSN 1542-7099, (Fall): 45-51.

Mair, J. and Marti, I. (2006). "Social entrepreneurship research: A source of explanation, prediction, and delight," *Journal of World Business*, ISSN 1090-9516, 41(February): 36-44.

Miller, E.J. and Rice, A.K. (1967). *Systems of Organization: The Control of Task and Sentient Boundaries*, ISBN 9780422729604.

Mohan, G. (2002). "Disappointments of civil society: The politics of NGO intervention in northern Ghana," *Political Geography*, ISSN 0962-6298, 21: 125-154.

Osborn, A. (1991). *Your Creative Power*, ISBN 9781569460559.

Peredo, A.M., and McLean, M. (2006). "Social entrepreneurship: A critical review of the concept," *Journal of World Business*, ISSN 1090-9516, 41: 56-65.

Rhodes, M.L. and Donnelly-Cox, G. (2008). "Social entrepreneurship as a performance landscape: The case of 'Front Line,'" *Emergence: Complexity & Organization*, ISSN 1521-3250, 10(3): 35-50.

Robinson, J.E. (2006). "Navigating social and institutional barriers to markets: How social entrepreneurs identify and evaluate opportunities," in J. Mair, J., Robinson, J. and K. Hockerts (eds.), *Social Entrepreneurship*, ISBN 9781403996640, pp. 95-120.

Waddock, S.A. and Post, J.E. (1991). "Social entrepreneurs and catalytic change," *Public Administration Review*, ISSN 0033-3352, 51(September/October): 393-401.

Wasserman, S. and Faust, K. (1994). *Social Network Analysis*, ISBN 9780521387071.

Arianna Dal Forno, Ph.D. (University of Trieste, 2001) is post-doctoral fellow at the department of Applied Mathematics at University of Turin. Her main research interests are mathematical modeling of organizations and microstructure of market architectures. She has been visiting post-doc at Harvard University. Her most recent publications have been published on Journal of Economic Interaction and Coordination, Journal of Artificial Societies and Social Simulation, Nonlinear Dynamics, Psychology, and Life Sciences, Decisions in Economics and Finance, as well as chapters in books.

Ugo Merlone, Ph.D. (University of Trieste, 1998) is associate professor of Applied Mathematics at University of Turin. His main research interests are mathematical modeling of organizations and bounded rationality agents interactions dynamics. He published several papers in peer reviewed journals, such as *Organization Sciences, Managerial and Decision Economics, Journal of Artificial Societies and Social Simulation, Nonlinear Dynamics, Psychology, and Life Sciences*, as well as chapters in books. He has been visiting scholar at University of Arizona, London School of Economics and Harvard University. Further details are available from http://web.econ.unito.it/merloneugo/.

Practitioner Paper Section

14. From Traditional Knowledge Management in Hierarchical Organizations to a Network Centric Paradigm for a Changing World

Kate Crawford[1], Helen Hasan[1], Leoni Warne[2] & Henry Linger[3]
1 Faculty of Commerce, University of Wollongong, AUS
2 Joint Operations Division, Defence Science and Technology Organization, AUS
3 Faculty of Information Technology, Monash University, AUS

At the beginning of the 21st century, all organizations need to address the continually changing social and economic landscape in which they operate. In this landscape organizations need to be responsive, flexible and agile and acquire the capability to leverage information and use collective knowledge to make appropriate decisions quickly and effectively. The practice of knowledge management allows knowledge workers to participate in dynamic processes that generate and use collective knowledge. However the complexity that arises from a continually changing global environment highlights the need for knowledge management to move in new directions both in practice and theory. This paper proposes approaches to knowledge management that incorporate concepts from complexity theory leading to the adoption of a network-centric paradigm in organizations, complementing or replacing traditional hierarchical bureaucracies.

Introduction

At the beginning of the 21st century, all organizations, whether government, industrial, commercial or civil, need to address the continually changing social and economic landscape, in which they operate. A central theme of this change is the emergence of information and knowledge as major elements of wealth creation processes including the growth of intellectual and social capital (Sveiby, 1997). The current corporate interest in knowledge is based on a realization that emerging economic theories, coupled with social and industrial restructuring, demand a more rigorous approach to the exploitation of knowledge, and knowledge making capabilities, as organizational resources (Drucker, 1998). As knowledge workers, modern employees are participating in dynamic processes that generate and use collective knowledge in a changing organizational landscape (Ivari & Linger, 1999; Pfaff & Hasan, 2007).

The practice of knowledge management (KM) is now distinguished from information systems and information management (Hart & Warne, 2005; HB189,

2004). As encapsulated in the Australian Standard (AS 5037, 2005), KM manifests itself in organizations through a variety of interpretations and implementations depending on the enterprise, the pressures for innovation and the market context. KM must contend with the increasing complexity that comes with the continually changing global environment, and the related need to negotiate, test, refine and share complex responses to the resulting challenges. This signals the need to re-evaluate organizational structures and processes to ensure that they appropriately enable these new activities at all levels. We propose approaches to KM incorporating concepts from Complexity Theory leading to the adoption of a network-centric paradigm in organizations, complementing or replacing traditional hierarchical bureaucracies.

In this paper we depict information as data in any media that is available and may be processed to be interpreted by people and thus potentially inform. Knowledge, on the other hand, can be distinguished as the human capability to interpret information and use it creatively, both individually and cooperatively, to add value to human activities and products. This accords with the Macquarie Dictionary definition of 'social capital' in terms of the investment in institutions, quality relationships and interactions that enhance wealth making processes. Issues raised by these changing environments are:

- Transformations in what constitutes wealth and what is valued;
- The shrinking distances around the world and global competition;
- The compression of time, which increases the tempo of our lives;
- The alteration in the distribution of power, including the power that is accrued by custodians of information.

Traditionally, wealth creation was based on land, capital and labour. Now, information and knowledge are major ingredients (Benkler, 2006). Creating wealth is about adding value by turning these new ingredients into products and exchanging them through open global markets. Information is a resource for management and decision-making. Information and Communications Technologies (ICT) have made information collection, storage, processing, retrieval, analysis and dissemination cheap, affordable and accessible. Zuboff (1989) argued, almost 20 years ago, that the information age carried with it, not liberation, but the threat that management, and perhaps the state, will use technology to invade more and more of the space previously claimed by workers and citizens for unsupervised, discretionary activity. However more contemporary ICT, exemplified by Internet-based applications, has the potential to significantly reduced control over emerging information and the knowledge derived from creative applications of new information (Pfaff & Hasan, 2007).

ICT are changing the space and time dimensions of value-adding activities and patterns of social capital investment by making information available at any physical place, reducing the dependence on location, and compressing process cycle times and decisions (based on information) about such processes (Coakes, 2002). Because information often changes rapidly and can be interpreted more or less effectively according to the knowledge of people using it, wealth creating processes and the opportunities that are associated with them now change more often and more rapidly. In such a changed landscape, organizations need to be responsive, flexible and agile. In other words, they now need to have the capability to leverage available information and to use their collective knowledge to make appropriate decisions quickly and effectively. This need has driven the KM movement in new directions for both practice and theory.

These changes have increased the complexity of organizational landscapes and the intricate processes carried out by people within them to a point where traditional hierarchical organizations no longer provide a platform for effective and efficient management and operations. The lack of success of traditional knowledge management initiatives highlights the need to adopt approaches that are informed by concepts derived from complexity theory, learning, action science and other theoretical positions that explicitly address complexity. An example of such an approach is the knowledge ecosystem concept on which the Australian KM Standard is based. In this paper, we propose a network-centric paradigm for knowledge management in complex organizational landscapes. The paper maps the journey the writers have made in defining this new paradigm. The paradigm is based on their collective case studies, from the military and community organizations and from the writers' reflections on the recursive nature of their own collaborative research as it negotiated its own levels of complexity, emergence, efficacy and tempo.

Complexity: The Dimensions of the Changing World

Using the Cynefin model of complexity (see Figure 1 below) the dimensions of the changed landscape can be characterized by:

- An increased shift towards the *unordered* half of the model;
- A reliance on the detection and leveraging of emergent patterns rather than pre-planning and design;
- Efficacy working comfortably and effectively in this environment;
- An awareness of, and adjustments to, variations in tempo. For example, changing culture is usually a long process whereas other changes happen rapidly.

Problems faced by organizations, whether in operations, structure or processes, can be explained in terms of these dimensions. This section will discuss how these aspects of the landscape shape the issues that confront organizations in the real world.

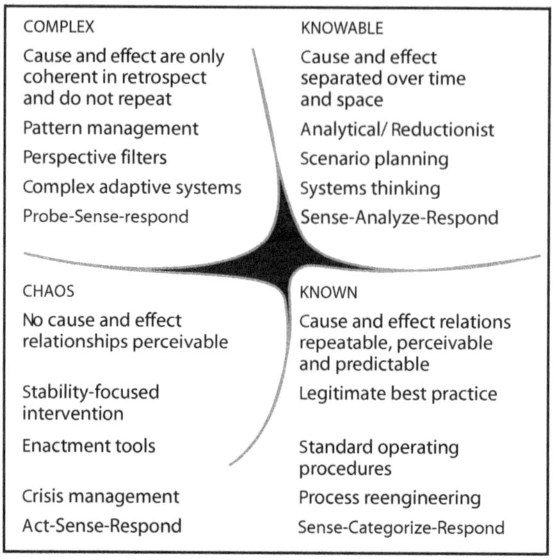

Figure 1 *The sense-making Cynefin framework (after Snowden 2002).*

The Cynefin framework is a knowledge space with five domains which set the context for collective decision making: two domains of order, the known and the knowable with the unordered domains of complexity and chaos. The domain of disorder provides a negotiable space for choices about strategic responses to new situations.

Firstly, the increased emphasis on dealing with complex situations for which established knowledge and practices are ineffective often produces feelings of fear and perceptions of chaos. In the face of these situations (Represented in the bottom left hand domain of the model) there are two common responses:

- Retreat to the established ordered position and reinforce the rules of engagement to avoid chaos and prevent unorder;
- Begin creative learning processes to explore the reasons for the apparent 'chaos' and understand the complex situation and use that understanding to leverage from emergent information and test the validity and effectiveness of response options.

Perceptions of, and chosen responses to, *unorder* are a reflection of working knowledge and interpretations of complex settings, personality differences and comfort levels in ambiguous situations requiring creative responses. The same environment can appear familiar and interesting, or *unordered* and chaotic to different people according to experience and world view. In traditional hierarchical organizations there is often an experience and knowledge gap between senior executives, who generally work in an ordered environment, and their younger knowledge working colleagues. The latter generally work at the periphery of an organization, have more recent qualifications, are more flexible about learning and consulting with each other, and are dealing with *unorder* by informally adding value to current knowledge and processes on a day to day basis.

Snowden (2002) identifies three generations of knowledge management. The first generation, clearly associated with increased ICT capabilities, focussed on timely information provision for decision support. The second generation, triggered by the SECI model (Nonaka, 1994), focussed on the tacit-explicit knowledge conversion as the one process of knowledge creation in organizations. The emerging third generation, associated with Snowden's work at the Cynefin Centre, uses complex adaptive systems theory to create a sense-making model of collective knowledge creation, disruption and utilisation that allows a pragmatic and conceptual alternative to the orthodoxy of scientific management. In the case studies presented below, it appears that first generation KM fitted comfortably into traditional hierarchical bureaucracies whereas maturation to the third generation follows the trend to more organic network-centric configurations where a knowledge ecosystem approach makes sense, as will be explained below.

The defining characteristics of the network-centric organization are flatter hierarchies; decentralised decision-making; greater capacity for tolerance of ambiguity; permeable internal and external boundaries; empowerment of employees; capacity for renewal; self-organizing units, and self-integrating coordination mechanisms (Daft & Lewin, 1993). In such organizations, knowledge is the most strategically important resource and organizational capabilities are the product of distinctive competencies in integrating and applying this knowledge. Thus active participation in communication is the pervasive, underlying force responsible for maintenance and dissemination of strategic capabilities based in knowledge. In this paradigm, there is distributed decision making and self-organization where attractors and barriers replace command and control. Participants are motivated (attracted) to participate with authority and initiative in activities around emergent aspects of the situation and have clearly understood boundaries for such actions. They also have responsibility to share the results of their participation in order to integrate strategic actions across the network (Allee, 2002). Tucker *et al.* (1996) point out that strategic capabilities result from new knowledge creation accomplished through a combination of individuals' tacit and objective knowledge, yet this collection of knowledge must somehow be aggregated and communicated at a collective level (Warne *et al.*, 2004).

In hierarchical organizations faced with rapidly changing contexts, an over emphasis by powerful senior executives on compliance, accountability, control of planning, and reinforcement of rules can slow the tempo of, and reduce, participation in, decision making and can also prevent those at the periphery of an organization from paying attention to emergent patterns and making the necessary adaptations.

In a complex and rapidly changing setting, successful and agile responses depend on capabilities for detection and leveraging of emergent patterns. Research (Whelton & Ballard, 2003) also indicates that these capabilities involve intuition and greater awareness (meta-cognition) of strategic development, purpose and the thinking underpinning new responses. In such a context, successful navigation depends on a non-linear process involving the creation of a mix of imposed limits or barriers and strategically placed attractors to explore, understand and eventually leverage emergent patterns and opportunities. The emergent networked business activity surrounding the You Tube video exchange phenomenon (http://www.youtube.com), including government participation to distribute drug education materials and the recent purchase by Google, is an example of these kinds of responses.

It has become apparent that in the current climate, awareness of change sometimes occurs rapidly and is leading to rapid changes in patterns of interaction, activity, awareness and capability for some participants in the emerging digital culture. There is now a categorisation of workers based on generational differences, known as baby-boomer, X and Y. According to Smola and Sutton (2002) a generation is "an identifiable group that shares birth years, age location and significant life events at critical developmental stages." While such generations share common historical and social life experiences, generational characteristics can be distinct from those purely due to age, maturity, experience and corresponding technical competence. However assumptions of ubiquity of emerging digital culture is particularly strong for young people many of whom have not only grown up with far greater access to information than previous generations, but also with developmentally powerful experiences of managing complex settings (for example, interactive multiplayer games), and capability and wide experience in using multiple media, often simultaneously, for wide ranging communication with different people in multiple networks. There is considerable evidence emerging from the field of neuroscience that such changes in interactive cultural experience change the course of brain development (Prensky, 2001a, 2001b).

The changes in experience, development and emerging values and capabilities also result in different world views. The differences in values and motivation between Generation Y (usually at the periphery of an organization and the Baby Boomers in the senior executive group is now well researched (e.g., Henry, 2007). These differences suggest that each group is likely to have a differing interpretation of and response to complex contexts and related barriers and attractors.

Organizational cultures fulfil needs for stability and often change slowly. However, as mentioned above, the gap between active participants in the emergent activity and those who conduct their lives according to pre-digital conceptions is ever widening. This causes tensions of several kinds. These include:

- A generation gap in digital culture that is acknowledged to be as wide as that between immigrant parents and their children growing up in a new culture (Prensky 2001a, 2001b);
- The tension (already mentioned) between emergent new agile patterns of network centric organizational activity and the well established hierarchical institutional platforms and legal regulation frameworks that have served developed countries for centuries;
- Changing conceptions of human rights and responsibilities (e.g., the emergent privacy debate);
- Changing patterns of commitment and participation as people manage in less stable, more global and rapidly changing landscapes.

The increasingly complex settings for knowledge work, and the changing capabilities, strategies, relationships and outcomes in a more *unordered* world, inevitably have an impact on the KM landscape.

The KM Landscape

The orientation of mainstream knowledge management has been organizational with an essentially top-down implementation that relies on the existing organizational hierarchy. Even if the stated objective is to "flatten the structure", knowledge management initiatives rely on senior "champions" in order to acquire the resources for the initiative and the authority for implementation. Yet knowledge production is explicitly seen as an individual activity. From an organizational perspective, this inherent contradiction is addressed by an emphasis on the sharing of knowledge and the transformation of tacit knowledge to explicit knowledge in parallel to the accumulation of organizational knowledge that is derived from personal knowledge. This effectively appropriates social production in that it applies Taylorist principles to "white collar" managerial work that traditionally has not been subjected to such treatment.

As a consequence much of KM has failed to deliver because those implementing KM initiatives have retained a bureaucratic perspective of work as performed by individuals in a formal organizational structure where knowledge is viewed as a static resource or asset that can be treated in much the same way as any other commodity. As long as organizations retain this perspective, the real nature of knowledge work remains hidden, and thus inaccessible to those who are trying to "improve organizational outcomes" through KM practices (Linger & Warne, 2001).

The concentration on formal organizational programs aimed at individual workers ignores the real nature of work practices that reside in a space between the organization and individual perspectives. It is our contention that revealing the nature of this hidden space holds the key to understanding knowledge work and is critical to successful organizational outcomes and learning.

Changing the focus of KM to work practices reveals how the work community and its members conceptualise the work they perform. It also exposes the synergistic roles of the community and its members in the processes of knowledge production and re-production (Burstein & Linger, 2003).

As an illustration of this change of KM focus we draw on the Australian KM Standard (AS 5037, 2005) which was produced in consultation with Australian businesses and defines knowledge management as follows:

> *A trans-disciplinary approach to improving organizational outcomes and learning, through maximizing the use of knowledge. It involves the design, implementation and review of social and technological activities and processes to improve the creating, sharing, and applying or using of knowledge. Knowledge management is concerned with innovation and sharing behaviors, managing complexity and ambiguity through knowledge networks and connections, exploring smart processes, and deploying people-centric technologies.*

The Standard's definition of KM is congruent with the notion that most knowledge is created through work in groups and networks that are not visible in the formal organizational structures and processes. The Standard also uses the concept of a knowledge ecosystem to represent the core organic nature of KM and provides a more relevant guide to KM for practitioners than the constrained, process-oriented approach currently prevalent in organizations. Figure 2 shows the Australian Standard's presentation of the elements, enablers and other factors of organizational KM as a knowledge ecosystem. Individual items within this ecosystem could comfortable reside in any one of the domains of the Cynefin model but the whole is a complex dynamic set of interactions between them. The ecosystem concept is strongly influenced by notions from Complexity Theory, as described above, where attractors and boundaries replace rules and control. It recognizes that every KM initiative is different, because of its unique context and strategic intent, and that the process is organic, informal and emergent rather than mechanistic and controlled. The knowledge ecosystem exposes the invisible aspects of knowledge work vital to the sustainable growth of modern organizations.

The invisible aspects of knowledge work relate to the upper two domains of the Cynefin model. The process of mapping the knowledge ecosystem makes some of the elements, processes, and characteristics of these domains more visible. Participation by businesses in the mapping process expands their awareness of these factors.

The evolving view of knowledge management has shaped our approaches to research and our interpretation of data in the case studies below. The very nature and conditions of work are changing. This necessitates a change in the paradigms used to design and interpret research.

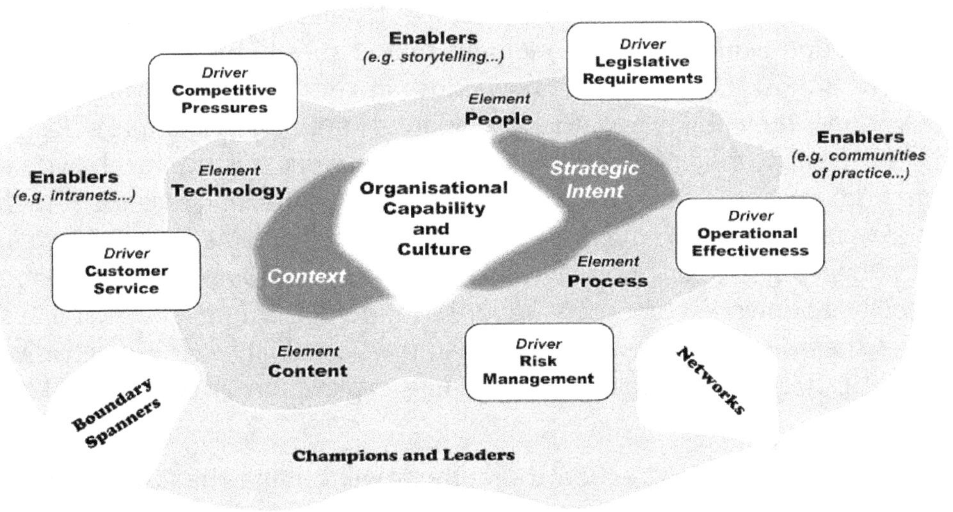

Figure 2 *A visualization of the Knowledge Eco-System from the Australian KM Standard (AS 5037-2005).*

Working in a Changing World

As discussed earlier, mainstream knowledge management has been largely concerned with the nature of knowledge in organizations, the knowledge processing capability of the organization, technology for knowledge processing and valuing knowledge in organizations. This perspective is essentially structural and hides the object of work that is the basis of knowledge management.

In the authors' view of the KM landscape, these approaches, which focus on the organization as a whole or the tacit knowledge of the individual, are not robust enough to service the complex changing world. In the authors' middle-out approach to KM the entities and context of KM form a complex set of interrelationships that are best described as an organic knowledge ecosystem rather than a machine or bureaucracy. Indeed, the development of this approach has revealed that the issue critical to organizational growth and learning is not what is now considered KM but rather the phenomenon of knowledge work.

The term 'knowledge worker' is attributed to Peter Drucker, who over forty years ago used the term to describe someone who adds value by processing existing information to create new information which could be used to define and solve problems (Drucker, 1959). Burstein and Linger (2003) have taken up Drucker's term "knowledge work" to refer to self-directed work practices of individuals and teams in almost every industry who continuously engage in processes that create and exploit knowledge. They go on to describe knowledge work as an activity system "located within the space defined by the *doing, thinking and communicating* dimensions. In order for actors to move seamlessly anywhere in this three-dimensional space, they need to have authority and responsibility that allows them to determine task outcomes." In our view, the purpose of knowledge work is a critical element that shapes behavior and outcomes. Likewise, Rogoski (1999) says of knowledge workers: "Their main value to an organization is their ability to gather and analyze information and make decisions that will benefit the company. They are able to work collaboratively with and learn from each other; they are willing to take risks, expecting to learn from their mistakes rather than be criticized for them." Haedrich and Maier (2004) also note that knowledge work is typically characterized by attributes such as mobility, flexibility, teamwork, computer-support and the use of intellectual abilities as well as specialised knowledge rather than physical abilities. It is creative work solving unstructured problems that require exploration or creation of knowledge.

In the following section the authors will outline the issues that have emerged from their own empirical investigations, outlining findings from their case studies. The issues are grouped around two main themes: *Situated Activity* as the work system that identifies the actors, intent, relationships, structures and tools; *Collaboration* as the social construct within which the activity occurs. Our intention is to highlight the complexity inherent in work practices that incorporate both the purposeful, productive and intellective aspects of activity systems.

In the case studies presented below, *situated activity* is described in terms of the following simple questions designed to reveal world views, relationship and strategic intent:

"**Where do you fit in?"** elicits information about shared or individual conceptions of identity, culture, values, power, systems thinking, boundaries, and ways of representing information that shape decisions about the scope of 'appropriate' behavior, strategic options, relationships and the quality of interaction and decision making.

"What are you trying to achieve?" focuses attention on the purpose(s), (object(s)) of situated activity, the priorities, commitment, roles and levels of participation that are required.

"How do you do your work?" invites descriptive accounts of the structure and organizing principles of the setting as well as information about the dynamics of critical communication, problem solving and production processes. This involves the tools by which activities are mediated by including technologies, institutional processes and structures as well as the context within which they occur.

As mentioned above, collaboration is the social construct within which activity occurs. *Collaboration in situated activity* involves feeling comfortable managing a series of effectors (some negative and some positive) and emergent outcomes in a complex setting rather than a pre-planned linear progression towards a predetermined outcome. In collaboration, the dynamics or quality of interaction and the shareability of information are critical factors. The level of trust between participants and their assessments about the reliability of emergent information or interpretations shape the extent that shared situated activity can become agile, responsive and aligned (and realigned) with maximally effective operational outcomes. Some understanding of the history, competence and interpretive perspective(s) (expertise) of collaborators is important. In this context, capabilities for intuitive thinking, complex data analysis, communication, informal leadership and strategic flexibility are at a premium.

Case Studies

Over the past decade the authors have conducted and published numerous studies of cases that illustrate the broad spectrum of organizational KM maturity. A representative selection of these will be presented and discussed in terms of current knowledge, situated activity, and collaborative activities. The case studies, all conducted in Australia, will cover the following domains:

1. The tertiary education sector, where hierarchical organizations persist despite early adoption of networking technologies;
2. The military, where moves towards network centricity are changing the traditional command and control culture;
3. Community networking in civil society enabled by Internet-based groupware;
4. A brief reflection on our own cross-institutional collaborative research.

Domain 1: Universities as Hierarchical Organizations

Universities in Australian were early adopters of organizational information and communications systems so that first generation KM flourished in the 1980s and 1990s. In her research (Hasan, 1999) chronicled 1990 projects in one university including a data-warehouse (DW) of student course information for line (department) managers, an information audit (IA) to determine sources and users of information, an Executive Information System (EIS) to present summarized performance information to management, an Intranet, managing documents on the web, and a campus-wide infrastructure (CWIS) to manage communications and access to computing resources. Early implementations historically functioned as replacements for previously paper based systems and processes and were oriented towards the needs of administrators. After six years the DW and IA had largely been abandoned. The EIS and CWIS were static and only the Intranet had expanded.

Subsequent developments included an automated form of web-based student enrolment. However one impact of these new systems has been to automate previously more flexible and negotiated transactions within the university. The negative consequences of the automation of student enrolment, reported by one academic student advisor, were the lack of personal contact, diminished control over student choices and having no feel for the type and number of students entering his program of study. There was also much tension evident between the market-driven focus of administrative staff and the teaching styles of academics manifest in continued arguments about quantity, versus quality, of students. The new technology enabled the easy processing of more students and allowed diverse forms of course delivery, including remote sites, but increased the complexity of the work of teaching staff with a lack of support in critical areas, such as moving subject delivery into the new flexible mode. With more work at the coalface, academic staff was disillusioned with the lack of understanding, coherence and direction from the top concerning teaching loads and work practices in a climate where they were being asked to also increase their research profiles.

In follow up research on the impact of information systems on universities (Crawford & Hasan, 2002) we found there was a disjunction between the purpose and function of the information systems and the more socially oriented, exploratory and knowledge making social capital development 'objects' of academic work. Apart from standard Internet communications there are no knowledge sharing systems which attempt to codify tacit knowledge, identify intellectual capital or encourage communications between knowledge workers, or connect experts with knowledge consumers. The less flexible, more rule bound automated arrangements appeared linked to new business imperatives and a push from administrative staff towards competition, regulation and cost efficiencies.

The current landscape of KM in these universities is that first generation, focussing on timely information provision for decision support, was embraced early but most have been frozen into legacy systems involving technology, business processes and structures.

These findings are consistent with our research in a second university (also reported in Crawford & Hasan, 2002) exploring the objects associated with social capital and capability development, IT implementation and its impact on academic staff. In this study, academic knowledge work was positively associated with email use ($p > .028$), internet applications ($p > .000$) and computer use ($p > .001$). Capabilities for knowledge work were also positively associated with agency or authority. In universities such self reported capabilities are strongly associated with seniority and administrative power in the still largely hierarchical structures.

The complexities of the uneven distributions of expertise and power and their impact on change were evident. The emerging knowledge system, and in particular the emerging knowledge management system associated with university administration, reflects the purposes of those with decision making power and these are not necessarily changing fast. As one staff member reported:

> *I'm okay because of my position, the system works well for old fogies like me, but post-grads and younger/ less senior staff have decreasing free time* (High Level Academic, Arts).

Administrative staff, regardless of their executive level, reported the lowest levels of personal agency of the group. The tensions among academic staff about the resulting conservatism are expressed eloquently as follows:

> *The University is losing the wrong people—the good people are going and the bad ones are staying... If you've already retired in the job, why would you bother retiring?* (Senior Academic, Science).

Applying the concepts of situated activity, together with collaboration Universities are very complex systems. Traditionally one of their central functions in society has been knowledge work and social capital development to create leaders in society able to assist in responding to challenges and change. With their long history also come forms of hierarchical organization or power and authority with their roots in medieval times. It is clear that the cultural history of technical development and its application in these organizations has had an unintended impact. The balance, in the day to day experience of people has changed. For many, there is less focus on knowledge work associated with research, innovation and creative activity and more on attending to rule based technologies and administrative processes that have evolved from industrial models and organizational processes. In seems to us that this experiment in the industrialization (including co-modification and increased automation) of knowledge work is unfortunate in an era where pressures for agile responses to rapid change have never been greater. Universities which should be leading the way in the knowledge era, appear in our case studies to be the least likely to move into the network-centric paradigm.

Domain 2: The Australian Defence Organization (ADO)

Conversely, the Australian Defence Organization which is stereotypically considered to be a rigid, traditionally hierarchical organization appears to be adapting well to the knowledge era and the network centric paradigm. The ADO includes the Australian Defence Force (ADF) and the government department that supports it.

In one study, over an extended period of four years, authors from the Defence Science and Technology Organization in Australia investigated the procedures that facilitate social learning and knowledge management in a number of different settings in the ADO (Ali *et al.*, 2001,2002; Warne, Hasan *et al.*, 2001, 2002). The term 'social' learning has been used to reflect that organizations, organizational units, and work groups are social clusters, and that learning therefore occurs in a social context. Social learning represents important processes that contribute to individuals' ability to understand information, create knowledge from that information and share what they know. Social learning is therefore intrinsic to knowledge management.

The research findings highlighted the importance of organizational and/or cultural values for effective social learning and knowledge management practices. In some cases, it was the absence of such values that made their importance clearer. Effective social learning was facilitated by the presence of a set of overarching values:

- *Empowerment*: autonomy to make individuals accountable and increase their sense of ownership of their role in the organization;
- *Cultural cohesiveness*: common identity, shared goals and a shared understanding;
- *Trust*: entailing mutual respect;
- *Forgiveness*: forgiving mistakes and creating knowledge from lessons learnt;
- *Commitment*: loyalty to the organization reciprocated by loyalty from the organization;
- *Openness of decision making*: transparent processes and information availability to employees at all levels of the organization;
- *Sharing of information*: information as an organizational asset not a source of an individual's power base

Apart from the overarching set of values, the research team identified additional sets of factors that support and enable effective social learning. These factors fall into two categories. The first, *Learning Capability Development*, refers to *characteristics in the environment* and provides a context in which the second category operates. This second category is referred to as *Enablers* and represents *processes and strategies* that, if present and effectively applied in an enterprise, can facilitate social learning. The learning capability is dependent on the priorities and objectives of the organization itself and the relative dominance, or perceived importance, of each of the *Values* in different research settings. However, the research also shows that the contribution of *Values* and *Enablers* to social learning is dependent on receptive and supportive organizational structures and processes. This research indicates that these vary across different sectors of the ADO possibly reflecting different cultural histories and different modes of operation and organization. Thus learning capability is nurtured by, and itself nurtures, organizational values that foster effective social learning.

A more recent analysis of this research (Warne, Hasan *et al.*, 2005) observes that technological tools of ever increasing sophistication are available for use in achieving the dissemination and sharing of data, information and knowledge across the organization. However, despite the existence and capability of these tools, knowledge management in many organizations all too often does not deliver the benefits sought from it, and the ideal inquiring organization is still a rarity. The ADO research suggests that organizational, human-related and cultural issues play a pivotal role in the success of knowledge management initiatives. Operating effectively as an ideal inquiring organization in times of change and uncertainty requires flatter organizational structures, a suitable technological infrastructure, relationships based on trust and many other cultural constructs as outlined in this paper.

What is pertinent about this case is that the ADO itself sponsored this work to understand the issues inherent in building learning, adaptive and sustainable organizations. Applying the concepts of situated activity, together with collaboration it indicated that, unlike the rigidity of Universities, change was taking place moving the ADO into at least the second generation of KM. Since the study described above, the focus of the military has shifted to the concept of Network-Centric Warfare (NCW). Although this focus arose out of technical considerations, it is now shifting to place much greater emphasis on the human and social aspects. The Australian *NCW Roadmap* isolated two dimensions of NCW. The first was the network dimension, referring to the physical systems providing connectivity between sensors, commanders and other parts of the network. The other dimension articulated by the *Roadmap* is the human dimension (Australian Department of Defence, 2005).

Similarly, a more recent study, based on interviews with ADF personnel returning from deployment in Iraq and Afghanistan, determined that under warfighting conditions, relationship building and cohesiveness are necessary ingredients for building situational awareness, achieving agility and, ultimately, securing successful mission outcomes. Furthermore trust is seen as the 'glue' that keeps networkers together and provides an underlying foundation for collaboration and the sharing of information. Data shows that often personal informal networks provide a means to verify situational awareness, often in preference to more formal means (Warne, Bopping *et al.*, 2005).

The importance of contextual factors and the relative inability of electronic communication methods and information systems to provide these came across strongly in the interviews. Moreover, the assumption that more data and information is necessarily better than less was also called into question. A person's understanding of their own situation and also that of the others with whom they communicate plays an important part in shaping how that communication takes place and, consequently, in its effectiveness in conveying intentions, meanings and implications. The technology, therefore, provides only the skeletal structure on which to build a successful communication infrastructure through which the ambitions of network-centric warfare can be achieved (Warne, 2006).

As the ADF moves into NCW, facilitated by the increased connectivity made possible by ICT advances, the human and social aspects of this work remain an important focus. In Australia, NCW is considered a key *capability enabler*, a means by which to enhance the overall warfighting effectiveness of the ADF. The Australian *NCW Roadmap* isolates two dimensions of NCW. The first is the *network dimension*, referring to the physical systems providing connectivity between sensors, commanders and those involved in engaging the adversary. The other dimension articulated by the *Roadmap* is the *human dimension* (Australian Department of Defence, 2005). It is this bilateral transition toward NCW that defines the ADF's adaptation to the demands of a 21st century context (Warne, 2006), and towards third generation Knowledge Management.

As Generation Y is moving up the ranks in the military, there are emergent changes in the use of Internet and social technologies in the field. Baum (2005) reports that, in Iraq, young platoon and company commanders were exercising their initiative in the face of a lack of training for the conditions they encountered. The younger officers had created for themselves, in their spare time, a means of sharing with one another, online, information that the Army did not control. These officers had been trained by officers of previous generations and equipped to fight against numbered, mechanized regiments in open-maneuver warfare. Instead of looking up to the Army for instructions, they were teaching themselves how to fight the war. This provides us with one example of the movement away from the hierarchical command-and-control culture towards one that is network-centric.

Domain 3: Community Networks

The concept of community has undergone considerable re-evaluation in the changing, more mobile, urban landscape which has seen the disappearance of traditional face-to-face neighbourhood communities and the emergence of many communities of interest and practice in the online environment. In whole sections of society neighbourhood support groups have disappeared and people are looking to their replacements on the Internet where web-based communities are not limited by location.

We are studying this phenomenon of emergent online community formation in our research (see for example Hasan & Crawford, 2003a,b) to complement the work we are doing in large bureaucracies which are attempting to become more network-centric. Many of the latter are seeking to emulate the formation of communities of practice that happen naturally in civil society. We approach this as both a social and a technical issue, indeed an inexorable interconnection of both aspects in a socio-technical system. The provision of suitable affordable technology is important but so are human issues of computer literacy, internet accessibility and the ability to act cooperatively. Even more significant is the understanding of how to participate in a community where much activity is conducted in a virtual space.

A technical tool, Eviva (see: www.eviva.com.au) has evolved throughout the research to support the purposeful activities of a wide variety of communities. Eviva is a web-based system that has been designed as a shared virtual space with tools that enable collaboration, coevolution, co-invention, and communication (Hasan & Crawford, 2006). During this evolution much learning has taken place among the members of the communities and the researchers. The knowledge gained from each stage has led to more sophisticated requirements for an online support tool appropriate for a wide range of such communities in a variety of settings. Our research approach has been to provide our online tool and knowledge of community behavior to actively set up and observe communities that could not afford their own technology. The first of these involved a community set up to promote awareness of new technologies (photonics) among the wider public through the activities of scientists with high school students. In these projects, subject-matter experts, students, teachers, parents, technologists and business people contributed to the community activity through:

- Initial workshops with input from all participants and including community-building exercises and heterogeneous project team formation. Their project was to create a website that could be used to inform other students about the new technologies;
- An online period of sustained creative activity as new materials are assembled and knowledge is exchanged by the teams online;
- Community celebrations where young people show their creative work and explain their new learning and interest to members of the community including politicians, local government officials and the media.

This experience was subsequently encapsulated in a socio-technical model, which was used to support a number of work/learning communities as case studies to provide data for the research. The model begins with a face-to-face workshop followed by a period online where learners, experts and instructors are linked. During this period the community of workers and learners undertake a self selected team-based, problem-solving project where experiential learning takes place through the generation of skills, ideas and solutions. Subsequent face-to-face meetings were found to be essential to celebrate achievement and sustain relationships among the community members.

The key issues emerging from our research in community settings were:

- A shared and inspirational project goal is an essential aspect of any truly collaborative project and essential for active online participation;
- The online system enabled convenient boundary crossing between groups who would not usually interact as team members;
- The workshops and meetings were an essential complement to online interaction;

- Experience of working in teams in collaborative and inclusive ways generates very high levels of voluntary participation and commitment and greater awareness of the benefits of collaborative team work;
- A limiting factor for older members of the community is their inexperience of and lack of capability in navigating and strategically using networked interactive communication tools;
- The complexity of coordinating, leading and inspiring creative team units is an area for further study. The Eviva system enables operational coordination among people who are not collocated and supports boundaries as these are required. However, the creation and maintenance of attractors that inspire meaningful participation appears to rest firmly in the social domain and depends on relationships, respect and shared understandings.

The most important thing learned from the project is that only experience of cooperative activities and working closely with other people engaged in meaningful cooperative projects provides the insight and confidence to build an environment that actually support such activities. The design of the system and the socio-technical methods used to involve people in projects evolved from our wider research. In the culture of many organizations, people have struggled to even imagine working in this way where the tool enables meaningful group coordination and provides evident (and thereby safe) boundaries for teams. Successful project teams also benefited from effective and flexible leadership processes. There appears to be a natural tendency for people to work in a network-centric way when there is a purpose, and support, to do so.

This research is continuing, taking a developmental approach to the social and technical support for work and learning in diverse communities (see Connery & Hasan, 2005; Hasan & Crawford, 2006). Applying the concepts of situated activity, together with collaboration we see these communities demonstrating, many of the attributes of third generation KM where there is a natural organic network-centric configuration consistent with a knowledge ecosystem. The challenge with these communities, many run by volunteers, is that, although they have a common focus and purpose, they are often ad hoc and lack the efficiency that comes with the structure of a formal organization. Despite the capability of online technology to support these communities, there is still a need to evolve human and social processes to sustain them.

Domain 4: The Authors' Distributed Research Team

The authors of this paper are members of a collaborative research team and have been working together for the past seven years. They are located in a variety of research institutions in different cities and use a combination of online applications and serendipitous meetings at conferences to support their collaborative activities. The nature of knowledge production and use has long been a topic of debate and academic research and this now needs to take into account the socio-technical nature of distributed research teams such as ours.

In Australia, research has been largely a publicly funded activity, and the Federal Government is now casting an increasingly critical eye over the way it is currently performed and evaluated. A series of Government moves are promoting inter-disciplinary research that aims to create and apply knowledge to address problems of national significance. This moves use from research where problems are set and solved in a context governed by the interests of a largely academic community (Mode 1 according to Baterham, 2003) to research where knowledge production is carried out within the context of application (Mode 2). It is intended to be useful to someone other than specifically the practitioner, be this industry, government or society generally; and this requirement is present from the beginning. Batterham (2003) describes Mode 1 as "discipline based; distinguishes between theoretical core and its conversion to application" while Mode 2 is "multidisciplinary, team based; constant flow between basic & applied; discovery occurs where knowledge is developed and put to use". The latter is a relevant description of our research approach. Applying the concepts of situated activity, we see ourselves as an emergent community of practice, engaging in dynamic purposeful activity and collaborating in an evolving socio-technical system; in other words a model of the network-centric paradigm.

The Picture Across these Domains

Table 1 attempts to interpret the activity situated in each of the domains presented here. This is not intended to be comprehensive but rather paint a picture of the diversity of complex organizational scenarios that coexist in a rapidly changing world. Taking the unordered perspective from the Cynefin complexity model of Figure 1, the picture reveals unexpected and contradictory emergent patterns. Universities, nominally bastions of knowledge, have the least flexible processes and structures. The military, once reliant on centralised command and control are now, enabled by technology, seeking to adopt more network-centric ways of working while retaining formal structures essential to coordinated operations. Communities in civil society, and to an extent our research group, have readily adopted new web-based ways of working and communicating to carry out meaningful activities with a common purpose outside traditional organizational constraints. These networks have many of the knowledge sharing cooperative characteristics desired by organizations. However while some communities thrive, many are not sustained, possibly because they lack the systemic knowledge management capability of a formal organization.

	University	Military	Community	Research Team
Where do you fit in? identity, values, power.	Disillusioned. Power shift from academia to business, loss of professional identity	Tensions between technology and people are starting to be addressed, changing values and changing paradigms of power	Ubiquitous use of Web-based communications across space and time, gaining power	Outside some institutional power pressures, peer support
What are you trying to achieve? purpose(s) priorities, commitment.	Market based, Customer focussed	Learning organization, effective operational capability	Focussing on community and global issues	Quality research with application to practice
How do you work? structure, context communication, production processes. Technologies.	Structured, accountable, performance indicators, rigid systems	Technology focussed with people awareness, C2 and with some team autonomy	Unstructured, flexible, networked, project/activity based	Unstructured, networked with constraints from member's institutional context
Trends.	Slow to change, becoming increasingly outmoded in many cases	Hybrid of traditional hierarch and network-centric	Self-organizing and networked with the need for new ways of managing for sustainability	More recognition of this mode of working

Table 1 *A summary of the case studies in terms of situated activity.*

Conclusion: Towards the Network Centric Paradigm

It can be argued that in post-industrial societies change is commonplace rather than the exception and recent years have witnessed a rapid increase in the pace of change. Our case studies illustrate how organizations respond differently to this unstable environment. In order to prosper we ideally need to make sense of the changing world in a way that is coherent and holistic. Interest in KM has grown in response to this need and has evolved through what Snowden (2002) names a series of three generations, until, in the third, this interest acknowledges the inherent complexity of contemporary organizations and approaches KM in the context of Complexity Theory. A framework is required where there is integration of diverse views with a conceptual, philosophical and pragmatic understanding of the bonds that bind the various components of an organizational system together. The knowledge ecosystem of the Australian KM Standard sets the basis for such a framework where dynamically interrelated elements permit the emergence of meaningful patterns in an unordered environment. This leads us into the concept of a network-centric configuration in which workers leverage information through the collaborative efforts of small and agile self-directed teams (Warne, Ali *et al.*, 2005). The capability to do this emanates from rapid developments of information and communications technologies which are driving and supporting the change from the industrial to the information age. The network-centric environment implies new ways of working, with consequences for the organization's infrastructure, processes, people and culture. One of the most challenging aspects of the network-centric paradigm is the need to moderate the organizational culture from one determined by a command and control, rule-based hierarchy to one which supports loosely-coupled, self-managed teams to make cooperative decisions through the sharing of knowledge.

One central limitation of traditional organizational frameworks is the life time of information is assumed to be long. Traditional information systems are designed on the assumption that storage and accessibility mean that approved information will eventually be used. The challenge we now face is that the value and life of information has been reduced by emerging practices and the convergence of information and communication technologies. In networked environments ongoing decision making involves sharing, analyzing, critical debate and negotiated transformation of meaning according to changing elements in the situation. In such a setting diversity and dissent are strengths that ensure that the scope of the decision making process is sufficient for the complexity of the context. Furthermore it ensures that all participants actively evaluate the quality of the information and interaction available as they make shared decisions about operational outcomes. In contrast, formal decision making processes, conducted in a hierarchical organization are often badly informed by out of date, or incomplete, or overwhelming amounts of information in organizational systems and cannot be made at a tempo that is fast enough for effective adaptation to rapid change. The result is a lack of coherence between the slow moving IT infrastructure, and related formal decision making processes, and the more ephemeral, ever changing and emergent shared needs and priorities of day to day human activity.

While technology is not the only or even the most important component of the network-centric paradigm, its constant evolution is providing tools for collaboration, independence and self organization that outstrip the pace of change possible in large monolithic organizations. Old technology (Enterprise Resource Planning (ERP) systems, databases, transaction processing) crystallized the centralised and hierarchical command structures with bureaucratic controls. New technologies (web-based communications and applications) enable anyone to set up a global virtual community or a business bypassing many traditional regulations, processes and structures. New group support, blogs, and wikis, readily taken up in civil society, could support the agility large enterprises need to adapt to change, but this requires a new paradigm.

Similarly, traditional organizational structures assume fixed roles and responsibilities over a long period. In many cases decision makers are not actually responsible for fashioning the operational response. In rapidly changing situations requiring creative and speedy adaptation, informal management roles must change according to the emergent demands of the situation. Management in networks is defined more holistically by the ability to grasp the complexity and changing demands of the situation and engender effective responses among participants. Networked knowledge management can be described in terms of varying connections and relationships between different agents according to the constraints and opportunities of the emergent context. Leadership and governance in such a context are not defined in terms of organizational authority or capital but in terms of capabilities to catalyze and monitor effective responses among the participants.

The network centric paradigm is a return to a recognition of the reality and value of human relationships, commitment, engagement and purpose, as the driving forces behind shared endeavour in any community. In such a model the different histories, capabilities, needs and purposes of participants and the quality of their relationships and shared interaction define the scope and possibilities for effective shared activity. This is aligned with Snowden's notions of the third generation of KM which involves a sense-making model of collective knowledge creation, disruption and utilisation that allows us to work in both the ordered and unordered quadrant of the Cynefin model. The need for this hybrid model emerges from our case studies. Ordered first generation KM in a traditional hierarchical bureaucracy is stifling growth and organizational learning. Information community networks exits in the unordered space but can fail due to lack of order. Third generation KM is consistent with organic network-centric configurations within a knowledge ecosystem perspective. However, as shown by the military case, it may have to exist along-side, or within, large ordered organizations that may retain some hierarchical command and control. Third generation knowledge management moves organizations to the network-centric paradigm.

References

Ali, I., C. Pascoe, L. Warne. (2002). "Yet another role for team building and work motivation-enabler of knowledge creation and knowledge sharing," *Proceedings of the Command & Control Research & Technology Symposium: C2Transformation Through Experimentation*, 11th-13th June, Naval Postgraduate School, Monterey, CA, http://www.dodccrp.org/events/2002_CCRTS/Tracks/Track_7.htm.

Ali, I., Warne, L., Agostino, K., and Pascoe, C. (2001) "Working and learning together: Social learning in the Australian Defence Organization", *Informing Science Conference IS2001: Bridging Diverse Disciplines Proceedings*, Krakow University of Economics, Poland, June 19-22, ISSN 1535-0703.

Allee V. (2003). *The Future of Knowledge: Increasing Prosperity through Value Networks*, ISBN 9780750675918.

AS 5037 2005 (2005). *Australian Standard Knowledge Management*, ISBN 9780733769306.

Australian Department of Defence (2005). *NCW Roadmap 2005*, Defence Publishing Office, Canberra, ACT.

Batterham, R. (2003). E-Science: A Frontier Technology for Achieving the National Research Priorities, http://www.glassearth.com/seegrid_pdf/seegrid_batterham.pdf.

Baum D. (2005). "BATTLE LESSONS: What the generals don't know," The NewYorker, January 17, http://www.newyorker.com/fact/content/?050117fa_fact.

Benkler, Y. (2006). *The Wealth of Networks*, http://www.benkler.org/wealth_of_networks/.

Burstein, F. and Linger, H. (2003), "Supporting post-Fordist work practices: A knowledge management framework for dynamic intelligent," *Information Technology & People*, ISSN 0959-3845, 16(3): 289-305.

Coakes E. (2002). "Knowledge management: A sociotechnical perspective," in E. Coakes, D. Willis and S. Clarke (eds.), *Knowledge Management in the SocioTechnical World*, ISBN 9781852334413. pp. 4-14.

Connery, A. and Hasan, H. (2005). "Social and commercial sustainability of regional web-based communities," *International Journal of Web-Based Communities*, ISSN 1477-8394, 1(3): 246-261.

Crawford, K. and Hasan, H. (2002). "Knowledge management in the knowledge industry: The challenge for universities," in G. Timbrell (ed.), *KM Challenge 2002: Advanced*

Models of Knowledge Management-Practical Approaches for Implementation, ISBN 9780733747922.

Daft, R.L. and Lewin, A.Y. (1993). "Where are the theories for the 'new' organizational forms? An editorial essay," *Organization Science*, ISSN 1047-7039, 4: i-vi.

Drucker, P. (1959). *Landmarks of Tomorrow*, ISBN 9781560006220 (1996).

Drucker, P. (1998). *Management's New Paradigms*, ISBN 9780887309984.

Hädrich, T. and Maier, R. (2004). "Modeling knowledge work," in P. Chamoni, P. Loos, W. Deiters, R.-D. Kutsche, K. Sandkuhl, N. Gronau, H. Müller-Merbach and B. Rieger (eds.), *Multikonferenz Wirtschaftsinformatik*, ISBN 9783898380508, pp.189-203.

Hart, D. and Warne, L. (2005) "Comparing cultural and political perspectives of data, information, and knowledge sharing in organizations," *International Journal of Knowledge Management*, ISSN 1548-0666, 2(2): 1-15.

Hasan, H. (1999) "The mediating role of technology in making sense of information in a knowledge intensive industry," *Knowledge and Process Management*, ISSN 1099-1441, 6: 72-82.

Hasan, H. and Crawford, K. (2003a). "A multifaceted approach to distributed communities of learning and practice," *Proceedings of ISGLOB03*, http://www.aueb.gr/ifip-isglob03/proceedings/10%20-%20Hasan.pdf.

Hasan, H. and Crawford, K. (2003b). "Socio-technical systems for knowledge mobilisation in communities," *Proceedings of ACIS2003*, ISBN 9781862726154.

Hasan H. and Crawford, K (2006). "The coevolution of an accessible but secure virtual space for collaborative activities," *19th Bled eConference: eValues*, Bled, Slovenia, June 5-7.

HB189 (2004). *Knowledge Management Terminology and Reading List: An Australian Guide*, ISBN 9780733757440.

Henry A. (2007). *Inspiring Tomorrow's Leaders Today: Breaking down the Generational Barriers at Work*, ISBN 9780977590513.

Iivari, J. and Linger, H. (1999). "Knowledge work as collaborative work: A situated activity theory view," *Proceedings of HICSS'32*, ISBN 9780769500010.

Linger, H. and Warne, L. (2001). "Making the invisible visible: Modelling social learning in a knowledge management context," Special Issue on Knowledge Management of the *Australian Journal of Information Systems*, ISSN 1039-7841, pp. 56-66.

Nonaka, I. (1994). "A dynamic theory of organizational knowledge creation," *Organization Science*, ISSN 1047-7039, 5(1), 14-37.

Pfaff, C. and Hasan, H. (2007). "Collaborative knowledge at the grass-roots level: The risks and rewards of corporate wikis," *Proceedings of PACIS*, ISBN 9780473124342.

Prensky, M. (2001a, September/October). "Digital natives, digital immigrants," *On the*

Horizon, ISSN 1074-8121, 9(5): 1-6.

Prensky, M. (2001b, November/December). "Digital natives, digital immigrants, part II: Do they really think differently?" *On the Horizon*, ISSN 1074-8121, 9(6): 1-6.

Rogoski, R.R. (1999). "Knowledge workers top company assets," *Triangle Business Journal*, ISSN 1060-5096, January 8, 14(19): 21.

Smola K.W. and Sutton C.D. (2002). "Generational differences: Revisiting generational work values for the new millennium," *Journal of Organizational Behavior*, ISSN 1099-1379, 23(4)363.

Snowden, D. (2002). "Complex acts of knowing: Paradox and descriptive self-awareness," *Journal of Knowledge Management*, ISSN 1367-3270, 6(2): 100-111.

Sveiby K. (1997). *The New Organizational Wealth*, ISBN 9781576750148.

Tucker, M.L., Meyer, G.D. and Westerman, J. (1996). "Organizational communication: Development of internal strategic competitive advantage," *Journal of Business Communication*, ISSN 0021-9436, 33(1):51-69.

Warne, L. (2006) "NetworkER centric warfare: Outcomes of the human dimension of future warfighting task," Human Factors Issues in Network-Centric Warfare, An International Symposium sponsored by TTCP Human Resources and Performance Group (HUM), Technical Panel, Sydney Australia 1-3 May, http://www.dsto.defence.gov.au/events/4659/.

Warne, L. (2000). "A sociotechnical approach to social learning analysis in the Australian Defence Force," in R. Lloyd-Jones and D. Willis (eds.), *The New Sociotech: Graffiti on the Long Wall*, ISBN 9781852330408.

Warne, L. Ali, I., Hasan, H. (2003) "Social learning through activity theory," in H. Hasan, I. Verenikina and E. Gould (eds.), *Information Systems and Activity Theory Volume 3 Expanding the Horizon*, ISBN 9780864187369, pp. 96-120.

Warne, L., Ali, I., Hasan, H. and Crawford, K. (2004). "From knowledge management foundations to network-centric organizations and communities," in G. Timbrell (ed.), KM Challenge, 2004: Driving Performance through Knowledge Collaboration, pp. 93-104, http://www.saiglobal.com/PDFTemp/Previews/OSH/as/sai-global/gb/GB005-2004.PDF.

Warne, L., Ali, I. and Hasan, H. (2005) "The network centric environment viewed through the lens of activity theory," in G. Whymark and H. Hasan (eds.), *Activity as the Focus of Information Systems Research*, ISBN 9781921047077, pp. 117-14.

Warne, L., Ali, I., Pascoe, C. and Agostino, K. (2001). "A holistic approach to knowledge management and social learning: lessons learnt from military headquarters," *Australian Journal of Information Systems*, ISSN 1039-7841, pp. 127-142.

Warne, L., Bopping, D., Ali, I., Hart, D. and Pascoe, C. (2005). "The challenge of the seamless force: The role of informal networks in the battlespace," *Proceedings of the 10th*

International Command and Control Research and Technology Symposium, http://www.dodccrp.org/events/10th_ICCRTS/CD/papers/215.pdf.

Warne, L., Hasan, H. and Ali, I. (2005). "Transforming organizational culture to the ideal inquiring organization: Hopes and hurdles," in J. Courtney, J.D. Haynes and D.B. Paradice (eds.), *Inquiring Organizations: Moving from Knowledge Management to Wisdom*, ISBN 9781591403104, pp. 313-336.

Whelton, M. and Ballard, G. (2003). "Project definition and wicked problems," *Proceedings of the 10th Annual Conference on Lean Construction*, http://www.leanconstruction.org/pdf/WickedProblemsinProjectDefinitionIGLC10.pdf.

Zuboff, S. (1989). *In the Age of the Smart Machine*, ISBN 9780465032112.

Kate Crawford has extensive experience as a researcher, developer and facilitator who works with clients to build self-governing, creative, agile and adaptive communities and organizations. She is the owner/director of a consultancy company, Eviva Pty Ltd, with a focus on knowledge mobilisation through social technologies and is an Honorary Fellow with the University of Wollongong, Australia.

Helen Hasan is an Associate Professor in Information Systems in the Faculty of Commerce at the University of Wollongong. She is a member of the Australian Standards Committees on Knowledge Management and Small to Medium Enterprises. She has published extensively in the areas of Human Computer Interaction, Decision Support Systems and Knowledge Management and more recently Network-Centric Organization.

Leoni Warne is a Science Team Lead within the Joint Operations Division of DSTO and she leads the Human Dimension Concepts Team. This team is responsible for researching and developing the human and organizational aspects of learning, knowledge mobilisation and network-centric warfare. She also holds an Honorary Principal Research Fellow appointment at the University of Wollongong and is Adjunct Professor of Information Systems at the University of Canberra. Leoni has published extensively both nationally and internationally.

Henry Linger is the Deputy Director of the Knowledge Management Research Program in the Faculty of Information Technology at Monash University. He has been a Research Associate at Defence Science and Technology Organization (DSTO) for the past 8 years. Henry conducts research in the area of knowledge work, knowledge management and the design of systems to support professional work. His research involves national and international collaborations addressing a broad range of domains including biology, immunology, epidemiology, meteorology, defence and clinical and management aspects of healthcare.

15. Framework for the Role of Self-Organization in the Handling of Adaptive Challenges

Saadia Mahmud
University of South Australia, AUS

This paper extends the concept of self-organization from the natural sciences to management and proposes a framework for the role of self-organization in the handling of adaptive challenges by enterprises. The process of self-organization is a characteristic of those complex adaptive systems that are far-from equilibrium, and results in the creation of order in a system by the internal interactions between agents leading to stronger adaptive capability. This paper presents a synthesis of the concept of self-organization suitable for management with communication as its central focus. Results from an empirical study in three Australian small and medium sized enterprises (SMEs) indicate that an adequate level of three key factors—trust level, open communication and strength of the value system in an enterprise—is needed for self-organization to occur.

Introduction

Managers today are faced by an increasingly turbulent and unpredictable environment with rapid changes in the market place (Morgan, 1988; Merry, 1995; Brown & Eisenhardt 1998; Pascale, 1999; White, 1999; Meyer, Gaba & Colwell, 2005). The changing reality for organizations has led to the long-held Newtonian paradigm in management being challenged by the view offered by chaos theory (Zimmerman, 1993; Wheatley, 1994; Tetenbaum, 1998; Tasaka, 1999). Chaos theory offers a view of the world as dynamic where change is the norm not the exception, and where prediction is impossible. Complexity theory contends that organization can arise spontaneously and is adaptive (Frederick, 1998). In the natural world "astonishingly simple rules, or constraints, suffice to ensure that unexpected and profound dynamical order emerges spontaneously", and this self-organization has been termed "order for free" (Kauffman, 1995: 74). Self-organization along with selection creates a system's capacity for adaptation to environmental conditions (Kauffman, 1993: 173).

A number of disciplines in the natural sciences have led the study into the concept of self-organization—for instance chemistry (Prigogine, 1968, 1976; Nicolis & Prigogine 1977; Prigogine & Stengers 1984), physics (Klimontovich, 1991; Shalizi, Shalizi & Haslinger 2004), and biology (Ulanowicz, 1979; Jantsch,

1980; Kauffman, 1993, 1995; Camazine *et al.* 2001). While the idea of self-organization has been taken up by management theorists (Weick, 1977; Foerster, 1984; Ulrich & Probst 1984; Drazin & Sandelands 1992; Comfort, 1994; Molleman, 1998, 2000; Biggiero, 2001; Stacey, 2001), there is a limited amount of empirical research—indicating a gap in the understanding of complexity science's application to the social world. Researchers have pointed out the need for appropriate consideration when applying concepts from the natural sciences to the social sciences. As Comfort (1994) notes "the concept of self-organization needs to be redefined and reinterpreted in order to assess both its presence and functions in the performance of social systems in rapidly changing environments." This paper extends the concept of self-organization from the natural sciences to management using a cross-disciplinary approach and examines this theoretical framework in light of empirical results.

Need for Applying Complexity Theory to Management

Numerous books aimed at management practitioners (Peters, 1989; Goldstein, 1994; Watson, 1994; Wheatley, 1994; Nonaka & Takeuchi 1995; Axelrod & Cohen 1999; Haeckel, 1999; Ralls Jr. & Webb, 1999; Wood, 2000) and the general public (Waldrop, 1994; Merry, 1995) have raised awareness for the potential application of complexity theory. However as Mitleton-Kelly (2003) notes, "by comparison there is relatively little work on developing a theory of complex social systems despite the influx of books on complexity and its application to management".

The dynamics of complex adaptive systems, such as their ability to evolve and learn over time, can provide organizations with insights into developing the capabilities to handle adaptive challenges. Organizations that can tap into the property of self-organizing or self-renewing systems have been termed adaptive organizations where the task determines the organization form (Dumaine & Anderson, 1991). A firm that can self-organise has the internal capacity for "spontaneously emerging structures", depending on what is required (Wheatley, 1994: 91).

Based on the laws of motion with a mechanical world view, there has been a diffusion of the concept of equilibrium from Newtonian physics, into neoclassical economics and then to organization theory (Meyer *et al.*, 2005). In the field of economics, the Newtonian paradigm interpreted the economy as "a closed, autonomous system, ruled by endogenous, mutually interdependent factors of a highly selective nature, self-regulating and moving toward a determinate, predictable point of equilibrium" (Weisskopf, 1979). Equilibrium in organization theory is "a condition in which all acting influences are canceled by others, resulting in a stable, balanced, or unchanging system" (Meyer *et al.*, 2005). In the Newtonian organization, "if nature or crisis upset this state the leader's role was to reestablish equilibrium" (Tetenbaum, 1998). In the Complexity paradigm the world view is of "Chaos"—a system that is 'far-from-equilibrium' and characterized by constant change and by never reaching equilibrium (Fuller & Moran, 2001). In the Newtonian view of organizations, the role of the leader is to maintain equilibrium, while in the complexity view, management is a social function and the role of the manager is as a facilitator of self-organization.

Self-Organization: From Natural Science to Social Science

Three characteristics referred to in self-organization process in the natural sciences can be related to the social sciences. These characteristics are emergence of complex behavior through iteration, rules guiding behavior, and the presence of attractors.

Self-organization in the natural sciences refers to the dynamic emergence of order, structure or patterns from the interactions of the elements inside the system. Iteration leading to complex behavior is found in various systems—including mathematical, biological and social systems. In chaotic systems the iteration of a few non-linear equations can lead to the creation of a range of possible behaviors. Iteration is referred to in the literature on self-organization in biological systems as the 'multiple, iterative interactions' (Camazine *et al.*, 2001: 488). In a similar sense, Luhmann (1986) refers to iteration in social systems as recursively produced and reproduced communication in a network. A social system has the ability to generate complex behaviors as a result of simple communicative events iterated over time. The link between communication and self-organization has been made by a number of researchers (Katz & Kahn, 1966; Simon, 1976; Jantsch, 1980; Comfort, 1994; Olson & Eoyang, 2001; Hammond & Sanders, 2002). Jantsch (1980: 196-197) notes:

> *... communication, however, is the key to complexity ... The world is full of vibrations originating in the manifold dynamics of dissipative self-organization. These vibrations are the communication media of a dynamic world.*

Besides iteration, social and biological systems have another similarity in the description of self-organization in the literature that relates to the importance of rules guiding behavior. In biological systems, interaction occurs between individuals executing rules of thumb (Camazine *et al.*, 2001: 488), while in human systems interaction occurs between participants having mutual acceptance of communicative rules (Scheidel, 1972: 45). The shared frame of reference is linked to the concept of psychological orientation (Mortensen, 1972: 127) which is an individual's attitude to communication, and the individual's predisposition to respond to message cues in characteristic ways. The term social schema is also used to denote this shared frame of reference. There is an apparent link between the social schema that lead to communication being a form of interaction and the schema mentioned by Gell-Mann (1995: 17) that govern the behavior of agents in a complex adaptive system (CAS). A shared frame of reference is a requisite feature of the communication needed if self-organization is to occur in human systems. Shared values and culture play an important role in creating a shared frame of reference among individuals. This strengthens the interaction in the complex adaptive system and therefore enhances the learning capability of the system.

The third characteristic of self-organization in natural sciences is the presence of "attractors". Long-term behavior of a chaotic system tends to settle down into a smaller number of behaviors or states. Attractors have been found in man-made systems like a Boolean network (Kauffman, 1995: 83); mathematical sets like the Mandelbrot set (Briggs, 1992: 78); and models of natural systems to predict weather (Gleick, 1988: 30). A strange attractor "consists of an infinity of points, in the plane ... these points correspond to the states of a chaotic system" (Ruelle, 1995: 195). Descriptions in management literature refer to strange attractors as binding or limiting the behavior of a system; for example, "if an attractor has multiple points of attraction within a finite space it is called a strange attractor and it limits a system's unstable behavior within those limits" (Mitleton-Kelly, 2003) and strange attractors represent "a basin of attraction that the system is drawn into, pulling the system into a visible shape" (Wheatley, 1994: 122-123).

A number of authors have suggested that values are the attractors in organizations. Thietart and Forgues (1995) refer to an organization in a chaotic state as being "attracted to an identifiable configuration", and Frederick (1998) interprets a corporation's strange attractor as its value system that permits change within constrained limits. On this view, a corporate system's phase space is the total range of value variables available to any given corporate dynamical system while its phase portrait is a set of swirling values that represents all possible behaviors (Frederick, 1998). Agents' behaviors in a system are governed by a few rules based on a small number of shared values (Lewin & Regine, 1999: 48).

The movement of a system from one attractor to another attractor is proposed as signifying the process of self-organization by Goldstein who notes that:

> *The idea of non-linear systems passing through the various attractor regimes is one of the startling discoveries of the new science of chaos and far-from-equilibrium thermodynamics ... the emergence of the new attractor signifies the onset of self-organization* (Goldstein, 1994: 67).

According to Goldstein (1994: 67), when an organization is under the sway of an "equilibrium attractor" it resists change, as opposed to the organization that can self-organize by moving from one attractor to the next.

Self-Organization in Management

One of the earliest references to the term 'self-organization' in management literature (Ashby, 1947) defines it as processes in which systems become more highly organized over time, and these are self-induced changes of organization, with no external order being imposed. A more recent description, similar to Ashby's concept, is that self-organization represents a system's affinity for evolving into modes of functioning exhibiting more complex and coherent patterns (Goldstein, 1994: 172). In synergetics (Bushev, 1994), an example of a self-organized process is the occurrence of regulated behavior that results in joint action to produce a product by workers receiving no external orders, but doing so by some kind of mutual understanding (Haken, 1978: 191). The description of self-organization by Haken (1978), is close to that of Molleman (1998) who considers self-organization as the local autonomy to make decisions on both the transactions to be realized and the way transformation processes are organized to achieve these transactions.

Autogenesis is a self-organizing perspective of organization that explains organization by "observation and categorisation of the interactions of independent actors whose behavior is governed by a system of recursively applied rules" (Drazin & Sandelands, 1992). There are three levels of structure in autogenesis, where the inner-most core is "Deep structure" or the tacit rules that govern action and interaction among actors, "Elemental structure" is the interactions among the actors, and "Observed structure" is the social facts constituted by the interactions. The tacit rules or deep structure guide the behavior of the individuals without the need for external order. The joint action of the individuals is the creation of organization.

All five descriptions of self-organization in management literature (Ashby, 1947; Haken, 1978; Drazin & Sandelands, 1992; Goldstein, 1994; Molleman, 1998) have the common theme of: a) creation of organization without external direction or external order; b) joint action by the constituents of the system for the achievement of a shared goal—higher organization, complex patterns, products, or transactions.

Research Methodology

The study sets out to test the contention, made by Kauffman (1993: 173), Axelrod (1999) and Goldstein (1994: 1) that organizations that have a greater ability to self-organise will adapt more effectively to environmental change than those with a lesser ability to self-organise. Based on a review of cross-disciplinary literature, self-organization is characterized for the author's study (Carapiet, 2006) as:

> *a collective process of communication, choice, and mutual adjustment in behavior based on a shared goal among members of a given system [without external order]* (Comfort, 1994).

The phenomenon of self-organization is studied in the specific context of 'an event'—an organization change or adaptive challenge faced in the past twelve months by the company. Event is used in this study in the sense of a 'significant experience'. Complex adaptive systems learn and evolve based on 'experiences'. Evidence of self-organization and the factors affecting self-organization is gathered in each case from four sources: interviews with agents; OCS questionnaires; documents; and observation. The measures of the ability to self-organize developed in the study are compared to the evidence of self-organization and indicators of adaptive performance for each case (Carapiet & Harris, 2007). The research design is a multiple-case study where the case study protocol has been followed consistently in each case.

Communication is the primary mechanism for the iterative interaction that drives self-organization. In this study, communication in each organization is assessed in two ways. Firstly, the interview protocol contained questions pertaining specifically to the communication processes in the company. Secondly, an instrument was used to gauge organization communication. The "Organizational Communication Scale" or OCS by Roberts and O'Reilly (1974) was selected as it has been shown to be to be reliable and valid across quite a variety of organization types including SMEs and has been used in recent studies.

In this study, the value system of a company has been linked to the measurement of self-organization. Interaction between the agents is according to a set of rules or 'schemata' (Gell-Mann, 1995), or by a "system of recursively applied rules" in organizations such as CAS (Drazin & Sandelands, 1992). Shared values create norms of behavior. Strong shared values lead to similar behaviors of individuals at various levels of an organization (McMillan, 2004: 88), learning (Senge ,1990: 182), and act as an attractor so that behavior remains within certain bounds (Frederick, 1998). According to Weick and Sutcliffe (2001: 124) a culture with a small number of shared values (three or four core values) that have been converted into norms of behavior will be coordinated, resilient and opportunistic.

The research path was cyclical and non-linear and ideas were revisited as new data was gathered. Refinement of conceptual insights emerged in the research. For instance, after the second case study the concept of the ASO Index was developed, and some of the outputs were subsequently discussed with the contact manager in the third case study during the interview that followed the fieldwork.

Research Participants

Three Australian SMEs (small and medium sized enterprises) participated as case study companies. The selected cases offered diversity in terms of industry, company size, and organization structure as shown below in Table 1.

Case study	No. of employees	Annual Revenue 2003 (AUDm)	Nature of Business	Company Structure	Interviews	Self-Organization
TE	200	50	System engineering	Business unit of larger company	8 Managers 1 CEO of affiliate co.	Evidence found
TC	90	15	Management training	Business unit of larger company	6 Managers	Evidence found**
TI	30	2	Software development	Stand alone business	4 Managers 3 Engineers	No evidence

** Concept based on a table to analyze variation in cases by Lapointe and Rivard (2005).*
*** Data indicates the occurrence of self-organization in 2005.*

Table 1 *Diversity in Case Companies*

These companies are referred to as Tech Enterprises (TE), Training Corporation (TC) and Telecom International (TI). Tech Enterprises is a system engineering company, Training Corporation is a management education training college, catering mainly to overseas students, and Telecom International is a software development company focusing on the telecommunications sector. Tech Enterprises and Training Corporation are 'stand-alone' business units forming part of larger corporations, whereas Telecom International is an independent company. A common element in all these cases is that the employees are highly qualified and that knowledge or information is a significant part of the value chain.

TI is the smallest company in terms of the number of employees and turnover, with 30 staff and annual revenue of AUD 2 million. TE generates AUD 50 million annual revenue with 200 staff. Both TE and TC are stand-alone units of a larger company, while TI is a stand-alone business. There was evidence of self-organization in TE in the past year, and in TC in 2005.

The relevance of the study of organizations as complex adaptive systems has been underscored with evidence of the extent to which contemporary organizations operate in far-from-equilibrium conditions. In all three case study companies, the year leading up to the fieldwork saw a major retrenchment, with change in senior management, and increased pressure on profitability.

The search for cross-case patterns would be more effective with a greater number of cases (Eisenhardt, 1989); however, the selected cases in this research project offered an opportunity to analyze the handling of an adaptive challenge in different enterprises, Given the project constraints—one researcher, limited resources, and time available, the number of case studies conducted in this research was therefore limited to three firms.

Data Collection

Four sources of evidence were collected in this study: interviews with agents, a questionnaire, observation, and documents. Data has been collected in each company by using the same interview protocol and questionnaire. The observation data has also been collected in two defined categories.

A total of twenty-one semi-structured interviews were conducted using the interview protocol. The number of agents interviewed is six to eight in each company. The use of a systematic protocol and questionnaire allows for future replication of this study and thereby contributes in satisfying the criteria of "transferability and confirmability" (Denzin & Lincoln, 1994: 13). Detailed notes were taken during the interviews, while observations were recorded as notes in the case file before and after the interviews. In order to avoid any inhibition in interaction, interviews were not audio-taped (Minichiello *et al.*, 1995: 98-101).

In the first case study company there was evidence of self-organization, therefore a second round of interviews was conducted with a supplementary interview protocol that provided a more 'in-depth' information. The interviews-data provided partial answers to the research questions of this study. The OCS discussed earlier, was another important source of data.

Qualitative Data Analysis

Data analysis techniques for the qualitative data (Miles & Huberman, 1994: 10-11) collected in this study include data reduction by writing summaries and coding responses to questions, data display by preparation of data matrices, as well as conclusion drawing and verification. Qualitative data collected mainly through semi-structured interviews was managed systematically by taking detailed notes. A clear indexing system has been used to distinguish among raw field notes, documents, and interpretive analytic materials. The field notes and other data collected by analyzing archives have been reduced by preparing data summaries. Data categorizing techniques (Dey, 1993) were used to aid analysis. Data display techniques, such as diagrams and matrices with text have been used to compress the data and to aid reflection on its meanings.

The organization data sources, both qualitative and quantitative (Yin, 1981), were used to gauge process outcomes for comparison with interviewee opinions. Triangulation was used by checking results with respondents, comparing interview comments with documentary evidence, and cross-checking the responses of different interviewees to a similar issue.

In order to analyze whether organizations with a greater ability to self-organise have a strong value-system based on a few core values, content analysis (Krippendorff, 1980) was used to analyze data of interviewee responses to questions on the organization culture, value system, and management control in their company. The data is in the exact words used by the interviewees.

The aim of the content analysis is to ascertain the strength of the value system "attractor" by measuring the presence of three core values: in my study they were, Honesty, Openness and Trust. The selection of these values was made for two reasons. Firstly, the words were mentioned by different respondents during the interviews and therefore indicated that these were 'shared values'. Secondly, there was mention of 'Openness' and 'Trust' in the literature (Eisenberg & Witten, 1987; Harkins, 1999: 14; Lewin & Regine, 1999: 3; McMillan, 2004: 88). Content analysis was a good method to use here as it is not only unobtrusive but effective in analyzing qualitative data.

Quantitative Data Analysis

The OCS uses a 7-point Likert scale to record responses to a number of questions. There is an ongoing debate about the data-analysis techniques suitable for data generated by Likert scale responses (Jamieson, 2004). The controversy surrounds the view of the data from the Likert scale as 'ordinal' or 'interval' (Knapp, 1990). Wang *et al.* (1999) note that "the controversy over treating ordinal scales as interval scales remains because of different views on the relevancy of measurement scales to permissible statistics".

According to Jamieson (2004) "Likert scales fall within the ordinal level of measurement, that is the response categories have a rank order, but the intervals between the values cannot be presumed equal". However a number of researchers use Likert data and analyze it as interval data, and I found a number of recent studies that use factor analysis on Likert data (Botha, 2005; Sarros *et al.*, 2005; Thomas-Sabado & Gomez-Benito, 2005). While the debate on whether

Likert data is ordinal or interval continues, for this study the OCS data generated using the Likert scale is treated as ordinal, and to avoid any distortion and incorrect interpretation of results the statistical techniques used are suitable for ordinal data. A consideration in the choice of non-parametric statistical methods is the small sample size (Wilcox, 2005), with a total of 21 respondents.

In the first instance, responses for each question have been analyzed on a case-to-case basis. Then for major findings in particular questions, cross-case comparisons are made. The statistical package SPSS has also been used for analyzing the data from the questionnaires. Responses to select questions have been mapped to variables used in SPSS analysis, with the statistical analysis of the quantitative data using SPSS being done in three ways; descriptive statistics of data in each case, Spearman rank correlation coefficient in each case to find significant relationships among the designated variables that comprise the measure of self-organization ability; and Categorical principal components analysis (CATPCA), through optimal scaling, is used on the combined data from all three cases to identify dimensions in the data, and its results have been compared with the other analysis conducted on the data.

Empirical Results and Framework for Self-Organization

Self-organization is the process where order emerges dynamically in response to a significant experience or event. Enterprises were identified that faced environmental pressures—such as competition, technological changes, and economic forces which challenged their ability to survive. Such turbulent conditions as "the space for creativity in an adaptive system" (Stacey, 1996: 97) were considered to be suited to the study of the self-organization process. It is anticipated that evidence of self-organization would correspond with an enterprise having higher adaptive performance reflected in better profitability and employee loyalty.

Evidence of self-organization has been documented in TE in response to the 'event', or adaptive challenge, faced by the company. The lack of external order was apparent in the self-organization process as the agents interacted with each other to achieve a mutual goal; namely, the removal of the Acting General Manager and the appointment of their own choice to the post. This process has been termed by the agents as a 'management coup'. The appointment of an outsider who threatened their ability to maintain the culture and work as they had before, disrupted the organization culture to which they were accustomed and resulted in a bottom-up change process. The self-organization process resulting in a restructuring of the management team is an example of high-order self-organization where the agents are highly capable and autonomous. TE is an organization that gives its employees a high level of autonomy. They have a saying "seek forgiveness not permission", which encourages employees to be entrepreneurial and to take risks. The extent of managerial control was one of the factors identified in the literature as affecting self-organization and the evidence in TE concurs with the view that a lower level of managerial control facilitates self-organization. Content analysis results indicate that TE has the strongest incidence of 'trust' as a value among the three cases, indicates the importance of this factor in self-organization.

Content analysis of the qualitative data from all three cases was conducted in order to ascertain the strength of the value system "attractor". Table 2 presents the results of content analysis conducted on the data from the interviews. The data tables consist of the exact wording used by the respondents interviewed in each case-study company—in response to questions on the organization culture of the company, its value system and its level of managerial control.

The content analysis has been conducted for two categories: with Category 1 being Honesty/Openness and Category 2 being Trust. Words and phrases conveying a similar meaning have also been included while counting the incidence.

The content analysis indicates a strong consistency of category 1 in Training Corporation's responses (83.34%), which is lower in Tech Enterprises (75%) and non-existent in Telecom International. Category 2 is highest in Case 1 (62.5%), lower in Case 2 (50%), while once again it is non-existent in Case 3. It would appear that Honesty and Openness are highly valued in Training Corporation and Tech Enterprises. It is significant that in Telecom International, none of the respondents used any word or phrase conveying the meaning of honesty, openness, or trust.

While the self-organization process was not evident in TC during the fieldwork, this study found that TC has a strong ability to self-organise. Evidence collected during the interviews with the agents in TC, and the quote of the General Manager of TC in a trade journal article in 2005, both support the view that the agents in TC self-organized to change the ownership of the company in

Case Study Company	**Incidence Category 1**	**Incidence Category 2**	**Consistency* Category 1**	**Consistency* Category 2**
Tech Enterprises	11	5	75%	62.5%
Training Corporation	8	3	83.34%	50%
Telecom International	0	0	0%	0%

**Consistency is the percentage of total respondents using category*

Table 2 *Result of Content Analysis*

2005. There was no evidence of self-organization in the third company TI and this was consistent with evidence of a low ability to self-organise.

Statistical analysis has been conducted on the quantitative data from all three cases as a joint data-set, using the module CATPCA or Principal Component Analysis of Categorical Data in the statistical package SPSS. Results from the CATPCA indicate that there is a significant pattern in the data that shows that in the first two case study companies' agents (TE and TC) cluster together in their responses and are in the quadrant that indicates a high desire to interact and a high level of open and honest communication. The agents in TI are scattered in their responses

Figure 1 below is a plot of the respondents (referred to here as objects labeled by the numbers 1-21) on the two dimensions of Open and honest communication (Dimension 1) and Desire to interact (Dimension 2). The numbers 1-8 represent the respondents in TE, 9-14 represent the respondents in TC; and 15-21 are representative of respondents in TI.

As Figure 1 shows, there is a cluster of the respondents in the lower left hand quadrant of the biplot. The quadrant represents a favorable position in both dimensions—Open and honest communication (Dimension 1) and a high desire to interact (Dimension 2). In the OCS, a score of 1 is the most favorable response for each variable analyzed here. The cluster consists of respondents from

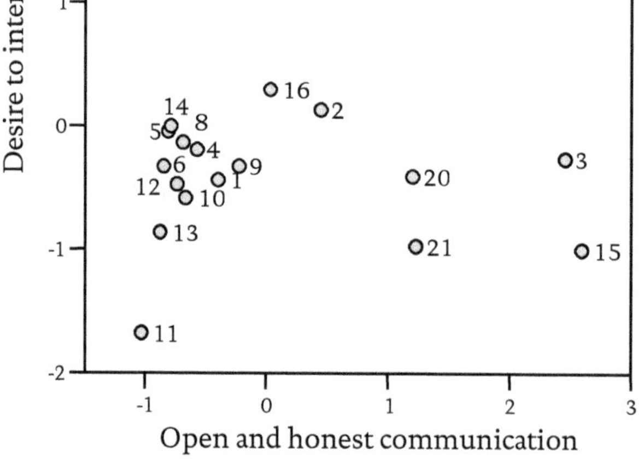

Figure 1 *Patterns in Data—Cluster of TE and TC respondents (respondents labeled by numbers)*

TE and TC only (objects 1, 4, 5, 6, 8, 9, 10, 12, 13 and 14). The ten respondents account for 71.5% of the respondents from the first two case study companies. Respondents from TI are scattered on the biplot away from the cluster.

CATPCA reduces data into a small number of dimensions that capture a high degree of the variance. Twelve variables were analyzed using CATPCA for the twenty-one respondents that filled out the OCS. Two dimensions were identified. Table 3 shows that both dimensions identified by the CATPCA have high reliability, and consequently account for a high variance in the data.

	Cronbach's	Variance
		Total
1	0.822	4.052
2	0.744	3.144
Total	0.939*	7.196

** Total Cronbach's Alpha is based on the total Eigenvalue.*

Table 3 *Model Summary—CATPCA*

The Cronbach Alpha indicates the internal consistency based on the average inter-item correlation. Alpha measured on a scale of 0-1 is high for both dimensions in Table 3 (that is, over 0.7). The variance in the total data-set accounted by each dimension is measured by the 'Eigen value'. As Table 3 shows, both dimensions have Eigen values over 1.0.

Dimension 1 can be described as "Openness and honesty in communication". In an organization where respondents trust their superiors as being generally fair, and trust their superiors to feel comfortable in having free discussion, there is no pressure to distort upward information, and consequently there is a good feeling about communication in the organization generally .

In the second dimension, the strong component loading is for the desire to interact with superiors (.811). All the variables that indicate a desire to interact—whether it be with immediate superiors, immediate subordinates or peers—have a positive covariance with the accuracy of information from superiors, and trust in the general fairness of the superior. Dimension 2 can be described as "Desire to interact". When superiors are considered as fair and providing accurate information, the respondents desire to interact with one another is better in all directions in the organization—upward, downward, and lateral.

Measures of the Ability to Self-Organize

This study led to the development of the ASO Index (Carapiet & Harris, 2007) which measures for the 'adequacy level of the factors' for self-organization. The qualitative data analysis results are supported by the measures of the ability to self-organise. Table 4 shows a comparison of the three case study companies on the basis of the ability to self-organize:

	TE (Case 1) N=8	TC (Case 2) N=6	TI (Case 3) N=7
ASO Index	2.06	1.89	2.79
TLCOM	1.5	1.67	3.0
CQCOM	2.61	2.11	2.44

All scores over 3 are 'Inadequate' according to the rating methodology

Table 4 *Self-Organization Measurement—Case Comparison*

The above table shows that Training Corporation has the best relative ASO (1.89) of the three organizations studied in this research. TLCOM measures trust level of communication and CQCOM measures the communication quality of the organization. Tech Enterprises has the best trust level among the three cases.

Based on the theoretical framework derived from earlier literature and the empirical results of this study, a framework for the role of self-organization in the handling of adaptive challenges has been developed illustrated in Figure 2. In this framework, enterprises face a number of adaptive challenges such as rapid changes in market, technology and within their own organization. In an Adaptive Enterprise, the agents interact in an iterative manner through communication.

Self-organization is the process where order emerges dynamically in response to a significant experience or event. The ability to self-organise is affected mainly by three factors (i) trust level between agents; (ii) open communication, and; (iii) a strong value system. When the level of factors is adequate, self-organization occurs. Value-based rules guide behavior of the agents who share a small number of core values. The enterprises that self-organise have a higher adaptive performance reflected in better profitability and employee loyalty.

An adequate level of factors was found in TE and TC which indicated an adequate ability to self-organise. Evidence of self-organization was found in the absence of external order in TE. A strong value system with three core-values—honesty, openness and trust was found in TE and TC. In TE a simple value-based rule was found that encouraged employees to be entrepreneurial and to take risks. This rule is to "seek forgiveness not permission". As an organization, employees at TE are encouraged to experiment and take initiative. This is consistent with earlier research that links self-organization with organizations capable of "experimentation" (Thietart & Forgues, 1995; Weick, 1977). In terms of indicators of adaptive performance, a rise in staff turnover, a fall in profitability and fall in

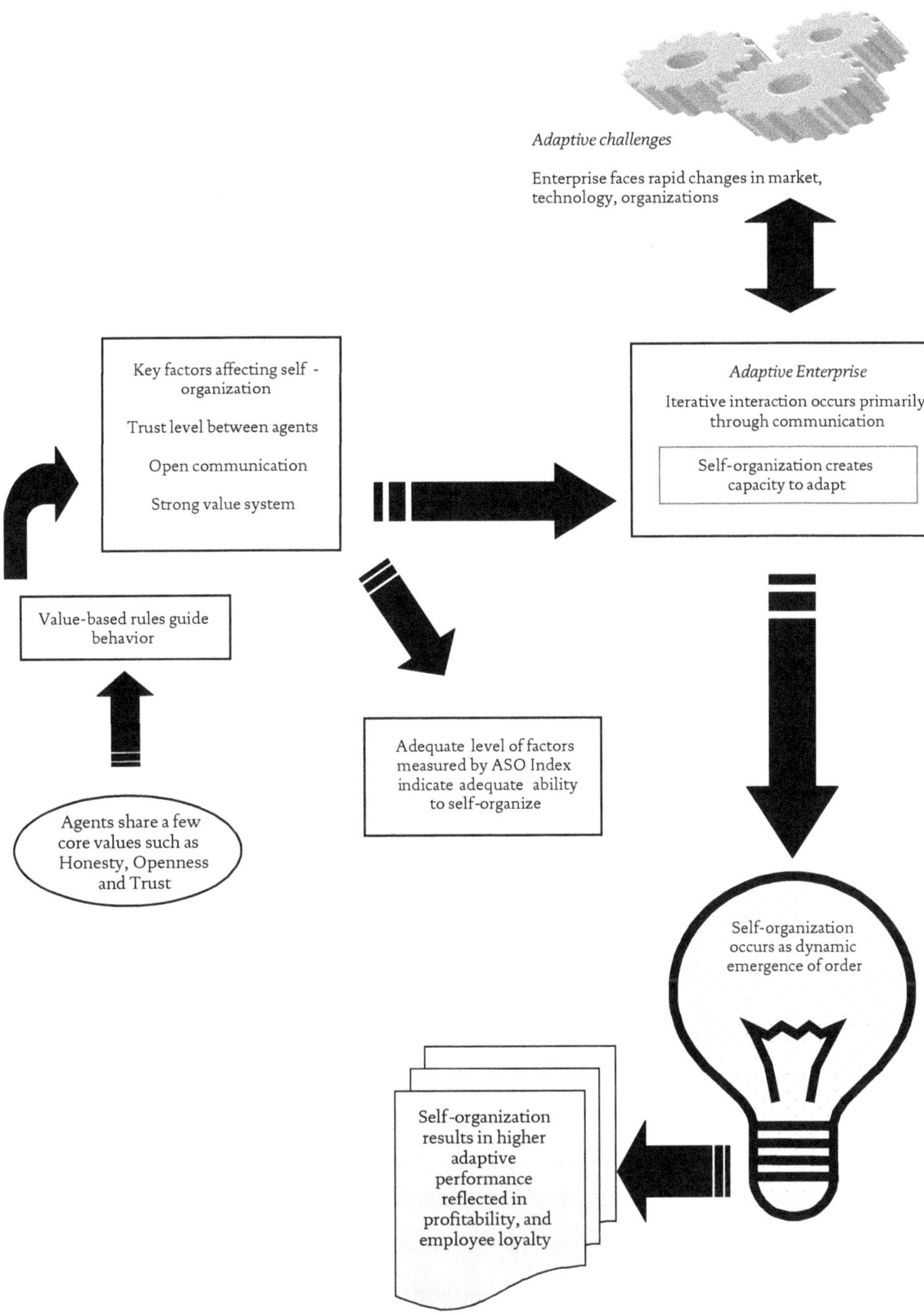

Figure 2 *Framework for Role of Self-Organization in Handling Adaptive Challenges*

employee loyalty each indicate problems in coping with adaptive challenges. All three indicators for TI, the enterprise with the lowest ability to self-organise, reflect a low level of adaptive performance.

Conclusion

Self-organization is the key for adaptive capability of complex adaptive systems. This paper extends the concept of self-organization from the natural sciences to management. Characteristics referred to in self-organization process in the natural sciences can be related to the social sciences such as the emergence of complex behavior through iteration, importance of rules in guiding behavior, and the presence of attractors. Based on the theoretical framework derived from earlier literature and empirical study results in this study, a framework is proposed for the role of self-organization in the handling of adaptive challenges by enterprises. Evidence of self-organization was found in two cases where an adequate level of three key factors affecting the ability to self-organise was present. The findings from this study has implications for organizations operating in a dynamic and unpredictable world. In order to improve the handling of adaptive challenges, organizations need to develop the ability to self-organise by having a high level of trust, open communication and a strong value system. The measure for the ability to self-organize (ASO Index) developed in this study is a first step to operationalize self-organization for management.

References

Ashby, W.R. (1947). "Principles of the self-organizing dynamic system," *Journal of General Psychology*, ISSN 0022-1309, 37: 125-128.

Axelrod, N.N. (1999). "Embracing technology: The application of complexity theory to business," *Strategy and Leadership*, ISSN 1087-8572, 27(6): 56-59.

Axelrod, R. and Cohen, M.D. (1999). *Harnessing Complexity: Organizational Implications of a Scientific Frontier*, ISBN 9780684867175.

Biggiero, L. (2001). "Self-organizing processes in building entrepreneurial networks: A theoretical and empirical investigation," *Human Systems Management*, ISSN 0167-2533, 20(3): 209-222.

Botha, D.F. (2005). "Towards an instrument for surveying knowledge management practices," *South African Journal of Business Management*, ISSN 0378-9098, 36(1): 1-6.

Briggs, J. (1992). *Fractals: The Patterns of Chaos - Discovering a New Aesthetic of Art, Science and Nature*, ISBN 9780671742171.

Brown, S.L. and Eisenhardt, K.M. (1998). *Competing On the Edge: Strategy as Structured Chaos*, ISBN 9780875847542.

Bushev, M. (1994). *Synergetics: Chaos, Order, Self-Organization*, ISBN 9789810212865.

Camazine, S., Deneubourg, J-L, Franks, N.R., Sneyd, J., Theraulaz, G. and Bonabeau, E. (2001). *Self-Organization in Biological Systems*, ISBN 9780691012117.

Carapiet, S. (2006). "Role of self-organization in the handling of adaptive challenges by enterprises," School of Management, PhD thesis, University of South Australia.

Carapiet, S. and Harris, H. (2007). "Role of self-organization in facilitating adaptive organization: A proposed index for the ability to self-organize," *Production Planning and Control: The Management of Operations*, ISSN 0953-7287, 18(6): 466-474.

Comfort, L.K. (1994). "Self-organization in complex systems," *Journal of Public Administration Research and Theory*, ISSN 1053-1858, 4(3): 393-410.

Denzin, N.K. and Lincoln, Y.S. (1994). *Handbook of Qualitative Research*, ISBN 9780803946798.

Dey, I. (1993). *Qualitative Data Analysis: A User-Friendly Guide for Social Scientists*, ISBN 9780415058520.

Drazin, R. and Sandelands, L.E. (1992). "Autogenesis: A perspective on the process of organizing," *Organization Science*, ISSN 1047-7039, 3(2): 230-249.

Dumaine, B. and Anderson, J. (1991). "The bureaucracy busters," *Fortune*, ISSN 0015-8259, 123(13): 36-42.

Eisenberg, E.M. and Witten, M.G. (1987). "Reconsidering openness in organizational communication," *Academy of Management Review*, ISSN 0363-7425, 12(3):418-426.

Eisenhardt, K.M. (1989). "Building theories from case study research," *Academy of Management Review*, ISSN 0363-7425, 14(4): 532-550.

Foerster, v.H. (1984). "Principles of self-organization: in a socio-managerial context," in H. Ulrich and G. Probst (eds.), *Self-Organization and Management of Social Systems*, ISBN 9780387134598.

Frederick, W.C. (1998). "Creatures, corporations, communities, chaos, complexity: A naturological view of the corporate social role," *Business and Society*, ISSN 0007-6503, 37(4): 358-389.

Fuller, T. and Moran, P. (2001). "Small enterprises as complex adaptive systems: a methodological question?" *Entrepreneurship and Regional Development*, ISSN 0898-5626, 13: 47-63.

Gell-Mann, M. (1995). *The Quark and the Jaguar: Adventures in the Simple and the Complex*, ISBN 9780805072532.

Gleick, J. (1988). *Chaos: Making a New Science*, ISBN 9780747404132.

Goldstein, J. (1994). *The Unshackled Organization: Facing the Challenge of Unpredictability through Spontaneous Reorganization*, ISBN 9781563270482.

Haeckel, S.H. (1999). *Adaptive Enterprise: Creating and Leading Sense-and-Respond Organizations*, ISBN 9780875848747.

Haken, H. (1978). *Synergetics: An Introduction - Nonequilibrium Phase Transitions and Self-Organization in Physics, Chemistry and Biology*, ISBN 9780387088662.

Hammond, S.C. and Sanders, M.L. (2002). "Dialogue as social self-organization: An introduction," *Emergence*, ISSN 1521-3250, 4(4): 7-24.

Harkins, P.J. (1999). *Powerful Conversations: How High Impact Leaders Communicate*, ISBN 9780071353212.

Jamieson, S. (2004). "Likert scales: How to (ab)use them," *Medical Education*, ISSN 0308-0110, 38(12): 1217-1218.

Jantsch, E. (1980). *The Self-Organizing Universe: Scientific and Human Implications of the Emerging Paradigm of Evolution*, ISBN 9780080243115.

Katz, D. and Kahn, R.L. (1966). *The Social Psychology of Organizations*, ISBN 9780471460404.

Kauffman, S.A. (1995). *At Home in the Universe: The Search for Laws of Self-Organization and Complexity*, ISBN 9780195111309.

Kauffman, S.A. (1993). *The Origins of Order: Self-Organization and Selection in Evolution*, ISBN 9780195058116.

Klimontovich, Y.L. (1991). *Turbulent Motion and the Structure of Chaos: The New Approach to the Statistical Theory of Open Systems*, ISBN 9780792311140.

Knapp, T.R. (1990). "Treating ordinal scales as interval scales: An attempt to resolve the controversy," *Nursing Research*, ISSN 0029-6562, 39(2): 121-123.

Krippendorff, K. (1980). *Content Analysis: An Introduction to its Methodology*, ISBN 9780803914988.

Lapointe, L. and Rivard, S. (2005). "A multilevel model of resistance to information technology implementation," *MIS Quarterly*, ISSN 0276-7783, 29(3): 461-491.

Lewin, R. and Regine, B. (1999). *The Soul at Work: Unleashing the Power of Complexity*, ISBN 9780752811857.

Luhmann, N. (1986). "The autopoiesis of social systems," in F. Geyer and J van der Zouwen (eds.), *Sociocybernetic Paradoxes: Observation, Control and Evolution of Self-Steering Systems*, ISBN 9780803997356, pp.172-192.

McMillan, E.M. (2004). *Complexity, Organizations and Change*, ISBN 9780415314473.

Merry, U. (1995). *Coping with Uncertainty: Insights from the New Sciences of Chaos, Self-Organization, and Complexity*, ISBN 9780275949105.

Meyer, A.D., Gaba, V. and Colwell, K.A. (2005). "Organizing far from equilibrium: Non-linear change in organizational fields," *Organization Science*, ISSN 1047-7039, 16: 456-473.

Miles, M.B. and Huberman, A.M. (1994). *Qualitative Data Analysis: An Expanded Sourcebook*, ISBN 9780803955400.

Minichiello, V., Aroni, R. and Sanders, G. (1995). *In-Depth Interviewing: Principles, Techniques, Analysis*, ISBN 9780582801011.

Mitleton-Kelly, E. (2003). "Ten principles of complexity and enabling infrastructures," in E. Mitleton-Kelly (ed.), *Complex Systems and Evolutionary Perspectives of Organizations: The Application of Complexity Theory to Organizations*, ISBN 9780080439570, pp. 23-50.

Molleman, E. (2000). "Modalities of self-managing teams: The 'must,' 'may,' 'can,' and 'will' of local decision making," *International Journal of Operations and Production Management*, ISSN 0144-3577, 20(8): 889-910.

Molleman, E. (1998). "Variety and the requisite of self-organization," *International Journal of Organizational Analysis*, ISSN 1055-3185, 6(2): 109-131.

Morgan, G. (1988). *Riding the Waves of Change: Developing Managerial Competencies for a Turbulent World*, ISBN 9781555420932.

Mortensen, C.D. (1972). *Communication: The Study of Human Interaction*, ISBN 9780070433953.

Nicolis, G. and Prigogine, I. (1977). *Self-Organization in Non-Equilibrium Systems: From Dissipative Structures to Order through Fluctuations*, ISBN 9780471024019.

Nonaka, I. and Takeuchi, H. (1995). *The Knowledge-Creating Company: How Japanese Companies Create the Dynamics of Innovation*, ISBN 9780195092691.

Olson, E.E. and Eoyang, G.H. (2001). *Facilitating Organization Change: Lessons from Complexity Science*, ISBN 9780787953300.

Pascale, R.T. (1999). "Surfing the edge of chaos," *Sloan Management* Review, ISSN 0019-848X, 40(3): 83-94.

Peters, T.J. (1989). *Thriving on Chaos: Handbook for a Management Revolution*, ISBN 9780330305914.

Prigogine, I. (1968). *Introduction to Thermodynamics of Irreversible Processes*, 3rd Edition, Interscience Publishers, New York.

Prigogine, I. (1976). "Order through fluctuation: self-organization and social system," in E. Jantsch and C.H. Waddington (eds.), *Evolution of Consciousness: Human Systems in Transition*, ISBN 9780201034394, pp. 93-126.

Prigogine, I. and Stengers, I. (1984). *Order Out of Chaos: Man's New Dialogue with Nature*, ISBN 9780434603954.

Ralls Jr., J.G. and Webb, K.A. (1999). *The Nature of Chaos in Business: Using Complexity to Foster Successful Alliances and Acquisitions*, ISBN 9780884155041.

Roberts, K.H. and O'Reilly, C.A. (1974). "Measuring organizational communication," *Journal of Applied Psychology*, ISSN 0021-9010, 59: 321-326.

Ruelle, D. (1995). "Strange attractors," in D. Ruelle (ed.), *Turbulence, Strange Attractors, and Chaos*, ISBN 9789810223106, pp. 195-206.

Sarros, J.C., Gray, J., Densten, I.L. and Cooper, B. (2005). "Chapter 8: The organizational culture profile revisited and revised: An Australian perspective," *Australian Journal of Management*, ISSN 0312-8962, 30(1): 159-182.

Scheidel, T.M. (1972). *Speech Communication and Human Interaction*, ISBN 9780673150059.

Senge, P.M. (1990). *The Fifth Discipline: The Art and Practice of the Learning Organization*, ISBN 9780385260947.

Shalizi, C.R., Shalizi, K.L. and Haslinger, R. (2004). "Quantifying self-organization with optimal predictors," *Physical Review Letters*, ISSN 0031-9007, 93(11): 118701.

Simon, H.A. (1976). "Communication," in *Administrative Behavior: A Study of Decision-Making Processes in Administrative Organization*, ISBN 9780029290002, pp.154-171.

Stacey, R.D. (2001). *Complex Responsive Processes in Organizations: Learning and Knowledge Creation*, ISBN 9780415249195.

Stacey, R.D. (1996). *Complexity and Creativity in Organizations*, ISBN 9781881052890.

Tasaka, H. (1999). "Twenty-first-century management and the complexity paradigm," *Emergence*, ISSN 1521-3250, 1(4): 115-123.

Tetenbaum, T.J. (1998). "Shifting paradigms: From Newton to chaos," *Organizational Dynamics*, ISSN 0090-2616, 26(4): 21-32.

Thietart, R.A. and Forgues, B. (1995). "Chaos theory and organization," *Organization Science*, ISSN 1047-7039, 6(1): 19-31.

Thomas-Sabado, J. and Gomez-Benito, J. (2005). "Construction and validation of the Death Anxiety Inventory (DAI)," *European Journal of Psychological Assessment*, ISSN 1015-5759, 21(2): 108-114.

Ulanowicz, R. (1979). "Complexity, stability and self-organization in natural communities," *Oecologia*, ISSN 0029-8549, 43: 295-298.

Ulrich, H. and Probst, G.J.B. (eds.) (1984). *Self-Organization and Management of Social Systems: Insights, Promises, Doubts, and Questions*, ISBN 9780387134598.

Waldrop, M.M. (1994). *Complexity: The Emerging Science at the Edge of Order and Chaos*, ISBN 9780140179682.

Wang, S., Yu, M., Wang, C. and Huang, C. (1999). "Bridging the gap between the pros and cons in treating ordinal scales as interval scales from an analysis point of view," *Nursing Research*, ISSN 0029-6562, 48(4): 226-229.

Watson, T.J. (1994). *In Search of Management: Culture, Chaos and Control in Managerial Work*, ISBN 9780415092302.

Weick, K.E. (1977). "Organization design: Organizations as self-designing systems," *Organizational Dynamics*, ISSN 0090-2616, 6(2): 30-46.

Weick, K.E. and Sutcliffe, K.M. (2001). *Managing the Unexpected: Assuring High Performance in an Age of Complexity*, ISBN 9780787956271.

Weisskopf, W.A. (1979). "The method is the ideology: From a Newtonian to a Heisenbergian paradigm in economics," *Journal of Economic Issues*, ISSN 0021-3624, 13(4): 869-884.

Wheatley, M.J. (1994). *Leadership and the New Science: Learning About Organization from an Orderly Universe*, ISBN 9781881052449.

White, M. (1999). "Adaptive corporations," in M.R. Lissack and H.P. Gunz (eds.), *Managing Complexity in Organizations: A View in Many Directions*, ISBN 9781567202854, pp. 281-291.

Wilcox, R.R. (2005). "Comparing medians: An overview plus new results on dealing with heavy tailed distributions," *Journal of Experimental Education*, ISSN 0022-0973, 73(3): 249-263.

Wood, R. (2000). *Managing Complexity: How Businesses Can Adapt and Prosper in the Connected Economy*, ISBN 9781861971128.

Yin, R.K. (1981). "The case study crises: some answers," *Administrative Science Quarterly*, ISSN 0001-8392, 26: 58-65.

Zimmerman, B. (1993). "The inherent drive towards chaos," in P. Lorange, B. Chakravarthy, J. Roos and A. Van de Ven (eds.), *Implementing Strategic Processes: Change, Learning and Cooperation*, ISBN 9780631185659, pp. 373-393.

Saadia Mahmud (formerly Dr. Saadia Carapiet) was awarded a PhD (July 2006) in the School of Management, University of South Australia for her thesis "Role of self-organization in the handling of adaptive challenges by enterprises". In the two decades since she graduated top of her MBA class at the University of Karachi, she has blended an interest in academic research with industry work. She read for an MPhil at the Judge Institute of Management at the University of Cambridge in 1993 after working in the banking and finance industry globally with Citibank N.A. Saadia continues to be involved with the finance industry and was awarded a Graduate Diploma in Financial Planning from the Financial Services Institute of Australasia, Australia in 2006. In addition to her role as Project Officer at the Learning and Teaching Unit, she currently teaches a number of courses at the University of South Australia including Contemporary Issues in Managing Change at the International Graduate School of Business, and Project Management at the School of Management.

16. Knowledge Transformation, Learning and Changes Giving Competitive Advantage in Aerospace Supply Chains

Christen Rose-Anderssen[1], James S. Baldwin[1], Keith Ridgway[1], Peter M. Allen[2] & Liz Varga[2]
1 University of Sheffield, UK
2 Cranfield University, UK

The paper investigates three levels of learning—adaptive, reactive and expansive—for the transformation of knowledge to enhance innovation and competitive advantage in commercial aerospace supply chains. A perspective of supply chains as complex Activity Networks is used for data analysis based on in-depth interviews in a global setting. Themes for the interviews were identified through rigorous literature research. The paper provides evidence of levels of learning in commercial aerospace supply chains. We found that a) adaptive learning brings a supply chain up to present industrial standards only, b) reactive learning makes the supply chain competitive, and c) expansive learning gives the supply chain potential for competitive advantage. By considering supply chains as the interaction of different work activities, the forces of change can be better understood. The findings may be useful to practitioners in understanding the importance of different levels of learning to supply chain sustainability.

Introduction

This paper identifies three levels of learning—adaptive, reactive and expansive—for the transformation of knowledge and the effects of learning for producing innovation and gaining competitive advantage.

Traditionally the large commercial aircraft manufacturer would define and specify exactly what their first tier suppliers should produce for them. This practice of high level of buyer dominance over suppliers has gradually changed, resulting in a worldwide reorganization of the industry. This is characterized by the commercial airframe manufacturers giving their 1st tier of selected suppliers directly below them, responsibility for complete systems (Giunta *et al.*, 2000).

Day and Atkinson (2004), argue that high impact suppliers have to commit themselves to cost reductions for long term business relationships, and create a culture of continuous improvement. However, to gain competitive advantage, more proactive responses are required. Nollet *et al.* (2005) argue that this is achieved through producing innovative solutions in terms of technology and products.

Recent developments in supply chain strategies are therefore based on a focus upon the sharing of skills and capabilities (Cousins & Spekman, 2003). In the context of creating technologically more advanced products, supply chains need to become learning communities with skills in creating and transforming knowledge (Preiss & Murray, 2005). In that sense, Easterby-Smith *et al.* (2005) highlight the absorptive capacity of a firm to appreciate and process external knowledge and the learning from past experiences. Engeström (1987), however, is more concerned about the learning and development experienced when moving away from the old. And this is about the creation of the new, i.e., innovation.

Risk-sharing partnerships combine these views in practice by bringing together the best expertise available to produce totally new concepts of aircraft design. These partnerships are the response to the need for sharing of skills, expertise and capabilities. This is driven by the fact that an airframe manufacturer cannot carry such huge investments in capital and expertise alone. The tight integration of risk-sharing partnerships can facilitate a totally new aircraft concept (Rose-Anderssen *et al.*, 2008). This paper therefore focuses on the levels of learning and knowledge transformation involved in the development of these practices.

An activity theoretical approach, viewing individual firms as work activities, and supply chains as complex networks of work activities is applied for analyzing the processes of learning and change. Activity Theory, which is an early form of evolutionary complex systems thinking, was formed by L. Vygotsky, A.N. Leont'ev and A.R. Luria in Soviet-Russia in the 1920-30s. The theory is a philosophical framework for studying different forms of human practices and their interaction. Complex Systems Thinking, however, was developed by many authors including I. Prigogine and his colleagues in Brussels and Austin, H. Haken and his colleagues in Stutgart during the 1970s, and later at the Santa Fe Institute in the US.

This paper is interested in the impact of levels of learning on supply chain change to gain competitive advantage. To further this end, the perspective of supply chains as networks of work activities is introduced first along with a model of knowledge transformation, learning and the creation of radical solutions. Following justification of the methodology, case data are discussed in terms of the model, leading to conclusions regarding the level of learning for transformation of knowledge to enhance innovation and competitive advantage in commercial aerospace supply chains.

Complex Activity Networks, Learning and Development

Johnson and Ford (2005) have written about the dynamic practices of supply chain network interventions and its effect on new product development. Their suggested intervention is to develop strategic alliances for sharing competencies and capital investments. In that sense, networks of work activities are collective attempts to share competencies and develop a common future.

A work activity is defined in Activity Theory as a developmental process connecting the individual and social levels through the human artefacts and the object orientation of the process. Work activities are characterized by the multi-voiced interaction created within them, and their potential for expansive transformation (Engeström, 2001). In these activities, there are continuous processes for creation of new instruments for change (Engeström, 1987).

The elements of the activity (Figure 1) are represented by the individual subjects of consideration, their activity community, the object and the human artefacts. There are three mediating artefacts. The instruments mediate between the individual and the object. The social rules mediate between the individual and the activity community. And the division of labour mediates between the community and the object.

Individuals have goals for their actions, whilst the object gives a collective purpose and direction to an activity (Hasu & Engeström, 1999). The object in this sense can be seen as the visual target or focus collectively being communicated and created by the community of a particular activity. It is important to make clear that an object is not the same as objectives, which are rigidly independent of individual conceptions, personal bias, thoughts and feelings. An object on the other hand is the vision that integrates the elements of the activity. Because of its imprecise nature the object needs to be reproduced more and more accurately in knowledge by the activity community for individuals to comprehend (Lektorsky, 1977). Essential to the development of the activity is the individuals' orientation towards the object.

The instruments are the concepts, language, technologies and strategies the individuals use to influence the object (Vygotsky, 1978). In that sense language is an essential instrument individuals are using to coordinate their actions with others (Blackler, 1993).

Evolutionary complex systems are governed partly by the invisible effects resulting from the interaction of individuals possessing a diversity of opinions and experiences within a system (Rose-Anderssen & Allen, 2008). Work activities are therefore difficult to observe and comprehend. Due to the diverse opinions and experiences people in a complex human system have, the system as a whole may potentially respond to challenges from the environment in multiple and unpredictable ways (Allen, 2001). In practice, this means that sufficiently complex activities will have a diversity of potential behaviors within them. This gives them the capacity to go beyond the marginal improvements characterized by routine behavior of a smooth and optimized terrain. That is, a complex work activity consists of a community of individuals with diverse perceptions, experiences and expertise. As such the community can explore more radically in a multidimensional "rough" performance landscape (Figure 2), and as a result produce radical solutions (Rose-Anderssen *et al.*, 2005). These are creative steps that are not foreseeable. Moving by only marginal changes in established routines, however, means that the community is merely 'hill-climbing' the smoother part of the terrain. This is just a rational extension of the present and will give marginal improvements only.

In this model of knowledge transformation and learning, there are three distinct and qualitatively different levels of learning in an activity (see Figure 3). Adaptive learning takes place when people adapt to practices developed by

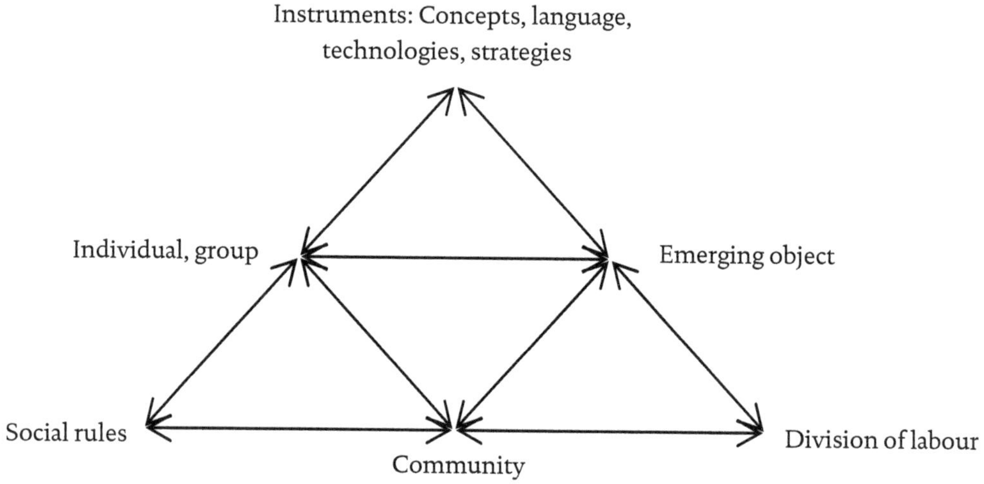

Figure 1 *Complex activity network (Engeström, 1987)*

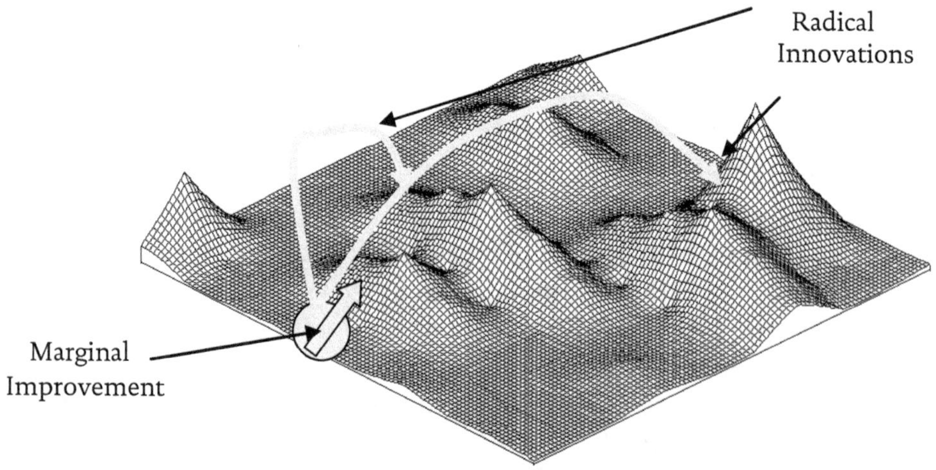

Figure 2 *Multidimensional Performance Landscape*

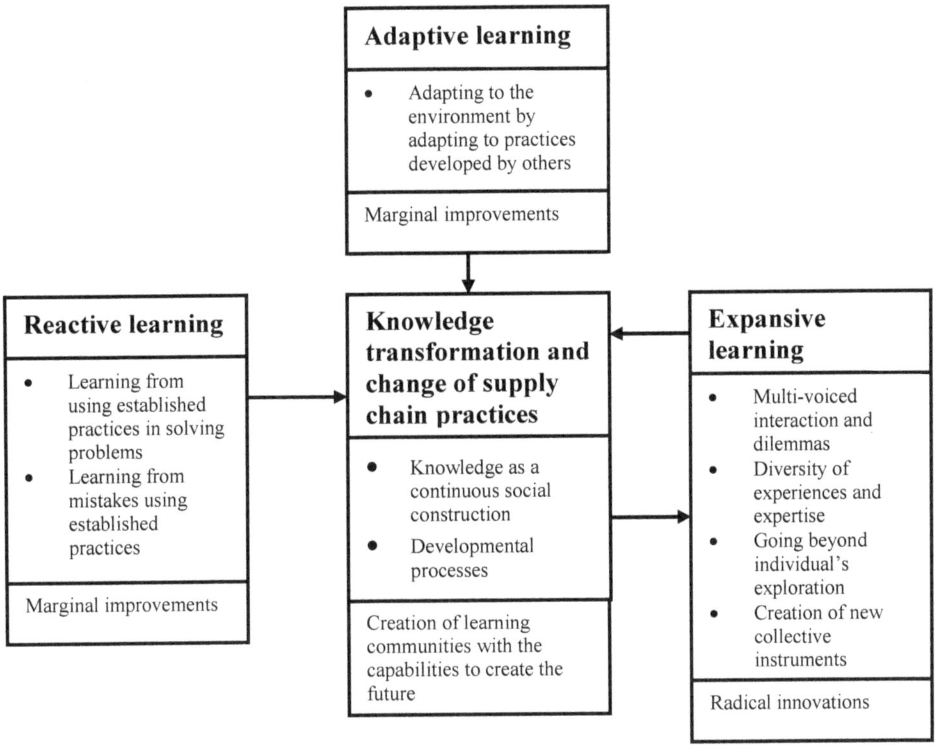

Figure 3 *Knowledge Transformation and Learning*

others. Adaptive learning is therefore about an awareness that things can be done differently and that these alternatives may serve in adapting to a changing environment. It means that there may be an opportunity to adapt to practices of more successful competitors or industries. Reactive learning occurs when using routine practices in solving problems. Reactive learning is therefore about taking corrective action to perceived mistakes and learning from that. These two levels of learning do not challenge the practices of the past. Neither of these two forms of learning requires the development of unknown approaches as they are still applied in the smoother and more familiar part of the multidimensional performance landscape. Supply chains operating according to these approaches are using single voiced interactions of established routines and practices.

A third form of learning takes place as an expansion of the given context (Engeström, 1987). This form of learning may occur at boundaries where people meet and interact to form new meanings that go beyond the limits of the individual person alone. This is the learning that takes place due to multi-voiced interaction between the individuals within the work activity. In that context they challenge the practices and perceptions of the past by moving into the rough terrain of the unknown. At a supply chain level, the multi-voiced interaction is extended to the boundaries between the firms.

Learning and Knowledge Transformation

Traditionally wisdom has been seen as something people possess in their head. In this perspective, learning is therefore associated with the acquisition of data accumulated by others (Gherardi, 2006). From this position, wisdom is a commodity that can be easily transmitted and controlled. And knowledge is about what an organization has accumulated over time. However, a more post-modern and complex systems position is taken where knowledge or knowing is seen as a dynamic process that is mediated, situated, provisional, pragmatic and contested (Blackler, 1995). From this position, knowledge is a continuous social construction in the collective infrastructure of knowing how and what to learn for change. The social construction of knowledge through learning has the potential of moving the activity beyond the rational and marginal extension of the present and into the rougher terrain of exploration and innovation. The static form of monumental knowledge transfer is replaced by a process of knowledge transformation. In this context the learning within a supply chain would move away from a risk-aversive position towards a risk-sharing one, and be searching opportunities for creating a new future.

Vygotsky's (1934, 1978) principles of proximal development is based on the difference in problem solving done by a child acting alone and what can be achieved with the guidance of an adult. In terms of work activities this means bringing together people of different experiences, expertise and responsibilities to solve a challenge of mutual interest (Rose-Anderssen & Allen, 2008). Through comparison and reflection these people are therefore being prepared to learn more and expand on what each of them could have achieved alone. This would mean going beyond each of the learner's previously acquired socially existing or documented forms of knowledge (Engeström, 1987). In the context of aerospace supply chains the success of the routines of the past are thereby challenged in order to produce more radical solutions in terms of practices, technologies and products.

Levels of Learning—From Adaptation to Creation of the 'New'

In work activities there is a move from individual action to the object of collective activity. In developing the object, the distinctive feature of human activity is the continuous creation of new instruments (Engeström, 1987). Instruments are here seen as the tools individuals are using at a non-conscious level in performing routine tasks and as models of thought when solving problem tasks.

Although the collective behavior of a work activity in principle should be characterized by a continuous process of change, the dynamics of change might be inhibited by the success of past practices, technologies and products. Following a success, for example, there is a tendency to simply use established routines again. The collective object that should be driving and guiding eventual changes to the activity has therefore been frozen. Thus new instruments that could transform the object are not created. Radical change of a work activity may therefore not be a natural choice of strategy for a single firm or from a supply chain management point of view. In such cases it might be more natural to keep to adaptive and reactive forms of learning.

To change from the successful routines of the past means a change of mind-sets. This might rely on significant disturbances within the market environment and therefore a need for change. In that case, the alternative to adaptive and reactive forms of learning is expansion which transcends the old context. The potential of an activity is that here people meet and interact to form new meanings that go beyond the evident limits of each of the people involved. The creation of the 'new' is not generated from the old but from the living movement of an activity leading away from the old (Engeström, 1987).

Situations of Dilemma, Exploration and the Creation of Radical Solutions

Important in activities are socially essential dilemmas which cannot be resolved through separate individual actions alone. Boundaries, therefore, are experienced as situations when individuals face a dilemma (Kerosuo & Engeström, 2003). This happens in interaction between individuals, and between their ideas. This typically happens in boundary zones between different activities. Creation of new instruments may resolve such dilemmas. This is facilitated through exploratory actions. Then tensions occur, leading to an analytical stage where new abstract models are produced which are tested and reflected upon, and eventually replace earlier ones (Blackler *et al.*, 1999).

Allen (2001) argues that because people have limited knowledge of what lies ahead they therefore cannot know precisely what actions to take at present. It is in the context of this dilemma of ignorance that learning, although never perfect, becomes important. It is in such situations of multiple choices where a diversity of perspectives and ideas are brought together in the supply chain that the exploration into more radical solutions can be made (Rose-Anderssen & Allen, 2008). This exploration into the unknown is achieved by challenging the routines and thinking of the past. These are characteristics of expansive learning processes. These expansive learning cycles, Engeström (1987) argues, should be viewed as partially destructive rejection of old perspectives and practices. These are important arguments as in several of the companies interviewed, interviewees realized that the industry as such has been conservative. It is therefore a key challenge to move away from the old through the creation of new instruments for change.

As the supply chain is a complex system of multiple work activities, learning can therefore be seen as a self-organization of collaborative relations and practices. The work activity therefore becomes an ongoing combination of intentions and emergence. Thus as individuals enter an activity, they face the challenge of acquiring the shared artefacts or established cultural means. At the same time they will face situations were they must formulate more desirable cultural means (Engeström, 1999). Due to the dualism between intentional action and emergence where individuals and their activity community are trying to change their situation, they are themselves transformed by these very actions. This is learning in doing. The connection of multiple work activities into a supply chain therefore complexifies learning and change. Supply chain management becomes a complex task of resolving networks of creative tensions that result from multiple dilemmas.

Methods of Data Collection

The research question for the paper is; what are the implications of different levels of learning and transformation of knowledge on competitive advantage? The model created from an activity theoretical point of view defines three levels of learning; adaptive, reactive and expansive. Empirical answers to this were sought through interviews.

Substantial literature research was conducted within several areas of relevance, i.e., on supply chains in general, on aerospace supply chains specifically, on the history of the commercial aircraft industry, and literature on evolution, learning, knowledge and change was conducted. Based on the interpretation of this data, open ended interviews were conducted. The interviewees were presented with a sheet of paper with the following key questions along with issues to expand on: *How have key practices changed and been introduced along the life cycle of an aircraft model and from aircraft model to aircraft model?* This was to be discussed in terms of the issues of: *Types of supplier selection, types of relationships with customers and suppliers; coordination and integration; training, learning and development (that is, how do people learn and therefore change practices?); quality assurance; risks; lean/agile practices; costs; offsets; technology; and future scenarios.* This paper focuses on the transformation of knowledge and learning issues elicited from the interviews.

In these interviews the interviewees were allowed to talk freely and tell their stories. Each interview became a stepping stone of comparison and reflection prior to entering the next interview session. The issues presented were the same each time. But the interviewer's questions on elaboration changed. This constant comparison and reflection on data is typical of the iterative and evolutionary process of discovering grounded theory (Glaser & Strauss, 1967).

The access to industry is always difficult to achieve. This limited the interviews to be conducted at one main airframe manufacturer only. However, the 1st tier suppliers visited for interviews were suppliers to both main competing airframe manufacturers. They could talk about their relationships to both companies. This made them part of a larger supply chain network. Some contacts in this research project were extraordinary helpful. That opened access to other companies. This meant the interviews could be conducted in Europe, Japan and the USA at the levels of airframe manufacturer and original equipment manufacturers (OEMs)/1st tier suppliers. The interviews were conducted in 7 companies with key senior managers and directors in each company. The interviews were conducted either as individual interviews or as focus group interviews. There were 8 individual interviews, one group of 2 interviewees, one group of 3 interviewees, and one group of 5 interviewees.

Focus group interviews produce an opportunity to observe interaction on a topic talked about. The advantage of individual interviews over focus group interviews is the control the interviewer has with closer communication with the interviewee (Morgan, 1997). Our experience, however, is that our focus group sessions became close conversations between the interviewees. The dialogue between interviewees was only interrupted when the interviewer intervened to ask for elaboration. The interviewees rather collaborated in an activity where the object was to explore and make sense of a specific issue. Due to the interviewees' complementary insights, the synergies produced were arguably an expansion of what the individuals could have produced alone. The atmosphere also became more relaxed than in the individual interviews where the individual may feel under constant pressure to perform.

Discussions on Knowledge Transformation, Learning and Development

In the following subsection the case is discussed from the point of view of the model of knowledge transformation and learning.

Inhibited Learning and the Restrictions of Old Supply Chain Routines

Prior to the intervention of Airbus into the world market for large airplanes a rather stable situation of slow evolution of change to practices and products existed. The airframe manufacturing could be argued to be in a context of one dominant voice. The division of labour was therefore based on a hierarchy of decisions with no collective object formation process, as there was one predetermined solution only. The airframe manufacturers produced the instruments, as strictly defined drawings and specifications. Similarly, in the case of military aircraft production a government imposed a strictly defined design on its supplier. In both cases learning, development and change were therefore inhibited. There were no significant disturbances that could initiate irreversible change. The rules of conduct within the companies and between supply chain tiers therefore became resistant to change.

> *Traditionally innovation with the suppliers has not been that great. We have done most of it and it is passed down as a detailed spec or a drawing in a sense* (Airframe manufacturer).

[The] aircraft industry in Japan was mainly for military aircraft. Our only customer is the Japan air-defence. [A] military contract is very unique. The only customer is [the] Japanese government, and the only supplier is us. You know, the changing is not that good for military contracts. Changing process experiences is difficult. If we try to change a process of [the] production, Japanese government must approve, inspect and review. They must approve the change process. This is very difficult process. So, historically our production system has not been changed for a long time (Japanese structural supplier).

Knowledge therefore was treated as if it was a commodity that is easily transmitted and controlled. It had served the success of the past. With past success there was apparently no need for a collective object formation process for change. Thus there was no creation of new instruments, learning and change.

Initial Dilemma Situation

As the U.S. airframe manufacturers were losing market share to the new European consortium of aircraft manufacturers in the mid 1970s, this triggered off models of strategic international relationships in order to utilise the best supplier expertise there was. The American companies were seeking more integrated systems and therefore looking for more to be done by the individual supplier company. In these new relationships they took into account market access, the suppliers' access to capital and the suppliers' unique technologies. Simultaneously suppliers had to commit themselves to cost reductions to achieve long term relationships with their customer. This was based on formal partnerships and partnership sourcing. It was a more collective channelling of competing interests into shared interests.

I would say the best learning [we] had was when they [the Europeans] were taking 50% of the market. There is always that when there is no competition you tend to become very arrogant. But as you enter into world competition you start trying to find the best players around the world (Airframe manufacturer).

A situation of creative tension therefore occurred when the US manufacturers lost 50% of the U.S. dominated world market for airplanes. This outsider intervention caused irreversible changes to the whole industry. At first, the responsive reaction from the US airframe manufacturers led to marginal improvements only as they sought out solutions mainly related to cutting costs. Secondly, these environmental disturbances were great enough to create more competitive awareness. This encouraged a practice of seeking out competence among international supplier with unique expertise.

There was therefore a creative tension or dilemma that had multiple and unpredictable paths to possible resolution. The success and speed of the resolution depended on the extent to which the supply chain was able to take advantage of the diversity of experiences and expertise into a collective object formation process, creating new instruments for change.

Knowledge Transfer and Adaptive Learning

Traditionally airframe manufacturers were in the dominant position of a high technology expert and the supplier would simply manufacture to strictly defined drawings and specifications.

There is a general consensus that the industry is still viewed as very conservative compared to, for example the automotive industry. But with the success of the automotive industry in improving their manufacturing practices and systems, and the increasingly competitive global aerospace industry, major changes were taking place. However, these changes were mainly coming from the top of the chains and emphasise the need for the supply chains to learn from the automotive industry and become lean.

> *Coming from automotive, I think there are different ways to do it. This is a pretty conservative industry, generally. The individual is also quite conservative. So the phase change tends to be slower than you would anticipate* (Mechanical systems supplier).

Adaptive learning was therefore manifested in the industry's attempt to embrace the lean practices associated with the automotive industry. Although the airframe manufacturer has very successfully implemented lean practices at their production sites, this was not true throughout their supply chains. They have therefore been trying to impose their demands as instruments for change. These demands were acknowledged by a key Japanese supplier. When these demands changed the mind-sets at the top of their company, they became willing to adapt to lean practices. In due course they would impose these practices on their own suppliers. However, their rigid social rules, en-cultured in resistance to changing previously successful practices of the past, had inhibited them from taking the initiative and seeking out lean practices experiences from within their own automotive division. Company behavior reflected an outlook of closed boundaries which failed to adapt external ideas and create new practices.

> *We are facing production problems right now, and they* [airframe manufacturer] *are, directing us to building under their instructions. So, we are now, you know, learning lean concept. How to do lean activities from them? Now we are learning. The biggest change in this century is the concept of the mind, you know. Our general manager and our executive, now they have changed their minds. We must be lean. So we are trying to change to be lean. The change has happened just recently. It's maybe two years ago,* [then] *we hadn't thought about that* (Japanese structural supplier).

However, in general, the use of e-portals for exchange of information and adapting new practices are shown to be successful instruments for supply chain integration.

> *We have developed portals in our system of design and engineering that enables us to work with our suppliers and researchers and partners at a completely virtual basis* (Airframe manufacturer).

> *We all have portal systems where we export our drawings in, where people can work on them more like real-time. So, it has for the last 10 years, the tools that have enabled the supply chains to become better integrated have improved phenomenally* (Equipment supplier).

Still, learning from fellow suppliers about lean production is bringing suppliers up to established industry standards only. Hence, the context is not changing much from the traditional practice of one dominant voice. That is, adaptive learning practices do not bring about enough disturbances for the continual formation of an emergent object. These practices are not boundary crossing and produce marginal improvements only. They are therefore helpful transformations in sustaining in the immediate future only.

By adapting to industry wide practices, it means that every competitor was responding to the challenges of the environment in the same way. They all adapted to pressure by adopting the marginal improvements of the same industry wide routine behavior.

Reactive Learning

Reactive learning within a supply chain was identified to take place in three ways. Firstly, the airframe manufacturer argues they have 'lessons learnt' exercises internally to learn from mistakes. Similarly they have random audits to assist suppliers to learn from their mistakes and correct their own work. These are quality improvement issues bringing products and practices up to expected standards only. Secondly, and more dynamically, a supplier is sharing and learning with suppliers to improve products.

> *Every quality reject is subject to a regroup cause analysis. And that is a team event were people from the shop floor, from purchasing, people from supplier, people from customer and also support crew, like engineers would be involved* (Electrical systems supplier).

Thirdly, good marketing principles were exercised when the airframe manufacturer started to listen to the needs of the airlines. This involvement with the airlines at an early stage of a new aircraft model made a particular aircraft take a huge part of the market compared to the competition. But still only took low risks and had a conservative approach to aircraft development. Compared to previous aircraft they made major improvements to the interior, to the cockpit etc. But these were, however, what they called, integrating point solutions only. They were renewals of some significant parts of old aircraft design rather than launching a totally new type of airplane. The approach therefore gave them marginal competitive abilities only. In practice the changes came about as repairing the single voice of past principles of design.

> *A decade ago really,* [we] *probably felt like it applied more to working together with our airline customers as a stronger connection. In driving the product design the airlines were embodied in the xx* [plane], *which we have huge part of the market compared to the competition. And I think we have learnt. We have been learning that aspect of collaboration* (Airframe manufacturer).

> *I think that the xx* [plane] *was not taking an existing system, but taking low risk and a conservative approach. It is not much different than the other aircraft we have produced. We made improvements with the cockpit, the navigation system and the seating capacity, storage, you know,* [the] *engine. It is a two engine aircraft. It is very powerful. They were what I call point solutions, moving to an integrated solution. We were very successful with* [that plane] (Airframe manufacturer).

These 3 slightly different approaches show good collaborative intentions. But the intentions did not facilitate instruments that are critical to the present, as new and radical perspectives were not brought in. The object remained frozen and they were repairing present tools only and were therefore merely hill-climbing in a smooth and familiar terrain. This made the supply chain competitive but did not bring about innovation and therefore industrial leadership. The example illustrates that whatever diversity of ideas existed, the inhibitions from embedded practices and old routines were able to stop the creation of critical disturbances for irreversible change to their design and production.

Major Dilemma Situation, Expansive Learning and Creating New Knowledge

Based on the positive previous experiences with early involvement of airlines, the airframe manufacturer started to spend more time with airline customers to make the passenger experience more rewarding. As part of the early involvement strategy they also started to share more with the 1st tier suppliers at that stage. In that sense the airframe manufacturer was moving from being risk-aversive to accepting risk in order to explore more and become innovative. They became dedicated to continuous development and trained their workforce to improve skills in order to become innovative. This was making them more open to communicating on mutual continuous improvement and sharing more with key suppliers, and they were trying to impose those philosophies down their supply chains.

> *We are an extremely risk-aversive company. And I think this is an indication [that] we have started to accept risk and become innovative* (Airframe manufacturer).

We [have] lately spent more time with our airline customers and trying to get to the ultimate customer. Why? In [order] *to figure out how to make the whole passenger experience more rewarding* (Airframe manufacturer).

The sustaining programme relationship with the supplier has changed over the years drastically. Today it is much more of an open communication, working together and continuous improvement. We are spending a lot more time educating and training (Airframe manufacturer).

A paradigm shift of philosophy occurred when the airframe manufacturer tried to bring together the best expertise and production capacity with the formalization of this through risk-sharing partnerships.

We learn and adapt based on trial and error. But because of that product cycle through our business, we also have to take significant risk to push the development (Airframe manufacturer).

Risk-sharing partners are also chosen for financial and sales reasons. In terms of financial reasons, the suppliers needed capacity to invest in innovation and thus producing a supply chain that has a competitive advantage. With sales, the suppliers are also expected to promote sales in the home countries of potential airline customers. Networks of maintenance and engineering suppliers are also built up in low cost economy countries to support potential sales in these countries. The aircraft manufacturer sees themselves as large scale integrators. Simultaneously they are trying to bring their suppliers up the value chain by trying to also make suppliers more responsible and efficient. The move from being satisfied by exploring within a familiar and optimized terrain into allowing the diversity of a multidimensional performance landscape of risk-sharing partnerships is therefore a bold step into inventing a future for a supply chain.

We need other companies to be risk-sharing partners. That could be financially. But also from a sales standpoint. Bringing other partners on board to help influence decision makers [in their countries]*, so that we can increase our sales* (Airframe manufacturer).

We also approached India to develop engineering skills, because we know that when we introduce commercial aircraft into a commercial market these airlines would need engineers [to] *help them maintain those airplanes* (Airframe manufacturer).

A major dilemma situation occurs due to the contradictions between the tradition of airframe manufacturer exercising customer power over suppliers and the concept of risk-sharing partnerships to enhance product innovation. Such partnerships are based on the dialogical principles of little power difference and closeness between potential partners. However, the airframe manufacturer exercises power as they replace previous contracts with suppliers by risk-sharing partnership contracts. From the moment the contracts have been signed power can be argued to have shifted to the endurable artefact, the contract. The contract from then on becomes an instrument for the development of sharing, creativity and innovation. At the same time, because of its strength, it might inhibit other looser supplier relationships potentially beneficial to the supply chain.

> *And if you look at the main contracts for example, for the new airplane, they have called back the terms we have with them. Pretty much straight through. And it is a partnership you know. They have become really a part of our business in the way their contract reflects. There is risk associated with that obviously* (Mechanical systems supplier).

This formal action is an attempt to open boundaries and encourage multi-voicedness. At this bifurcation point a path for new behavior is initiated. Thus several dilemma situations occur. These are shifts of roles from being suppliers and being told what to do, to becoming risk-sharing partners exercising expert power and having a voice and becoming responsible. These are moves from the marginal innovative potential based on previous experience and financial ability to actually stretch boundaries by implementing innovative solutions.

> *Before it was; 'here is your contract, this is what you do'* (Airframe manufacturer).

> *So as opposed to previously, we would have supplied subsystems and components. But* [now] *our responsibility is to integrate that and give it to the airframe manufacturer. It is that integration the airframe manufacturer previously have done themselves* (Mechanical systems supplier).

Finally, the outcome of the risk-sharing partnerships has so far shown unique innovative strength in terms of a conceptually new product. By bringing together the best expertise and financial strength available it has therefore facilitated dialogue followed by expansive learning and innovation beyond expectation. Still the partners are on the move of good intentions to continually improve the implementation in praxis. They are still learning through adaptation, reaction and exploration at all levels of the supply chain.

> *And some of the things that I* [have] *read, that has truly made me proud about us in the last six months is, throughout the world, our new airplane is recognized as an extraordinary high technology advanced aircraft. People believed that we have stepped out a very innovative solution. We have stretched the boundaries of technology, offered a unique solution that has never been offered to the industry before. And we have always felt that internally. The industry is concerned because of the innovation. We are an extremely risk aversive company. And I think this is an indication we have started to accept risk and be innovative. Quite honestly we have to be global. We have to have those relationships. So globally is, I think those supplier relationships are really important. If we want to be a large system integrator then we have to trust other people. And the only way we can trust other people*

is [in] relationships. It can not just be a business relationship it has to be a risk sharing relationship (Airframe manufacturer).

The financial power and expertise of every partner made it possible to collectively explore, learn and develop the new aircraft concept. The risk-sharing partnership facilitated the collective object formation process through exploring, learning and creating the new.

Conclusions: Knowledge Transformation and Levels of Learning

In this paper, the intervention started off when the airframe manufacturer tried to make sense of the more rapidly changing global market. This posed a major dilemma, and the airframe manufacturer realized that it could not be resolved either technologically or financially, by them alone. The new instrument chosen for resolving their dilemma was therefore risk-sharing partnerships. Once this had been decided it represented a major bifurcation point in the evolution of their supply chains. Thereby the old system of vertical domination became obsolete and a collective movement of partners was chosen to explore new opportunities. And the exploration of new opportunities was based on the inherent expertise accumulated by each partner. The challenge was however, how to develop this further. The new instrument of risk-sharing partnerships facilitated a new division of labour by sharing of diverse expertise for exploring new solutions.

These ideas of adaptive, reactive and expansive learning correspond to ideas used in earlier work by Allen *et al.* (2006) on "cladistic" evolution of organizational forms. In describing organizations as being bundles of practices, skills or techniques, the evolution of the industry was seen as an evolutionary tree created by the successive adoption of new practices. However, for many organizational forms the new practice might well come from other organizations in the industry and would correspond to "adaptive" or "reactive" learning. However, sometimes a new practice would emerge from within a particular organization and then it would correspond to "expansive" learning. In the cladistic papers, however, there is no distinction made between practices that are coming from elsewhere and those that emerge from within. Al that matters is whether or not the new practice is adopted or not. In some ways the detailed reflection here resulting from the interviews in the aerospace sector throw more light on the processes that lead organizations to create a new practice, which may diffuse across the industry.

It is essential to appreciate that all three forms of learning act together. As individuals or groups interact in a common activity they adapt successful practices from each other. They react to external changes and their own mistakes. They codevelop new practices. The collective learning therefore can be argued to be fluctuating between the three forms of learning as the context of the activity changes. The context changes due to imperfect and varied human

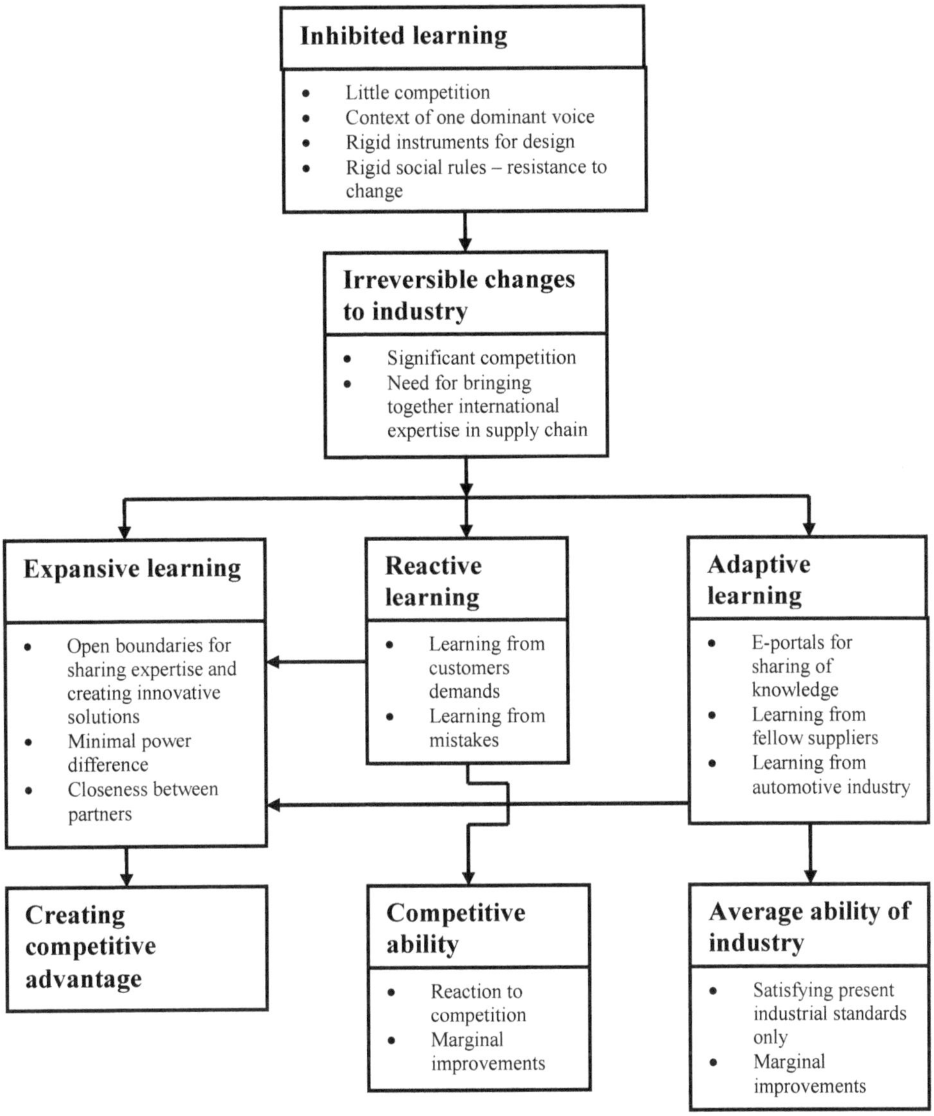

Figure 4 *Implications of Levels of Learning for Aerospace Supply Chains*

interaction. The simultaneously occurring collective learning that is developing new meanings enhances the collaborations skills for implementing new practices.

A simple rule is that adaptive learning in supply chains is about getting practices up to the present quality standards of the industry or comparable industries. Reactive learning is about getting practices that makes the supply chain able to compete. And expansive learning is about getting practices that gives the supply chain a potential for winning. In the long run, however, they need to act together.

Due to successful learning and development with one airframe model and its supply chain, the respective practices and systems are transferred as routines for other airframe supply chains. These then become established as new supply chain forms. In that sense the airframe manufacturer's success becomes the strategies for change for entire supply chains. This illustrates a clear path dependency of practices due to successful previous implementations.

> *It is much easier for us to take new technology,* [and] *get it on a new airplane model than to retrofit new technology back on an existing model... We are starting to have a better understanding of technology capability. I think we're on getting them on the next airplane programme. We are learning that from the* [new aircraft] *and the learning that comes out* [from that experience] (Airframe manufacturer).

However, the success of new supply chain forms in the long run become resistant to change until their inner contradictions exceeds stability and people engage themselves in the similar reflective processes that created them. The facilitation of continuous practices of collective reflection therefore is likely to become the essential instrument for innovative solutions and sustainability of a supply chain and its member firms.

The model of knowledge transformation through adaptive, reactive, and expansive learning could be useful in research when focusing specifically on changes to an existing supply chain. In terms of more interactive research with industrial partners the model could be applied to a supply chain in the process of being established. An understanding of the implications of the differences between the three concepts of learning could assist the management involved in the processes of supply chain change. Subsequently they could apply the learning mechanisms most appropriate at the different stages of change.

References

Allen, P.M., (2001). "Knowledge, ignorance and the evolution of complex systems," in J. Foster, and S. Metcalf, (eds.), *Frontier of Evolutionary Economics: Competition, Self-organization and Innovative Policy*, ISBN 9781840645255, pp. 315-350.

Allen, P.M., Strathern, M. and Baldwin, J.S. (2006). "Evolutionary drive: New understanding of change in socio-economic systems," *Emergence: Complexity and Organization*, ISSN 1521-3250, 8(2): 2-19.

Blackler, F. (1993). "Knowledge and the theory of organizations: Organizations as activity systems and the reframing of management studies," *Journal of Management Studies*, ISSN 0022-2380, 30(6): 863-884.

Blackler, F. (1995). "Knowledge, knowledge work and organizations: An overview and interpretation," *Organization Studies*, ISSN 0170-8406, 16(6): 1021-1046.

Blackler, F., Crump, N. and McDonald, S. (1999). "Managing experts and competing through innovation: An activity theoretical analysis," *Organization*, ISSN 1350-5084, 6(1): 5-31.

Cousins, P.D. and Spekman, R. (2003). "Strategic supply and the management of inter- and intra- organizational relationships," *Journal of Purchasing and Supply Management*, ISSN 1478-4092, 9(1): 19-29.

Day, M. and Atkinson, D.J., (2004). "Large-scale transitional procurement change in the aerospace industry," *Journal of Purchasing and Supply Management*, ISSN 1478-4092, 10(6): 257-268.

Easterby-Smith, M., Graca, M., Antonalopoulou, E. and Ferdinand, J. (2005). "Absorptive capacity: Tales from the field," EBK Publications, Working Paper 2005/08 ESRC.

Engeström, Y. (1987). *Learning by Expanding: An Activity-Theoretical Approach to Developmental Research*, ISBN 9789519593326.

Engeström, Y. (1999). "Activity theory and individual and social transformation," in Y. Engeström, R. Miettinen and R-L. Punamäki (eds.), *Perspectives on Activity Theory*, ISBN 9780521437301, pp. 19-38.

Gherardi, S. (2006). *Organizational Knowledge: The Texture of Workplace Learning*, ISBN 9781405125598.

Giunta, A. (2000)."Large Firms and subcontracting relations in commercial aircraft industry," in A. Giunta, A. Lagendijk and A. Pike (eds.), *Restructuring Industry and Territory: The Experience of Europe's Regions*, ISBN 9780117023802, pp. 37-56.

Glaser, B.G. and Strauss, A.L. (1976). *The Discovery of Grounded Theory for Qualitative Research*, ISBN 9780202302607.

Hasu, M. and Engeström, Y. (2000). "Measurement in action: An activity-theoretical perspective on producer-user interaction," *International Journal of Human-Computer Studies*, ISSN 1071-5819, 53(1): 61-89.

Johnson, T. and Ford, D. (2005). "At the receiving end of supply network intervention: The view from an automotive first tier supplier," *Journal of Purchasing and Supply Management*, ISSN 1478-4092, 11(4): 183-192.

Kerosuo, H. and Engeström, Y. (2003). "Boundary crossing and learning in creation of new work practice," *Journal of Workplace Learning*, ISSN 1366-5626, 15 (7/8): 345-351.

Lektorsky, V.A. (1984). *The Dialectic of Subject and Object and some Problems of the Methodology of Science*, Moscow: Progress Publishers.

Morgan, D.L. (1997). *Focus groups as Qualitative Research*, ISBN 9780761903420.

Nollett, J., Ponce, S. and Cambell, M. (2005). "About strategy and strategies in supply management," *Journal of Purchasing and Supply Management*, ISSN 1478-4092, 11(2/3): 129-140.

Preiss, K.J. and Murray, P.A. (2005). "Fashions of learning: improving supply chain relationships," *Supply Chain Management: An International Journal*, ISSN 1359-8546, 10(xx): 18-25.

Rose-Anderssen, C., Allen, P.M., Tsinopolous, C. and McCarthy, I., (2005). "Innovation in manufacturing as an evolutionary complex system," *Technovation*, ISSN 0166-4972, 25: 1093-1105.

Rose-Anderssen, C. and Allen, P.M. (2008). "Diversity and learning for innovation: Dialogue for collaboration," *Journal of Management Development*, ISSN 0262-1711, 27(3): 307-327.

Rose-Anderssen, C., Ridgway, K., Baldwin, J.S., Allen, P., Varga, L. and Strathern, M. (2008). "The evolution of commercial aerospace supply chains and the facilitation of innovation," *International Journal of Electronic Customer Relationship Management*, ISSN 1750-0664, 2(1): 63-84.

Vygotsky, L.S. (1934). *Thought and Language*, ISBN 9780262220293.

Vygotsky, L.S. (1978). *Mind in Society: The Development of Higher Psychological Processes*, ISBN 9780674576285.

Christen Rose-Anderssen was a Research Associate at the Advanced Manufacturing Research Centre with Boeing, University of Sheffield. He was engaged in the ESRC research project Modelling the Evolution of the Aerospace Supply Chain. He previously worked as a Research Officer in the project New Product Development as a Complex System of Decisions at the Complex Systems Research Centre, Cranfield University. He worked as a naval architect and manager in the shipbuilding industry in Northern Europe for many years. He worked as a consultant in shipbuilding in Asia and as a manager in the Norwegian offshore engineering industry.

James S. Baldwin is a Lecturer at the School of Management, University of Sheffield. He was engaged in the ESRC research project Modelling the Evolution of the Aerospace Supply Chain. He has previously researched into the sustainability of complex systems with an emphasis on manufacturing in South Yorkshire. He is also involved in research on the evolutionary classification and modelling of industrial ecosystems.

Keith Ridgway, OBE is head of the research institute, Advanced Manufacturing Research Centre with Boeing, University of Sheffield. He established the AMRC to carry out research in manufacturing technologies directly related to aerospace industry. He previously held the position as Professor of design and manufacturing and was Director of the Ibberson Technology Transfer Centre, Department of Mechanical Engineering, University of Sheffield. He was awarded the OBE in 2005 for services to UK manufacturing industries.

Peter M. Allen is head of the Complex Systems Research Centre, School of Management, Cranfield University. He worked for 20 years with the Nobel Laureate, Ilya Prigogine in Brussels. For almost 30 years he has been working on the mathematical modelling of change and innovation in social, economic, financial and ecological systems, and the development of integrated systems models linking the physical, ecological and socio-economic aspects of complex systems as a basis for improved decision support systems.

Liz Varga is a Research Officer at the Complex Systems Research Centre, School of Management, Cranfield University. She was engaged in the ESRC research project Modelling the Evolution of the Aerospace Supply Chain. She works part-time for the Society of Information Technology. She also provides management consultancy and programme management to a number of public sector organizations.

17. Sustainable Supply Networks: A Complex Systems Perspective

Liz Varga[1], Peter M. Allen[1], Mark Strathern[1], Chris Rose-Anderssen[2], James S. Baldwin[2] & Keith Ridgway[2]
1 School of Management, Cranfield University, ENG
2 The University of Sheffield, ENG

A complex adaptive systems perspective is used to examine the sustainability of the supply network in the commercial aerospace manufacturing sector. A framework for the analysis of coevolutionary dynamical change is used which examines the structure, integration methods and process dynamics within the supply network. Multiple methods are used for data collection from 8 firms in the sector. The framework identifies 14 management implications related closely to the sector's current heterarchical supply network archetype. The management implications address known environmental factors for the sector and the broader techno-economic paradigm.

Introduction

Significant changes have been observed in the structure and dynamics of supply networks over the last 20 years, driven by changes in the environment and by the firms within the supply networks. This paper recognizes firms, supply networks, and the environment as complex adaptive systems (Choi *et al.*, 2001). A paradigm shift to a complex adaptive systems view and a coevolutionary dynamical epistemology of supply networks holds promise for future research addressing Daft and Lewin's (1993) observation that firms are changing cataclysmically with respect to networks and that a new theory of supply networks is needed (Salancik, 1995). This shift also gives primacy to 'organizational becoming' (Tsoukas & Chia, 2002) and a new way to approach change.

The paper starts with an introduction to the tenets of complexity science, and how they apply to supply networks. It is followed by a description of the market and environmental context of jetliner manufacturing in order to place the firm and the supply network with the current eco-historical regime (Garnsey & McGlade, 2006). This includes a brief history of the demand for jetliners, followed by the identification of relevant global technological change, ending in a discussion of the current environment influences for the industrial sector: global connectedness, technology know-how, customer expectations (for quality, cost, and security), competition and 'regulations and legislation'. The case study methodology is described next followed by the case study results which are presented in the order of structure, integration methods and process dynamics of the supply network. Results from the case study, synthesized with the current environment raise the most relevant management implications for the sustainability of commercial aerospace supply networks. A discussion follows with suggestions for further research.

Multi-Layer Complex Adaptive Systems

Complex adaptive systems are the structuring mechanisms, integration methods and associated process dynamics which adapt, have emergent properties and are open to their environments (Varga, 2009). In complex systems terms, the environment is merely another layer in a nested system in which each system takes every other system as its environment in a process of coevolution whilst adapting to their environment. The environment (or landscape) is changed as a result of changes in the systems that constitute the environment (Kauffman, 1993). A complex system does not act in isolation (Barabasi, 2002).

Emergence occurs at different layers in the supply network, enabling and constraining the potential for new emergent properties at the next layer (Fuller & Warren, 2006). Existing structures are the consequence of the irreversibility and path-dependency of the supply network and of the organizations, past and present, within the supply network. Experimentation is important (Allen, 1988) as innovations and evolutions fit within the wider milieu of the social, cultural, environmental and technological of their own history: the *"eco-historical regime"* (Garnsey & McGlade, 2006).

Each organization's networks are idiosyncratic and have followed a path dependent process (Gulati & Gargiulo, 1999) conferring competitive advantage as they are not easily imitated or substituted. As a result of these relationships, dynamic network constraints and benefits occur (Gulati *et al.*, 2000), e.g., benefits of lock-in to a profitable network or lock-out of a failing network. Conversely, the constraints can act disadvantageously, e.g., lock-out of profits and lock-in to a failing network.

Small differences in market share can be amplified and develop into much larger differences (Arthur, 1994) so long as self-reinforcing processes, that is positive feedback, dominates self limiting processes or negative feedback which act as a self-regulatory mechanism and the key to equilibrium (Capra, 1996). One action may have varying effects on different parts of the complex system and may result in varying degrees of feedback, driving virtuous or vicious cycles (Holland, 1998).

The supply network has matured through a number of observable stages (see Table 1): from connecting intra-organizational components of inbound materials and outbound products; to dyadic (two-sided) supplier relationships in which each organization attempts to manage immediate suppliers; to dyadic chains which extend the relationships of the organization to both to customers' customers and suppliers' suppliers; to supply chain management in which all organizational supply chains are managed holistically; to integrated business networks which manage multiple businesses that create products and service packages; to demand chain communities, which manage multiple enterprises practising agility to customer demand. This paper adds the supply network as a complex adaptive system in which coevolution occurs between the firms and the supply network, in the context of the environment.

Supply networks are recognized as innately dynamic, responsive to the environment and constituted from the complex adaptive systems at the levels below. They are complex adaptive systems, which have variations and are selected for in particular environments as they coevolve with other complex adaptive systems.

The Environment of Commercial Aerospace Manufacturing

This section describes the environment of commercial aerospace manufacturing by examining the products of the sector, the global techno-economic paradigm and current environmental influences.

The Products: Jetliners

Modern day commercial airliners or jetliners are used to carry large numbers of passengers on medium and long-haul journeys across the globe. The firms examined in this paper produce jetliners carrying at least 100 passengers. These planes are powered by turbofan jet engines which are relatively quiet and fuel-efficient when compared to first generation of jetliners such as the de Havilland Comet and Douglas DC-8. Such jetliners were not manufactured until the end of World War II, the first being the Lockheed Constellation in 1943, and the Convair 880 and Boeing 707 not until the late 1950s.

By 1977, over 30 years ago, the Boeing 727 fleet had carried its one billionth (1,000,000,000) passenger, a first achieved by a commercial aircraft. Today the number is in excess of 4 billion. The 727 was the first best-selling jetliner in the world when orders passed 1,000 in 1972, although the Boeing 737 has exceeded the final total orders for the 727 which ceased production in 1984 (Freeola, 2008).

Primes, that is those organizations which produce jetliners, are focused in the US and Europe. Some small amounts of production occur in Brazil and Russia. China plans to produce an jetliner with 150 seats in 2010, but are first concentrating on smaller regional planes. China produced a large jetliner, the "Yunshi," which had its maiden flight in 1980 but production was stopped as it failed to gain foothold (Asia Times, 2006).

Stage of Development	Type of Trading Relationship	References
Intra-Business Chain	Internal supply chain integrates business functions involved in the flow of materials and information from inbound to outbound ends of the business	(Harland, 1996)
Dyadic Relationship	The management of dyadic or two party relationships with immediate suppliers; Extended to downstream distribution channels and upstream production chains; Structure and scope of supply chain consists of material and information processing units: demand, value-adding transformation and supply	(Harland, 1996) (Womack *et al.*, 1990; Womack, 2002) (Davis, 1993)
Dyadic Chain	The management of a chain of businesses including a supplier, a supplier's suppliers, a customer and a customer's customer, and so on	(Harland, 1996)
Supply Chain	Management of multiple company relationships (SCM); Managing and coordinating multiple business activities across functions and firms, and viewing the supply chain as a single entity, rather than as a set of separate functions.	(Lambert *et al.*, 1998) (Mentzer *et al.*, 2001) (Larson & Rogers, 1998) (Christopher, 2005, 1992)

Stage of Development	Type of Trading Relationship	References
Integrated Business Network	The management of a network of inter-connected businesses involved in the ultimate provision of product and service packages required by end customers	(Harland, 1996)
Demand Chain Communities	Demand driven, agile, multi-enterprise organizations, increasingly complex with various inter-relationships between companies, growing number of participants which does not remain constant throughout product life cycles; unlike old models where customer orders were delivered from on-hand inventories.	(Hewitt, 2000) (Lummus & Vokurka, 1999) (Bowersox *et al.*, 2000)
Complex Adaptive Supply Network	Complex Adaptive System; emergent and coordinated; embedded in its environment and eco-historical regime; non-linear effects and feedback; multi-layer coevolutionary system with structure, integration and dynamic	(Surana *et al.*, 2005) (Choi *et al.*, 2001) (Kauffman, 1993) (Garnsey & McGlade, 2006) (Holland, 1998) (Arthur, 1994) (Varga, 2009)

Table 1 *Developments in Supply Network Explanation*

The Boeing and Airbus fleets dominate commercial aerospace manufacturing. Deliveries to March 2008 from Boeing (first was the Douglas DC-9 in 1969) number 16,713 (Boeing Website, 2008) and Airbus (first was the A300-B4 in 1974) number 5,545 (EADS Airbus, 2008). Airbus has delivered more jetliners than Boeing since 2003 and orders for aircraft for 2008 show that Airbus continues to maintain a lead in production with a net order book of 756, whilst Boeing's is 661 (Crump, 2009).

Only a small number of jetliner models seating at least 100 passengers have been introduced over the last 50 years by what are now only a handful of organizations: Boeing (US), Airbus (Europe), Embraer (Brazil), Ilyushin and Tupolev (Russia). The jetliners are distinguishable by size: wide body/2 aisle and narrow/single aisle. By the third quarter of 2008, a total of 5,808 wide-body jetliners have been delivered, over 50% being Boeing planes. Three-times as many narrow-body or single aisle jetliners have been produced (16,745), a third of which are Boeing 737s. The jetliners which continue in production are easily identified in Table 2 as those with a to-date which indicates the cumulative count at that point.

Variants of commercial jetliners sometimes exist for the transport of freight. Other adaptations include that for VIP corporate use, typically where the jetliner contains fewer seats. A third variant is government use, where it is modified for different types of use, such as airborne tankers, air ambulance and reconnaissance, as well as troop carrying purposes. These variants are not of direct interest in this paper however the versatility of many jetliners has contributed to their longevity, for example, the Boeing 737.

Global Techno-Economic Paradigms

Three generations of globalization are used to describe the economic progress towards our current environment. Described variously as dominant "logics" of production/techno-economic paradigms (Tuomi, 2007), Schumpeter's waves (Schumpeter, 1942) or Kondratiev's macroeconomic cycles (Kondratiev, 1984), these paradigms describe the dominant or standard ways of economic development over different time periods. Their relevance is to highlight new generic technologies driving renovation and innovation leading to new economic models through waves of creative destruction (Schumpeter, 1942). These globalization paradigms are consolidated in Table 3. The first globalization included four "logics" of production: water power, steam power, electricity; and oil.

The earliest dominant "logic" was the harnessing of water power via canals in the period from the 1770s. This coincides with Schumpeter's 1st wave in which renovation was triggered by water power and the use of iron. The standard ways of working moved to mechanized methods.

The second dominant "logic" was steam power. This drove the investment in railways, increased the scale of production and the widespread use of

Original organization	Organization today	Wide-body/ long-haul Model	dates	number delivered	as at if still produced	Narrow-body/ medium-haul Model	dates	number delivered	as at if still produced
United States & Canada									
Boeing	Boeing	747	1970 -	1409	30/09/2008	707	1958 - 1979	1010	
		767	1982 -	968	30/10/2008	717	1999 - 2006	156	
		777	1995 -	741	30/10/2008	737	1968 -	5857	30/09/2008
		787	2009 -	0	31/12/2008	757	1983 - 2004	1050	
Consolidated Vultee Aircraft Corporation	Boeing					Convair 880	1959 – 1962	65	
McDonnell Douglas & Douglas Aircraft Company	Boeing	DC-10	1971 – 1988	386		DC-9	1965-1992	976	
		MD-11	1990 – 2000	200		MD-80	1980 – 1999	1191	
						MD-90	1995 – 2000	116	
Lockheed	Lockheed Martin	L-1011 TriStar	1972 – 1984	250		Constellation C-69/ Super Constellation C-121	1943-1958	856	
South America									
Embraer (Brazil)	Embraer					E-jets: E-170, E-175, E-190, E-195	2002 -	446	30/09/2008
Europe									
Airbus	EADS	A300	1974 - 2007	561		A318	2003 -	65	30/09/2008
		A310	1983 - 2007	255		A319	1996 -	1126	30/09/2008
		A330, A340	1993 -	882	30/09/2008	A320	1988 -	2016	30/09/2008
		A350	2011 -	0	30/09/2008	A321	1994 -	483	30/09/2008
		A380	2007 -	22	31/12/2008				
British Aircraft Company (BAC) & British Aerospace	BAE Systems					BAC-111	1963 – 1989	244	
						Concorde	1976 - 2003	20	
						Bae-146/Avro RJ	1983 - 2003	387	
Fokker (Holland)	none					F100	1986 – 1997	283	
Vickers-Armstrongs	BAC					VC10	1962 - 1970	54	
Russia									
Ilyushin	Ilyushin	Il-86	1980 – 1994	106		Il-62	1967 – 1993	276	
		Il-96	1992 -	28	30/09/2008				
Tupolev	Tupolev					Tu-204/ Tu-215	1995 -	68	30/09/2008
						Tu-334	2009 -	0	30/09/2008
Totals				5808				16745	

Table 2 *Jetliner Models and Deliveries*

Dominant "logics" of production/ techno-economic paradigms	"Standard way of doing business," utilizing the new generic key technology	Schumpeter's Waves
Canals and availability of water power, new infrastructures from 1770s	Mechanized work	**Schumpeter I** renovation triggered by availability of waterpower and use of iron
Steam and railways; universal postal service and the emergence of telegraph networks	Increased scale of production, which in turn was based on standardization of mechanical components; increased use of telegraph	**Schumpeter II** use of steam, the production and wide use of steel
Age of Steel, Electricity and Heavy Engineering, starting from 1875	Cheap steel, which, among other applications, was used to build steam-engine based steel ships, worldwide railways, electrical networks for illumination and industrial use, as well as bridges and tunnels that complemented the new transport technologies	**Schumpeter III** wide use of the electricity; development of internal-combustion engines; start of chemical industries from 1900s
Age of Oil, Automobile and Mass Production from 20th century End of 1st globalization	Associated with key new technologies such as the automobile, electricity and the availability of cheap oil, the standardization of products, utilization of scale benefits, realization of the emerging economic potential required the development of infrastructures such as highway networks, airports, oil ducts, and worldwide telecommunications; Mid 1960s—international telephony	**Schumpeter IV** rise of petrochemical products from oil and gas; development of electronics and aviation industries
Age of Information & Telecommunications (or the Information Society) 2nd globalization	Since the invention of semiconductors and general-purpose microprocessors, the economic system has increasingly been organized around efficient use of information and communication technologies; exchange of documents with rich content in early 1990s; direct dialing and email messaging	**Schumpeter V** growth of digital networks, software and the new media
3rd globalization	Broadband communications network, global division of labour; internet-based business models, real-time virtual service	

Table 3 *Globalization Paradigms (adapted from Tuomi, 2007 and Schumpeter, 1942)*

steel which is described by Schumpeter's 2nd wave. Standardization of mechanical components was a feature of the increased production. Universal postal services and an emerging telegraph network were also characteristic of this era.

The third "logic" was the age of steel, electricity and heavy engineering from around 1875. Cheap steel was used to build ships, railways, bridges and tunnel. The wide use of electrical networks for lighting and industrial purposes completed these developments. Internal combustion engines and the start of the chemical industries typify this 3rd Schumpeter wave.

The fourth "logic" was the age of oil and the new technologies of the 20th century such as automobile production. Products became standardized and scale benefits were extracted as functionally diversified and hierarchical organizations appeared. Investment in infrastructure such as highways, airports, oil ducts and worldwide telecommunications was typical in the West. The rise of petrochemical products from oil and gas, the development of electronics and aviation industries characterizes Schumpeter's 4th wave. After the 1940s, jet airplanes enabled managers to travel regularly to distant plants.

The 2nd globalization is known as the age of Information Technology or the Information Society. The invention of semi-conductors and general purpose microprocessors has shifted the economic system to the use of information, enabling the exchange of rich content via direct dialing and email messaging. This is the start of Schumpeter's 5th which highlights the growth in digital networks and software.

We are now in the 3rd globalization or Schumpeter's 5th wave, defined by broadband communications networks, global division of labour, internet-based business models and real-time virtual service. Continued miniaturization is enabling the end-customer to perceive an increasing value during flights with the accessibility of increased multi-media technologies.

Environmental Influences

Five types of environmental influences are evident in jetliner production: global connectedness, technology know-how, customer expectations (for quality, cost, and security), competition and 'regulations and legislation'. The following sections discuss some examples of these major changes.

Global Connectedness

The transition from a national-regional logic to a global-local one (Longhi, 2005) is evidenced in changes at Toulouse, the European capital of the commercial aerospace industry. The increasing use of information communication technologies (ICT) is uniting global networks across different time periods, enabling the sharing of information and the use of previously untapped resources in the development, production and servicing of aircraft (Ho *et al.*, 2003).

Technological Know-How

The dominant technologies in jetliner production are the jet engine and composite materials. The jet engine was the technology which enabled the creation of jetliners. Modern jetliners are powered by turbine engines which operate efficiently at much higher altitudes and with greater reliability than piston engines. They also produce less vibration and noise. Many second generation jetliners such as the BAC-111, Boeing 727 and Tupolev Tu-154, used a rear-engine T-tail configuration. Whilst this configuration is still used on some short and medium haul planes, design of jetliners has converged to a low-wing design with engines mounted in under-wing pods. Access is quicker and easier for maintenance compared to tail-mounted engines and it enables a lighter wing structure.

Composites are advantageous due to their light weight, and so good for fuel consumption, and strength, however they are more complex to manufacture than most metal structures and are difficult to repair and to detect faults. Collaborative knowledge sharing in the recent developments in composite materials shows the value of out of sector knowledge for aerospace (McAdam *et al.*, 2008). In the recent A380, composites form 25% of the airframe by weight, including the central wing box although the fuselage is aluminium.

Customer Expectations

At the start of the 21st century, the severe knocks to the global commercial aerospace industry following 9/11 (New York), SARS (originating in China) and the Iraq war caused industry lay-offs and consolidation as customer demand for air travel waned. Terrorism acts such as 9/11 create incredible shocks to aircraft orders, although traditionally the success of jetliner models and the firms producing them has depended on the severity and frequency of accidents. Nearly all jetliner accidents make the news for the very reason that they transport the public. Accidents with fatalities are reported (see for example Airsafe (2008)) giving details of crash events and crash rates based on number of flights. Safety and the role of the civil aviation authority of each nation in maintaining safety standards during civilian flight (and the power to ground aircraft) are critical to customer confidence in selecting flight travel.

Air traffic has now fully recovered from the impacts of 9/11 and North American and European airports are facing increased demand and urgent need for expansion of runways and terminals but this is being resisted largely as a result of environmental opposition. Dramatic economic growth and increased aviation demand has been achieved in emerging market countries. In China, this growth is led by government investment and economic partnerships in manufacturing and exports. In India, public-private partnerships, government liberalization and high-technology industries drive the growth. In the Middle East, government investment in infrastructure is funded by oil and gas revenues. Each of these emerging areas are able to respond aggressively to demand, building airport facilities quickly on a massive scale. These changes are "*shifting the future of aviation from West to East*" (Assa & Denton-Brown, 2008).

Social and economic benefits of air transport, including improved international cooperation and increased consumer choice are persistent drivers increasing demand. Other social changes, such as the growth in obesity due to energy dense food, motorized transport and sedentary lifestyles (Foresight, 2007) will influence the design of future planes.

Competition

Government protectionism is observed in two ways. Statutory instruments or laws are sometimes created, supporting industry restructuring and consolidation, enabling them to compete globally, but also preventing sales, and loss of intellectual property rights, to other nations. Secondly, governments have awarded contracts to local firms, rather than sought to achieve best value, although this approach tends now to be more related to defence contracts. Historically, national prestige was attached to developing jetliners and bringing first generation designs into service. There was also a strong nationalism in purchasing policy. In 2008, we find Boeing's bid for a U.S. Air Force contract has been won by a non-U.S. based consortium. Nationalism was overcome by global competition despite Boeing's reliance on global networks to complete its contracts (Epstein & Crown, 2008).

Industry consolidation and tiering (reduction in direct suppliers) is likely to take place in the UK and increased sourcing from low cost economies is predicted (DTI Aerospace and Defence Directorate, 2003). The location of developing skills and knowledge, particularly in great numbers in China and Russia will be a source of competition for the West.

Regulations and Legislation

The effects of flight travel continue to influence the industry. The need to reduce CO_2 emissions and effects on global warming, have become global political issues. But remedies to reduce emissions appear to have a negative effect as world-wide demand for air service including freight transport which is expected to double by 2010 (Air Transport Action Group, 2008). Political interventions include carbon-offsetting, proposed restrictions on airport growth and routes, and rising air passenger duty. Other environmental concerns range from aircraft noise, to fuel consumption and the need for energy efficiency, improved infrastructure and land use. Compliance is required by firms with a range of environmental legislation (SBAC, 2001). Legislation in the form of "The Civil Aviation Act" implements, amongst other things, commitments to sustainable aviation and protection of passenger interests as defined in the Future of Air Transport White Paper (Department for Transport, 2005).

Air pollution and toxic waste monitoring and intervention continue to dominate the agendas of environmental protection agencies. The airline industry is responsible for about 11 percent of greenhouse emitted by the U.S. transportation sector. Boeing estimates that biofuels could reduce flight-related greenhouse-gas emissions by 60 to 80 percent by blending algae fuels with existing jet fuel (Gonzalez, 2007). An informal collaboration between Boeing and leading biofuels makers, for example in Brazil and New Zealand, is testing biofuel options for aviation.

The primes have programmes of investigations which are focussed on improved product efficiency. Four concept designs were being examined by Boeing (Gates, 2006) codenamed after the well-known Muppets. All four design designs have rear-engine layouts and concentrate on reducing fuel usage. Two of the concept planes target low emissions and low noise.

Summary

The major challenge facing the aviation industry is sustaining growth which meets the needs of modern society and also safeguards the environment for future generations (SBAC, 2001).

The Case Study

This section presents the research methodology and the case study results.

Research Methodology

A case study approach has been chosen because it is well suited to new research areas (Eisenhardt, 1989) or when new perspectives are sought or there is little knowledge of the phenomenon (Patton, 1990; Patton, 1987). A multiple case approach was selected because the unit of analysis, the supply network, straddles firms in many tiers. The adoption of a single case or a one firm perspective would provide too narrow a view. Nevertheless, the multiple case approach is not intended to be a macroscopic study and has only limited generalisability (Hamel *et al.*, 1993; Yin, 1984). Multiple interview methods of data collection are used for each case. Ragin's 'case oriented comparative methods' (1987) and later fuzzy-sets (2000), demonstrates that case-oriented research using multiple methods finds reasons for deviating cases, and so creates a rich dialogue between theory and evidence. The use of multiple methods is typical for theory building (Eisenhardt, 1989).

Three types of interview were used: open-ended, semi-structured and structured (using repertory grid technique). Open-ended interviews are unstructured and free-ranging but within the overall scope of the supply network, with the primary purpose of understanding the meanings interviewees attach to aspects of the supply network without being influenced by the researcher's assumptions (Easterby-Smith *et al.*, 1994). Semi-structured interviews attempted to delve into the rationale for organizing the supply network in its current form. The repertory grid technique creates qualitative interview data and matrices of quantitative data conjointly (Goffin, 2002). Within a rep grid design, it must be decided what the columns will be beforehand. Two rep grid designs were used: the first with customers and the second with suppliers, in order to look down-stream and up-stream respectively, in the supply network.

Case selection was driven by the need to ensure adequate variety of cases in particular at different tiers of the supply network and different types of firm, intending to incorporate extreme and polar types (Pettigrew, 1990). Variety is more important than the frequency (Ragin, 1987) as notions of sampling are less relevant in case study research. This variety should create adequate possibilities for deviation which should be accounted for in any theory developed from the cases, so replication logic applies, rather than sampling logic (Eisenhardt, 1991). Nevertheless, firm selection was driven more by opportunism than planned theoretical or selective sampling. As Pettigrew (1990: 274) puts it: "there is an intentional or design component in the process of choosing and gaining access to research sites, but the practicalities of the process are best characterized by the phrase 'planned opportunism'". In all, 8 firms who are significant in commercial aerospace manufacturing were the subject of the case studies.

Data was collected from interviewees largely on a face-to-face basis. Only 4 interviews were telephone interviews and these were semi-structured in nature. Most interviews were recorded and all interviews were typed-up into electronic format (MS Word) providing a permanent record of the research.

Content analysis was managed by using ethnographic coding techniques (Strauss & Corbin, 1990). Open coding was applied independently to each interview transcript, coding sentence by sentence. Corroboration of codes and categories was achieved via independent coding and via the Centering Resonance Analysis (CRA) tool, which uses concordancing to arrive at frequently used words and words used frequently together (Williamson *et al.*, 2004). In addition, another coder was asked to code two interview transcripts in order to identify significant mismatches or bias in the coding process. The coder's results agreed at around the 90% level thus inter-rater reliability gives adequate assurance that coding was consistent (Rungtusanatham *et al.*, 2003).

The semi-structured interviews collected information relating to specific categories, which themselves had been decided by a combination of literature review and the coding of open-ended interviews. Coding of semi-structured interviews therefore sometimes required a little 'unravelling' of the orderly semi-structure, to create an improved coding categorization. Constructs and principal components identified in the qualitative and quantitative data from structured repertory grid interviews was incorporated into the analysis.

Once the coding was completed from the different methods of collection, a comparison was made against literature. Those categories which were found using multiple interview methods and were verified in extant literature were selected for this paper. The whole process from research design, through data collection, analysis and comparison with literature enables grounded theory to be developed (Pandit, 1996).

All data collected protected the privacy and confidentiality of the individual respondents. This was declared prior to interview and maintained by good records management after the interview. The focus of the research was not to make a critical investigation nor was there intention to expose unethical behavior or malpractice. Confidential information provided regarding specific suppliers was anonymized in order to protect the commercial confidentiality of these firms.

Additional information was sourced via two streams. The first was information in the public domain, including publications from the firm, and comment and news, via the internet. The second was via documents, usually charts, organ-o-grams and MS PowerPoint slide shows provided by employees of the firms being researched.

Analysis of Case Studies

Three dimensions were uncovered in these case studies: 1) structuring, 2) integration, and 3) coevolutionary dynamics (Varga, 2009). Prior to examining the detail of these dimensions, we define sustainability. Sustainability for the commercial aerospace manufacturing sector is measured for the purposes of this paper, as the maintenance of current total firm revenue relating to the production of jetliners. This definition permits flexibility in three ways. First, manufacturing capacity may be located anywhere in the world and so revenue is not nationalistic allowing competitive positions to change globally and the sector to remain sustainable. Second, if more jetliners are produced at lower cost, then total revenue of the sector can be maintained whilst also meeting demand for more flights. Third, it is assumed that current and future environmental demands can be met by new technologies, better designs and new methods whilst maintaining quality and safety requirements without incurring price rises.

Structuring of the Supply Network for Sustainability

Structuring is the first dimension of supply network sustainability. There are two key themes which define these structural mechanisms. The first theme is that of globalization and the second is related to product complexity-cum-miniaturization.

Globalization

The primary theme in the structuring of the commercial aerospace manufacturing supply network is increased globalization. Global connectedness for a variety of purposes is achieved over a wide geographical dispersion of suppliers. There is a tangible global division of labour as noted in the current techno-economic paradigm. Factors relating to globalization include access to skills for innovation, low costs for manufacturing, and access to global markets for sales.

Access to skills: The diminishing and narrow diversity of aerospace engineers in the West is in contrast to the East where there are young and growing skills, is likely to move the locus of innovation to the East as noted in current environmental influences. Very few new individuals enter the aerospace industry in the West. Engineering graduates choose different industries, so existing aerospace engineers move between firms. Skills need to be found globally.

MI (Management Implication):1—Develop relationships, educational, R&D, etc., around the world, promoting own product training, researching new technologies and allowing non-linear effects (Lorenz, 1963) to work.

Low cost manufacturing: The availability of products around the world (and the knowledge that these products exist and can be transported) means that the lowest price can be sourced from increased locations. But lower costs have in the past been associated with firm immaturity with respect to other capabilities, such as quality and delivery precision.

MI:2—Outsourcing can create resilience to demand fluctuations but the cost is resources to manage and develop suppliers to attain quality and other performance standards. Outsourcing also creates diversity in the supply chain and can lead to innovation and positive feedback.

Access to global markets: The loyalty of airlines to home country manufacturers has changed over the last 30 years. Nowadays, there is global competition for local customer business. There is no longer national loyalty. Competition is global so primes need to sell globally, but government restrictions can act as constraints. For example, the exports of some commercial jetliners have required export licences which have not been readily granted to some countries in the past. In terms of environmental influences, there has been investment in infrastructure in the East and growing competition.

MI:3—Know the implications for export of the components and technologies used in the jetliner. Some global markets which will form the long-term market for jetliners may have under-developed transport infrastructure and industrial skills for the service and maintenance of jetliners. These are opportunities for developing relationships and shaping future markets, but it is likely that these countries will demand a level of in-country manufacture. The implication is that the next generation of jetliners is likely to use the same manufacturing and assembly plants and will produce for other nations. The balance of cooperation (and permissible emergence through positive feedback) and coordination (control managed by negative feedback) is a strategic issue for the design and operational performance of the supply network (Choi *et al.*, 2001).

Product complexity-cum-miniaturization

The second theme in the structuring of the commercial aerospace manufacturing supply network is increased product complexity and miniaturization leading to increased specialization of suppliers into product systems or specialist products, and a move away from the generalist firm. Product complexity has also driven increased consolidation of suppliers. Finally, the primes have demanded complete systems of their first tier suppliers, leading to systematization, and the redefined role of the 1st tier as systems specialists and OEMs. As the first tier produce systems, so component design work has shifted up the supply network. In summary, three factors are relevant: specialization, consolidation and systematization.

Specialization: Increased specialization has driven the requirement for an increased customer base, providing global resilience during market demand fluctuations. The effect on suppliers, particularly suppliers of key commodities, is that they need to be diversified into other markets for their commodities so that they do not become liable to bankruptcy. "We used to be our own supplier in a lot of ways. Primes owned many more manufacturing facilities then and do much less manufacturing now. Nowadays, primes prefer their suppliers to have other customers to help spread the long term involvement in business risk". This industry diversification means new customers are able to exert power, creating pressure for timely delivery of the commodities demanded by the aerospace firms.

MI:4—Control over suppliers is no longer possible as the specialist systems suppliers could now have much bigger customers. Relationship and partnership development is needed and new ways established for mutual collaboration and integration.

Consolidation: Demand fluctuation and the reduction of in-house manufacturing has led to vertical disintegration and increased consolidation of firms. Consolidation has led to less diversity and less choice in the supply network; the network has become more dense. This strengthens the market position of the consolidated firm and reinforces the supply network heterarchical archetype in which collaboration is key as control is diminished. When the consolidated firm is brought on board to a new supply network, the firm is likely to stay in the supply network for the life-time of the product. "When you choose that supplier you pretty much marry to him on that programme. So the supplier selection has evolved, and is almost all upfront. We essentially take into account market access. We take into account access to capital market, access to technology..." Once a product/model is established, the organization is locked in, or vertically integrated, to a supply network (Gulati *et al.*, 2000). However at the lower part of the supply chain at the commodity level it is much more competitive. Short relationships come under more evaluation.

MI:5—Consolidation of firms can lead to a fragmented organization with multiple businesses producing similar products; reorganization along product line rather than business line would improve specialization and variety.

Systematization: Increased product complexity has led to simplification by the notion of systematization. Suppliers who were previously 1st tier (to the primes) are now consolidated or working for the new 1st tier systems' integrators to produce integrated systems. Integration work previously carried out by the primes has moved to the 1st tier diminishing their detailed knowledge of system design. The role of systems integrator is a competitive position of engineering and integration expertise. Functional complementarity of network partners can be achieved in which specialist skills and capabilities are provided by relevant network members (Parkhe, 1991). Suppliers further down the chain will become increasingly commoditized and will have to compete on price. Systematization is an example of a nearly decomposable system in which the rates of interaction between components at one level of the hierarchy are higher than those between different components (Simon, 2002; Levin, 2002).

MI:6—Systematization improves the rate of evolution making the system fitter, so is to be encouraged in order to find higher performing bundles of components. IPR ownership moves to the OEM but advanced design and build skills are needed for integrated systems.

Integration of the supply network for sustainability

Integration is the second dimension of supply network sustainability. Integration is concerned with protocols, methods and enablers that permit two or more firms to communicate and deliver their obligations. There are three factors in the integration of the supply network for sustainability; the first is that of standard interfaces, the second is related to technology, the third is that of quality standards.

Standard interfaces: The requirement for boiler-plate or standard product interfaces is essential. It enables suppliers to work to the same interface standards in the certainty that parts from different suppliers will be compatible. It also avoids the need for bespoke equipment and so keeps cost lower. Paradoxically, it is allowing a wider mix and match that is allowing customization to take place in a very agile manner.

MI:7—Firms should contribute to the shared development of standard interfaces, learning from other industries as necessary. Project teams could be constituted from the customer, integrator and supplier to develop solutions creating integrated designs, rather than bi-lateral ones.

Technology: The ability to use a global ICT infrastructure and shared systems is as important for product development as it is for production planning and delivery. Networks and distributed computing are continually improving general purpose technologies and resources allowing shared systems to be accessed from any location around the world. Broad-band and real-time communications is a distinct feature of the 3rd globalization/Schumpeter's 5th wave. IT is now dominating supply chain management (Ho *et al.*, 2003). Primes have developed portals in their design and engineering system that enables them to work with suppliers, researchers and partners on an integrated basis, enabling high-speed data transfer of digital information supporting virtual working in a fairly complex environment of 3D CAD. Computing tools are incredibly powerful and driving significantly changing relationships with suppliers. "Suppliers need to be organized for efficiency of manufacturing but also for information and collaboration". The use of technology to automate production can reduce resource requirements and act as an alternative to outsourcing to LCEs.

MI:8—Technology integration of a global nature attempts to overcome cultural, language and distance differences. Skills are needed to use the appropriate technologies, but also to prevent information/knowledge sharing leakages and vulnerabilities. Technological solutions may provide alternatives to outsourcing.

Quality Standards: Maintaining and improving quality standards is critical within the sector and drives supplier selection. Expectations of supplier quality performance have never been higher, particularly of risk-sharing partners, who may be supported by supplier development programmes. Suppliers are likely to be classified by quality standard, so that the highest quality suppliers are offered new work first. Certificated aerospace manufacturers are limited in numbers, have very high performance and very high cost.

MI:9—There is a cost to the firm associated with quality dependent on the skills of the supplier. Quality needs to be learned, particularly by new firms in the supply network. Quality performance is a key component of continuous improvement.

Coevolutionary Dynamics of the Supply Network for Sustainability

The third and final dimension of supply network sustainability looks at the process dynamics which occur within the structure of the supply network and using the available integration methods. These process dynamics relate to inter-firm dynamics which enable the firm to coevolve with the supply network, in the context of the environment.

The dynamics of the commercial aerospace supply network appear to be Janusian—having two faces—one related to the formative stages of the supply network and the second related to the operational supply network. Five key factors emerge, each of which has two faces: these are innovation, market responsiveness, information sharing, supplier performance and risk mitigation.

Innovation: Radical innovation occurs for new products in new supply networks. Nevertheless suppliers will invest R&D funds on technology research continuously, although it may not be usable for some time. Engineering designs largely stagnate once a jetliner is in production; a focus on production process technologies can create a competitive edge for manufacturers. The creation of 'blurred technologies' for improvements in current production are desirable. New jetliner designs must increasingly accommodate requirements for lower CO_2 emissions, improved fuel efficiency and other environmental influences, some of which are regulatory or legislative. These are environmental influences, some of which are regulatory or legislative. Increased terrorism activity increases the need for product security innovation measures. Increasingly, new product designs must be fit for sustainable production.

MI:10—Process innovation is key in operational supply networks, whereas product innovation is key for new supply networks. Both can contribute to increased sustainability by addressing energy and resource use, and improve flight safety.

Market Responsiveness: Customer demand for jetliners halved over three years then recovered in the following three years (2001-2007). In operational supply networks this was achieved through vertical disintegration, supplier consolidation and supplier specialization and diversification. Demand for new jetliners is created from product concepts using strategic marketing methods. The order book for commercial aerospace can have a 2 year backlog. Although demand has been high and growing since 2006, there is a global financial crisis which is likely to have some impact on commercial air travel. Demand fluctuations may be a recurrent theme in commercial aerospace manufacturing.

MI:11—Structural agility is more easily met through heterarchical or temporary (programme) structures helping to meet demand fluctuations. Diversifying into other related sectors or industries may be an option.

Information Sharing: Information sharing is largely translucent in product design as opposed to transparent in the operational supply network. Translucency is sometimes a consequence of sharing product designs before they are finalized but can be intentional, as too much transparency may give away information to competitors, although OEMs are particularly aware of this risk. In operational networks, mutual sharing, mutual sharing of information is believed to give better warning of fluctuations in demand and avoid some of the bull-whip effect, however premature or incomplete sharing of information can also cause production delays.

MI:12—Mutual and timely information sharing and appropriate dialogue is perceived as beneficial. Reliance and completeness assessments of information could be shared.

Supplier performance: Development of suppliers influences not only the performance of the supplier but the performance of the supply network. In particular the adoption of quality management techniques, Just-in-Time, Kanban, and lean techniques to name a few, can reduce rework, cut stock holding levels, improve responsiveness to demand fluctuations and improve the supplier's chances for selection in new supply networks. The resource costs associated with the support of learning and development in 'low cost economies' are high and demand long-term commitment particularly from 1st tier organizations. "So for instance with the Chinese we had to teach them a lot about quality. But with the Russians we are teaching about schedules. Depending on which part of the world you have got into or what maturity of the supplier you are dealing with depends on how much you have to teach them to be able to be an important supplier... Some of that is a long time commitment. It takes a lot of resources and energy."

MI:13—Benefits of supplier development accrue to the entire supply network although costs are disproportionately distributed. Increased quality performance has often been driven by industrial change, particularly from automotive manufacturing.

Risk mitigation: When the firm cannot exercise direct control (via ownership) over the supplier, then methods are required to ensure that the risks of failure are mitigated. Limited numbers of suppliers are available for most parts and whilst double sourcing is possible, it is not always cost effective for the management of operational risks. In terms of suppliers, their risk of failure or bankruptcy is a risk to the Prime, so risk in managed by understanding mutual interests and responsibilities in a partnership capacity. Partnering also helps to influence sales and to create shared investments. Firm dominance over suppliers can be asserted in a collaborative way rather than in a control manner.

MI: 14—There needs to be a balance between retaining risk and cost to mitigate risk. Partnerships can spread risk and improve collaboration. 'Templates' which achieve defined outcomes can be a useful collaborative technique which provide sufficient flexibility.

A summary of the findings in presented in Table 4.

Discussion

The supply network of the commercial aerospace manufacturing sector is dominated by a duopoly of Boeing and Airbus. The current supply network archetype is that of a heterarchy. The issue of influencing independently owned firms within a heterarchy requires coordination mechanisms. Management implications identified in the case study were strongly connected to coordination and collaboration as a means to achieve stated objectives.

Mulford and Rogers (1982) define coordination as the establishment of decision rules between two or more firms to deal collectively with the shared tasks in their environment, whereas cooperation is focused on the joint achieve-

Dimension	Theme	Factor	Management implication
Structural Mechanisms	Globalization	Access to skills for innovation	MI:1 Develop relationships, educational, R&D, etc. around the world, promoting own product training, researching new technologies, allowing non-linear effects to work.
		Low cost manufacturing	MI:2 Outsourcing can create resilience to demand fluctuations but the cost is resources to manage and develop suppliers. Outsourcing also creates diversity in the supply chain and can lead to innovation and positive feedback.
		Access to global markets	MI:3 Know the implications for export of the components and technologies used in the jetliner. Develop relationships with LCEs and help shape future markets.
	Product complexity-cum-miniaturization	Specialization	MI:4 Control over suppliers is no longer possible as the specialist systems suppliers could now have much bigger customers. Relationship and partnership development is needed and new ways for mutual collaboration and integration.
		Consolidation	MI:5 Consolidation of firms may require reorganization along product line rather than business line to improve specialization and variety.
		Systematization	MI:6 Systematization improves the rate of evolution making the system fitter, so is to be encouraged in order to find higher performing bundles of components. Advanced design and build skills are needed for integrated systems.
Integration Methods	Standard interfaces		MI:7 Firms should contribute to the shared development of standard interfaces, learning from other industries as necessary. Project teams could be constituted to develop solutions creating integrated designs, rather than bi-lateral ones.
	Technology		MI:8 Technology integration skills are needed to use the appropriate technologies, but also to prevent information/knowledge sharing leakages and vulnerabilities. Technological solutions may provide alternatives to outsourcing.
	Quality Standards		MI:9 There is a cost to the firm associated with quality dependent on the skills of the supplier. Quality needs to be learned, particularly by new firms in the supply network.
Coevolutionary process dynamics	Innovation		MI:10 Process innovation is key in operational supply networks, whereas product innovation is key for new supply networks.
	Market Responsiveness		MI:11 Structural agility is more easily met through heterarchical or temporary (programme) structures helping to meet demand fluctuations. Diversifying into other related sectors or industries may be an option.
	Information Sharing		MI:12 Mutual and timely information sharing and appropriate dialogue is perceived as beneficial. Reliance and completeness assessments of information could be shared.
	Supplier performance		MI:13 Benefits of supplier development accrue to the entire supply network although costs are disproportionately distributed. Increased quality performance has often been driven by industrial change, particularly from automotive manufacturing.
	Risk mitigation		MI: 14 There needs to be a balance between retaining risk and cost to mitigate risk. Partnerships can spread risk and improve collaboration. 'Templates' which achieve defined outcomes can be a useful collaborative technique which provide sufficient flexibility.

Table 4 *Summary of Management Implications by Coevolutionary Dimension and Factor*

ment of firm goals. The need to coordinate assumes that cooperation is needed between firms. Soft or intangible capabilities are particularly important competencies for strategic operations (Lewis, 2003). Cooperation heightens the need for communication, and for information technologies and associated software to support that communication (Castells, 1996). The need to cooperate and leverage complementary competencies within the network becomes essential (Yusuf *et al.*, 2004). Collaboration happens globally; Boeing collaborates with European aerospace firms to jointly design airframe components. But these endeavours are huge, expensive undertakings beyond the means of one firm alone. So the efficiencies of collaboration are significant enough to overcome the "logistical hassles, security issues and general mistrust that tend to isolate U.S. companies" (Pastore, 2004: 1).

Compared with organizational hierarchical relations, the network is more loosely coupled, relies more on self-organizing processes and has greater competitive pressures (Ring & Van de Van, 1994). However, strong ties may improve the probability of oligopolistic coordination more than weak ties (Galaskiewicz & Zaheer, 1999). This emphasis on longer term relationships reduces market focus and which would otherwise exist in a supply network (Cohen & Agrawal, 1999) but this is mitigated somewhat by unequal distribution of costs and benefits between the supply chain partners making inter-company cooperation difficult (Kärkkäinen *et al.*, 2003).

Whilst the initial conditions at the time of creation of an alliance have an influence on the development of the alliance (e.g., Hamel 1991), some alliances evolve in a punctuated equilibrium manner due to changes in the environment (Gulati, 1998) and other exogenous factors such as industry competition. Organizations and the networks to which they belong are dynamic and need to adapt and combat nonlinearities such as the bull-whip effect which spirals between tiers in the supply network. Thus each firm in the supply chain directly and indirectly affects the performance of all other supply chain members as well as ultimate supply chain performance (Cooper *et al.*, 1997) in a non-equilibrium manner.

The rise of the information age and greatly reduced information communication costs is changing coordination mechanisms among partners in the supply network (Coase, 1998), increasing collaborative work within teams on high capacity networks (Tapscott, 1996), enabling continuous information flow in an integrated supply network (Lambert & Cooper, 2000) and providing new opportunities for customers to connect to supplier and to reduce transaction costs and risks (Lewis & Talalayevsky, 2004). Supply chain integration of legally and spationally separated firms is shown as a vital tool for competitive advantage (Yusuf *et al.*, 2004). A supply network can be defined as a group of semi-independent firms which collaborate in "ever-changing constellations" in order to achieve some business goal related to the collaboration (Akkermans, 2001; Tapscott, 1996).

Further Research

The research identified the growing importance of product service and 'flight by the hour' as a likely future business model. The transformations required to the current supply network archetype and its structural mechanisms, integration methods and coevolutionary process dynamics would be valuable research for practice and policy. Further research which incorporates a broader industry survey would be useful for practice.

If environmental concerns heighten and accelerate the need for new products then research into whether the existing supply network archetype is able to deliver considerably more jetliners per year would be valuable research. Currently the rate of change in the environment is running at a slower rate than change in the firm and supply network. If this is reversed then according to the Red Queen effect (Van Valen, 1973; Kauffman, 1995; Barnett & Hansen, 1996) the firm and the supply network will not be able to evolve unless it continues to develop just in order to "maintain its fitness" relative to the systems it is coevolving with. This speeding up of environmental change relative to socio-economic change is interesting for practice and research.

A comparative research pathway would be to examine and compare alternative supply network archetypes for the purposes of sustainability, such as the market, hierarchy, Kieretsu and 4PL ™ (Varga, 2009). Finally, the research methodology and/or the three dimensional analysis framework could be attempted in another industry or sector to examine the coevolution of adjacent levels of complex adaptive systems.

Acknowledgement

This work was supported by 1) the Economic and Social Science Research Council (ESRC) grant RES-000-23-0845 "Modelling the Evolution of the Aerospace Supply Chain" 2005-2008; and 2) by the EC grant for the QosCosGrid project grant number: FP6-2005-IST-5 033883.

References

Air Transport Action Group (2008). "Air transport and the environment," http://www.atag.org/files/AirTransportEnvironmentFlyer.pdf.

Airsafe (2008). "Fatal plane crash rates for selected airliner models," http://www.airsafe.com/events/models/rate_mod.htm.

Akkermans, H. (2001). "Emergent supply networks: system dynamics simulation of adaptive supply agents," *System Sciences: Proceedings of the 34th Hawaii International Conference* on 3-6 Jan.

Allen, P.M. (1988). "Evolution, innovation and economics," in G. Dosi, C. Freeman, C. Nelson, G. Silverberg and L. Soete, *Technical Change and Economic Theory*, ISBN

9780861879496, pp. 95-119.

Arthur, W. B. (1994). *Increasing Returns and Path Dependence in the Economy*, ISBN 9780472064960.

Asia Times (2006). "China renews airliner manufacturing plan," http://www.atimes.com/atimes/China_Business/HC14Cb07.html.

Assa, O. and Denton-Brown, J. (2008). "The future of airport growth," http://www.sovereign-publications.com/bechtel.htm.

Barabasi, A. (2002). *Linked: The New Science of Networks*, ISBN 9780738206677.

Barnett, W.P. and Hansen, M.T. (1996). "The red queen in organizational evolution," *Strategic Management Journal*, ISSN 1097-0266, 17: 139-157.

Boeing Website (2008). "Boeing commercial deliveries," http://active.boeing.com/commercial/orders/index.cfm?content=userdefinedselection.cfm&pageid=m15527.

Bowersox, D.J., Closs, D.J. and Stank, T.P. (2000). "Ten mega-trends that will revolutionize supply chain logistics," *Journal of Business Logistics*, ISSN 0735-3766, 21(2): 1-16.

Capra, F. (1996). *The Web of Life: A New Scientific Understanding of Living Systems*, ISBN 9780385476751.

Castells, M. (1996). *The Rise of the Network Society*, ISBN 9781557866172.

Choi, T.Y., Dooley, K.J. and Rungtusanatham, M. (2001). "Supply networks and complex adaptive systems: Control versus emergence," *Journal of Operations Management*, ISSN 0272-6963, 19(3): 351-366.

Christopher, M. (1992). *Logistics and Supply Chain Management Strategies for Reducing Costs and Improving Service*, ISBN 9780273034155.

Christopher, M. (2005). *Logistics and Supply Chain Management: Creating Value-Adding Networks*, ISBN 9780273681762.

Coase, R. (1998). "The new institutional economics," *American Economic Review: Papers and Proceedings*, ISSN 0002-8282, 88(2): 72-74.

Cohen, M.A. and Agrawal, N. (1999). "An analytical comparison of long and short term contracts," *IIE Transactions*, ISSN 0740-817X, 31(8): 783-796.

Cooper, M.C., Lambert, D.M. and Pagh, J.D. (1997). "Supply chain management: More than a new name for logistics," *International Journal of Logistics Management*, ISSN 0957-4093, 8(1): 1-13.

Crump, E. (2009). "Airbus still the world's biggest maker of planes: Final 2008 orders total set to edge Boeing," http://www.dailypost.co.uk.

Daft, R.L. and Lewin, A.Y. (1993). "Where are the theories for the 'new' organizational forms? An editorial essay," *Organization Science*, ISSN 1047-7039, 4(4): i-.

Davis, T. (1993). "Effective supply chain management," *MIT Sloan Management Review*, ISSN 1532-9194, 34(4): 35-46.

Department for Transport (2005). "Civil aviation bill," http://www.dft.gov.uk/about/strategy/whitepapers/air/cab/civilaviationbill.

DTI Aerospace and Defence Directorate (2003). "Aerospace and defence technology report 2002/3," http://www.berr.gov.uk/files/file10022.pdf.

EADS Airbus (2008). "Airbus orders and deliveries," http://www.airbus.com/en/corporate/orders_and_deliveries/.

Easterby-Smith, M., Thorpe, R. and Lowe, A. (eds.) (1994). *Management Research: An Introduction*, ISBN 9780761972846.

Eisenhardt, K.M. (1989). "Building theories from case study research," *Academy of Management Review*, ISSN 0363-7425, 14(4): 532-550.

Eisenhardt, K.M. (1991). "Better stories and better constructs: The case for rigor and comparative logic," *Academy of Management Review*, ISSN 0363-7425, 16(3): 620-627.

Epstein, K. and Crown, J. (2008). "Globalization bites Boeing," *Business Week*, ISSN 0007-7135, 4076: 32.

Foresight (2007). "Social foresight," http://s3.amazonaws.com/foresight/20.pdf.

Freeola (2008). "Boeing 727," http://www.century-of-flight.freeola.com/Aviation%20history/coming%20of%20age/Boeing%20727.htm.

Fuller, T. and Warren, L. (2006). "Entrepreneurship as foresight: A complex social network perspective on organisational foresight," *Futures*, ISSN 0016-3287, 38: 956-971.

Galaskiewicz, J. and Zaheer, A. (1999). "Networks of competitive advantage," in S. Andrews and D. Knoke (eds.), *Networks In and Around Organizations: Research in the Sociology of Organizations*, ISBN 9780762304738, pp. 237-261.

Garnsey, E. and McGlade, J. (2006). *Complexity and Coevolution: Continuity and Change in Socio-economic Systems*, ISBN 9781845421403.

Gates, D. (2006). "Clean engines, wings that fold: Boeing dreams of futuristic jets," *Seattle Times*, ISSN 0745-9696,

Goffin, K. (2002). "Repertory grid technique," in D. Partington (ed.), *Essential Skills for Management Research*, ISBN 9780761970088.

Gonzalez, A. (2007). "To go green in jet fuel, Boeing looks at algae," *Seattle Times*, ISSN 0745-9696,

Gulati, R. (1998). "Alliances and networks," *Strategic Management Journal*, ISSN 1097-0266, 19(4): 293-317.

Gulati, R. and Gargiulo, M. (1999). "Where do interorganizational networks come from?" *American Journal of Sociology*, ISSN 0002-9602, 104: 1439-1493.

Gulati, R., Nohria, N. and Zaheer, A. (2000). "Strategic networks," *Strategic Management Journal*, ISSN 1097-0266, 21(3): 203-215.

Hamel, G. (1991). "Competition for competence and inter-partner learning within international strategic alliances," *Strategic Management Journal*, ISSN 1097-0266, 12: 83-103.

Hamel, J., Dufour, S. and Fortin, D. (1993). *Case Study Methods*, ISBN 9780803954168.

Harland, C.M. (1996). "Supply chain management relationships, chains and networks," *British Journal of Management*, ISSN 1045-3172, 7: S63-80.

Hewitt, F. (2000). "Demand satisfaction communities: New operational relationships in the information age," *International Journal of Logistics Management*, ISSN 0957-4093, 11(2): 9-20.

Ho, D.C.K., Au, K.F. and Newton, E. (2003). "The process and consequences of supply side virtualization," *Industrial Management & Data Systems*, ISSN 0263-5577, 103(5): 423-433.

Holland, J. (1998). *Emergence: From Chaos to Order*, ISBN 9780201149432.

Kärkkäinen, M., Holmström, J., Främling, K. and Artto, K. (2003). "Intelligent products: A step towards a more effective project delivery chain," Computers in Industry, ISSN 0166-3615, 50(2): 141-151.

Kauffman, S.A. (1993). *The Origins of Order*, ISBN 9780195079517.

Kauffman, S.A. (1995). "Escaping the red queen effect," *The McKinsey Quarterly*, ISSN 0047-5394, 1: 119-129.

Kondratiev, N. (1984). *The Long Wave Cycle, translated from The Major Economic Cycles*, originally published in Russian in 1925, ISBN 9780943940076.

Lambert, D.M. and Cooper, M.C. (2000). "Issues in supply chain management," *Industrial Marketing Management*, ISSN 0019-8501, 29(1): 65-83.

Lambert, D.M., Cooper, M.C. and Pagh, J.D. (1998). "Supply chain management: Implementation issues and research opportunities," *International Journal of Logistics Management*, ISSN 0957-4093, 9(2): 1-16.

Larson, P.D. and Rogers, D.S. (1998). "Supply chain management: Definition, growth and approaches," *Journal of Marketing Theory and Practice*, ISSN 1069-6679, 6(4): 1-5.

Levin, S.A. (2002). "Complex adaptive systems," *Bulletin (New Series) of the American Mathematical Society*, ISSN 1088-9485, 40(1): 3-19.

Lewis, I. and Talalayevsky, A. (2004). "Improving the interorganizational supply chain through optimization of information flows," *Journal of Enterprise Information Management*, ISSN 1741-0398, 17(3): 229-237.

Lewis, M.A. (2003). "Analyzing organizational competence: implications for the management of operations," *International Journal of Operations & Production Manage-*

ment, ISSN 0144-3577, 23(7): 731-756.

Longhi, C. (2005). "A French revolution: Technology management in the aerospace industry the case of Toulouse," *International Journal of Technology Management*, ISSN 1741-5276, 29(3/4): 194-215.

Lorenz, E.N. (1963). "Deterministic nonperiodic flow," *Journal of the Atmospheric Sciences*, ISSN 0022-4928, 20: 30-141.

Lummus, R. and Vokurka, R. (1999). "Defining supply chain management: A historical perspective and practical guidelines," *Industrial Management & Data Systems*, ISSN 0263-5577, 99(1).

McAdam, R., O'Hare, T. and Moffett, S. (2008). "Collaborative knowledge sharing in composite new product development: An aerospace study," *Technovation*, ISSN 0166-4972, 28(5): 245-256.

Mentzer, J.T., DeWitt, W., Keebler, J.S., Min, S., Nix, N. W., Smith, C. D. and Zach, Z. (2001). "Defining supply chain management," *Journal of Business Logistics*, ISSN 0735-3766, 22(3): 1-26.

Mulford, C.L. and Rogers, D.L. (1982). "Definition and models" in D.L. Rogers and D.A. Whetten (eds.), *Interorganizational Coordination: Theory, Research and Implementation*, ISBN 9780813809861.

Pandit, N.R. (1996). "The creation of theory: A recent application of the grounded theory method," *The Qualitative Report*, ISSN 1052-0147, 2(4).

Parkhe, A. (1991). "Interfirm diversity, organizational learning, and longevity in global strategic alliances," *Journal of International Business Studies*, ISSN 1478-6990, 22(4): 579-601.

Pastore, R. (2004). "A failure to collaborate?" *CIO*, ISSN 0894-9301, 17(19): 14.

Patton, M.Q. (1987). *How to Use Qualitative Methods in Evaluation*, ISBN 9780803931299.

Patton, M.Q. (1990). *Qualitative Evaluation and Research Methods*, ISBN 9780803937796.

Pettigrew, A. (1990). "Longitudinal field research on change: Theory and practice," *Organization Science*, ISSN 1526-5455, 1(3): 267-292.

Ragin, C.C. (1987). "Case oriented comparative method," in C.C Ragin (ed.), *The Comparative Method: Moving Beyond Qualitative and Quantitative Method*, ISBN 9780520066182, pp. 34-52.

Ragin, C.C. (2000). *Fuzzy-Set Social Science*, ISBN 9780226702773.

Ring, P.S. and Van de Ven, A.H. (1994). "Development process of cooperative interorganizational relationships," *Academy of Management Review*, ISSN 0363-7425, 19(1): 90-118.

Rungtusanatham, M.J., Choi, T.Y., Hollingworth, D.G., Wu, Z. and Forza, C. (2003). "Survey research in operations management: Historical analyses," *Journal of Operations*

Management, ISSN 0272-6963, 21: 475-488.

Salancik, G.R. (1995). "Wanted: A Good Network Theory of Organization," *Administrative Science Quarterly*, ISSN 0001-8392, 40(2): 343-349.

SBAC (2001). "Air travel challenge: Greener by design," http://www.foresight.gov.uk/Defence%20Aerospace/Air_Travel_Challenge_Aug_2001.pdf.

Schumpeter, J.A. (1994). *Capitalism, Socialism and Democracy*, originally published in 1942, ISBN 9780415107624.

Simon, H.A. (2002). "Near decomposability and the speed of evolution," *Industrial and Corporate Change*, ISSN 0960-6491, 11(3): 587-599.

Strauss, A. and Corbin, J.M. (1990). *Basics of Qualitative Research: Grounded Theory Procedures and Techniques*, ISBN 9780803932517.

Surana, A., Kumara, S., Greaves, M. and Raghavan, U. N. (2005). "Supply-chain networks: a complex adaptive systems perspective," *International Journal of Production Research*, ISSN 1366-588X, 43(20): 4235-4265.

Tapscott, D. (1996). *The Digital Economy: Promise and Peril in the Age of Networked Intelligence*, ISBN 9780070633421.

Tsoukas, H. and Chia, R. (2002). "On organizational becoming: rethinking organizational change," *Organization Science*, ISSN 1047-7039, 13(5): 567-582.

Tuomi, I. (2007). "Learning in the age of networked intelligence," *European Journal of Education*, ISSN 1465-3435, 42(2): 235-254.

Van Valen, L. (1973). "A new evolutionary law," *Evolutionary Theory*, ISSN 0093-4755, 1: 1-30.

Varga, L. (2009). *The Coevolution of the Firm and the Supply Network: A Complex Systems Perspective*, unpublished PhD thesis, Cranfield University.

Williamson, J., Dooley, K.J. and Corman, S. (2004). "Using text mining to create actionable knowledge: Application to network failure incident reports," in Karagiannis, D. and Reimer, U. (eds.), *Practical Aspects of Knowledge Management*, ISBN 9783540003144 (2003), pp. 71-81.

Womack, J.P. (2002). "Lean thinking: Where have we been and where are we going?" *Forming and Fabricating*, ISSN 1075-3699, 9(9): L2-L6.

Womack, J.P., Jones, D.T. and Roos, D. (1990). *The Machine that Changed the World: Based on the Massachusetts Institute of Technology 5-Million-Dollar 5-Year Study on the Future of the Automobile*, ISBN 9780892563500.

Yin, R.K. (1984). *Case Study Research: Design and Methods*, ISBN 9780803920583.

Yusuf, Y.Y., Gunasekaran, A., Adeleye, E.O. and Sivayoganathan, K. (2004). "Agile supply chain capabilities: determinants of competitive objectives," *European Journal of Operational Research*, ISSN 0377-2217, 159: 379-392.

Liz Varga is Senior Research Fellow at the Complex Systems Management Centre (CSRC), a cross-school centre for research, teaching and practitioner-led community for the investigations of complexity science in the real world. The track record for CSRC includes a wide range of research projects including working with: an ESRC funded joint project with Sheffield University 'Modelling the Evolution of the Aerospace Supply chain'; European Union Sixth Framework project 'Quasi-opportunistic supercomputing for complex systems in grid environments'; OMEGA, a multi-university consortium carrying out aviation and environment research. Liz has also contributed to a European Union project on inter-disciplinary in the field of Corporate Social Responsibility, developing an interest personal construct psychology, and using repertory grid in various organizational settings. Her interests are in the coevolution of socio-economic systems and how we might understand these better via agent-based modelling. Liz also lectures on the MSc for Business Performance and provides consultancy services to the public sector.

Mark Strathern is a Senior Research Fellow with the Complex Systems Research Centre. At present he is part of an inter-university team working on the EPSRC CASCADE project modeling smart electrical grids. Recently Mark has worked on a three-year EC project ,QosCosGRID, on the development of grid computing techniques for modeling complex systems Previously he was part of the team working on 'Modeling the Evolution of the Aerospace Supply Chain' and he was also part of the team on the ESRC NEXSUS project to understand and model sustainability in complex socio-economic networks. He has also been involved with the modeling and teaching of sustainability within the EU fishing industries, and has had a collaboration with the Tyndall Centre working on formal and mathematical methods for use in the climate change arena. Mark has worked for nearly two decades at Cranfield modeling complex systems, particularly complex socio-economic systems: including the modeling of financial trading systems, banking customer services, the effects of climate change on food security, the Senegalese economy, the water quality in the Rhone valley, Bayesian approaches to helicopter gearbox monitoring and water recycling. Before joining Cranfield Mark worked for a number of years as a consultant in Information Systems and has held positions in accountancy and marketing. Mark has a BSc in mathematics and a PhD in Modeling Organisational Evolution and Change.

Chris Rose-Anderssen was a Research Associate in the three-year project 'Modelling the Evolution of the Aersopace Supply Chains' at the Advanced Manufacturing Research Centre with Boeing, University of Sheffield. His contribution in this project was the construction of a cladistic classification scheme of supply chain evolution, and the analysis of supply chain change in terms of risk-sharing partnership development and innovation, knowledge transformation and learning, integration and communicative interaction; which he disseminated in 5 journal articles and 6 conference papers.

James Baldwin's research interests range from theoretical model development of industrial ecosystems; sustainable industrial development at the regional level; guidelines for classification development for sustainable manufacturers and sustainable industrial networks; the simulation of evolutionary and innovation processes in manufacturing firms, supply networks and clusters; and management decision-making and the evolutionary consequences. James serves on the Editorial Board of the international scientific referee journal *Progress in Industrial Ecology: An International Journal*, and has also participated in the review process on special issues for both the *Journal of Cleaner Production* and *Benchmarking: An International Journal*. He has been invited to guest-review articles for *Journal of Industrial Ecology*, *Journal of Environmental Management* and *International Journal of Global Environmental Issues*; and, *Innovation: Management, Policy and Practice*. James was a member of the 2005 scientific committee of the '11th Annual International Sustainable Development Research Conference'. He also chaired a special stream on 'Industrial Ecology, Systems Analysis and Evolution'.

Keith Ridgway OBE, FREng is Research Director and HEFCE Business Fellow at the Advanced Manufacturing Research Centre with Boeing (AMRC). In January 2001 Professor Ridgway established the AMRC to carry out research in manufacturing technologies directly related to the aerospace industry. In addition to research on manufacturing technologies Ridgway has also worked on the translation of theories from biological science into manufacturing management, most notably the work on cladistics and evolutionary modelling with McCarthy, and Baldwin. Professor Ridgway is actively involved in a wide range of AMRC research projects, funded by the EPSRC, ESRC, EU framework 6, DTI and industry. In June 2004 Patricia Hewitt announced a grant of £4.5 million to support the development of the University of Sheffield Composites and Advanced Materials Technology Centre CAMTeC with Boeing. This will form part of the National Composites Network and will be integrated with the AMRC to form the Advanced Manufacturing Institute at the University of Sheffield. Professor Ridgway has 123 publications including 31 journals articles.

18. Social and Ecological Transitions: Winemaking in California

Greig Tor Guthey[1] *& Gail Whiteman*[2]
1 Department of Liberal Studies, California State University San Marcos, USA
2 Sustainability and Climate Research Centre, Erasmus University, HOL

With rising interest in sustainability, ecology is an increasingly important dimension of organizational research. Yet few empirical studies integrate local ecology into coevolutionary approaches where firms are key actors, and fewer still approach the question of sustainability and organizations from a systems perspective. In this paper, we ask how organizations can effectively participate in efforts to increase sustainability from a systems perspective. We develop an interdisciplinary framework for understanding firm-ecology relationships and then explore how this framework sheds light on regional planning and industrial practice in northern California's wine industry.

Introduction

Porter (2006) argues that we need to take a systems perspective to adequately describe how firms affect natural systems and how they can contribute to increased sustainability. This is a new area for research. While studies on organizations and the natural environment have grown considerably over the last few decades (Bansal & Gao, 2006; Jermier *et al.*, 2006), only a few use systems theories (e.g., Allen, 1997; Boons, 2008; Loorbach *et al.*, forthcoming; Whiteman *et al.*, 2004). Instead, organizational research tends to focus on whether or not it pays to be green (Berchicci & King, 2007), on the various strategic approaches to framing environmental issues (Etzion, 2007; Jermier *et al.*, 2006), on the impact of regulation on firm behavior, and on measuring the environmental performances of organizations in terms of waste production, resource use, or the adoption of ISO 14001 (Bansal & Gao, 2006). But persistent sustainability problems—such as climate change or unsustainable agriculture—require more radical and structural changes within and between organizations (Rotmans, 2005) as they relate to the needs and constraints of the local ecology (Allen, 1997; Boons, 2008; Guthey, 2004; Lockwood, 2007; Whiteman *et al.*, 2004).

From a complex systems perspective, sustainability and resilience emerges from the coevolution of social, ecological and economic systems (Allen, 1997; Porter, 2006; Korhonen & Seager, 2008). As Seager (2008: 447) writes, "[t]he locus of study in sustainability science is on *the interaction between human and natural systems*" (italics in original). Capturing this interaction requires a multidisciplinary approach to organizational research which moves beyond a linear search for eco-efficiencies (Allen, 1997; Korhonen & Seager, 2008; Loorbach *et al.*, forthcoming). In this paper, we look to the fields of ecology and geography for theoretical tools that can help managers and researchers think about complex organization-nature relationships. We also empirically examine how organizations can effectively participate in efforts to increase sustainability from a systems perspective given the needs and constraints of a specific local environment.

In section one, we briefly review the growing body of work in organizations and ecology. Section two identifies how approaches from ecology can help clarify organization-nature relationships (e.g., Holling, 1986; Folke *et al.*, 2002). In section three, we draw upon research from economic geography to propose that organizations can usefully be viewed as powerful systemic actors that coproduce social understandings of "place" that has direct implications for sustainability. We then present empirical findings in section four to illustrate how organizations (including firms) have coevolved with a specific place through a case study of winemaking and regional planning in northern California.

Data for the case study was collected using documentary analysis, interviews, and observation in order to show how firms embracing environmental values can generate innovations in production processes and markets (Guthey, 2004). In Napa and Sonoma County, firms have contributed to an emergent set of policies concerning ecological protection, agricultural land preservation and promotion for more than 40 years with the result that firms and other stakeholders seem to be coproducing not just wine but also the 'place' associated with its production.

Ecology in Organizational Research

Although corporate environmentalism has progressed from reactive responses in the early years to more proactive business strategies (see, e.g., Etzion, 2007; Hoffman, 1997; Jermier *et al.*, 2006), firms continue to focus more on reducing unsustainable firm-level behavior rather than on increasing the sustainability of the system via radical change across actors and levels (Ehrenfeld, 2005). This is surprising given that Starik and Rands (1995) provided an early framework for studying the relationship between the organization and other levels of analysis (political-economic, social-cultural, ecological and individual). However, there are only a few empirical studies that can provide direction in this area (e.g., Allen, 1997; Boons, 2008; Loorbach *et al.*, forthcoming; Whiteman *et al.*, 2004).

That companies coevolve with other institutions and actors is well established in organization studies (Rodrigues & Child, 2003; Lewin & Volberda, 1999). And while the majority of coevolutionary studies "overlook" the natural environment due to their focus on institutional linkages, there is a growing body of work attempting to link ecology and organizations in new ways. For instance, a number of studies examine how industry in the Netherlands is adopting a systemic approach to sustainability. Boons (2008) describes how a network of firms in the Port of Rotterdam developed an emergent self-governance approach to link their production processes and reduce environmental waste and energy use following selective pressure by the Dutch government. Before this initiative, individual environmental managers were not fully aware of their collective impact on the local ecology. Bass (2008) found that local embeddedness (in cognitive, structural, cultural, political, spatial and temporal dimensions) within industrial systems in Rotterdam facilitated learning which helped businesses radically innovate. Similarly, in the densely urbanized social-ecological context of the Netherlands, Loorbach *et al.* (forthcoming) illustrate how businesses can coevolve with government-backed societal transitions in areas such as sustainable mobility. The key is to encourage social processes amongst innovative front-runners, who can then begin to self-organize and experiment with new approaches to sustainability. This is similar to Clarke and Roome's (1999) case study of a Canadian utility. These researchers identified the important role of learning action networks in their case study of an energy company's efforts to develop more sustainable practices. In their view, learning action networks enabled a broad set of stakeholders to engage in a continuous process of learning, action and change that was interactive and reflexive. As all of these studies show, there is tremendous value in studying emergent social processes in the coevolution of organizations that share or operate within related social-ecological systems.

Importantly, Allen (1997) and Whiteman *et al.* (2004) also highlight the need to better understand local ecological processes in order to evaluate organizational effects on the larger system. For instance, Allen (1997) argues that cities and regions are complex systems in which spatial self-organization occurs through the interaction of economic, demographic, and environmental factors. He argues that modeling can be used to identify the range of potential alternatives that arise from policy decisions over the long term. There is an "unsustainable hidden reality to urbanization," he argues, which means that "decision makers are increasingly divorced from the reality of the natural system that really supports cities" (Allen, 1997: 19).

By contrast, Whiteman *et al.* (2004) explored the social-environmental impacts of multinational companies in high-latitude eco-regions in order to conceptualize how oil, natural gas or forestry companies interact with "socially and economically mediated ecosystems" (p. 371). They suggest more generally that business theory and practice need to move beyond an unduly narrow social or economic concept of organizational survival. Instead, they propose that regional ecosystem resilience is a more accurate measure of sustainable corporate governance because it shifts the objectives of the corporate board away from short-term individual firm benefits to the long-term viability of the nested social-ecological systems which firms rely upon and collectively shape.

Despite these examples, the concurrent examination of social and ecological processes within organization studies is not common. For instance, Loorbach *et al.* (forthcoming) do not explicate in detail specific ecological processes related to sustainable innovations other than more general descriptions of system constraints such as CO_2 emissions, air pollution and traffic congestion. In a related way, Whiteman *et al.* (2004) do not identify how ecological processes interact with existing social processes used and enacted by boards of directors. Yet a combined approach which examines the coevolution of social and ecological processes is required in order to understand the complexity of nested social-ecological systems. In this paper, we address this gap by exploring how and where social and ecological processes coevolve, and to do so, we first introduce aspects of systems thinking from ecology and economic geography.

Complex Systems as Dimensions of "Real" Ecologies

Ecologists like C.S. Holling (1986) explore sustainability by identifying the resilience or vulnerability of *social*-ecological systems. A social-ecological system (SES) is the combination of natural ecosystems and social systems. An SES is believed to be resilient to the degree that it can absorb shock or 'disturbance' and either bounce-back to its original healthy state or coevolve to another stable state. Resilience is "the capacity of a system to absorb shocks while maintaining function" (Folke *et al.*, 2002).

From this perspective, the emergence of 'adaptive management' practices which anticipate complexity and surprise is encouraged. Adaptive management in conservation or natural resource management emphasizes flexibility, learning and the need to loosely couple local knowledge on ecological processes with institutional knowledge of social processes amongst organizational actors in a cross-scale manner (Folke, 2007). This literature perceives nature, society and economic organizations as part of several nested systems (Holling, 1986; Folke *et al.*, 2002; Whiteman *et al.*, 2004). Resilience is thus related to the adaptive capacity of a complex SES to adjust and adapt, and not simply to the ability of individual organizations to bounce back from social or economic crisis as would be the case when terms like ecology and environment are understood as metaphors for a set of institutions.

While resilience theory has proven robust in conservation management (Folke *et al.*, 2002; Ostrom, 2008), it has yet to conceptualize a concrete role for firms in social-ecological systems. In general, resilience theory conceives of corporate activities as belonging to the 'economic' sphere of systems theory. The literature has yet to study how the actions of individual firms, and industries collectively, are integral coevolving parts of social-ecological systems. This is an important gap to address in ecology, because as Chapin and Whiteman (1998) argue, firms often actively seek ways to create disturbance in the social system to achieve competitive advantage over their peers. Such disturbances—while useful in creating competitive space for action—have unrecognized and perhaps unintended impacts on the resilience of local and global ecologies. Furthermore, how do firms and other system actors become aware of their own embeddedness (Whiteman & Cooper, 2000) in local ecologies and coevolve within specific locations with unique ecological characteristics as well as social and economic developmental paths and prospects? The opportunity is to use conceptual and practical tools from the field of geography that embrace the necessary relationships between ecology and industry, and not just the impacts of one on the other.

Complex Systems as Places

The geographic concept of 'place' offers potentially useful ways of bridging the gap between our use of organizational coevolution and ecology's framework of social-ecological systems using complexity thinking. 'Place' is a rich term that incorporates the natural, cultural, social, political and economic dimensions of a given space and seeks to encapsulate not just local or industrial activity there but also larger scale influences and connection. Importantly, place is not a static concept but is considered an open and historically contingent process in a constant state of change through interaction and renegotiation of power geometries that extend from endogenous local dynamics to global processes (Pred, 1984; Massey, 1994; Paasi, 2002). "No economic landscape, place or region—however defined—stands still. Every economic landscape, place or region is a composite of simultaneous diversity and successive variation, of unfolding processes and historical sedimentation" (Pred, 1985: 7).

In this way, geographers understand organizations and societies as producers of place in ways that neither organization scholars nor ecologists often consider. Places are the result of, and also are themselves always, unique human-environment relations located in a specific location. A classic example of the production of place as a consequence of changing social relations is Cronon's (1983) *Changes in the Land.* This study shows how the beliefs, understandings, and uses of nature by North American colonists led to a different kind of ecology from that of Native Americans. English settlement involved a large-scale transition in eastern North American ecology because the practices, beliefs, and physical landscapes of Native peoples differed vastly from and were supplanted by settlers' understandings and uses of nature. Langston makes a similar observation about the Blue Mountains of Northeastern Oregon and Southeastern Washington; settlement physically changed the pre-settlement landscape and ecology of the Blue Mountains.

In a more metropolitan setting, Guthey *et al.* (2003) found that unique sets of consumers in the San Francisco Bay area are leading the way to new locally-based understandings of food and environment which drives innovation in food industries like dairy. Moreover, beliefs and ideas about nature influenced the sustainability of the system as a whole via the development of new institutions and industrial practices. From across these studies, local cultural patterns and beliefs about nature, environment and lifestyle influence the evolution of local ecology in complicated ways. In the following case study, we map out how a new understanding of place and ecology was tied to the development of the California wine industry.

Organization-Ecology Relationships in California

Sustainability raises clear challenges for firms and industries alike but we have a number of examples of firms embracing or at least incorporating substantive interest in their impacts on the local environment that seem to extend beyond the usual individualized approach such as reducing paper waste through recycling or adopting ISO 14000 production criteria. In this case, we explore how organizations have been actively involved in planning for improved sustainability of economy, society and nature over a forty year period. In short they have not only been parts of a complex organizational system, but also critically aware that they are dimensions of a complex adaptive place.

The history of winemaking in California is often considered one of transition from one form of wine making—bulk production—to another form of wine making—premium production. An interesting facet of this transition is how long and how many different actors including the local ecology have been involved in the industry's transformation. The development of truly premium wine production has taken several decades of research and development involving state-funded university research, significant private investment and entrepreneurship, and fundamental change in people's understandings of human-environment relationships. These changes have led to further innovations in policy making, environmental management, and thus increasing attention to ecology and industrial practice in regional planning. Thus the northern California industry is an example of a complex social-ecological system. Together with a range of other organizations and nature, the industry has coproduced a 'place' with a unique set of beliefs, practices and conventions that are in a continuous process of revision (Guthey, 2008).

Napa County and Sonoma County are part of the nine county San Francisco Bay metropolitan area, a region with a Mediterranean climate generally characterized by a brief 3 month rainy season followed by a 9 month dry season with a high level of biodiversity. It is covered in mixed oak woodlands with chaparral and an understory of mixed grasses and shrubs (Hilty & Merenlender, 2004). In wetter and cooler sections, large stands of conifers (e.g., redwoods *(Sequoia sempervirens)* and Douglas fir *(Pseudotsuga menziesii))* persist. In fact, Sonoma County is known as 'The Redwood Empire.' For the past 150 years, the region has been actively farmed for timber, truck and nursery crops, stone fruits, wheat, dairy, and most notably wine grapes with consequent impacts on ecosystem functioning. But while the ecology has changed over this period, it also forms the basis for ongoing environmentalism and restoration efforts. Of particular concern is the potential for landslides and slumping of vineyard development as a consequence of heavy winter rains which risks damage to public water resources including local streams and rivers with habitat for steelhead and salmon. Development affects dozens of other rare and protected species of plants and animals; approximately 81 special status plants and 68 special status animals—species identified under various protective federal, state, and local laws—are thought to live in Napa County alone (Napa County, 2004). Over the past several decades, a regional consensus has emerged which seeks to integrate environmental protection, agricultural preservation, and smart growth.

Wine firms as Producers of Place

Firms were important players in building resilience into the agricultural industry over this time period. The problem facing the region from a winemaking perspective in the 1960s was that landowners were essentially speculating that housing development would be more profitable than the existing wine industry since the industry was locked into a low road with respect to quality and markets. Thus wine making was considered at the time a low performing industry and many people were hoping to get out of the business. As a result, housing development became the primary threat to a sustainable agricultural economy in the 1960s.

The influx of entrepreneurs into winemaking in this region after the 1950s promised a different sort of wine industry from the essentially bulk industry prior to 1960. These people were part of a broad social movement in the United States during the 1960s and 1970s, known as "The Back to the Land Movement," and associated with a migration of large numbers of people from cities and suburbs into the countryside. In Napa County, many people who attempted this migration were experienced managers who left positions in larger organizations out of frustration with bureaucratic process. They had a vision of small-scale farming as an improved form of enterprise, improved because it was supposed to be more innovative, smaller scale, and closer to the land. They also wanted to live in a place without the complications of urban life and with stronger emotional attachments and public engagement in community life, in short, a stronger sense of place. Thus from the beginning, producers were palpably aware that they were attempting to produce a 'place' that had social and ecological robustness.

In pursuing the seemingly diverse goals of agricultural production and communal living, the new cadre of wine producers helped transform an historically unregulated rural production space into a tightly regulated agricultural place geared to production of fine wines. The distinction between space and place is critical. Hudson (2006: 255-284), for example, argues that space suggests limited connection to a locale while place involves multiple connections and emotional ties to locale. In this case, firms helped induce both new forms of regulation and new understandings of nature-society relationships in Napa County and later Sonoma County. In fact, protecting the land from other types of development that detracted from the agricultural economy was absolutely fundamental to the success of the premium wine industry.

The Region as a Complex Adaptive Place

Agricultural preservation in Napa County became possible because of the combination of industry interest combined with local voters, state legislation and a broad statewide planning movement in the 1960s. In the same year that environmentalists floated a proposal for making Napa Valley a nationally protected vineyard district, the California Assembly passed the California Land Conservation Act which empowered county tax assessors to assess land based on its use value rather than its full market value which is generally higher (Tryphonopoulos, 1967). In this way, state legislators hoped agricultural land might be preserved by shielding farmers from taxation based on their property's full development value. In order to gain these tax benefits, landowners enter into contracts that bar landowners from subdividing or otherwise developing their property for 10 years or more. In order to gain access to these benefits, counties had to implement agricultural zoning, which set the stage for changes to regional development policy in Napa County.

In a county like Napa—a place largely in favor of agricultural preservation but also suspicious of government—large lot agricultural zoning combined with state tax benefits provided a way around the suspicions of "nationalization" implicit in the national vineyard proposal (Sokolow, 1990; Medvitz, Sokolow & Kemp, 1999). The implementation of the agricultural preserve was nevertheless a difficult political project because it conflicted with deeply held beliefs about property rights and existing interests in subdividing agricultural land which held a majority vote on the county board of supervisors. The entire valley floor was zoned for one-acre subdivision. Yet many winery interests were not convinced this was the best alternative including firms like Schramsberg Vineyards, Volker Eisele Family Estates, Martini Winery, and Robert Mondavi.

Progressives also worked closely with a much broader community of interests. For example, San Francisco resident Dorothy Erskine helped to catalyze support for their cause. Erskine believed "Land is a resource, not a commodity subject to speculation and mindless use" (Walker, 2008: 136). She also owned a second home in Napa County which gave her insider credibility. And she was instrumental to convincing recent winemaker Jack Davies to work on policy to protect agricultural land in Napa County. In fact, Davies was not a conservation activist until after having dinner at her house at which point he became convinced that "winemaking and the land were inextricably linked" (Conaway, 1990: 82-83).

Thus outsiders and insiders worked together to convince Napa County public officials and growers to support the development of a locally designated agricultural preserve using state law. Among the local officials important in the agricultural preserve effort were large landowner, grape grower, and Planning Commission member Andrew Pelissa; former County Supervisor Lowell Eddington who represented grape growers at the Napa County Planning Commission and Board of Supervisors; Board Member and Republican Dewey Anderson; County Tax Assessor George Abate, County Administrator Al Harberger; and Planning Commission Chair Felix Vanderschoot (Eisele, 1999: 106). Important grape growers, such as Louis and Louis Peter Martini and one of the Galleron brothers, eventually stood up in support of the proposal as well.

This coalition of urban activists, local officials, and winemakers and farmers eventually succeeded in convincing public officials that preserving agricultural land was good policy, but the agricultural preserve was neither a silver bullet to the question of agricultural land conversion, nor a uniform victory for proponents. Supporters originally wanted to have a 40 acre minimum parcel size. The Board of Supervisors reduced the minimum lot size to 20 acres in order to maintain majority support for the rezoning. They included new rules to allow landowners to sell off large portions of the property while still maintaining a house and one acre parcel, thus enabling a creeping residential estate growth in the preserve. While path breaking, the strategy has led to further subdivision, additional ecological constraints due to habitat fragmentation and increased sedimentary loading, and thus potentially decreased ecosystem resilience. Nevertheless the agricultural preserve was an important step forward, and was a step led in many ways by wine entrepreneurs in view of their efforts to build a new sense of place and a renewed wine industry focused on premium wine production.

Their effort signaled what appears in historical perspective to be an emergent process of additional policies and institutions to promote sustainability in the region. The agricultural preserved passed in 1968. By the early 1970s, the county began to plan for agriculture instead of housing. Voters then passed measures to limit population growth and to place land use decision making authority with the voters. More recent efforts include the Napa Green Certified Land Program developed by vintners, local agencies and conservation groups, a California Sustainable Winegrowing Program, a state-wide producer driven effort to promote sustainability in collaboration with multiple stakeholders, and new incentives in the Napa County General Plan for landowners who "incorporate environmentally sustainable practices that avoid or mitigate significant environmental impacts" (Napa County, 2008: CON-33). In short, sustainability involves many different scales, actors and institutions.

Discussion and Implications

Northern Californian efforts to preserve agricultural land and improve environmental quality present a number of important insights into organizational relationships with nature. Organizations must take into account the unique dynamics of place in understanding and developing approaches to negotiating change. Our case study shows the importance of social embeddedness (Baas, 2008; Granovetter, 1985) and ecological embeddedness (Allen, 1997; Whiteman & Cooper, 2000) within a distinct place. In this case, a number of leading firms, firms organized into trade associations, and voluntary working groups have been able to influence regional policy and industrial practice over a long period. These initial steps towards creating a more sustainable agricultural economy were driven by entrepreneurship that envisioned a certain kind of wine industry and lifestyle in the wine region. The interesting point here is that the initial effort led to a cascade of policies, practices, public support and activism that also proved important in maintaining long term system resilience. Moreover local political actors, including firms, have been critical to generating the momentum to overcome rigidity in the existing social-ecological system so that overtime a movement focused on agricultural and natural resource conservation could emerge and begin to build a different sense of human-nature relationship in the region (Baas, 2008). The case also shows the importance of new emergent governance institutions to maintain multiple moments of stability within the shifting dynamic of local place (Boons, 2008).

The case helps deeper our understanding of systemic change within specific place-based economic activities that are transitioning towards greater sustainability. There were multiple forces and interests vying for influence over development policy in the region. And it was only through fundamental changes in beliefs and understandings of nature (and careful and strategic policy development) that a more sustainable basis on which to build a renewed wine industry emerged. The mechanisms for engaging in this process of policy development and change came from understanding and connecting with place at a deep intellectual level on the one hand, and also from engaging in a long term effort to build a new affective sense of place among multiple stakeholders. This was accomplished, in part, by the creation of local stories on the need for sustainability that in turn contributed to, and reinforced, this evolving sense of place (Baskin, 2008). In this process, firms played central roles in coalition building, research, and network development that ultimately increased the resilience of the regional agricultural economy as a whole.

The results of our study may have implications for many other firms which operate within complex nested systems linked together by specific local places. Our case study demonstrates that organizational actors exist within, change, and are changed by constantly shifting SESs (Holling, 1986; Folke, 2007; Whiteman *et al.*, 2004). Research on conservation management suggests that

1965	The California Land Conservation (Williamson) Act passed by state legislature.
1968	Napa County Agricultural Preserve passed by County Board of Supervisors.
1973	Goal One of Napa County General Plan becomes "Plan for Agriculture"
1976	The Land Trust of Napa County and the Sonoma Land Trust founded to promote private land conservation through tax easements, in fee-title ownership and donations.
1980	Napa County voters pass Measure A limiting population growth to 1% annually in unincorporated Napa County.
1990	Napa Voters pass Measure J requiring public referenda for all changes in agricultural and open space land use through 2020.
	Sonoma County citizens vote to create the Sonoma County Agricultural Preservation and Open Space District, a public sector land trust funded through a quarter-cent sales tax extended by voters in 2006 to 2031.
1991	Napa County Board of Supervisors enacts Hillside Ordinance/Conservation Regulations to address erosion control and stream setbacks.
1995	Napa Sustainable Winegrowing Group formed to address integrated pest management practices and to promote stewardship that optimizes ecological stability, reduces pesticides, and promotes soil health.
1999	Fish Friendly Farming certification program begun in Sonoma County to promote best practices for maintaining viability of salmonid species.
	Napa River Benthic Macroinvertibrate Study initiated to provide baseline and longitudinal data on the Napa River's health.
	Sonoma County Board of Supervisors approves Vineyard Erosion and Sediment Control Ordinance.
2000	Napa County Watershed Task Force releases "Phase II Final Report" detailing the findings and recommendations for managing county watersheds.
2002	Board of Supervisors establishes Watershed Information and Conservancy of Napa County
	Napa County Vintners and Grape Growers initiate Napa Green, a version of Fish Friendly Farming for the Napa River
2005	Napa River Macroinvertibrate study shows both incredible richness and areas of concern along 30 different segments of the Napa River.
2008	Napa County General Plan revision includes incentives for landowners to implement environmentally sustainable practices.
	Napa County voters approve Measure P which extends Measure J to 2058 with minor revisions.

Table 1 *Institutional Development in Napa and Sonoma County, 1965-2008. Sources: Guthey (2004), Jones (2008), Marcus (2004), Medvitz et al. (1999), Napa County (1983), Napa County (2004), Napa County (2008), Sonoma County (2008).*

"[b]uilding social-ecological resilience requires understanding of ecosystems that incorporates the knowledge of local users. Thus the ecological ignorance of some contemporary societies undermines resilience" (Folke *et al.*, 2002: 8). Our findings extend this concept beyond conservation management and suggest that local users with a strong understanding of the cultural, ecological, economic and social dimensions of a specific place will be important for building future sustainability (Baas, 2008; Whiteman & Cooper, 2000). Conceptualizing nature-firm coevolution is thus a critical step in developing adaptive capacity. It can for example yield additional insight into sustainability because we can then identify firm roles and more holistically theorize stakeholder relations with local ecology and local 'places.'

Two shifts seem to be critical for researchers and practitioners alike in the context of sustainability and social-ecological systems. The first involves reconfiguring our understandings of economic performance (Korhonen & Seager, 2008; Whiteman *et al.*, 2004). As Folke *et al.* (2002) write "a fundamental challenge is to change perceptions and mind-sets, among actors and across all sectors of society, from the over-riding goal of increasing productive capacity to one of increasing adaptive capacity, from the view of humanity as independent of nature to one of human and nature as coevolving in a dynamic fashion with the biosphere." Adaptive capacity refers to the ability of an entire SES to adjust in some kind of sustainable manner. Of course, this raises the question of exactly how organizations might develop this capacity; hence the second necessary shift.

The second shift involves focusing and building the capacity within organizations for incremental adjustment, or multidisciplinary learning, that can lead to better outcomes over time for both nature and society (Boons, 2008; Clarke & Roome, 1999; Loorbach *et al.*, forthcoming; Seager, 2008). The resilience of complex social-ecological systems is related to the capacity of the entire system and its individual managers *to learn and adapt* to changing social and ecological processes (Holling, 1986; Korhonen & Seager, 2008; Whiteman & Cooper, 2000). Resilience is not defined solely as a recovery from social or economic crisis using existing practices, nor is it solely the recovery of the natural ecology alone. There has to be emergence of new practices, new social and ecological processes and new understandings of nature-society interactions. This involves institutional learning (Rodrigues & Child, 2003) and adaptive experimentation with the many relationships between ecology and organizations (Holling, 1986). Conceived in this manner, the wine industry is a type of organizational field (DiMaggio & Powell, 1983) that is not only embedded in local and global social-ecological systems but also shapes these systems in distinct ways

Theory from geography and ecology can thus enrich the practice of researchers and managers alike by encouraging them to stretch their understanding of inter-firm networks to include organization-nature relationships emerging in specific places over time. As Lockwood (2007) noted, ecology is critical to organizing sustainable futures. And the challenges we face will involve complex interactions with a set of historically conditioned and place-influenced policies and practices that are emergent and necessarily linked to local ecologies. From this perspective, we see the challenges of sustainability as both deeply socially and ecologically embedded (Whiteman & Cooper, 2000), and subject to multiple levels of influence (micro-level, meso-level, and macro-level). But importantly, the challenges and opportunities arise out of and are created by firms, organizations, people and nature in place. Future organizational studies which seek to examine sustainability from a multidisciplinary perspective would be enriched with a simultaneous analysis of how social and ecological embeddedness in local place contributes or detracts from such activity. Integrating these dimensions of place into organizational theory provides ways of analytically conceiving of firms in relation to place and nature in a more systematic manner and provides a richer understanding of organizational performance within complex social-ecological systems.

References

Allen, P. (1997). "Cities and regions as evolutionary complex systems," *Journal of Geographical Systems*, ISSN 1435-5930, 4: 103-130.

Baas, L. (2008). "Industrial symbiosis in the Rotterdam Harbor and Industry Complex: Reflections on the interconnection of the techno-sphere with the social system," *Business Strategy and the Environment*, ISSN 1099-0836, 17, 330-340.

Bansal, P. and Gao, J. (2006). "Building the future by looking to the past: Examining research published on organizations and environment," *Organization & Environment*, ISSN 1086-0266, 19: 458-478.

Baskin, K. (2008). "Storied spaces: The human equivalent of complex adaptive systems," *Emergence: Complexity & Organizations*, ISSN 1521-3250, 10 (2): 1-12.

Berchicci, L. and King, A. (2007). "Postcards from the edge: A review of the business and environment literature," in J.P. Walsh and A.P. Brief (eds.), *Academy of Management Annals*, ISBN 9780805862201.

Berkes, F. and Folke, C. (1998). *Linking Social and Ecological Systems: Management Practices and Social Mechanisms For Building Resilience*, ISBN 9780521785624.

Boons, F. (2008). "Self-organization and sustainability: The emergence of a regional industrial ecology," *Emergence: Complexity & Organizations*, ISSN 1521-3250, 10(2): 41-48.

Boons, F. and Roome, N. (2005). "Sustainable enterprise in clusters of innovation: New directions in corporate sustainability research and practice" in S. Sharma and J.A. Aragón-Correa (eds.), Corporate Environmental Strategy and Competitive Advantage, ISBN 9781845420055, pp. 259-285.

Chapin, F.S. and Whiteman, G. (1998). "Sustainable development of the boreal forest: Interaction of ecological, social, and business feedbacks," Ecology & Society, ISSN 1708-3087, 2(2): 12, http://www.consecol.org/vol2/iss2/art12.

Clarke, S. and Roome, N. (1999). "Sustainable business: Learning-action networks as organizational assets," *Business Strategy and the Environment*, ISSN 1099-0836, 8: 296-310.

Conaway, J. (1990). *Napa: The Story of an American Eden*, ISBN 9780618257980.

Cronon, W. (1983). *Changes in the Land: Indians, Colonists and the Ecology of New England*, ISBN 9780809001583.

DiMaggio, P. and Powell, W. (1983). "The iron cage revisited: Institutional isomorphism and collective rationality in organizational fields," *American Sociological Review*, ISSN 0003-1224, 48: 147-160.

Ehrenfeld, J.R. (2005). "The roots of sustainability," *MIT Sloan Management Review*, ISSN 1532-9194, 46: 23-25.

Eisele, V. (1999). "Twenty five years of farmland protection in Napa County" in A. Medvitz., A. Sokolow, and C. Lemp (eds.), *California Farmland and Urban Pressures: Statewide and Regional Perspectives*, Davis: University of California Agricultural Issues Center.

Etzion, D. (2007). "Research on organizations and the natural environment, 1992 - present: A review," *Journal of Management*, ISSN 0149-2063, 33: 637-664

Folke, C. (2007). "Social-ecological systems and adaptive governance of the commons," *Ecological Research*, ISSN 0912-3814, 22: 14-15.

Folke, C. Carpenter, S. Elmqvist, T. Gunderson, L. Holling, C.S. Walker, B. (2002). "Resilience and sustainable development: Building adaptive capacity in a world of transformations," *ICSU Series on Science for Sustainable Development*, ISSN 1683-3686, 3:1-74.

Granovetter, M. (1985). "Economic action and social structure: The problem of embeddedness," *American Journal of Sociology*, ISSN 0002-9602, 91: 481-510.

Guthey, G. (2004). *Terroir and Politics of Agro-Industry in the North Coast Wine District*, ISBN 9780542008184.

Guthey, G. (2008). "Agro-industrial conventions: Some evidence from the northern California wine industry," *The Geographical Journal*, ISSN 0016-7398, 174(2): 138-148.

Guthey, G., Gwin, L. and Fairfax, S. (2003). "Creative preservation in the California dairy industry," *Geographical Review*, ISSN 0016-7428, 93: 171-192.

Hilty, J. A. and Merenlender, A. M. (2004). "Use of riparian corridors and vineyards by mammalian predators in Northern California," *Conservation Biology*, ISSN 0888-8892, 18(1):126-135.

Hoffman, A. (1997). *From Heresy to Dogma: An Institutional History of Corporate Environmentalism*, ISBN 9780804745031.

Holling, C. S. (1986). "Resilience of ecosystems: Local surprise and global change" in W.C. Clark and R.E. Munn (eds.), *Sustainable Development and the Biosphere*, ISBN 9780521323697, pp. 292-317.

Jermier, J.M. Forbes, L.C. Benn, S. and Orsato, R.J. (2006). "The new corporate environmentalism and green politics," in S.R. Clegg, C. Hardy, T.B. Lawrence and W.A. Nord (eds.), *The SAGE Handbook of Organization Studies*, ISBN 9780761949961, pp. 618-650.

Jones, J. (2008). "Measure P passes with ease," Napa Register, 5 Nov, http://www.napavalleyregister.com/articles/2008/11/05/news/local/doc491106264b4fc273281050.txt.

Korhonen, J. and Seager, T. P. (2008). "Beyond eco-efficiency: A resilience perspective," *Business Strategy and the Environment*, ISSN 1099-0836, 17, 411-419.

Lewin, A.Y. and Volberda, H. (1999). "The coevolution of new organizational forms," *Organization Science*, ISSN 1526-5455, 10: 519-534.

Lockwood, D. (2007). "Ecology is not rocket science," *Emergence: Complexity & Organizations*, ISSN 1521-3250, 9(1-2): 107-119.

Loorbach, D., van Bakel, J., Whiteman, G. and Rotmans, J. (forthcoming, 2009). "Business strategies for transitions towards sustainable systems," *Business Strategy and the Environment*, ISSN 1099-0836,

Marcus, L. and Associates (2004). *Fish Friendly Farming: History*, http://www.fishfriendlyfarming.org/history.html.

Massey, D. (1994). *Space, Place and Gender*, ISBN 9780816626175.

Medvitz, A., Sokolow, A. and Lemp, C. (eds.) (1999). *California Farmland and Urban Pressures: Statewide and Regional Perspectives*, Davis: University of California Agricultural Issues Center.

Napa County (1983). "Napa County General Plan," *Napa: The County*.

Napa County (2004). "Napa County Baseline Data Report," Napa County Conservation, Development and Planning Department, prepared by Jones & Stokes (Oakland, Ca) and EDAW (San Francisco, CA), http://www.napawatersheds.org/docs.php?ogid=10079.

Napa County (2008). "Napa County General Plan Conservation Element," http://www.co.napa.ca.us/GOV/Departments/DeptPage.asp?DID=8&LID=1790.

Ostrom, E. (2008). Keynote presentation. Resilience 2008 Conference, Stockholm, April.

Paasi, A. (2002). "Place and region: Regional worlds and words," *Progress in Human Geography*, ISSN 1477-0288,

Peltoniemi, M. (2006). "Preliminary theoretical framework for the study of business ecosystems," *Emergence: Complexity & Organizations*, ISSN 1521-3250, 1: 10-19.

Porter, T. B. (2006). "Coevolution as a research framework for organizations and the natural environment," *Organization & Environment*, ISSN 1552-7417, 19: 479-504.

Pred, A. (1984). "Place as Historically Contingent Process: Structuration and the Time-Geography of Becoming Places," *Annals of the Association of American Geographers*, ISSN 0004-5608, 74 (2): 279-297.

Pred, A. (1985). "Interpenetrating processes: Human agency and the becoming of regional spatial and social structures," *Papers of the Regional Science Association*, ISSN 0486-2902, 57: 7-17.

Rodrigues, S. and Child, J. (2003). "Coevolution in an institutionalized environment," *Journal of Management Studies*, ISSN 0855-0352, 40: 2137-2162.

Seager, T.P. (2008). "The sustainability spectrum and the sciences of sustainability," *Business Strategy and the Environment*, ISSN 1099-0836, 17, 444-453.

Sokolow, A. (1990). *The Williamson Act: 25 Years of Land Conservation*, UC Agricultural Issues Center.

Sonoma County (2008). Sonoma County Agricultural Preservation and Open Space District Annual Report 2007-2008, http://www.sonomaopenspace.org/docManager/1000001235/District%20Report%2008.pdf.

Starik, M. and Rands, G. (1995). "Weaving an integrated web: Multilevel and multisystem perspectives of ecologically sustainable organizations," *Academy of Management Review*, ISSN 0363-7425, 20: 905-935.

Tryphonopoulos, N. (1967). *An Investigation of the Economic Structure of a Small Area: Napa County*, California, Department of Agricultural and Resource Economics, University of California, Berkeley.

Walker, R. (2008). *The Country in the City: The Greening of the San Francisco Bay Area*, ISBN 9780295988153.

Whiteman, G. and Cooper, W. H. (2000). "Ecological embeddedness," *Academy of Management Journal*, ISSN 0001-4273, 43: 1265-1282.

Whiteman, G. Forbes, B.C. Niemelä, J. and Chapin, F.S. (2004). "Bringing feedback and resilience of high-latitude ecosystems into the corporate boardroom," *Ambio*, ISSN 0044-7447, 33: 371-376.

Greig Guthey is Assistant Professor of Public Policy and Planning in the Department of Liberal Studies at California State University San Marcos. He completed his Ph.D. in Geography at the University of California, Berkeley in 2004. He has published in *The Geographical Journal* (2008, Agroindustrial Conventions: Winemaking in Northern California), *Conservation Biology* (2004, Land Trusts: Who is conserving what for whom?) and *Geographical Review* (2003, Creative Preservation in the California Dairy Industry). His research interests center on sustainability, regional development and innovation in agriculture and food industries.

Gail Whiteman is an Associate Professor at the Rotterdam School of Management, Erasmus University, the Netherlands. She received her PhD in Management from Queen's School of Business in Canada in 1999. Currently, she is co-founder and Director of the Sustainability and Climate Research Centre at RSM. Dr. Whiteman's research focuses on the roles of companies and managers in the resilience of complex social-ecological systems. She takes a multidisciplinary approach and works closely with management theorists, geographers and ecologists. She has published on sustainability issues in *Nature, The Academy of Management Journal, Ecology & Society, Organization Studies, Human Relations, Organizations and the Natural Environment, Ambio: A Journal of the Human Environment, Business Strategy and the Environment*, and numerous book chapters.

19. Collective Case Study Method and Fractal Geometry: Instrumental and Intrinsic Cases in Organizational Research

Donald L. Gilstrap
University of Oklahoma, USA

In case study methods of organizations, researchers are often limited to the aggregation of individual cases within the context of the organizational case. Borrowing from Stake's (1995) use of instrumental and intrinsic case studies, this paper presents a fractal geometry case study method. For the purposes of this article, on site interviews of seventeen librarians who work in a research institution were conducted to learn more about their experiences with organizational change. Instrumental case studies of these individuals, or rather those cases that respond to other phenomena, were performed and analyzed at the micro level. A clustering technique, serving as a fractal seed, was also incorporated to draw out themes that highlighted the interconnections of individuals. These cases were then recursively integrated into an emergent framework of the intrinsic case of the organization. The use of this method suggests that observations of individuals, and the subsequent meaning they generate at the micro level, reflect the complex interconnections of these cases. At the same time, this method suggests that the recursive integration of individual cases contributes to the understanding of the complex organization at the macro level.

Introduction

Case study research methods of organizations have been debated with some frequency in the social and behavioral sciences. In addition to the case findings, epistemological differences are represented in the disparate approaches performed by case study researchers. Moreover, there are ontological differences between case studies thought to be traditionally scientific and those case study methods that are purely qualitative. As an example, some scholars approach case study research as a means to generalize the findings of a case or a collection of cases to the population at large. Arguing that this must be avoided in qualitative research, however, Stake (2005) focuses primarily on the specificity of cases and how their uniqueness contributes to further understanding. "Case study has been too little honored as the intrinsic study of a valued particular, as it is in biography, institutional self-study, program evaluation, therapeutic practice, and many lines of work" (Stake, 2005: 448).

In case study methods of organizations, researchers are often limited to the aggregation of individual cases within the context of the organizational case. Conversely, narrative research methods are often chosen to focus on the individual but may sometimes lack coherence in their connection to the case of the organization. Borrowing from Stake's (1995) use of instrumental and intrinsic case studies, this article presents a complexity science research methodology, using a fractal lens for case study method. For the purposes of this article, on site interviews of seventeen individuals who work in an academic unit of a research institution were conducted to learn more about their experiences with organizational change. Instrumental case studies of these individuals, or rather those cases that respond to other phenomena, were performed and analyzed at the micro level. A clustering technique, serving as a fractal seed, was also incorporated to draw out themes that highlighted the interconnections of individuals. These cases were then recursively integrated into an emergent framework of the intrinsic case of the organization. It is the purpose of this paper, therefore, to show how this fractal case study method suggests that observations of individuals, and the subsequent meaning they generate at the micro level, reflect the complex interconnections of these cases. At the same time, this method highlights the recursive integration of individual cases and how they contribute to the understanding of the complex organization at the macro level as a fractal form.

Theoretical Orientation

The theoretical orientation of this paper draws from the scientific and social sciences research of complexity theory while grounding qualitative case study method in the works of Stake (1995, 2005, 2006). Stake suggests three main types of case studies: intrinsic, instrumental, and collective. Intrinsic case studies are used to understand the particular aspects of one or more cases. A researcher might utilize instrumental case studies, however, to understand another case or another issue. And collective case studies are used to draw thematically from several cases. Additionally, instrumental cases do not rely on a priori hypotheses to develop intrinsic themes; rather case study themes can emerge during the course of data collection and analyses in a continual process of interpreting and reinterpreting data. Moreover, the use of case study method enables researchers to find "interactivity" (Stake, 2005: 452) and connectedness between the individuals participating in case study while transferring the experiences of the participants to the reader through the development of a story or stories that describe the cases; or, as Stake (2005) suggests, description of "the case in sufficient descriptive narrative so that readers can experience these happenings vicariously and draw their own conclusions" (p. 450).

Integrating instrumental with intrinsic cases into collective case study can be a large task, however. Stake (1995) admits that researchers in the social and behavioral sciences might attempt to "serve too many audiences" (p.

135) when they combine the two methods. However, as Stake (2005, 2006) also notes, the use of multiple methods of case study can be viewed as a nonlinear approach to qualitative research:

> *There is no hard-and-fast line distinguishing intrinsic case study from instrumental, but rather a zone of combined purposes... I call this multiple case study or collective case study. It is instrumental study extended to several cases. Individual cases in the collection may or may not be known in advance to manifest some common characteristic. They may be similar or dissimilar, with redundancy and variety each important. They are chosen because it is believed that understanding them will lead to better understanding, and perhaps better theorizing, about a still larger collection of cases* (Stake, 2005: 445-446).

Therefore, the use of these two methods in a fractal framework can be beneficial to qualitative researchers.

The generation of a fractal model of case study method draws from the theoretical framework of chaos and complexity theories. Expanding upon the works of Fournier d'Albe and Stoney (1907) on electro-magnetic theory, the iterative mathematics of Fatou (1906) and Julia (1918), and from Lévy (1948) on stochastics, Mandelbrot (1975, 2004) expanded his own work on economic theory to create visual representations of complex geometrical patterns in mathematics now known as Mandelbrot sets. Important to this discovery are the bounded conditions of self-similarity between individual agents of a complex function viewed at the micro level in relation to the macro structure. Fractal images represent the simultaneous iteration of real with imaginary numbers, known as complex numbers. A fractal mathematical process contains two or more values—one representing the real and one representing the imaginary—while incorporating the use of a seed upon which each iterative calculation returns recursively until the computation either collapses or goes on into infinity. This use of complex mathematics reveals that, rather than the one dimensional linear points on a plane represented in Euclidian geometry, fractal iterations represent two dimensional iteratively created planes that are observable by the eye (Gribben, 2004; Mainzer, 2005; Mandelbrot, 1975) (see Figure 1). Moreover, this view of planes folding in on themselves in self-similar and attractor patterns is further understood in three dimensional images where each iteration adds further complexity to the full phase space of a fractal process (Peitgen, Jürgens, & Saupe, 2004). Equally, the distance represented as height in a three dimensional view of a Mandelbrot set further explains the difference between one dimensional real numbers and two dimensional complex numbers.

In Mandelbrot's (1975, 2004) view, the re-discovery of fractals provided a springboard for new types of experimentation in mathematics and subsequent re-analyses of how we view the world. While mathematics had been, for many years, subject to the empirical linearity of Euclidean geometry, Mandelbrot

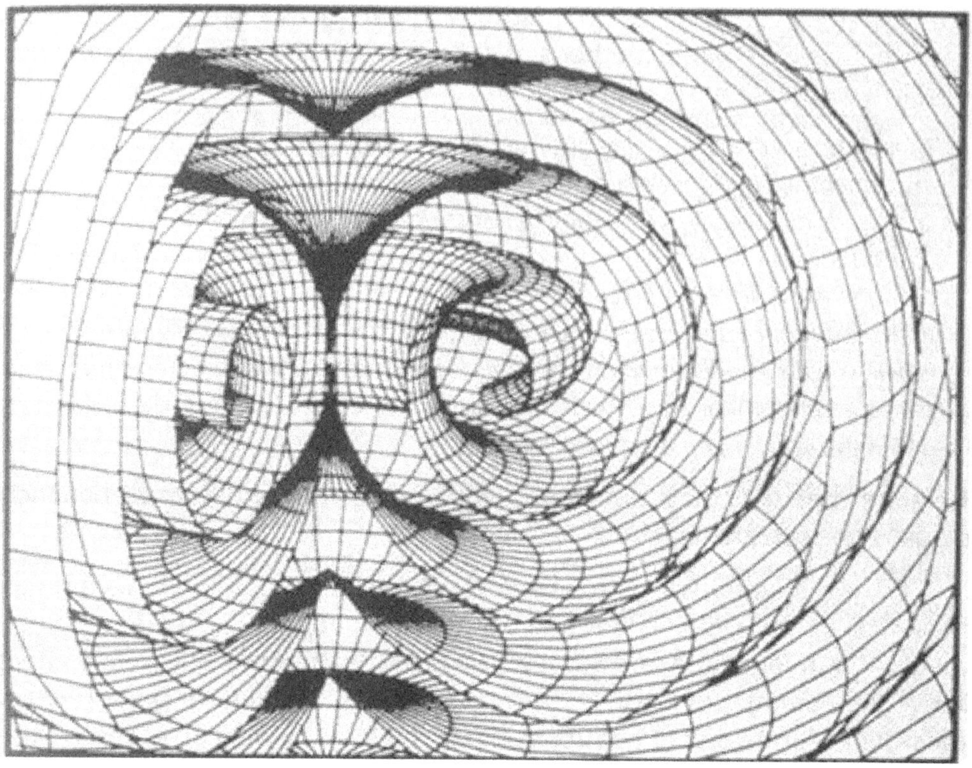

Figure 1 *Fractal Representation of Planar Space (http://www.quantumconsciousness.org/ultimatecomputing/ulti_files/image010.jpg)*

(2004) believed that fractal geometry provided mathematicians with the ability to begin to "see" mathematics in new ways.

> *Altogether, my lifetime scientific work rescued the verb "to see" from the figurative meaning to which both common usage and hard quantitative science had reduced it, and restored its concrete meaning, whose instrument is the eye... It suggests that a strong interest for mathematics is widespread among humans, but only if its links with nature and the eye are not actively suppressed, but, instead, brought out and celebrated* (Mandelbrot, 2004: 3, 10).

As an example, what appeared originally to Julia (1918) and Fatou (1906) as Julia sets disconnected from parent Mandelbrot sets later was shown by Mandelbrot (1983, 2004) to be a reliance on specific variables chosen a priori in experimentation. By allowing the computations to choose their own course, the appearance of the connectedness of Julia and Mandelbrot sets emerged in visual representation. In some ways, Mandelbrot (2004) notes, this new, observable object forced mathematicians and other scientists to question discrepancies in empiricism, representing "a historical change and a return to a far less ideological view of their craft" (p. 36). Furthermore, the fractal nature of the Mandelbrot set yields the strange attractor patterns found in chaotic and dissipative structures which have also been studied in organizational dynamics (Gilstrap, 2005, 2007a).

Variance emerges in fractal systems in non-linear ways that can prove challenging to linear mathematical models. As is the case with the example of Julia sets, self-similarity takes on a paradoxical nature in that three-dimensional representations of fractal patterns reveal simultaneous boundary conditions and unique internal attractor patterns. As Mainzer (2005) states, "the dynamics of a system converge neither to a fixed point nor to a closed curve but to a bounded region of a state space that is filled up by irregular and nonperiodic trajectories of self-similarity." Furthermore, fractal images contain layers of complexity where the strange attractor pattern migrates from plane to plane rather than plot to plot (Gribbin, 2004). In effect, Mandelbrot sets reflect the associations of real, imaginary, and complex numbers found in Julia sets (Mainzer, 2005). This concept of *seeing* mathematics in new ways, therefore, extends a metaphorical, if not epistemological, framework for describing how humans interact in complex, scale independent, and self-similar ways in organizations.

The use of fractal representations has also been expanded to qualitative research in the social and behavioral sciences. Scholars such as Davis (2004, 2005) and Fleener (2002) have investigated the use of fractals to generate visual frameworks for understanding human dynamics. Davis (2005), as an example, has described the recursive process inherent in fractal patterns that emerge at both the micro and macro levels:

> *Fractal images are produced by establishing a rule, applying that rule to generate a result, reapplying the rule to that initial result, and continuously reapplying the rule to results as they are determined in an unending, reiterative process… fractal images don't get simpler when you magnify them, nor more complicated when you pull back from them. Said differently, there is a scale independence to complex phenomena… Closing in on any aspect of this web, or pulling back from the image, reveals another web that is similar in detail and form to this image. Each moment is a node, connecting and collecting a new weave of associations. There is no line joining birth to death. Rather, life is presented as a complex unfolding that is always renewed, always brimming with detail* (Davis, 2005: 124-126).

In a similar manner, the instrumental case of the individual can represent the micro observation of the fractal's pattern. Each of these instrumental cases holds a unique pattern yet has self-similarity with other instrumental cases. When focusing on fractal geometry at the macro level, therefore, the intrinsic case of the organization emerges where this self-similarity is reflected throughout the fractal image.

As Davis (2004) has argued, fractal geometry provides us with a representation of both the intersubjective and the interobjective. Intersubjectivity is an epistemology of the social construction of reality, knowledge, and humanistic interpretivism. In Davis's (2004) view, intersubjectivity is largely a response to the ontology that truth and knowledge exist independent of the observer. In case studies similar to this research study, intersubjectivity in group dynamics can be described as:

> *How a 'common' sense of the situation emerges among the participants while engaged in a cooperative activity in a work-like context... Thus, at various points of the cooperation, the participants must make adjustments that enable them to align their private views of the situation... In work-like environments, the co-construction of intersubjective space is as much about how the participants develop their representational practice to avoid jointly focused, sequential grounding interaction* (Alterman, 2008: 816).

Interobjectivity borrows from an ecological epistemology grounded in "mutually affective relationships between phenomena and knowledge of phenomena" (p. 101). Similar to the case of Mandelbrot's (2004) description of planar phase spaces, Davis (2004) contends that an interobjective philosophy does not espouse an observer-less truth, but rather seeks out qualitative spatial representations of interconnectedness. As is the case with this research study, the author projects observations of the fractal structure of the case of the individual and the organization in a coherent system of interrelationships. However, the methodological proclivities of scholars often place the intersubjective and interobjective in conflict with one another. Likewise, the epistemologies of qualitative and quantitative research might even "weave between induction and deduction in an iterative process," (Bruce, 2007: 52), similar to qualitative research on group dynamics where simultaneous emergence and reductionism are used to analyze phenomena (Gilstrap, 2007b). Davis (2004) therefore suggests that fractal geometry provides us with a framework for understanding how the two discourses can coexist while being philosophically contradictory and simultaneously complimentary in the iterative branching of methodological thought (Figure 2).

Expanding upon this use of fractal thinking, Peitgen, Jürgens, and Saupe (2004), provide great detail of the Sierpiński Gasket as a fractal representation of infinite recursion (Figure 3). The gasket starts with a three-dimensional, equilateral triangle. The center point of the bottom line of the triangle is used to form the top point of a new equilateral triangle within the original. Once this triangle has been formed, it is then removed from the original triangle. This process then repeats recursively with an infinite number of new triangles being formed with each iteration.

In the social and behavioral sciences, Fleener (2002) provides us with an example of the Sierpiński Triangle in explaining curriculum development (see Figure 3). Fleener (2002) presents a visual representation of the interconnections of curriculum processes when interpreting Doll (1993): the Sierpiński Trian-

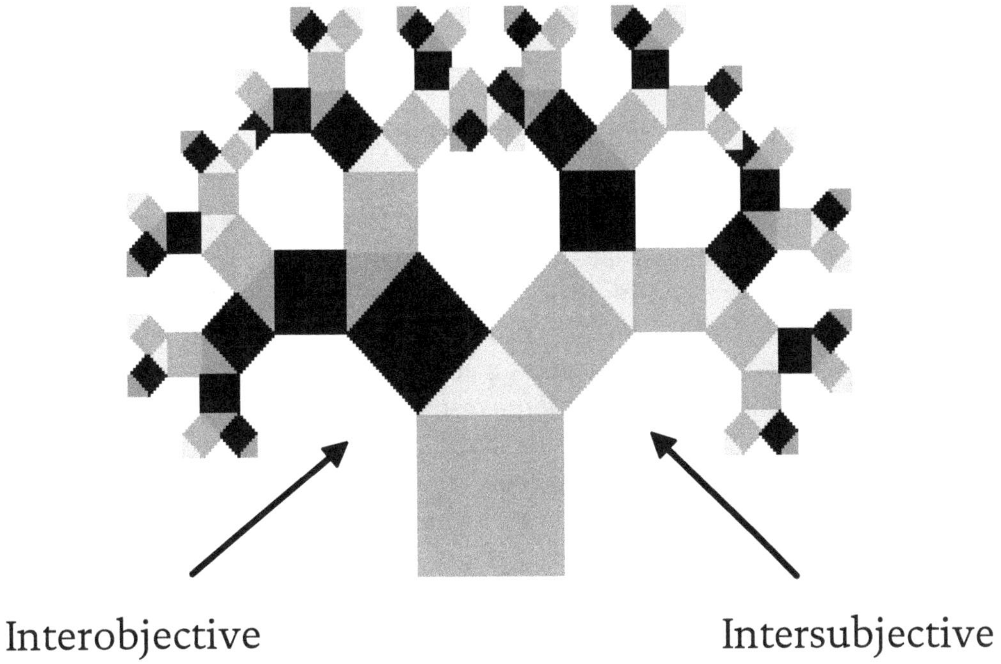

Figure 2 *Intersubjective and Interobjective Branching Discourses (Adapted from http://www.cevis.uni-bremen.de/fractals/index.html)*

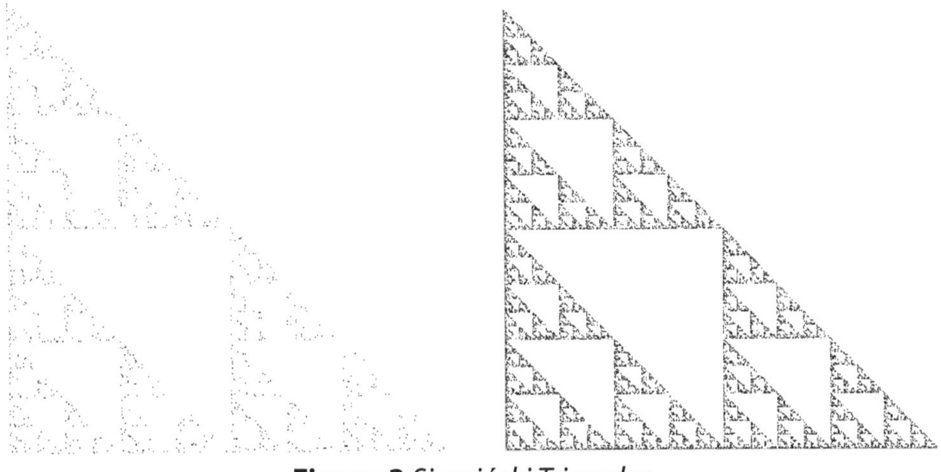

Figure 3 *Sierpiński Triangles*

gle shows the emergence of self-similarity in further definition through several subsequent iterative processes. By representing curriculum development with a fractal model, the metaphor generated reveals that curriculum processes can be simultaneous outcomes and developments. Likewise, Smitherman (2005) uses a fractal approach to curriculum development by introducing a fractal seed into preliminary classroom discussion to which she returns recursively as the conversations become more complex. In ways similar to these examples, a fractal case study method encourages a generative framework for understanding the cases of individuals, how they interact with each other, and how the instrumental case of the individual contributes to the intrinsic case of the organization.

Method and Analysis

In qualitative case study research, Stake (2005) contends that the sampling of individuals should primarily be purposive. In this research, a study was conducted to investigate librarians' experiences with change at a large research library (Gilstrap, 2007c). Administrators at several research libraries were contacted to identify an institution that could exemplify the transformative change process. Based on their identification of particular institutions, a preliminary survey was sent to all librarians at a large research library, asking each to discuss their experiences with change in their library. Thirty librarians responded, approximately fifty percent, and seventeen librarians were then selected purposively for on site interviews based on the richness and depth of their responses. Since institutional review board policies prohibit disclosure of the institution and participants, further detail about the selection of the institution and participants cannot be provided in order to protect their anonymity. The term East Coast University (ECU) Libraries and the names of the participants, therefore, are pseudonyms used to refer to the organization and the individuals.

On site interviews consisted of a clustering technique and semi-structured interviews. The use of "clustering" was first introduced by Rico (1983) in creative writing, and the technique has been expanded for use in the social and behavioral sciences by Karpiak (2000). Participants were provided with a sheet of paper containing the phrase: "change in my library." They were then asked to write down short ideas or mental pictures they thought of while contemplating this phrase. The use of the clustering technique served as a method that allowed the participants to describe the first things that came into their minds (Figure 4). This researcher then allowed participants as much time as they needed to present their individual mental maps that surrounded the main phrase.

After participants appeared to be finished with their clustering map, this researcher then turned on the audio recorder and asked the participant to describe what s/he was thinking about each of the concepts s/he had written down on the paper. The researcher encouraged the participants to go into as much detail about their thoughts and feelings regarding each cluster as each person wished. When the participant felt s/he had responded sufficiently to each of the clusters, the researcher then asked the participant to provide a summary of his or her thoughts about the contents of the clusters. Participants were given as much time as they needed to describe the concepts they had written down which usually lasted around fifteen minutes.

After participants finished their clustering exercises, the researcher moved to the second portion of the interviews. Questions used for the semi-structured interviews were developed based on the main research questions for

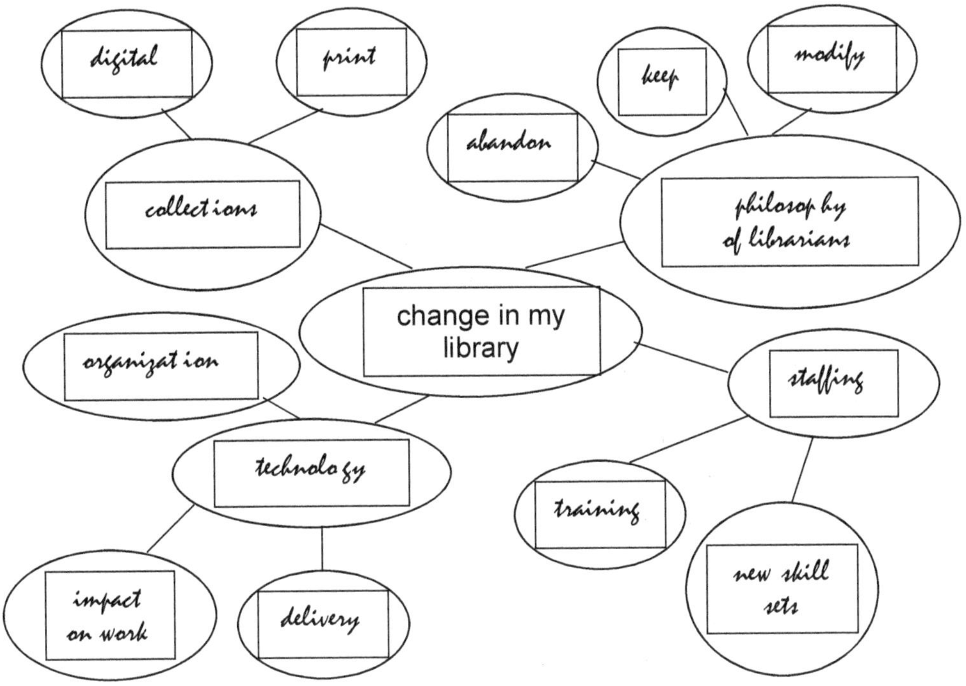

Figure 4 *Clustering Technique Used during Interviewing*

this study and were asked of each of the participants. Each interview lasted approximately 60-90 minutes. Consequently, the main questions from the structured template were used, but follow up questions of a semi-structured nature were also asked based on both the participant's and the interviewer's interpretations of the questions and responses that emerged during the interview. When participants felt they had answered all of the questions sufficiently, they were given an opportunity to add any other information to the interview.

Instrumental case studies were then coordinated through collective case method into a complex systems framework for intrinsic case study, looking at the interactions and connectedness of individuals operating within an organization (Stake, 1995, 2005). Since the instrumental case studies of each individual are woven into the fabric of the intrinsic case study, the use of the fractal case study method allowed for the voices of the participants to rise through the aggregation of data themes. While analyzing the data provided from these interviews, it is also important to note the role that the researcher plays in observation and interpretation of case study participants. As the work of Heisenberg (1944) showed in quantum theory, the role of the observer influences that which is observed. Davis (2004) also expands upon this paradox in social and behavioral science research, noting that "observed and observer are not two forms separated by space. It's more the converse" (p. 145). In like manner, the role this researcher played in the analysis and interpretation of data collected from study participants suggests a shared space between the observer and the observed, much like the intricacies that emerge when seeing fractal geometry with the eye as contended by Mandelbrot (2004). This interconnection between the instrumental and intrinsic case studies provides a framework for the complex web of associations and patterns that emerge at both the individual and organizational levels. As Alterman (2008) argues, "intersecting sets of internal representations cannot account for the commonality of the situation. Rather, the focus should be on the organization and flow of social interaction that *produces* a common understanding" (p. 817). In like ways, this view of multiple case study analyses reveals similarities between the fractal patterns in chaotic and complex systems and those inherent in the relationships of human experiences (Figure 5).

The following image represents a fractal map of one theme that emerged during intrinsic case study analysis of the ECU Libraries as an organization. The findings of these instrumental case studies suggest that the experiences of individuals converge toward and diverge away from intrinsic case study themes in ways that are very similar to the chaotic structures associated with a fractal

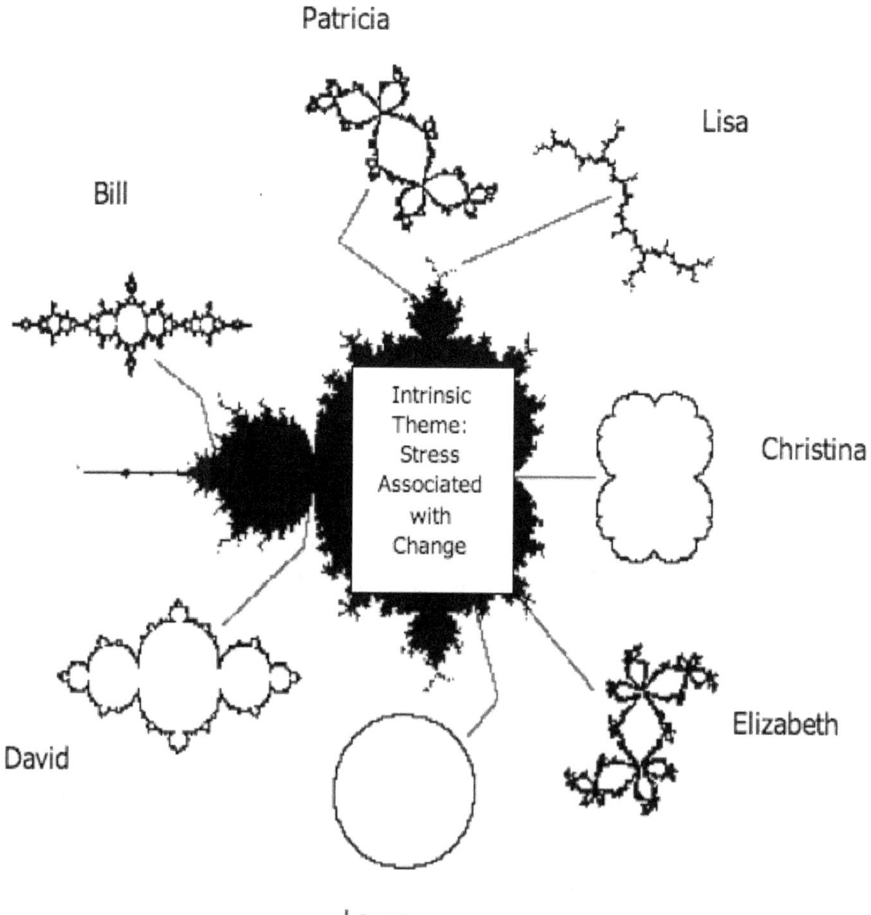

Figure 5 *Fractal Representation of Instrumental Case Studies as Julia Sets Relating to Mandelbrot Set of Intrinsic Case Study Theme (Adapted from http://hypertextbook.com/chaos/)*

image. In the case of several instrumental case studies, for example, a reflection of similarities to a general theme of the intrinsic case is observed. At the same time, the experiences of any one individual under study contribute to characteristics that are unique to that instrumental case. As Stake (1995) notes in case study research, instrumental cases are used to help understand the emergence of other phenomena. The complexity of instrumental cases represented as fractal images at the micro level therefore emphasizes how both unique and interdependent relationships contribute to the emergence of complex themes at the macro level of the intrinsic case.

Equally, when viewing these instrumental case studies through the lens of fractal geometry, analysis of data collected from interviews using the clustering technique shows how the development of themes at the individual level serves as an emergent quality of the themes generated for the intrinsic case study (Figure 6).

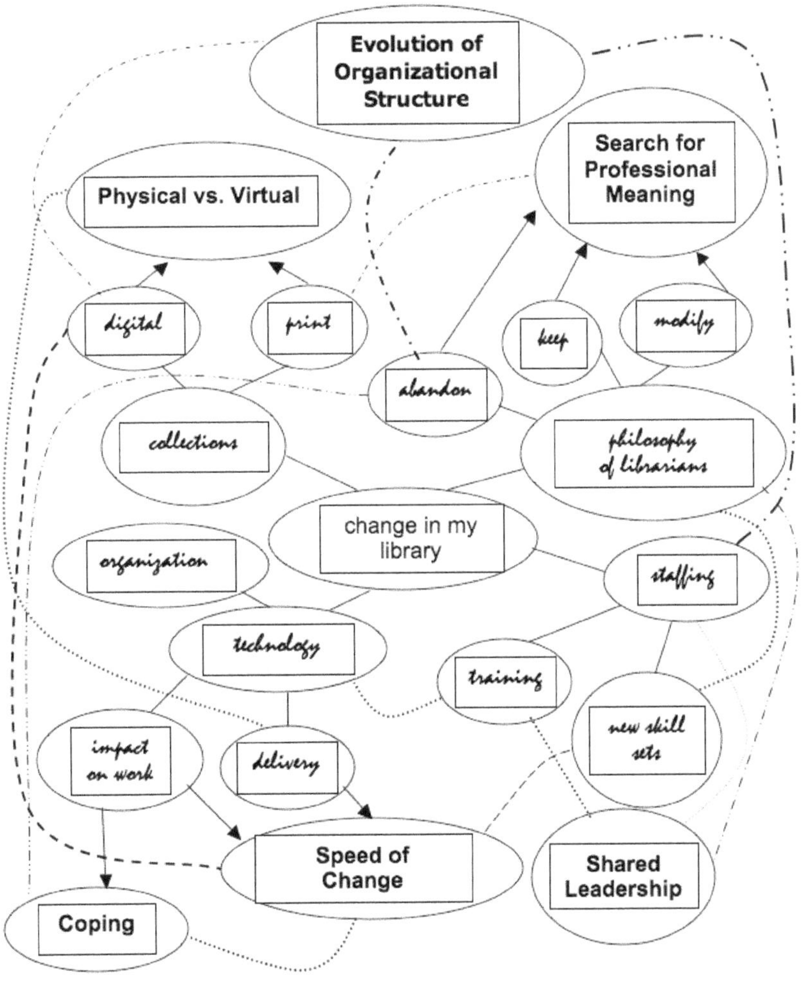

Figure 6 *Concept Map of Intrinsic Themes Emerging from Instrumental Case Clustering*

Fractal Case Study Method: Instrumental Case Studies Integrating with an Intrinsic Case Study Theme

The following section provides a working example of how the fractal representation of the instrumental case studies can integrate into the intrinsic case study. As Denzin and Lincoln (2005) in qualitative research, Stake (1995, 2005, 2006) in case study research, and Davis (2004, 2005) and Doll (1993) in complexity science research have all addressed, richness in description is needed to capture the specificity, nuances, and interconnectedness of a particular research study. Although space in this article will not permit for the inclusion of this research study in its entirety, it is important to note that several additional instrumental case studies, as well as intrinsic case study themes, are included in the full research study.

There are two sections incorporated in the following pages. This first section details the analysis of data collected from the study's participants during their interviews. Reinforcing that the clustering technique (see Figure 4) served as a fractal seed for generating emergent qualities of the semi-structured interviews, each case that follows is an instrumental case which helps to explain the individual and his or her connection to the intrinsic case of the organization (Figure 5). Much like a playbill, the characters are introduced in order of appearance with short descriptions of the instrumental cases of the individuals. This provides a representative example of how the Julia sets which accompany the case are both filled in and bounded by the cases of each of these individuals. The instrumental cases are then used to understand the intrinsic case of the organization. The second section details the descriptions emerging from recursive analysis of the instrumental cases while generating themes used to encapsulate the intrinsic case of the organization. The Mandelbrot set provides a fractal representation of how these unique individuals provide self-similarity within the larger intrinsic case. This theme, in particular, focuses on how these individuals' experiences with change contribute to the overall stress within the larger far-from-equilibrium system.

Filled Julia Sets: Instrumental Case Studies of Individuals

The Case of Bill

Bill has worked as an administrator in the ECU Libraries for many years. He is nearing retirement, and he reflects on the significant changes that have taken place at the library over the past 15 years. Bill has been through both of the reorganizations at the ECU Libraries, and he provides a broad understanding of the decision-making processes and organizational structure of the library. He also recounts how business models are becoming more common place in research libraries, and, although he supports them, he questions their longevity in higher education. Bill is near retirement, and he focuses more deeply now on how he may be able to contribute to the future of the ECU Libraries by enabling staff in his area to move forward after he is gone.

The Case of Patricia

Patricia has worked for the ECU Libraries for many years and most recently has served as an administrator in a technical area. Patricia is deeply committed to the educational mission of the university and in guiding the people in her area toward positive professional growth. She sometimes questions whether the future of librarianship is certain, however. Patricia consequently encourages all librarians working in research libraries to step back from their everyday routines and find ways to implement change that is concurrent with the shifts in higher education practices.

The Case of David

David has worked at the ECU Libraries for many years and now serves in an administrative capacity in a technical area. He is nearing retirement and questions whether research libraries are evolving and adapting fast enough to survive the future of the twenty-first century. David reflects critically on the profession and the changes that have taken place at his own library, searching for much deeper meaning associated with change and his connection to his life's work.

The Case of Lisa

Lisa is still in the beginning stages of her career and has been with the ECU Libraries for only a short time. However, she brings a great wealth of knowledge and experience, having worked at other academic libraries while currently working as one of the library's administrators. Lisa was not employed at ECU during the first major organizational restructuring and does not comment on this experience, but she identifies herself as one of the people who actively promotes change within the ECU Libraries. However, she struggles with the stress brought on by the rapidity of change in her library and recognizes this symptom in her other colleagues.

The Case of Elizabeth

Elizabeth is new to the profession of academic librarianship and new to the ECU libraries. Her primary responsibility is working in an educational capacity, and she brings a wealth of external professional experiences, having served as an educator before changing careers. Elizabeth encourages further change to take place at the ECU Libraries and challenges professional thinking among all academic librarians who hold on to traditional ideologies that primarily protect and force people to use books. She fears a future where librarians do not advocate for a futuristic role of librarianship. She also predicts that if librarians do not confront these philosophical issues at an individual level, the profession may be reduced to a servile technical role or become obsolete altogether.

The Case of Laura

Laura is a mid-career librarian who has in recent years come to ECU to serve as a manager in a public service area. She is an advocate for incorporating technologies that reflect contemporary practices in society. Additionally, her experiences in other libraries and in her participation in professional groups leads her to question whether organizations like the ECU Libraries, which are instituting transformational changes, are anomalies in the larger scale of research libraries. Since she is many years away from retirement, she fears that the traditional practices engrained in the profession at large will lead to a bleak future for librarians.

The Case of Christina

Christina is relatively new to the ECU Libraries. Although she primarily serves the library in an educational role, her previous experiences in technology development have also been utilized by the library. Christina has little tolerance for the traditional practices of the profession and strongly encourages librarians to continue to make the shift to a completely electronic environment. She has a strong desire to focus on competencies in the workplace and learn new skills. Christina provides the ECU Libraries with mechanisms for positive feedback in her continual challenges to traditional librarianship.

Intrinsic Case Study Theme and the Mandelbrot Set: Stress Associated with Change

The concept of stress associated with rapid change causes the librarians at ECU to respond with different types of emotions. Some grow despondent, others look for opportunities within the change process, and still others find particular coping methods that work best for their individual experiences. Many librarians identify with the stress that comes as a result of constant change. Librarians worry that they are not keeping up with the needed technological shifts, because there are so many skill sets required to manage hi-tech research libraries. And this angst is exacerbated if a person does not have enough knowledge to assist in decision-making or to make the correct decisions. Bill, who has worked with the conversion of print to digital resources for many years, notes for example, "it's overwhelming in some ways—it's often like you're running to catch up." As a result, librarians rely more on the collection of people with different knowledge and skill sets to make recommendations that will influence individual decision-making. Patricia, who oversees a technical services unit, expands on this example, noting that she has learned to delegate more responsibility for decision-making to her staff while relying on them to provide her with information. By letting go of control as an administrator, this process not only lowers her own stress, but her staff generally come back with more information, adding value to the decision-making process.

An overwhelming sense of stress associated with the volume of change at the ECU Libraries also takes place in the organization. Librarians comment that the amount of scenarios upon which decisions can be made are far more now than there were just a few years ago. As an example, Lisa, who oversees work in a public services area, states that she feels an increased amount of anxiety that goes along with determining whether you have made the correct decision:

> *It's like you're constantly searching the landscape to make sure there isn't something else ahead of you that you need to be on top of. And for me, I get over 100 email messages a day, and a lot of people have demands on my time and attention. It is hard to keep up with that. At what point do you say, "we're kind of maxed out on what we can do?"*

As another example, librarians recognize emotional responses similar to these among staff whose tasks are no longer valued. This increases their own stress, because they realize that these changes must be made to maintain the relevance of the library to the university community.

Sometimes these emotional responses can be accompanied by feelings of despair. As an example, Elizabeth, who works in educating students about changing library resources, states that the stress associated with competition with the private sector leads her to feelings of hopelessness in her ability to keep up:

> *I think our library has been having to face how difficult it is just to stay in the present—not able to get ahead—just keeping things on. Because library users just do not understand and have absolutely no sympathy for things not working.*

When a computer quits working, users must wonder why someone just does not simply replace it with a new one. Moreover, library users have little sympathy for system failures and network bottlenecks, and they do not understand what goes on behind the scenes to make all this technology work in research libraries. And as Elizabeth comments, in a research library there probably should be little sympathy.

Technology, conversely, can be viewed as contributing to increased productivity in the workplace. Many librarians at ECU comment that technology has enabled them to increase the volume of work they oversee on a daily basis. But some librarians question if this productivity can sometimes actually be unhealthy for a sustained career in research libraries. As an example Laura, who oversees a large public services unit, states that, "pretty much from the moment I step through the door to the moment I go home, it's non-stop." Moreover, librarians are also finding that the lines between their professional and personal lives have significantly blurred due to the technology they use. Since ECU librarians have access from their homes to all the technology they use at work, several joke that finding enough time to sleep is now becoming more problematic.

Many librarians characterize the ECU Libraries as "non-stop" work and also identify the resulting stress in less positive ways. They note that their typical day starts when they hit the door and does not let up until they leave. Some librarians comment that the work of teams also takes up a large portion of time when they could be working on individual projects. Additionally, librarians identify that the stress in their work lives spills over into their private lives where they must sometimes juggle between family and work. As an example, Christina, who works in an educational capacity, states that her commute to work intensifies this stress, as she has to be away from her children for longer periods during the day. She jokes about these high levels of stress, stating, "you might as well take up smoking, you know?"

The experiences of other librarians also supports the constancy of stress brought on by working at the ECU Libraries. Librarians comment that the stress brought on by so many changes can be exhausting. Moreover, they are not sure that the university has mechanisms in place to allow librarians to relax periodically and use creativity serendipitously to stumble upon solutions. David, as an example, is coming closer to retirement which causes him to reflect on how tiring all of the changes have been. And he highlights a key theme that can perhaps be attributed to the stress associated with research libraries in general which pertains to support of both a print and a virtual environment:

> *I find myself getting exhausted by it all; by the constant uphill battle with it. It became obvious to me a long time ago that it wasn't going to be either a brick or a byte library, but it was going to be a brick and a byte library... And we were going to have to maintain them both with the same amount or less staff. The same or less budget. Quite frankly... it burns you out after a while.*

Conclusion

This article is intended to describe the use of fractal case study method to understand individuals and their connection to their organization. In some ways the fractal case study method described might reflect narrative inquiry in its rich description of individual, instrumental cases. Likewise, the generation of themes for the intrinsic case might resemble similar methods used in phenomenological research. But rather than copying commonly used methods, much like Davis' (2004) discussion of intersubjective and interobjective discourses, fractal case study method extends narrative inquiry and phenomenology in that it provides for interconnectivity and interrelationships between instrumental and intrinsic cases. Equally, as Davis, Sumara, and Luce-Kapler (2000) have argued, fractal geometry provides many advantages for looking at social and behavioral science research through the lens of self-similarity. "Each layer or body can be simultaneously seen as a whole, a part of a whole, or as a complex compilation of smaller wholes" (p. 73). In the previous cases, the fractal case study method reflects this self-similarity in that each instrumental case study of the individual can encapsulate that individual intrinsically, instrumentally in his or her connection to others, and collectively in the development of the intrinsic case of the organization.

As Davis (2004) describes, the fractal geometry of social and behavioral sciences research is not intended to be an exact science. In like form, the fractal representation of case study method cannot reflect in exact manners those forms generated by computer simulated models. In the human and behavioral sciences, recursive analyses generated by a fractal seed cannot, in most cases, eventually collapse in on themselves as is the case with complex numbers in Mandelbrot set construction (Peitgen, Jürgens & Saupe, 2004). Likewise, as Gribbin describes with the human body:

> *Of course, the analogy breaks down at the extremes. Systems within the human kidney do not actually branch infinitely many times, just very many times; and going in the other direction, we do not find kidneys embedded with superkidneys, and so on forever, but that each system is self-contained. Nevertheless, the similarity between many living systems and fractals is more than merely an analogy... Near-fractal self-similarity is a pervasive feature of the bodies of living organisms* (Gribbin, 2004: 108).

Further discussion is also needed to explore the connections between qualitative research and such fractal issues as prisoner sets, imaginary numbers, and planar spaces (Mainzer, 2005; Mandelbrot, 1983, 2004; Peitgen, Jürgens, & Saupe, 2004). And in the social and behavioral sciences, this might even traverse time-space parameters in the study of the imaginary (Korth, 2007). At the same time, however, we know that fractal models exist in nature without collapsing in on themselves or dissolving, as is the case with Cantor sets. It is also known that multi-fractal models do exist in mathematics but still have yet to be represented in three-dimensional form due to their immense complexity (Mandelbrot, 2004; Peitgen, Jürgens, & Saupe, 2004). In like manner, the possibilities of exploring multi-dimensional fractals created from numerous seeds might present a more complex perspective of qualitative research. In addition to the use of fractal geometry to develop case study methods, these concepts generate great potential for further research on the human as a fractal geometry of nature, as well as his/her connection with others and with the world in fractal form.

References

Alterman, R. (2008). "Representation, interaction, and intersubjectivity," *Cognitive Science*, ISSN 0364-0213, 31(5): 815-841.

Bruce, C. (2007). "Questions arising about emergence, data collection, and its interaction with analysis in a grounded theory study," *International Journal of Qualitative Methods*, ISSN 1609-4069, 6(2): 51-68.

Davis, B. (2004). *Inventions of Teaching: A Genealogy*, ISBN 9780805850390.

Davis, B. (2005). "Interrupting frameworks: Interpreting geometries of epistemology and curriculum," in W.E. Doll, J.M. Fleener, D. Trueit, and J. St. Julien (eds.), *Chaos, Complexity, Curriculum, and Culture: A Conversation*, ISBN 9780820467801, pp. 21-75.

Davis, B., Sumara, D. and Luce-Kapler, R. (2000). *Engaging Minds: Learning and Teaching in a Complex World*, ISBN 9780805862874.

Denzin, N.K. and Lincoln, Y.S. (2005). *The Sage Handbook of Qualitative Research* (3rd ed.), ISBN 9780761927570.

Doll, W.E., Jr. (1993). *A Post-Modern Perspective on Curriculum*, ISBN 9780807732175.

Fautou, P. (1906). "Sur les solutions uniformes de certaines équations fonctionnelles," *Comptes Rendus Mathematique*, ISSN 1631-073X, 143: 546-548.

Fleener, M.J. (2002). *Curriculum Dynamics: Recreating Heart*, ISBN 9780820455402.

Fournier d'Albe, E.E., and Stoney, G.J. (1906). *The Electron Theory: A Popular Introduction to the New Theory of Electricity and Magnetism*, ISBN 9780548637425.

Gilstrap, D.L. (2005). "Strange attractors and human interaction: Leading complex organizations through the use of metaphors," Complicity: An International Journal of Complexity and Education, ISSN 1710-5668, 2(1): 55-70.

Gilstrap, D.L. (2007a). "Dissipative structures in educational change: Prigogine and the academy," *International Journal of Leadership in Education*, ISSN 1360-3124, 10(1): 49-69.

Gilstrap, D.L. (2007b). "Phenomenological reduction and emergent design: Complementary methods for leadership narrative interpretation and metanarrative development," *International Journal of Qualitative Methods*, ISSN 1609-4069, 6(1): 95-113.

Gilstrap, D.L. (2007c). *Librarians and the Emerging Research Library: A Case Study of Complex Individual and Organizational Development*, Doctoral dissertation, University of Oklahoma, Norman, OK.

Gribbin, J. (2004). *Deep Simplicity: Bringing Order to Chaos and Complexity*, ISBN 9781400062560.

Heisenberg, W. (1944). *Die Physikalischen Prinzipien Der Quantentheorie*, ISBN 9781114248472.

Julia, G. (1918). "Mémoire sur léiteration des fonctions rationnelles," *Journal de Mathématics Pures et Appliquées*, ISSN 0021-7824, 4: 47-245.

Karpiak, I.E. (2000). "The ésecond callé: Faculty renewal and recommitment at midlife," *Quality in Higher Education*, ISSN 1353-8322, 6(2): 125-134.

Korth, B.D. (2007). "The leaps of faith in social science: Study of the imagined in the discourse of the real," *International Journal of Qualitative Methods*, ISSN 1609-4069, 6(1): 69-94.

Lévy, P. (1948). *Processus Stochastiques et Mouvement Brownien: Suivi Déune Note de M. Loéve*, ISBN 9782876470910.

Mainzer, K. (2005). *Symmetry and Complexity: The Spirit and Beauty of Nonlinear Science*, ISBN 9789812561923.

Mandelbrot, B. B. (2004). *Fractals and Chaos: The Mandelbrot Set and Beyond*, ISBN 9780387201580.

Mandelbrot, B.B. (1975). *Les Objets Fractals: Forme, Hasard et Dimension*, ISBN 9782082111881.

Mandelbrot, B.B. (1983). *The Fractal Geometry of Nature*, ISBN 9780716711865.

Peitgen, H.-O., Jérgens, H., and Saupe, D. (2004). *Chaos and Fractals: New Frontiers of Science* (2nd ed.), ISBN 9780387979038.

Pickover, C.A. (ed.) (1996). *Fractal Horizons: The Future Use of Fractals*, ISBN 9780312125998.

Rico, G.L. (2000). *Writing the Natural Way*, ISBN 9780874779615.

Smitherman, S. (2005). "Chaos and complexity theories: Wholes and holes in curriculum," in W.E. Doll, J.M. Fleener, D. Trueit, and J. St. Julien (eds.) *Chaos, Complexity, Curriculum, and Culture: A Conversation*, ISBN 9780820467801, pp. 153-180.

Stake, R.E. (1995). *The Art of Case Study Research*, ISBN 9780803957671.

Stake, R.E. (2005). "Qualitative case studies," in N.K. Denzin and Y.S. Lincoln (eds.), *The Sage Handbook of Qualitative Research* (3rd ed.), ISBN 9780761927570, pp. 433-466.

Stake, R.E. (2006). *Multiple Case Study Analysis*, ISBN 9781593852481.

Donald L. Gilstrap is Associate Professor and Associate Dean of Libraries at the University of Oklahoma, Norman, OK, USA. His research and scholarship includes complexity theory in leadership and management, research methods, and nonlinear dynamics and learning during organizational change. He can be reached at gilstrap@ou.edu.

Complexity and Philosophy

Albert Einstein once pointed out, "The results of scientific research very often force a change in the philosophical view of problems which extend far beyond the restricted domain of science itself" (Einstein writing with Leopold Infeld). This is certainly true with regard to contemporary research into complex systems since, by stimulating a re-thinking of many of our fundamental assumptions about the nature of our organizations as systems, complexity theory is radically altering the sundry folklore and outmoded conceptual freight underlying traditional models of organizations. By re-thinking assumptions, the varied sciences of complex systems are inspiring a re-look at age old philosophical and metaphysical issues.

The new insights offered by complexity theory, therefore hold the promise that Schopenhauer once gave voice to, "...the corrected, extended, and more thorough knowledge of nature is the very knowledge that always undermines and finally overthrows the metaphysical assumptions that until then have prevailed." That is why *E:CO* includes a section committed to offering fresh and insightful papers on the interface of complexity and philosophy. In this section, researchers and scholars grapple with foundational questions emerging from the debate between modernist and post-modernist perspectives, a debate which has led to intriguing and fresh speculations in ontology, epistemology, cognition, semiotics, ethics, and aesthetics. Again, these articles have been selected through their recognition that new constructs are not merely ends in themselves, but can offer new methods with pragmatic benefits.

Jeffrey Goldstein

20. Unpredictability and Nonlinearity in Complexity Theory: A Critical Appraisal

Stephen Kemp
University of Edinburgh, UK

This piece explores potential problems with the focus on unpredictability and nonlinearity within complexity theory. Whilst not completely rejecting the application of ideas of nonlinearity and unpredictability within the social sciences, I argue that greater empirical and conceptual care is needed. The arguments made are illustrated by a critical examination of cases from John Urry's Global Complexity, including the dominance of the petroleum-fuelled car in the 20th century and the prevalence of wild-fires in Malibu. Empirically speaking, I argue that claims about particular instances of nonlinearity and unpredictability in the social world must be backed up by appropriate evidence, rather than analysts simply assuming that all social phenomena have these characteristics. Conceptually speaking, I suggest that care needs to be taken to distinguish genuinely unpredictable phenomena from those that are simply poorly understood at the present time. I also argue that predictability should be seen as a matter of degree.

Introduction

It's arguable that the emphasis in complexity theory on unpredictability and nonlinearity did not surprise social scientists, especially social theorists, but reinforced what they already believed. The existence of nonlinear and unpredictable phenomena may have come as a shock to those natural scientists who bought into a (roughly) Newtonian picture of a deterministic universe.[1] But the vast majority of social theorists, and some empirical social scientists, already believed that the social world was not predictable and could not be understood by undertaking linear operations on quantitative information about social phenomena. Defenders of interpretive social thought, (e.g., Winch, 1990 [1958]) social constructionists (e.g., Gergen, 1985), and even quasi-naturalists such as critical realists (e.g., Bhaskar, 1979), have argued against the possibility of scientific prediction in the social domain, and have preferred instead to focus on qualitative change and the richness of unquantifiable meaning. This is not to say that complexity theory adds nothing to previous understandings of the social world, and at the very least complexity theory brings its own vocabulary and explanatory repertoire to account for the social world. Furthermore, complex-

1. Presumably the shock was somewhat diminished for those who had already accepted indeterminism within the quantum domain.

ity theory offers concepts that are held to apply to both the social and natural worlds, rather than arguing for a clear-cut division between the two in the manner of interpretive social thought[2].

Nevertheless, given the extent of common sympathies between complexity theory and its precursors, it may be that certain shared claims have not been as rigorously interrogated as they might otherwise have been. If many are already convinced that the social world is an unpredictable place, and one in which social changes and developments rarely, if ever, have a 'linear' character, then claims by complexity theorists in these areas may have been given a relatively easy ride. In this article I intend to build on the small amount of work of those who have questioned, to a greater or lesser extent, complexity theory's treatment of prediction and linearity in the social world (see McLennan, 2003, 2006; Thompson, 2004). My aim is not to attempt a demonstration that the social world is, contrary to complexity theory, fully predictable and linear in character. Rather, my intention is to argue two things: firstly, that notions of unpredictability and nonlinearity need to be used in a more careful and analytically precise way then they are in certain existing complexity arguments; and, secondly, that we need to carefully attend to the ways in which empirical cases actually do, or do not, fit in with ideas of unpredictability and nonlinearity as formulated in complexity theory.

Given the limitations of space, I will focus only on the work of one prominent exponent of complexity theory in the social sciences, John Urry, and particularly on his *Global Complexity* (2003). In *Global Complexity*, Urry strongly defends the relevance of complexity theory to the social sciences, and puts it to work in order to account for phenomena usually dealt with under the rubric of 'globalization', as well as applying it to numerous other examples. Urry's book offers a useful basis for discussion both because it makes these linkages between complexity theory and empirical examples, and because Urry draws heavily on ideas of nonlinearity and unpredictability when analyzing the social world. In this article I shall consider his treatment of these concepts in turn, and offer some critical commentary on the extent to which they successfully account for the empirical examples that Urry presents.

Nonlinearity, Small Causes and Large Effects

Although 'nonlinear' is a term used extensively by chaos and complexity theorists, there is some degree of dispute about its meaning. Strictly speaking, 'linear' and 'nonlinear' are terms that are used to distinguish two types of mathematical equation; whereas linear equations can be solved by breaking them down into smaller pieces then recombining, this is not possible

2. Critical realism could be argued to share complexity theory's emphasis on continuity between the natural and social domains, although some hiatus between the two is often implied (see for example Archer, 1998).

with nonlinear equations (see McLennan, 2006: 145; Kiel & Elliott, 1996: 4). However, many less technically-minded discussions of chaos and complexity theory use the term to cover processes where there is something like a lack of proportionality between (changes to) the input and the outcome. One oft-discussed type of case is that of 'sensitivity to initial conditions' in which the minutest difference in the starting conditions of particular kinds of system (such as those whose trajectories are governed by nonlinear equations) may lead to large-scale differences in the state of that system when an extended period of time has elapsed. The most famous instance of this is Lorenz's discovery of the sensitivity of weather systems to initial conditions, often referred to as the 'butterfly effect' (Gleick, 1988). Lorenz argued that the minutest alteration in the initial state of a weather system, such as the tiny amount of turbulence generated by a butterfly flapping its wings, can lead to a vastly different outcome—a storm occurring in a far-away city—than that which would have occurred if the butterfly had decided to have a rest—a sunny day in that city.

In *Global Complexity*, Urry argues that the principle of nonlinearity can be used to explain the emergence and persistence of certain technical forms such as the QWERTY keyboard and the petroleum-fuelled car (Urry, 2003: 54-5). Urry does not offer a rigorous definition of the term 'nonlinear' (see McLennan, 2006)[3], but appears to be drawing on the broad idea that for some phenomena there is a lack of proportionality between inputs and outputs. In the realm of causal explanation this involves the claim that 'there are often massive disproportionalities between causes and effects' such that 'Causation can indeed flow from contingent minor events to hugely powerful general processes' (Urry, 2003: 7, 54). Insofar as this is a coherent notion, it suggests that small, apparently accidental or insignificant causes can come to have a major influence on the development of a system.

We will consider the example of the petroleum-fuelled car rather than the QWERTY keyboard here because its rise and persistence has had arguably more significant consequences. As we know, petroleum came to be the power-source for the vast majority of cars in the 20th century. Looking back at the history of this development, Urry suggests that in the 1890s petroleum was inferior to two other means of powering cars, electricity and steam. In his article 'The "System of Automobility"', Urry discusses what happened next:

---

3. McLennan argues that the terms linear and nonlinear are misused by Urry (and complexity theorists more generally) because their strict mathematical meanings are not adhered to (McLennan, 2006: 145). Whilst I have some sympathy for this argument, the problem I am concerned with here is not the appropriation of the aura of the 'hard' sciences by social scientific complexity theories, so much as evidential and conceptual issues in the usage of 'nonlinear' that actually has been adopted by social sciences, that relating to the proportionality of causes.

> *Petroleum-fuelled cars were established for small-scale, more or less accidental reasons, partly because a petrol-fuelled vehicle was one of only two to complete a "horseless carriage competition" in Chicago in 1896* (Urry, 2004: 32).

On this view, the petroleum-fuelled car became dominant not because of its technical superiority but through chance events. This dominance was possible because small-scale, contingent causes were magnified through processes of 'lock-in' in which 'an astonishing array of other industries, activities and interests came to mobilize around the petroleum-based car' (Urry, 2003: 55). Once one path of development came to be backed by enough of the players within the industry and in supporting industries, it became extremely difficult to produce competitive alternatives because a whole network of institutions and businesses—such as petrol stations, petroleum producers, and car part manufacturers—came to have a vested interest in the persistence of a particular technological form rather than its opponents. On Urry's view, small scale chance events led to one power-source rather than others becoming dominant and also unchallengeable through nonlinear processes of reinforcement present in institutional lock-in.

I want to make two general kinds of argument about Urry's analysis, one of which relates to the empirical basis for his claims, and the other of which is a more theoretical point. Beginning from the empirical, I would argue that there are problems with Urry's claim that the dominance of the petroleum-fuelled car is the result of 'more or less accidental reasons' (Urry, 2004: 32). For one thing, Urry offers little or no evidence in support of this claim, offering only one reference to back up his historical account in 'The "System of Automobility"' (2004). For another thing, if we look at relevant sources on the competition between different ways of powering cars (Motavalli, 2000; Cowan & Hultén, 1996; Foreman-Peck *et al.*, 1995), the picture looks different to the one painted by Urry. The most opposed interpretation to Urry's is that put forward by Foreman-Peck *et al.* (1995) who argue that the petroleum car eventually won out in its contest with other power sources because it had the best match between its technical characteristics and those features that were desired by consumers. As with Urry, Foreman-Peck *et al.* refer to road races and endurance trials for cars, but present these as not minor shows or entertainments, but as important means by which technological advantages could be displayed (Foreman-Peck *et al.*, 1995: 15). Although they admit that steam-powered cars out-performed petrol cars on-and-off in contests of speed, they argue that the wider demand was not for the fastest cars but for those that were reliable and convenient, factors which endurance tests were a better proxy for (Foreman-Peck *et al.*, 1995: 15-6). Indeed Foreman-Peck *et al.* suggest that the superiority in reliability and convenience of petroleum cars was clearly apparent prior to their mass production, with some notable manufacturers abandoning steam for petrol before this point. They thus challenge the view that the dominance of the petroleum-fuelled car was a product of the chance reinforcement of one power source rather than another through institutional lock-in.

Other writers, such as Motavalli (2000), and Cowan and Hultén (1996), do not challenge the 'lock-in' thesis as a partial means to explain the dominance of petroleum-powered cars over steam- and, particularly, electric-powered cars during the 20th century. Nevertheless, this does not mean that they attribute the dominance of petroleum-powered vehicles to chance, accidental events. For one thing, these writers are clear that there were only three plausible sources of power for cars at the end of the 19th and beginning of the 20th century—steam, electricity and petroleum. Even were it to be established that petroleum had no intrinsic merits over the others, its status as a serious contender at the time does not seem to be challenged. Thus we already have a narrowing of the field of possibility that was not the result of chance or accident *per se*. Further than this, both Cowan and Hultén and Motavalli acknowledge that at the time the mass-production of cars was undertaken, petroleum-based cars had clear advantages over their competitors. In their analysis, Cowan and Hultén list the problems faced by cars using each of the three power sources at around the turn of the century, and point out that for the petroleum-powered car, all of them had been solved by 1912, except their noisiness. By contrast, both the steam car and electric car seemed to have serious remaining problems—the steam car's consumption of 'immense amounts of water' and the electric car's limited range and low top speed (Cowan & Hultén, 1996: 68). Motavalli, who is extremely sympathetic to alternative, non-petroleum-based ways of powering cars, also acknowledges the limitations of alternative forms of power at the time the petroleum-powered cars were being mass-produced (Motavalli, 2000: 9).

Summarizing these points, I would suggest that the empirical analyses considered here throw Urry's analysis into question. Even if we only accept what is agreed by all of these writers, it does not make sense to say that the dominance of petroleum-fuelled motor cars is due to minor, accidental, factors. Petroleum was one of only three plausible means to power cars at the end of the 19th century, and the major problems with petroleum-powered cars were resolved before those of steam- and electricity-powered cars. As such, Urry's claim that this is an example of a nonlinear complex phenomenon, in which accidental causes lead to hugely significant outcomes, is thrown into question.

My second overall point about Urry's use of this example is that even if he were to challenge other accounts and provide the empirical evidence to establish that the dominance of the petroleum car was the result of small-scale, accidental causes, there would need to be a degree of refinement in how such a case is talked about. Both Urry and other commentators refer to the idea that in some cases *a small cause has large effects* (e.g., Urry, 2005: 6; Hayles, 1991: 11). However, this is problematic because it misdirects attention by focusing only on one of the causes that leads to the final outcome in any given case. To see this, let us start with the classic 'chaos theory' example of a putatively small cause that has large effects, the butterfly effect. As we saw above, this involves the claim that the tiny amount of turbulence generated by a butterfly flapping its wings can lead to a vastly different outcome, say a storm occurring in a far-away city instead of the sunny weather that would have occurred if the butterfly had decided to have a rest. Is the butterfly's wing-flapping a small cause that generates a large effect? I would argue that this is not the case. After all, the small cause of the butterfly's wing-flapping does not cause, all by itself, the large effect of a storm occurring. Rather, the butterfly's wing-flapping is one of a range of atmospheric causal influences that interact and ramify over time, resulting in a storm. It is thus more accurate to say that two systems whose end-states are the results of many causes, may end up in markedly different end-states even if their starting conditions differ only by one small causal influence[4]. This is not merely a pedantic point, in my view, because it emphasizes the importance of attending to the various other causes that contribute to any outcome. If one believes that small causes have large effects, one does not need to seriously consider the other causes that contributed to these effects.

To be fair to Urry, his discussion of the domination of the petroleum-car does point towards the kinds of processes which reinforce the early advantages that the petroleum car might have had, insofar as there are learning processes in both constructing and using petroleum cars and their support systems which make it less effective for organizations to shift to other modes once they 'back' the petroleum car (Urry, 2003: 55). However, stronger emphasis needs to be given to these processes in my view, and further examination. Even if the complexity-based explanation was correct, Urry would need to accept that the dominance of the petrol car, as an 'effect', would not be anything like suffi-

4. Perhaps some of the confusion here arises from cases where there is just one iterating cause, where the state of a system depends on just one cause repeated over time. A simple system of this kind would be one in which the key variable capturing the system state was X, and where for every time interval that elapsed a cause operated such that the value of X was squared. The equation for such a case would be $X_{(T+1)}=X_{(T)}^2$. This equation means that the value of X at some particular time (Time 'T+1') is the square of the value of X at the time interval before (Time 'T'). As Harvey and Reed point out, in situations of this kind, slight initial differences in the value of X will lead down the line to much larger differences because of the repeated squaring of the value (Harvey & Reed, 1994: 379). It is correct to say that it might only take a very small causal input to alter the initial state of the system, captured in the initial value of 'X'. It is also correct to say that there will be a marked difference in how the system develops over time even if this initial value was altered only minutely, for example the difference between $X_{initial}=2$ and $X_{initial}=2.0001$. However, this does not mean that the small cause altering the start value of X has had a large effect. Rather, the difference between the system that results depending on the two different values of $X_{initial}$ is the result of the *combination* of the small cause altering $X_{initial}$, and the many instances of the cause that produces a squaring of the value of X.

ciently explained by reference to the 'cause' of it completing a race in Chicago in 1896. Rather, it would be explained by a variety of causal factors including initial causes, but also including the institutional processes that created and supported the large-scale dominance of the petrol car against competitors, as well as those social and natural features that allowed the petrol car to continue to be viable (the existence of vast reserves of oil in the world, for example). These other causal features may well turn out to be somewhat distinctive, or at least only one class of phenomena amongst others, insofar as not all technological fields are dominated by one form of technology[5]. As such, I would suggest that we should dispose of the summary claim that a small cause can have large effects, and offer more precise formulations that encourage attention to be given to all of the relevant causal influences at work[6].

Unpredictability

The other aspect of complexity discourse that I would like to examine here is the argument that complex phenomena are intrinsically unpredictable in character. Complexity discourse shares an emphasis on unpredictability with chaos theory, although arguably not to the same extent. Chaos theorists often argue that the trajectories of chaotic systems are inherently unpredictable in the sense that it is not possible to precisely predict where the system will be in state-space at any particular point in time.[7] Complex orders are frequently held to have a degree of stability, but to be periodically subject to unpredict-

5. There do seem to be technological fields where there are multiple viable competitors at any one time. One example from the past few years is that of music in which CDs, vinyl records, tapes and MP3-files coexist as formats. This is not to say that all of these forms will persist, but that they have meaningfully persisted for a certain period of time along with one another. Contrasts between this kind of case and the apparently almost utter dominance of the petroleum car need to be examined. Of course, a further issue is where one draws the line with competitors; if aeroplanes, trains and boats are considered competitors then the petroleum car is not utterly dominant.

6. It is worth briefly noting that the causal processes involved in the 'butterfly effect' and 'lock-in' are somewhat different. In the case of the butterfly effect, the difference in outcomes is based on the vast number of causal interactions that take place in order to generate the weather. Weather systems with slightly different initial conditions may develop in quite strongly divergent ways. However, this divergence is not predictable given our current capacities of measurement and calculation. As such, it is not possible to predictably generate bad weather at some part of the globe by slightly altering the conditions at another point. In the case of 'lock in', the divergence of outcomes is based on the way in which one development path or another is reinforced over time by surrounding developments. Insofar as the lock-in theory is correct, it should be possible to manipulate initial conditions in order to predictably generate the preferred path of development.

7. For an argument about the limitations of unpredictability in chaotic systems see Brown (1996).

able developments in which self-organizing processes will reformulate the system and its structure (Harvey & Reed, 1994). Thus, key writers such as Prigogine have tended to place an emphasis on the ultimate unpredictability of complex systems, their tendency to bifurcate such that they develop in one of two possible directions, and the impossibility of predicting which direction will be taken (Prigogine & Stengers, 1984: 160-170)[8]. Proponents of complexity theory in the social science have been particularly enthusiastic to embrace this emphasis on unpredictability chiming in as it does with social theoretical concerns after positivism which have variously emphasized the role of agency, contingency and unintended consequences in social life (see for example Giddens, 1984; Bhaskar, 1979).

Urry is certainly keen to find unpredictable phenomena in the social world (see for example Urry, 2003: 59-72, 95, 105), and here I want to consider how analytically robust two of Urry's examples are. The first example makes a link between the nonlinear character of a phenomenon and its unpredictability, dealing with the ecological relationships between settlers and the environment in Malibu, United States. This is a case that Urry draws from Mike Davis' fascinating book *Ecology of Fear: Los Angeles and the Imagination of Disaster* (2000). Urry presents the example as follows (Urry, 2003: 34-5). Malibu can be considered the 'wildfire capital of North America' in that it suffers incredibly intense fires with relative frequency. The most significant factor in causing this is 'nonlinear relationship between the age structure of vegetation and the intensity of fires that are generated' (Urry, 2003: 34). For example, 50 year old vegetation of the kind present in Malibu does not burn 2 ½ times as intensely as 20 year old vegetation, but 50 times as intensely. Instead of being burnt off when it is young, however, the vegetation is often given the chance to age significantly because of the policy of 'total fire suppression'. The result of this is that when blazes do get started they are extreme in character, frequently burning down numerous properties as well as putting the lives of residents, fire-fighters and others in the locale at risk. Summing up his discussion, Urry states:

> *This example shows that certain kinds of cause can generate huge and unpredictable change while other examples would show that external causes could generate almost no significant effects* (Urry, 2003: 35).

In response to Urry's analysis, I want to suggest that it is seriously unhelpful to describe the presence of intense forest fires in Malibu as unpredictable. Certainly, an emphasis on their unpredictability is quite out of the spirit of Davis's analysis (see Ch. 3 of Davis, 2000). For one thing, Davis presents the move to 'develop' Malibu in the 20th century as misguided from the start, because of the:

---

8. It is worth noting that other complexity theorists, while acknowledging the lack of strict predictability of complex systems, argue that there are attractors in the state-space of complex systems which draw them towards a limited set of possible configurations (see for example Kauffman, 1993).

...uncanny alignment of its coastal canyons with the annual "fire winds" from the north: the notorious Santa Anas... The San Fernando Valley acts as a giant bellows, sometimes fanning the Santa Anas to hurricane velocity as they roar seaward... Add a spark to the dense, dry vegetation on such an occasion and the hillsides will explode into uncontrollable wildfire (Davis, 2000: 100).

Davis's argument is that the combination of geography and climate of the region makes it predictable that there will be fires, particularly at certain times of year.

As Urry reports, Davis further argues that the combination of a 'total fire suppression' policy and the nonlinear increase in burning intensity leads to particularly severe fires in the Malibu area. However, unlike Urry, Davis makes no connection between nonlinearity and unpredictability. Rather, Davis' analysis is based on the fact that knowledge of the nonlinear increase in burning intensity actually makes outcomes predictable in some sense. In other words, given the nonlinearity in question, it is precisely predictable that a policy of total fire suppression will lead to hugely intense fires[9]. Furthermore, these intense fires could be largely avoided if there were controlled burn-offs every few years, as these would prevents the vegetation getting to an age where it predictably burns with high intensity (Davis, 2000: 144)[10]. Another predictable outcome that can be generated from Davis' analysis is that comparable environments, such as parts of Baja California, will fare differently to Malibu because of their differing policies on fires. In the uplands of Baja California, where there is no attempt at fire suppression, there are more fires but they are 'patchier' and less intense, because high intensity vegetation does not build up (Davis, 2000: 103). The overall point I want to make here is that Urry misdescribes this kind of example because of his belief that nonlinearity is intrinsically linked to unpredictability. The intensity of Malibu wildfires is not unpredictable given the relevant knowledge.

It is important to clarify that in claiming certain aspects of Malibu wildfires to be predictable, I am not suggesting that we can be *absolutely certain* about the timing and placement of their occurrence and their precise characteristics (exact temperature and so on). However, if this level of detail and precision is necessary to be able to deem some phenomenon to be predictable then this would exclude many if not all social scientific cases, as well as many natural scientific ones. Indeed, absolute certainty of this kind is not even present in natural scientific laboratory controlled situations, given that scientists can never be certain that they know all of the possible confounding conditions including

---

9. Of course, those who set up the total fire suppression policy may not have known about the nonlinear relationship between vegetation age and burning intensity. But this limitation to their knowledge does not make the situation unpredictable in principle, just unpredicted at that point in time.

10. Davis does note that the small possibility of break-away fires would expose the fire departments to potential lawsuits (Davis, 2000: 144). However, if there was a general policy of controlled burn-offs, the material would not be there for fires of high intensity.

local contingent ones (See for example Collins, 1985)[11]. Rather than placing all social scientific phenomena in the 'unpredictable' category, I would suggest that there are some cases in which we can have confidence that, under specified conditions, certain outcomes are likely to occur within a given time and space range. We can thus distinguish relatively predictable events, such as the intense character of Malibu wildfires, from completely unpredictable events, such as the numbers in the next lottery draw, rather than putting these together in the same category—'not predictable with certainty'.

The other putative example of unpredictability that I would like to consider here arises from Urry's discussion of networks and fluids that are global in character. Urry introduces his analysis of networks and fluids as part of his exploration of new ways to analyze global spaces. Urry's intention is to analyze the emergent properties of the global order without placing old-fashioned and outmoded concepts like 'nation' and 'region' at the centre of the analysis (Urry, 2003: 42-6). According to Urry, globally integrated networks are not nationally bounded, but operate transnationally 'across multiple and distant spaces and times' (Urry, 2003: 57). These networks make fairly stable and predictable links between people and objects around the globe, and are employed by organizations like Microsoft, McDonalds and Greenpeace to allow efficient communications, successful delivery of products and profits, and so on.

Global fluids are, according to Urry, less stable than globally integrated networks, and interact with other fluids, networks and organizations in unpredictable ways. These fluids include travelling peoples, global brands, social movements, the world's oceans, environmental and health hazards and automobility (Urry, 2003: 59-74). Urry argues that fluids are the result of many local actions which can interact and ramify to create 'global waves'. Urry states:

> *The "particles" of people, information, objects, money, images, risks and networks move within and across diverse regions forming heterogeneous, uneven, unpredictable and often unplanned waves... Such waves demonstrate no clear point of departure, deterritorialized movement, at certain speeds and at different levels of viscosity with no necessary end state or purpose* (Urry, 2003: 60).

Thus global fluids are seen by Urry as unpredictable, and as often lacking a clear start and end point.

Urry is aware that earlier writers within the sociological tradition, such as Durkheim, have argued against modes of representation that attempt to capture the 'flux' of life without abstracting from this to arrive at more stable, fixed conceptualizations (Urry, 2003: 59). However, Urry defends his approach in a

---

11. One does not have to agree with every aspect of Collins' analysis to accept his point that local contingencies of experimental set-up can have an impact upon any particular attempt to run an experiment.

broadly realist fashion by arguing that concepts of flow, liquidity, waves and so on are necessary to accurately describe what global society is like. The defence of these concepts is, for Urry, simply a matter of representational adequacy.

Whilst I would accept that, in principle, fluids, flows and unpredictable waves may exist, it is important to remember that identification of phenomena as having these characteristics is potentially fallible. Furthermore, as John Holmwood has argued, accounts of particular phenomena that emphasize their indeterminacy—unpredictability, unevenness, no clear point of departure, no necessary end state or purpose—have similar characteristics to confused accounts of determinate phenomena (Holmwood, 1996; see also Thompson, 2004). Putting this another way, a process that appears to be erratic and unpredictable in its operation may actually be a relatively determinate process that is currently being misunderstood. The relevance of Holmwood's contention to the present case is backed up, in my view, by the lack of analytical rigour present in Urry's discussions of some examples of global fluids. By categorizing all 'travelling peoples' as one fluid, for example, it is hardly surprising that this fluid seems unpredictable, heterogeneous and so on. However, a more careful separation of some of the constituents of that fluid might allow a more coherent picture to emerge, insofar as, for example, the 'flows' of refugees and transnational capitalists are likely to have different characteristics and logics (see McLennan, 2003: 555). In another of his examples, that of automobility (the car system), Urry himself undermines the notion that he is analyzing an unpredictable fluid with no clear starting point. For one thing, Urry clearly attributes the beginnings of the system of automobility to North America, implying that there is a straightforward, identifiable point of departure of this transport wave. For another thing, Urry goes on to suggest that societies become 'locked in' to the system of automobility in a way that seems 'impossible to break from' (Urry, 2003: 69). These phrases are hardly redolent with notions of unpredictability and fluidity. As such, some of Urry's claims about the unpredictability of 'global fluids' seem rather overstated. Furthermore, in some cases it may be that it is analytical and conceptual problems which gives social phenomena the appearance of indeterminacy, rather than this being inherent in the phenomena themselves.

Conclusion

The purpose of this piece has been to critically reflect on the concepts of nonlinearity and unpredictability, concepts which are commonly invoked by complexity theorists. The aim has been not to debunk complexity theory, or the work of John Urry which I have focused upon, but to use Urry's stimulating discussion of global complexity as a basis for thinking about how complexity theory might become more analytically precise in its use of notions like nonlinearity and unpredictability. This has involved both challenging the standard formulation that 'small causes have large effects' and arguing that there is a range of levels of predictability, rather than a simple contrast between completely predictable and totally unpredictable. As well as this, I have been raising the question of the empirical justification of claims about nonlinearity and unpredictability as constitutive features of the social and natural worlds. Two dimensions are important here. The more narrowly empirical issue is that claims about the nonlinearity and/or unpredictability of certain phenomena, such as the dominance of the petrol engine or the spread of wildfires in Malibu, need to be carefully justified by empirical evidence, rather than being accepted too swiftly by those enthusiastic to find complexity everywhere in the social and natural worlds. The wider conceptual component of the argument is that the apparent indeterminacy and unpredictability of phenomena may be signs of the limitations of existing concepts and theories rather than indicators of the true nature of the phenomena in question. I say *may* be signs here because it cannot be demonstrated that all phenomena can be rendered determinate and predictable, and complexity theorists have made interesting arguments which account for why certain phenomena may not have this character (see for example Cilliers, 1998). However, there is a danger in always operating with an assumption that the social world and its constituents *must be* intrinsically fluid and unpredictable. Those who are convinced that the social world has this form may fail to examine whether the apparently intrinsically confusing and unpredictable nature of the phenomena they study is actually an artefact of their own limited categories. For these reasons, a careful analytical and empirical approach to issues of nonlinearity and unpredictability is essential.

References

Archer, M. (1998). "Introduction: Realism in the social sciences," in M.S. Archer, R. Bhaskar, A. Collier, T. Lawson and A. Norrie (eds.), *Critical Realism: Essential Readings*, ISBN 9780415196314.

Bhaskar, R. (1979). *The Possibility of Naturalism: A Philosophical Critique of the Contemporary Human Sciences*, ISBN 9780855277727.

Brown, T. (1996). "Measuring chaos using the Lyapunov exponent," in L. Kiel and E. Elliott (eds.), *Chaos Theory and the Social Sciences: Foundations and Applications*, ISBN 9780472106387.

Cilliers, P. (1998). *Complexity and Postmodernism: Understanding Complex Systems*, ISBN 9780415152877.

Collins, H. (1985). *Changing Order: Replication and Induction in Scientific Practice*, ISBN 9780803997172.

Cowan, R. and Hultén, S. (1996). "Escaping lock-in: The case of the electric vehicle," *Technological Forecasting and Social Change*, ISSN 0040-1625, 53: 61-79.

Davis, M. (2000). *Ecology of Fear: Los Angeles and the Imagination of Disaster*, ISBN 9780330376556.

Foreman-Peck, J. Bowden, S. and McKinlay, A. (1995). *The British Motor Industry*, ISBN 9780719026126.

Gergen, K. (1985). "The social constructionist movement in modern psychology," *American Psychologist*, ISSN 0003-066X, 40(3): 266-275.

Giddens, A. (1984). *The Constitution of Society*, ISBN 9780745600062.

Harvey, D. and Reed, M.(1994). "The evolution of dissipative social systems," *Journal of Social and Evolutionary Systems*, ISSN 1061-7361, 17(4): 371-411.

Hayles, N.K. (1991). *Chaos Bound: Orderly Disorder in Contemporary Literature and Science*, ISBN 9780801422621.

Holmwood, J. (1996). *Founding Sociology? Talcott Parsons and the Idea of General Theory*, ISBN 9780582291652.

Kauffman, S. (1993). *The Origins of Order: Self Organization and Selection in Evolution*, ISBN 9780195058116.

Kiel, L. and Elliott, E. (1996). "Introduction," in L. Kiel and E. Elliott (eds.), *Chaos Theory and the Social Sciences: Foundations and Applications*, ISBN 9780472106387.

McLennan, G. (2003). "Sociology's complexity," *Sociology*, ISSN 1469-8684, 37(3): 547-564.

McLennan, G. (2006). *Sociological Cultural Studies: Reflexivity and Positivity in the Human Sciences*, ISBN 9780230008854.

Motavalli, J. (2000). *Forward Drive: The Race to Build "Clean" Cars for the Future*, ISBN 9781578050352.

Prigogine, I. and Stengers, I. (1984). *Order Out of Chaos: Man's New Dialogue With Nature*, ISBN 9780434603954.

Thompson, G.F. (2004). "Is all the world a complex network?" *Economy and Society*, ISSN 0308-5147, 33(3): 411-424.

Urry, J. (2003). *Global Complexity*, ISBN 9780745628172.

Urry, J. (2004). "The 'system' of automobility," *Theory, Culture and Society*, ISSN 0263-2764, 21(4/5): 25-39.

Urry, J. (2005). "The complexity turn," *Theory, Culture and Society*, ISSN 0263-2764, 22(5): 1-14.

Winch, P. (1990/1958). *The Idea of a Social Science: And its Relation to Philosophy*, ISBN 9780415054317.

Acknowledgements

I would like to thank the British Academy for providing an Overseas Conference Travel Grant to the World Sociology Congress 2006, at which a version of this paper was presented. Thanks also to Angus Bancroft for his comments on this paper.

Stephen Kemp is a Lecturer in Sociology at the University of Edinburgh, UK. Having pursued undergraduate and masters level studies at Massey University, New Zealand, he completed his doctorate at the University of Edinburgh. His published papers include 'Critical Realism and the Limits of Philosophy', which appeared in *European Journal of Social Theory* in 2005, and 'Saving the Strong Programme: A Critique of David Bloor's Recent Work', which appeared in *Studies in History and Philosophy of Science* in 2005. He has a particular interest in debates about the epistemology and ontology of the social and natural sciences.

21. Complexity and the Phenomenological Structure of 'Surprise'

Darren Stanley
University of Windsor, USA

Surprising, unexpected events happen all the time which can be thought of and addressed in a variety of ways. On one hand, surprise can be something that is not desired, something suppressed or controlled for. Or, it can be something that is embraced, sought out or encouraged. Conceptually speaking, conceiving surprise in this fashion is not uncommon. Still, there seems to be an important piece missing from the many discourses on surprise. This paper offers some a possible framework to understand the *experience* of surprise in relation to a more complexified framing of the lived-experience. Specifically, by drawing upon principles from the complexity sciences, this paper considers the lived human experience of surprise as an emergent phenomenon that arises from a complex system.

Introduction

The phenomenon of surprise is typically understood to be a special event that happens *to* people where the surprising event may be thought of as a good or a bad experience. Moreover, for many people who manage or direct the actions of others in an organization (like many workplaces), surprises are seen to be unwelcome moments that sometimes evoke a feeling of discomfort, prompting the need to know more, plan better, or design better systems to avoid the possibility of any unexpected surprises (McDaniel *et al.*, 2003; Weick & Sutcliffe, 2001). Certainly, it is possible to attend to many different kinds of organizations to find examples of how human expectation and certainty, as well as, unexpectedness and surprise, affect and shape our knowledge and actions within those organizations (Stacey, 1992). It is not, however, the aim of this paper to explore organizational contexts for an understanding of surprise. Rather, it is the idea that surprise and unexpectedness are particular *kinds* of *experience*. In the complexity science literature, however, the *experience* of surprise is not, by and large, something that is discussed, but is taken up in slightly different ways—often, for instance, psychologized, framed by the notions of bounded rationality and a lack of information (Simon, 1991). Nevertheless, it is the aim of this paper to suggest some connections that can be made between the experience of surprise and a complexity science interpretation of surprise. As such, in this paper, I outline a view of "surprise" and "unexpectedness" framed by the philosophical branch of phenomenology, further interpreted through the lens of complexity and some of its attendant principles. In doing so, this paper seeks to illuminate the notion of surprise by tending to the common ground of complexity and lived-experience.

Phenomenological Research: Finding A Methodological "Middle Ground"

Phenomenological research might appear to be antithetical to the work of science. After all, whereas matters of methods and technique, which are quite antithetical to the spirit of phenomenological research, are not fitting in any purely prescriptive sense (van Manen, 1990), science requires some attention to methods of study and tools and techniques: scientific techniques can create a certain illusion of control and knowability, but then difficulties and tensions can arise. For example, as Barritt (1983: 4) and her colleagues write, "techniques of collecting information are more important than what the collection is all about–that is, the thing one is trying to understand. There is too much talk about things which matter very little–numbers, methods, contexts stripped of their meaning in the name of research–and too little about the important events–the real stuff of life...". To suggest, however, that science is the same as methodological reductionism, in the derogatory sense of the word "reductionism", is short-sighted. In fact, any sort of methodology—the scientific method, for example—in a prescriptive sense, as when one follows a recipe, is bound to fail to capture in any real sense the messiness of doing research, particularly research about lived-experience. Still, not only is a phenomenological "method" apparently paradoxical in that, traditionally speaking, it is not a prescriptive method of inquiry at all, but it often requires some attention to other forms of inquiry to understand lived-experience.

To some, it may appear that the work presented here is not particularly phenomenological—perhaps I may even be "tainting" the field. To be sure, the work presented here is more in line with an evolved framing of the field of phenomenology. Thus, given the usual need to announce one's allegiance to a particular conceptual framework, this particular inquiry draws upon a "naturalized phenomenological" approach to understanding surprise. Following the lead of researchers and scholars like Jean Petitot, Francisco Varela, Bernard Pachoud and Jean-Michel Roy (1999), my approach here is more in line with the shared intellectual space of contemporary phenomenology, the cognitive sciences, and complexity science (see Petitot *et al.*, 1999; Thompson, 2007; Varela, 1999). As such, the apparent mixing up of theoretical frameworks here is actually not so unorthodox. In fact, phenomenological inquiry is one which actually attends to the *emergence of a phenomenon*, a notion very much associated with complexity science.

The essence of such an event—an emergent event—is what makes, as Dastur (2000) describes, the "impossible possible". Moreover, this notion of the "essence" of such an event or phenomenon is in line with what Merleau-Ponty (1962) describes as the meaning of phenomenology. In other words, any phenomenon is always and already an emergent phenomenon. It is a happening, an event—a playful moment when the possible emerges, comes out, or appears to us. As such, no amount of conceptualizing can tell us what the human experience of surprise might be. Of course, pure experience is not enough for an understanding of how and why we are or are not surprised. That is, to be clear, other perspectives—in this case, complexity science—will be drawn upon here for further insight.

Experiencing Surprise: A Phenomenology Of Unexpectedness

U*nexpectedness* as a lived human experience is not some phenomenon limited to one particular type of experience. What is it that one recognizes in a moment of unexpectedness? When one is shocked to hear news about the death of someone that one knows well, is that said to be an experience of unexpectedness? Is my work colleague who appears startled whenever I approach her with a question, who appears so completely unaware of my presence, experiencing unexpectedness? What about other experiences? Would it be like having guests suddenly show up at your door unannounced? Or what about a surprise birthday party? Is unexpectedness any of these things: what is the lived experience of unexpectedness?

In the realm of human experience, one might say that we are seldom surprised. But, why do we not experience surprise in every moment of life? Surely, the experience of surprise might be taken to be a rare event—perhaps even something special. Moreover, the experience of surprise seems to come unannounced, and, just as suddenly, it disappears. That is, it is unforeseen, where one could not have "seen it coming". To be clear, however, it is the immediate *experience* of surprise that does not happen all of the time, and this should not be confused with a surprising event. What could be surprising for one person may not be so for another.

Framed in this way, when we are surprised, it seems to sneak up on us. Interestingly, the etymology of the word surprise has something insightful to say to us about this notion of "sneaking up on" ("surprise, n", 1989) . As van Manen (1990: 58) writes: "Being attentive to the etymological origins of words sometimes put us in touch with the original form of life where the term still had living ties to the lived experiences from which they originally sprang." In this spirit, then, the word "surprise" is closely related to such notions as "a sudden unexpected attack", "to come upon unexpectedly", "to take unawares", "the feeling or emotion excited by something unexpected, or for which one is unprepared" or something "akin to astonishment and wonder, caused by an unexpected occurrence or circumstance." There is, as its French roots suggest, a sense of being "overtaken" (*sur* + *prendre*).

Etymologically, then, the idea of being surprised is, as the OED suggests, about being "taken over" or "attacked" all of a sudden. And, it is an unexpected event or happening for which we might be unprepared. Yet, this is paradoxical since we must be prepared, neurologically, to be surprised (Gazzaniga, 1998; Nørretranders, 1998)! Nontheless, there is still a sense that surprise can be by design. That is, surprises can be deliberately planned events to catch *someone else* "off guard". In this manner, certain aspects of the planned surprise must be kept away from the other. Hence, the surprise must remain hidden or invisible to the other, although, to be clear, it is the lived human experience of surprise that cannot be planned.

Certainly, it would appear that sometimes surprises are hidden from us, however, the unexpected need not be hidden at all. In fact, the unexpected happening can be there in front of us all the time (Merleau-Ponty, 1962). But why is it that we might not see it? The difference may be a distinction between seeing and looking. In looking, we are paying attention, and there is an act of intentionality. When we are merely seeing, the world fades into a background of over-familiarity (Berger, 1973).

Embodying Surprise: An Illusion of Stability and Expectation

To the western mind, the idea that consciousness plays such a small role in our "experience of experiencing" may be hard to believe. Indeed, some may believe that consciousness is limited, but the idea that consciousness contains so little information may be a bit surprising. Since the mid-20th century, psychologists and communication theorists have known that large quantities of information as sensory inputs are discarded, prior to the conscious perceptions of human beings (Nørretranders, 1998). So, do human beings see the world as it is? Generally speaking, the answer is "no".

To be sure, human beings are bathed in a sea of sensory possibilities, and research suggests that what human beings can be consciously aware of are several magnitudes of difference away from one another (see Nørretranders, 1998, for a full list of references). In fact, while the details may differ from one piece of research to another, the results are generally in the same order of magnitude with the number of sensorial possibilities being around ten-million bits per second compared to the very narrow bandwidth of 40 bits per second of conscious perceptual possibilities. In spite of so much information being "filtered out," the sensations that we do have still affect how human beings think and act at a non-conscious level (Davis *et al.*, 2000). This is, quite possibly, a feature for many different kinds of organizations of various scales, including the biological, but also the sociological, cultural and ecological. The difference between what is perceived and what is sensed cannot be overstated enough. But it does beg the question: If one is not and cannot be aware of every sight, sound and smell, then what is it that human beings see, hear and smell that happens on a non-conscious level?

As it turns out, figuratively and literally, we don't see what we don't see (Maturana & Varela, 1992). And, yet, human perception of the world appears so seamless. Human beings have blind spots and sensorial equivalents to our visual blind spot that remain outside of human consciousness unless certain deliberate "tests" are done to show these "limitations". But why do human beings experience a seamless world as opposed to one filled with gaps? The answer is perceptual completion or apperception.

Perceptual completion is a subject-level phenomenon which presents a kind of "filling in" of information by the brain to make up for some kind of absence of information (Thompson *et al.*, 1999). Moreover, all of this happens before we know it, in a manner of speaking. That is, the brain seems to get things done before a person knows it (Gazzaniga, 1998). This kind of predictive perception offers up, from past experiences, a possible or to-be-expected perception. In other words, through experience, human beings become primed or prepared to deal with life's unknowns. Paradoxically then, one must prepared, in some sense, to be surprised. As such, the experience of surprise suggests a certain complicity. Thus, surprise is an emergent phenomenon that arises in the interactions between one's self and the world.

Theorizing "Surprise" Using Complexity-Related Frames

What I aim to show here are two possibilities for thinking about the nature of surprise and unexpectedness through the conceptual frames of catastrophe theory and self-organized critical systems. It must be born in mind that these two "theoretical" frames–in as much as they could be coherent theories–are not the same because they are informed by different assumptions. The field of complexity is, after all, an interdisciplinary field that is shaped and guided by a diversified collection of tools and views. In this light, the idea of presenting a unified frame for thinking about the nature of surprise and unexpectedness is not troublesome. Certainly, the emerging field emphasizes a diversity of views anyway.

In *Reality Rules*, John Casti (1992: Vol. 2: 259) notes: "One of the great challenges to both science and philosophy is to provide a rational account of the uncertainty we perceive in the events of daily life." He remarks, however, that although a variety of theories, including classical probability and fuzzy logic, have addressed this issue, these domains have made unsatisfactory progress in answering this challenge. Each of these domains, moreover, has suggested that uncertainty is not associated with randomness. It is, rather:

> *...that the uncertainty we feel over everyday events and situations cannot usually be attributed to the influence of a random mechanism, but appears to stem from an inherent vagueness, or lack of information, either in the linguistic description or other circumstances surrounding the situations we find ourselves in* (Casti, 1992: 259).

As has already been suggested, it is not so much that there is some inherent vagueness or a lack of information in what is known that marks human experience. Rather, there is a great deal of re-interpretation or apperception that happens prior to human experience, especially in times when the brain itself remains uncertain when faced with something novel or unexpected (Austin, 1999). In fact, while a great deal of "information" may be filtered out while unconsciously assimilated as "exformation", the perceptual completion of such a filtering of information draws upon a particular history of past events as particular expectations of the world (Thompson *et al.*, 1999).

Returning to Casti (1992), one might consider the proposition that surprise is some result arising from some discrepancy between the behavior of some open system through its interactions within the larger context of that system (the "outside") and the behavior of a closed system to those same interactions. In other words, the discrepancy arises through some difference between what is experienced and what could have been expected–even on an unconscious level. Last, Casti suggests that complexity arises from a potential of bifurcation which is a notion that seems fitting for a catastrophic-inspired model of surprise. The question of surprise, as an emergent phenomenon, however, cannot be accounted for in such a model. That is, the "novelty" of a surprising moment of unexpectedness cannot be addressed. In this manner, surprise as an emergent phenomenon would be better addressed through the notion of self-organized criticality.

A Catastrophe Model

Phenomenologically, the quality of "suddenness" is a particular lived existential (van Manen, 1990). In this case, it is the temporality or "lived-time" aspect of surprise that proves to be a helpful category for reflecting on the nature or essence of unexpectedness. Certainly, one cannot separate this lived existential from the others (i.e., corporality, spatiality and relationality). Still, it can be differentiated, and it is used here as one aspect that could inform a general shape for a catastrophic model for surprise and unexpectedness.

With this idea of suddenness in mind, it is fitting to consider a theory that examines phenomena that unfold discontinuously (although not necessarily continuously discontinuous). In this manner, catastrophe theory, as a theory about discontinuous change governed by local interactions, may prove helpful. In fact, many physical events display discontinuous change: cell differentiation, turbulence, and buckling in iron structures of buildings, although models involving the behaviors of living systems have also been studied, e.g., Thom and Zeeman (Casti, 1992). In this case, it is a lived-experience—a human emotion—that is the subject for this model: while modeling physical systems may be easier, attempting to model surprise as a complex phenomenon could open up other possibilities for thinking about and understanding this deeply-felt human experience.

Original Ref—E:CO Vol. 11 No. 2 2009 pp. 46-53

Self-Organized Criticality

Of course, the world is always changing: this may be an obvious statement to make. Nonetheless, human beings sometimes seem to forget this, and thoughts and phrases like "I was caught off guard" and "Why is this happening?" point to a rather mystifying thought. That is, we seem to *forget* that the world is continually changing—as are we. This is a rather puzzling paradox. Still, what complexity scientists believe about how the human brain emerges can illuminate the tension in this apparent paradox.

If the world is continually changing, then why are there times when we do not seem to notice change? How are we affected by imperceptible change? Is there something significant or special concerning those changes that we can perceive as opposed to those that are imperceptible to us? I argue that there is nothing special or privileged about change that is perceptible to human beings, as opposed to imperceptible change. That is, all perceptual change arises from the same dynamics: yes, in some cases, dramatic effects can happen. But more times than not, mostly smaller and more frequent imperceptible change happens.

Conceived as a phenomenon that emerges at times of critical transformation, surprise could be patterned in the same fashion as two canonical examples of self-organized criticality: sand piles and earthquakes. Moreover, I shall refer to the experience of "surprise" as a particular moment of unexpectedness that happens at some critical moment of transformation to distinguish it from other experiences of unexpectedness. To do so, aligns such a conceptual framework for surprise with the complexity sciences as a particular perceptual discrepancy between expectation and experience. This allows me to make the claim that every moment is a moment of surprise. This shift in meaning is important for framing a theory of surprise in the light of the complexity sciences; however, it most certainly does not hold up to any phenomenological study of surprise.

With this in mind, it can be said that the world is full of surprises. Every moment is a surprising moment. Now. Now. And, now, again. Did you notice anything surprising? Naturally, to use a cliché, I wouldn't be surprised if you said "no". But, again, why are we not surprised (in every moment)? Surprises happen *for* us, and not necessarily *to* us. In this manner, to be surprised means attending to, and responding through, our current lived embodied experiences in the world, rather than reacting to some apparently large, significant human event.

As such, the world cannot necessarily *cause* us to be surprised without our partial collusion—particularly when we are inattentive to it and the world fades into a background of over-familiarity. Thus, if the world is full of surprises, some are filtered out: in fact, the human mind must necessarily forget some things and the brain must *learn* to forget (Davis *et al.*, 2000). The brain has developed in such a way as to bring forth a mind that functions in the world with particular expectations. This allows human beings to constantly adapt to a changing world. Otherwise, to use a computer metaphor, the human brain would "overload".

At this point, I still wish to hold onto Casti's notion of surprise. That is, surprise is some result arising from some discrepancy between the behavior of some open system through its interactions within the larger context of that system and the behavior of a closed system to those same interactions. The measure of such a discrepancy will be important for describing further this self-organized critical model of surprise, although I will not—or rather cannot—say what such a thing would be. That is, as Einstein suggested, some things that matter cannot be measured, while other things that can be measured do not matter. Moreover, although the measures of such discrepancies have been useful and used in classical probability studies of surprise, I will refer to such measures in more of a conceptual manner. That is, I will use a particular conceptual *idea* for a measure of a surprise by suggesting that surprise could be conceived as a particular geometrical structure: a fractal.

Readers familiar with the concept of a fractal structure will recall that such structures have a quality of *scale invariance* on a number of different scales, temporal and/or spatial (Bar-Yam, 1997). This means that the appearance of some structure of the fractal will resemble other larger or smaller parts. Indeed, it may resemble the whole fractal itself. Sometimes fractals are described as being self-similar objects; however, it should be born in mind that only mathematical fractals can be self-similar, in a sense that a part of a fractal can be *identical* to the whole. Fractal-like structures in nature are best described as scale-invariant structures where, for example, the branch of a tree may *resemble* the larger tree itself in all of its detail, although the two are not identical. But the two different scales do arise from the same underlying principles of dynamics for complex phenomena.

The emerging thinking behind self-organized critical phenomena has shown that many phenomena evolve and emerge at some "critical state" where change in the phenomenon, brought about by some disturbance, can lead to an event of any size (Jensen, 1998). Earthquakes and avalanches tend to be the canonical examples for such phenomena. And, for ages, human beings have thought that major events involving phenomena like earthquakes and avalanches were special events. In fact, it is the same evolution to such a state that gives rise to a large avalanche that brings about smaller ones. The dynamical interactions create a self-organizing system capable of change at any time. The size of such changes, however, will vary in a distribution referred to as a power law.

Phrased mathematically, the power law represents a way of stating a law of distribution of particular events through the observation of statistical measures (Schroeder, 1991). Of course, what can be observable depends upon our capacity to perceive. For instance, earth tremors happen constantly, but so many remain of them lie below our own thresholds of perception. Thus, when a tremor is felt, it is perceived to be a special event. Similarly, I claim that we are surprised more often that we realize according to the same kind of dynamical fractal pattern except that our own physiology is complicit in removing or filtering out surprising possibilities through mechanisms such as perceptual completion, for example, which renders the world a bit more stable.

The point, here, is that surprises happen constantly and come in a variety of "measures"—in many sizes, small and large. More specifically, the number of times for and the measure of a particular measure of surprise are related and expressed as a power law. Graphically, such distributions looks like straight lines on a double-log coordinate scale: as the size of the measure increases, the fewer of such measures there will be, and, thus, the collection of data points resembles a straight with a negative slope. If, for example, one counts all of the animals on the planet, size is an indicator of number—the bigger the animal, the fewer there will be. Analogically, we experience more surprises which, if they could be measured, are small in nature compared to surprises that "stand out more" or, in a more common sense manner, that seem like surprises that are a lived emergent experiences.

Certainly, the notion of emergence is a popular one in the complexity literature (Bar-Yam, 1997). A variety of definitions abound involving concepts such as interactions, parts of a system, and something new. As a working definition here, *emergence* is about a novel phenomenon that arises from the interactions of parts of a system and is not found in the parts themselves. Emergent phenomena, thus, are inherently unpredictable. Moreover, emergence cannot be understood by attending to the parts of a larger system nor their interactions. Emergence is something more. That is, emergence is not readily *understood* from the self-organizing interacting parts of a system that give rise to a new phenomenon. In other words, the level of observation is important to understanding emergence.

A self-organized critical view of surprise then has this to say: "surprise" is an emergent phenomenon that manifests itself at a level different from the interactions of (at least two) different systems, as well as the systems themselves. Through the interactions, the possibility of surprise is brought forth through some "push" into novelty. At some perceptual threshold, the experience of "unexpectedness" is felt: below that threshold, surprises of "all sizes" or 'measures" continue to happen, although they remain outside of the narrow bandwidth of human consciousness. Such sub-liminal surprises, nevertheless, become a part of who we are as we are a product of our historically-situated experiences. In this manner, human beings become prepared for further surprises. In other words, surprise is possible because of our embodied history, and, when prompted, a surprising moment happens for us, and not to us, because of who we are.

Concluding Thoughts

While many types of organizations attempt to remove the possibility of being surprise, surprise is actually something that will happen in spite of our efforts to eliminate it. Becoming rigid and stable, in fact, is no way to adapt to change; however, encouraging diversity and possibility through interactions in dynamic ways and in dynamic spaces suggests a more promising way to deal with a constantly changing world.

To be sure, surprising, unexpected events happen all the time. Depending upon the particular view that a person or organization may have, these surprising moments can be thought of and addressed in a variety of ways. On one hand, surprise can be something that is not desired, something suppressed or controlled for. Or, it can be something that is embraced, sought out or encouraged. Such a conceptual view of surprise can have some profound effects upon an organization. Those effects, however, are quite different depending upon the "nature" of the organization.

Conceptually speaking, conceiving surprise in this fashion is not uncommon. Still, there seems to be an important piece missing from the many discourses on surprise. This paper, however, offers some speculation for framing the *experience* of surprise in relation to a more complexified framing of the lived-experience. And, it is hoped that some greater attention might be made for considering phenomenological aspects as an important, if not necessary, attempt to understand the notion of surprise through the science of surprise itself.

References

Austin, J.H. (1999). *Zen and the Brain: Toward an Understanding of Meditation and Consciousness*, ISBN 9780262011648.

Bar-Yam, Y. (1997). *Dynamics of Complex Systems*, ISBN 9780813341217.

Barritt, L.S. (1983). *A Handbook for Phenomenological Research in Education*, OCLC 65708913, Ann Arbor, MI: School of Education University of Michigan.

Berger, J. (1973). *Ways of Seeing*, ISBN 9780140216318.

Casti, J.L. (1992). *Reality Rules: Picturing the World in Mathematics*, ISBN 9780471184355.

Dastur, F. (2000). "Phenomenology of the event: Waiting and surprise," *Hypatia*, E-ISSN 1527-2001, 15(4): 178-189.

Davis, B., Sumara, D.J. and Luce-Kapler, R. (2000). *Engaging Minds: Learning and Teaching in a Complex World*, ISBN 9780805837858.

Gazzaniga, M.S. (1998). *The Mind's Past*, ISBN 9780520213203.

Jensen, H.J. (1998). *Self-Organized Criticality: Emergent Complex Behavior in Physical and Biological Systems*, ISBN 9780521483711.

Maturana, H.R. and Varela, F.J. (1992). *The Tree of Knowledge: The Biological Roots of Human Understanding*, ISBN 9780877736424.

McDaniel, Jr., R.R., Jordan, M.E., and Fleeman, B.F. (2003). "Surpise, surprise, surprise! A complexity science view of the unexpected," *Health Care Management Review*, ISSN 0361-6274, 28(3): 266-268.

Merleau-Ponty, M. (1962). *Phenomenology of Perception*, ISBN 9780415045568.

Nørretranders, T. (1998). The User Illusion: Cutting Consciousness Down to Size, ISBN 9780140230123.

Petitot, J., Varela, F.J., Pachoud, B., and Roy, J.-M. (eds.) (1999). *Naturalizing Phenomenology: Issues in Contemporary Phenomenology and Cognitive Science*, ISBN 9780804736107.

Schroeder, M. (1991). *Fractals, Chaos, Power Laws: Minutes from an Infinite Paradise*, ISBN 9780716723578.

Simon, H.A. (1991). "Bounded rationality and organizational learning," *Organization Science*, ISSN 1047-7039, 2(1): 125-134.

Stacey, R.D. (1992). *Managing the Unknowable: Strategic Boundaries between Order and Chaos in Organizations*, ISBN 9781555424633.

"surprise, n" (1989). *The Oxford English Dictionary*, OED Online: Oxford University Press, http://dictionary.oed.com/cgi/entry/50243394 (subscription required).

Thompson, E. (2007). *Mind in Life: Biology, Phenomenology, and the Sciences of Mind*, ISBN 9780674025110.

Thompson, E., Noe, A., and Pessoa, L. (1999). "Perceptual completion: A case study in phenomenology and cognitive science," in J. Petitot, F.J. Varela, B. Pachoud and J.-M. Roy (eds.), *Naturalized Phenomenology: Issues in Contemporary Phenomenology and Cognitive Science*, ISBN 9780804736107, pp. 161-195.

van Manen, M. (1990). *Researching Lived Experience: Human Science for an Action Sensitive Pedagogy*, ISBN 9780920354421.

Varela, F.J. (1999). *Ethical Know-How: Action, Wisdom, and Cognition*, ISBN 9780804730334.

Weick, K.E. and Sutcliffe, K.M. (2001). *Managing the Unexpected: Assuring High Performance in an Age of Complexity*, ISBN 9780787956271.

Darren Stanley is Assistant Professor in the Faculty of Education, University of Windsor, in Windsor, Ontario, Canada, where he teaches in the areas of mathematics education and curriculum theory. His diverse research interests lie at the interstices of complexity, learning, and healthy learning organization and include democratic schools and eco-justice scholarship and activism.

22. Making Room for Affordances

Hugo Letiche[1] *& Michael Lissack*[2]
1 University for Humanistics, NLD
2 Institute for the Study of Coherence & Emergence, USA

This is a conceptual paper about 'affordances'. It is inspired by Gregory Bateson (1972) who argued that consciousness is a person/environment interactive process; we will focus on how relationships between environments and organisms lead to perceived possibilities, actions, and cognition. Both the relationships between environments and persons, and the relationships between persons and environments count. The connection between world and consciousness is dynamic. There is a mutual causal link between circumstances and organisms. We argue that the world via affordances presents itself to consciousness. Emergent possibilities afford; complexity affords. Scott Kelso's 'complementary relating of contrarieties' affords. Affordances are the dynamic reciprocal relationships between animate persons and their environments. Affordances are in-between—their cognition is situated and contextual. Affordances are the a next frontier for organization studies.

Introduction

Profoundly influenced by Micheal Foucault and Jacques Derrida, most social complexity theory has focused on the knowing languaged-based subject (Cilliers, 1998). In Foucault's terms, (social) cognition is grounded in the historically specific episteme. Complexity scholars have asserted that the contemporary episteme is complex—that is, it is emergent, dynamic and resembles a strange attractor. The episteme characterizes the dominant way of seeing, and operates as a very socially and economically powerful hermeneutic. The contemporary episteme, Foucault argued in the mid-twentieth century, has centered on 'discipline' and 'power'—i.e., it focused on the ways of structuring physical and mental existence prerequisite to the development of technology and industrialization, physical and social science, bureaucracy and globalization, material wealth and post-Fordist capitalism. Derrida focused specifically on how texts emerge from one another, refer and defer to one another, and form complex webs of signification. Text-based consciousness (which includes art, music, mathematics, etcetera—i.e., a great variety of forms of text)—it is asserted, is the only form of consciousness that we (can) know. Text is emergent, dynamic and operates in webs of relationship and pre-assumptions. While a social complexity theory grounded in Foucault and Derrida is revelatory; it is very perception and consciousness directed. The danger is that 'world' gets lost in some sort of consciousness studies. Without wanting to trash the Foucault-Derrida interpretation of complexity theory, in this article we focus on 'affordances'—i.e., not on language, but persons and circumstances. It is our conviction, with Bateson, that complexity involves a multi-dimensional ecology of world and consciousness, objects and perception, opportunities and language.

We examine here the concept of 'affordances'—the assumed mechanism(s) whereby complexity presents itself as (weak) signal(s) to consciousness. The "World", in the form of affordances, invites response by subjects. Affordances act as attractors drawing humans into action. Humans are not in a universe of dead material (hyle), but live in a world of active subject-world inter-relationship(s). The world acts, makes occur and initiates possibilities. 'Affordance' is a word for this activity.

This article does not debate epistemological or research methodological issues. Obviously experience would remain unheard of and unanalyzed if it could not be named. As Karl Weick keeps repeating, what actually is cognitized and named undergoes sensemaking after-the-fact (Weick, 1997). But as Bateson insisted, language and existence (i.e., knowing and world) are two dimensions (or two sides of the coin) of the same existence. Of course, the world that we can know at any one point is deeply determined by our assumptions, thought processes and mental characteristics; but our mental structure(s) have to do with the materiality of the brain, the makeup of our environment and the history of our universe. Weick's one-sided attention to language based cognition entails putting the material world into the abstract, and produces a mentalist phantasm. Narrative(s) are language-based cognition and language is not the physical, material and living 'world'. The naming of the 'affordances' is not the same thing as the affordances themselves. But it is really meaningless to debate whether affordances give rise to experience, or human perceptual and social assumptions define the affordances. It is both at once. Assumptions, models and speech, resonate with situations, organizations and events. Affordances point toward opportunities and actions—affordances give direction to the potentially unlimited space(s) of (proximate) development. Affordances facilitate the human movement towards world and action. *Affordances are what points to the adjacent possible.*

The theory of affordances rejects the idea of the incommensurability of the known and knower. It embraces Bateson's ecology of the mind (1972), asserting that the known and the knower are mutually originatory, generative and epistemologically primary. Affordances are about subject/object relationship, interaction and complementarity. Affordances are about opportunities, dangers, and possibilities that call organism, consciousness and environment, to activity and sense-making. Getting the balance right of the 'something out there' and the 'consciousness of the actant' may be a philosophical nightmare, but we do it every day. Opportunities for action only exist if there is an actant to whom they appear. Bodily-embedded being is attuned to certain environments. Affordances have to "fall neither too far to the object side nor too far to the subject side of perception" (Sanders, 1997: 104). Our challenge is to get that balance and dynamic right in social complexity theory. First, we will examine the original 'ecological psychology' tradition of affordances as kicked-off by James Gibson; and then we will discuss Scott Kelso's complexity based reinterpretation of the idea under the name of 'complementary relating contrarieties' (Kelso, 2008).

Affordances in Ecological Psychology

Affordances occur when self and other, perceiver and perceived, objects and persons meet in actionable combinations. Affordances are relationships—in affordance there is connection, interaction and bonding. Mutually determinative relatedness between life and environment, self and world produces coevolution through action. Persons and situations, information and environments, activities and identities, interact in what is afforded. In particular, we are interested in organizational affordances, or in how physical, individual, and economic environments come together in collective activity. Such activity is emergent; it is neither mechanistic nor volunteeristic. Affordances are neither *solely* subject driven, nor situationally determined. Affordances are complex and interactional; they mobilize subject and object, consciousness and environment into activity.

Affordances invite participation, action and response. When circumstance invites reaction, context demands a response, or the situation offers opportunity, something is afforded. In affordance, perception, information, and activity are related in a manner that seems to beg for action. Affordances are not just labels—i.e., the product of a subject's naming something. Nor are affordances retrospective—i.e., a quality of reality identified after-the-fact. Affordances are prospective—context invites action, environment points to activity. In the ascribed sense-making of Karl Weick, action is not prospective, circumstance does not invite action and the environment is not proactive. In affordances, world, situation and location, point to action, shout for response and offer opportunities for attainment.

The term "affordance" was first coined by the perceptual psychologist J.J. Gibson (1977, 1979); it referred to actionable properties between the world and an actant (person or animal). To Gibson, affordances are relationships. They exist: they do not have to be visible, known or desirable. Affordances entail the possible relationships amongst actors and objects; they are properties of the world. For instance, affordances are what objects or things offer people to be done with them. Affordances are bestowed by the environment. They are what it offers, provides and supplies. Affordances invite activity, reaction and point to possibilities. An affordance is a relationship between something in the world and the intentions, perceptions and capabilities of a person or persons.

When used in this sense, the term 'affordance' refers to the perceived and actual properties of things—primarily those properties that determine just how the thing could possibly be used. A chair affords support and therefore affords sitting. A chair can also afford supporting things and can therefore afford being used as a 'table', or it can afford being used as a stepladder, or it can afford being used as decoration, or as an art object (among many other possibilities). Affordances extend across users and vary with them. "An affordance is an action possibility available in the environment to an individual, independent of the individual's ability to perceive this possibility" (McGrenere & Ho, 2000). But if the connections between perceiver and perceived, i.e., humans and their environment, are fortuitous or absurd; then affordances merely produce senseless illusions, hyper-real absurdities, and postmodern disorder. Sense-making can be illusionary, or at its worse, nonexistent. Asserting that an affordance exists does not necessarily mean an avowal of any positive mutuality in perception between self and world. Affordances are adjacent possibilities, but not all possibilities should be acted upon.

Gibson's ecological psychology studied perception as interaction between living beings (mainly humans) and (their) environment(s). The four distinguishing features to his theory are:

1. Perception is an adaptive process that steers physical (biological) and social (cultural) activity;
2. Perception—for instance, of structured light and sound—specifies opportunities or 'affordances' for (inter-)action;
3. Perception reveals (realist) 'information' in dynamic interaction with the perceptual environment of surfaces, objects, places and others;
4. Attunement to (potential) affordances is regulated via evolutionary design, individual history, intentionality and context (Based on McArthur & Baron, 1983).

Gibson's point of departure is his rejection of traditional psychology's effort to explain perception as stimulus/response or as a communication and computational achievement. The problem with stimulus/response theory is that it really has nothing to say about perception. There is no awareness or mind in the S/R schema—the conscious relationship of a self to a world is not thinkable in S/R terms. Communication and information processing model(s) conceive of perception as a process between sender, signal and receiver. How 'world', 'circumstance' or 'reality' are transformed by a 'sender' into 'signals', and/or how these 'signals' are re-transformed by a 'receiver' into 'knowledge' 'information' or 'awareness', is here an unsolved and unsolvable dilemma. How does a particular circumstance get encoded into 'signals' and how does another person 'decode' those 'signals', back into a 'representation'? Gibson dismisses 'representation' as a confusing and needlessly complicated concept. His assertion is that the circumstance is directly perceived. Perception is of, and in the world. Perception does not depend on mentalist abstractions such as 'representations'.

> *An affordance cuts across the dichotomy of subjective-objective and helps us to understand its inadequacy. It is equally a fact of the environment and a fact of behavior. It is both physical and psychical, yet neither. An affordance points both ways, to the environment and to the observer* (Gibson, 1979/1986: 129).

Gibson insists on 'direct perception'. It is asserted that perception occurs directly, instantly, and non-inferentially. 'World versus perception', and subject / object' dualism are rejected. Commonly it is assumed that world and perception are two very different orders of things. For Gibson, perception is direct and immediate. He assumes the "ontological homogeneity of perceptual experience and reality" (Oytam & Neilson, 2007). The only world that exists to be known is the world that is perceived. There is not some sort of *Being* or truer reality behind perception. Ontological dualism—for instance, between the surface and depth levels of experience, or between 'throwness' and 'Being'—is rejected. The authenticity of perceptual lived-existence is incorrigible. Gibson embraces 'epistemological realism' in a culture that has accepted skepticism, and wherein the 'seen' is thought to be mere 'appearance' and the 'experienced' is only 'doxa', and the 'lived' is 'hyperreal'. For Gibson, to exist is to perceive—existence is the stream of consciousness. Mind and sensible objects are in the act of perceiving, a single phenomena.

The inner sense of place and direction in Gibson is part and parcel of human ambient existence—because humans move constantly about, they inevitably experience an ever-changing stream of perceptions—i.e., an existence that is continually moving about. Perception is not static or stable; it is direct and immediate. For Gibson, experience and ontology are the same thing. Sensible reality just *Is*. Gibson's ontology is a radical phenomenological ontology wherein motility and temporality play a crucial affording role. Time and space are not idealized pre-structures of reality; motility and duration are direct experiential qualities of lived circumstance. In the 'real', there is spatial as well as temporal movement. Time and space as abstract pre-conscious qualities of mind, destroy the person ~ world dynamism: for Kelso (see below) '~' means mutually related and inextricable connected. In Gibson's theory of affordances, perception or (what he calls) 'information' is unequivocally perceived as if world, event, or circumstance directly addressed themselves to the subject. 'Information' is 'invariant'—i.e., the relationship of the 'here to there' of the landscape just *Is* (Gibson, 1979 / 1986).

In Gibson, 'affordances' are specific aspects of the physical and cultural environment that directly address themselves to the user. The user is characterized by a particular form of attunement that has coevolved in relationship to the environment. The subject <> environment relationship is co-defined; humans are suited to sight, the seen world informs human existence. Self <> world are co-determinant. Humans are defined by how they inhabit and contribute to the sustenance of their environment. Gibson's point was that perception should not be conceived of as psychology was wont to do—that is as a passive internal matter of representation. Perception is not a matter of reactive senses merely receiving physical sensations. It is a living action of an open system that by its very nature is attuned to interact with its environment. Gibson insisted that perception is coevolutionary. For instance, perception is a quality of motion; and motion makes perception necessary.

For Gibson the relationship of a species to its affordances is dialectical. The species is characterized by what it makes of its affordances (or environment), and its affordances (or environment) provide it with its niche or opportunities for living and existing. An opportunity is only proven to be an opportunity if a species actualizes it. For Gibson affordances exist on the species and not on the individual level. An affordance is a possibility of perception, action and existence, whether or not that opportunity is presently actualized. But at this point, Gibson's theory runs up against an epistemological barrier—how does the individual researcher know if something is or is not an affordance? If affordances are unknowable, the theory becomes fairly irrelevant. How could an individual observer claim species wide awareness? The answer is that he or she cannot do anything of the sort. Gibson limited his research into affordances, to basic, very abstract perceptual themes. On this level of universality, he could avoid epistemological dilemma. But epistemological problems became very visible when for instance Gaver (1993) and Norman (1988) applied the theory of affordances to design, or when Costall (1995) and Gombrich (1999) tried to analyze social and cultural factors.

Gibson tried to overcome the subjective/objective divide by studying the material conditions of human activity. An affordance, for him, is a physical attribute; such as, rocks afford: throwing, catching, injuring, collecting, etcetera. But surely a subject never responds to all the possible affordances at once. Affordances always stand in relationship to an actor and human actors are inevitably social. Affordances are not pre-social or universal, but are part and parcel of concrete events and interactions (Costall & Still, 1989). Culture is not located in the head, but in the everyday affordances of practice. The socially constructed environment hides some affordances, emphasizes others, and continually creates new opportunities for action. As Alan Costall has argued, there is no conflict between affordances and culture; quite the opposite, the socio-culturally structured environment, provides humanity with a large portion of its affordances. Gibson wanted to illustrate the subject's involvement in the materiality of the everyday world. In Gibson meaning as 'information' is not generated either by subject or object, but in the space in between them. But Gibson's only described individual consciousness confronted by an affording environment (Costall, 1995). There's no explicit acknowledgement in Gibson of social affordances or of how persons collectively produce and consume what they afford one another.

Gibson limits himself to relationships with the physical environment and objects. Tools, culture, and humanly mediated circumstances, get short shrift.

Social Affording / Affording the Social

Affordances explain how perception and activity entail mutuality between actor and environment. The focus is on how living beings live in and thanks to their relationships with circumstance. They are in a world that supports, stimulates, and nourishes them. The mental and physical worlds are not separate domains. They merge:

> *You are a jazz musician playing in a small improvisational group, a quartet, say. You are the undisputed leader of the group. While others may be featured from time to time, you are the one that generally controls the flow of the music. A number of points describe your situation.*
>
> 1. *Action is unavoidable. As the music progresses, you are continually playing, either as the featured player or, occasionally, as part of the back-up. Playing is action. Even if you are not playing, during a brief respite after a long solo, you are still in action, since you are not playing and leading the group at that time also has consequences for the group's playing.*
> 2. *Detached reflection about action is impossible. It is not possible to simply analyze your actions during the course of playing. Actions must take place immediately in order to respond to the group's playing. After the song is over, you may of course think to yourself that you should have played a series of notes differently, taken a faster tempo, and so on.*
> 3. *Action effects are unpredictable. Even if it were possible to step back from the situation and analyze potential outcomes of your actions, the effects of these actions cannot be predicted. If you suddenly change the nature of your playing, perhaps you change rhythm or key, the rest of the group may conceivably be inspired by this innovation and play with renewed rigor. It is also conceivable that the group may dislike your changes, losing their playing energy. You must simply … go with the flow.*
> 4. *Stable representation of the situation is impossible. There is no musical score, no written notes. You are simply improvising. After the song ends, you may be able to represent the situation: this solo did this, followed by that solo doing that, and so on. During the song, however, such an analysis is impossible, since you do not have a complete picture of the situation.*
> 5. *Representation is interpretation. Even though it is possible to form a representation of the situation after the fact, after the song is over, this representation is still a relativist interpretation. After a really good set, you might want to attempt to write down some of the music for future*

reference. Your transcription of the music is surely influenced by your current experiences and mood state. If another member of the group, less enamored of the set than you, were also to write down the music, his score would undoubtedly be different from yours. Each of you has the potential to interpret the song differently (Zahorik & Jenison, 1998).

The person (i.e., the jazz musician) is 'thrown' into the situation of the jazz performance. The person is part and parcel of circumstances. For the person, 'world' exists as 'readiness-to-hand'—i.e., the musical instruments, knowledge of jazz melodies, stage and lights, etc., are all absorbed into the activity-at-hand. This metaphor extends well beyond Jazz.

Circumstances are not analytically transparent to the participant; a circumstance is an event, interaction and involves direct participation. Via a deliberate effort, it is possible to reflect on our position in the physical and social environment. But normal 'knowledge' is 'active'—that is grounded in processes of lifelong and continuous interaction. Most of the time, the subject is linked to direct encounters with circumstances and practical tasks. Via 'affordances', we live the quotidian reciprocity of self and surroundings. 'Affordances' are what the social and physical environment 'affords the animal, what it provides or furnishes, either for good or ill [and thus] it implies the complementarity of the animal and the environment' (Gibson, 1979/1986: 127). Reciprocal relations coevolve: they form ecosystems or niches of physical and social existence. Put in physical terms—certain places, offering a knee high surface of support, afford sitting. In mental terms, the musician who starts to play a favorite (jazz) musical theme affords improvisation.

Our world affords looking; it invites 'visual' recognizance. It supports visual continuity, wherein the visual appears to give a more 'universal' or 'consistent' version of reality, than do sound, odor or touch. The visual appears to be 'laid out before our eyes' and to be 'made available to our senses'. In the perceptual ecology of humans, circumstance or 'reality' is given or offered to be perceived. Erik Jonssen (2002) has persuasively described such spatial and visual affordances. But his concept of 'affordance' is not (entirely) Gibson's. In his extensive description of the 'sense of direction', Jonssen has argued that humans orient themselves via implicit spatial cognitive maps. He asserts that people constantly understand themselves in their physical surroundings via an inner sense of where they are. They non-consciously know where other things are in relationship to where they now are. All of this is felt as a tacit sense of place. It is not analytic or cognitive, but relational and perceptual. It is a directly felt sensation of presence— 'here' and 'this place' are spontaneously experienced via implicit inner navigation. Maps and compasses require abstract comparisons between cognitive models and circumstances. Inner navigation is direct, unmediated and immediately perceived. Such an inner implicit sense of direction, affords human being-in-the-world.

In Garver (1993) and Norman (1988), the application of affordances in a very un-Gibsonian manner looses the quality of radical mutuality:

> *... the term affordance refers to the perceived and actual properties of the thing, primarily those fundamental properties that determine just how the thing could possibly be used. A chair affords ('is for') support, and, therefore, affords sitting* (Norman, 1988: 9).

Norman sees 'affordances' as the agency of materials and things—'when affordances are taken advantage of, the user knows what to do just by looking: no picture, label or instruction is required' (Norman, 1988: 9). Norman writes about things like stove controls, functions of telephones, light switches and doorknobs. On stoves it is rarely apparent which control is to be used for which burner. Telephones tend to have very few buttons and very many functions, whereby users do not know how to activate the functions. Often it is very confusing which light switch turns which light on or out. Whether a door has to be pushed, pulled, or (often the worst of all) slid aside is mostly unclear from the fixtures. Norman believes that good design should be self-evident; objects should unambiguously afford their functions.

In Gaver (1993), affordances become a form of (social) materialism—i.e., the world's imprint on the human. Affordance refers to 'facts of the physical world' wherein 'individual activities are shaped by the environment'. Gaver wants to reveal 'how individuals ... orient to the relevant environmental attributes.' 'Affordances' are primarily facts about the 'constraints of our physical environment'. The objects are determinant. Affordances, as mutually formative processes, are not relevant here. Gaver examines 'social interaction', describing how a lawful physical world exerts material influences on behavior, without making use of 'convention' or culture as explanatory principles. Design he argues needs to focus on the material features relevant for behavior. He illustrates his approach by comparing the affordances in information transfer on paper, in comparison to those using electronic media. While paper combines the storage and display functions; in electronic media they are separate. Layers of significance—for instance, via marginal notes or marks of wear and tear—are spontaneously added, one upon the other, in a paper document; but not in an electronic one. With a (paper) book, one can see how far one is and one can literally feel the pages. Originality and authenticity are well served by paper. On the other hand, electronic documentation is much more changeable or adaptable, and supports co-presence and collaborative action. The differences in materiality, supposedly, provide different sets of affordances. Gaver downplays cognitive structures, such as memory, problem-solving and culture, in order to stress physical factors. He focuses on how social activities are 'embedded in and shaped by the material environment'. His 'ecological psychology' rejects mutualism and embraces a materialism wherein the physical is <u>the</u> determining aspect. Norman (1993) and Gaver (1988) escape the epistemological complexity of mutualism, by reducing affordances to functional or material first-causes, and by relegating consciousness either (Norman) to an epiphenomena, or by trying (Gaver) to avoid it altogether. Ecological psychology without mutual causality becomes just another materialism.

Costall (1995) and Gombrich (19999) make exactly the opposite move. They assume that human affordances are always social affordances, and thus that no concept of affordances can exist without a theory of social hermeneutics. Affordances are inherently interpretative; and interpretation is textual, cultural and illuminative. Gombrich insists that an ecological psychology has to be a psychology of flow or activity. Perception is understood as mutual action—i.e., a process of human interrelationship and of influencing. Affordances teach us that perception is action—a process of defining possibilities, opportunities and options. Perception is not passive or merely spectator-ial. Gibson's theory of 'invariant information' is applied in a very limited way. Gibson makes use of a very particular concept of 'information'. In concrete circumstance—i.e., in the world experienced as 'readiness-to-hand'; some aspects of the situation are directly perceived. Aspects of the condition that are characteristic of it, or that define its affordances, are called 'information'. 'Information' exists on a species level and is circumstantially invariant. Affordances are not particular or moment bound; they are general and ecological:

> *... Gibson's assertion that we do not really see the individual aspects of things but their invariants perceived in the round is easily refutable. ... Whether we are painters or merely nature lovers, we have the right to examine and admire the ... view without wanting to become aware of affordances or indeed of our exact position in relation to these vistas. Enjoying the magic of a sunset we cannot be concerned with invariants. We may respond to its beauty, and this response can also be evoked by a painting, which may arouse a mood rather than a potential for active exploration. Gibson was right in drawing our attention to this latter function and purpose of our senses, but his account may still have to be supplemented by doing justice to the undoubted emotional effects of seeing, effects which may also be wrongly considered to be purely subjective* (Gombrich, 1999).

Perception is an act and not an automatic or passive response. Perception is an achievement and not a reflex. Gibson understands perception as a successful interaction between the living being and its environment. Perception is a key aspect—for instance, in determining for humans' their ecological niche. Perception provides a flow of 'information'—that is, a stream of possible actions, probable opportunities and the anchorage (order and certainties) needed to be effective. Making contact with the world by exploring its affordances takes many different forms. The tabletop's affordance may be for depositing things, but it can serve as a platform from which to address a crowd, or be an art object just to be looked at. Gibson's realism is very problematic—how can one know what is and is not 'information'?

For Gombrich, the object or tool that is 'grasped' with hands or eyes and is experienced as ready-at-hand is appropriated in terms of its affordances. There is no representation or ideation involved; there is a direct flow of perception, motion and action. If that flow breaks down—i.e., the tool doesn't do what it was supposed to do, or the appearance suddenly takes on an unexpected aspect or significance; then the affordance has to be reviewed. Most often, non-analytical ready-at-hand cognition suffices—things just 'are'. Most of the time, existence unfolds within the unproblematic perceptual affordances of everyday experience, but not all the time. When everyday affordances do not suffice, then aesthetic analytical and (self-)conscious thought has to take over. Blind response to affordances may suffice much of the time, but some of the time new events and remarkable circumstances demand more self-aware forms of perception.

The concept of 'social affordances' provides a unit of analysis, which potentially encompasses both the perceiver and the situated action, and which does justice to the human world. But, what is the nature of the middle ground wherein perceiver and perceived actually meet?' To answer requires dismissing the mechanistic world-view, because in it 'mechanisms do not act unless put into action by an external agency' (Reed, 1996). But the problem for ecological psychology is: 'How do persons put themselves into motion?' Can one retain the Gibson-ian sense of 'information' and 'affordance', but also acknowledge autonomous agency? Perceivers are not simply perceivers. They have a 'will' and 'do' things. Affordances may indicate opportunities for action, but that does not alone explain activity. Which opportunity is seized upon and which is ignored? Which affordances define the human or social envelop of opportunity? Posing the question in this manner, parallels Vygotsky's investigation of the 'zone of proximate development' (Vygotsky, 1962). Re-stated in Gibsonian terms, the 'zone of proximate development' is characterized by learning whose affordances originate in the physical and social environment. Perception and cognition are processes of change and activity. These processes are emergent—i.e., coevolve within the temporal flow. Reed liberates (Vygotsky's) activity theory from its somewhat narrow socioeconomic concept of (historical) change by understanding affordances as inherently emergent. Cognition occurs inside an eventful and changing world. Cognition is neither cause nor effect, but both at once. Social affordances are cognition in the world. Perception, information and consciousness, all come together in affordances.

To make the analysis more concrete: the financial crisis of 2008 to the present is illustrative of the working of affordances. According to most observers it was the all-pervasive belief that housing prices could only rise that afforded easy lending and lax regulation, which in turn allowed the housing bubble to inflate. The belief in always-rising prices allowed lenders—supposedly rationally—to look only to the value of the underlying asset (which the belief set asserted could only rise), rather than to the ability of the borrower to make payments. The same beliefs allowed regulators to be unconcerned when stories arose in the media regarding "liars' loans" (loans made to people who blatantly could not afford them and who lied on their mortgage applications). The belief in rising prices afforded the bullish lack of unease or of controls. Beliefs and stories provided a context that afforded resulting actions.

Complexifying Affordance(s)

The theory of affordances tries to provide a theory of perception that is not lopsidedly directed to mind or world, consciousness or brain, self or object. We believe that social complexity theory needs the concept of affordances to describe 'complexification' and to fill-in its coevolutionary theory of change. Social complexity theory is directed to the self-complexifying nature of individual and/or shared consciousness, and demands whether complexification is a material and/or inherent quality of life. Forewarned of the conceptual pitfalls of social affordances, we nonetheless (re-)assert that there are internal affording relations between perceiver and perceived. These affordances are "equally a fact of the environment or a fact of behavior" (Gibson, 1979). Affordances produce experienced, perceived properties that create the possibility for certain kinds of behavior. But properties, at a higher level of complexity, cannot be directly ascribed to (lower level) component parts. Key properties do not belong to any single aspect of the organism or organization, but to the relationships between the parts in the whole.

Affordances are emergent. The whole happens—but it can be unpredictable, dynamic and creative. Organization is emergent; it is not a mere collection of pre-defined components. In organizing, the parts of the organization develop in relationship to one another. These parts only take on their identity through their relationship(s) to the whole—the parts and whole are mutually one another's causes and effects. Development of the parts depends on the development of the whole, and vice versa. Obviously the parts do not add up to the whole; the organization or organism is more than the sum of its parts. The organized whole—at whatever complexity or aggregation level—is semi-autonomous. Start-up ventures and entrepreneurs are afforded the opportunity for reward and failure by their (investors') belief in the prospects for success. But profits can depend on labor markets, exchange rates, taxation and regulation, technology, the economic climate, competitors' behavior, etcetera—organizational success of failure occurs in very complex ecologies, however simple the affordances may appear to be at any given moment.

Every aggregation level has its own rules. The genes of the living thing or the atoms of the object do not determine the living thing's identity or the object's characteristics. Context is crucial:

> *The function of the part depends on its context within the whole. One sees this on several levels involving the gene. First, what a gene "does" depends on its context. The function of a gene or gene product depends on its interactions*

with other genes and gene products. β-Catenin can be an adhesion protein in the liver or a transcription factor in the skin. Lactate dehydrogenase can be an enzyme in the muscle or a structural crystalline in the eye. ... the whole determines the function of the parts just as the parts determine the function of the whole (Gilbert & Sarkar, 2000: 6).

The environment can make all the difference in the organism's or organization's development. The living actant can make a big difference to its environment. The general law determines the specific instance, and the specific instance determines the general law. Actions and actants are linked in complex interacting systems.

Whatever may be the theoretical power of Gibson's ecological psychology and concept of affordances, in comparison to a theory of emergence, it is rather static. Affordances just *are*; consciousness and world coevolve of themselves. There is no agency in any of this.

Scott Kelso's theory of the 'complementary relating of contrarieties' provides us with a contemporary process theory of affordances that is more dynamic. Kelso adds a more explicit complexity perspective to Gibson. He focuses on the 'metastability' of relatedness, characterized by "the simultaneous realization of two competing tendencies: the tendency of the components to couple together and the tendency for the components to express their intrinsic independent behavior" (Kelso, 2008: 186). Kelso argues that "informationally coupled, self-organizing dynamical systems, like the brain and its complementary relation to mind," require the metastability of 'complementary relating of contrarieties' (Kelso, 2008: 187). Affordances provide this 'complementary relating of contrarieties'. They encompass both sides of any duality, such as: organism and environment, nature and nurture, mind and body, friend and enemy, living and dying. These terms are all 'mutually related and inextricably connected'. Kelso uses the '~' to express the complementaries between two factors: integration ~ segregation, local ~ global, individual ~ collective, part ~ whole, competition ~ cooperation, creation ~ annihilation, convergence ~ divergence, dwell ~ escape, states ~ tendencies, symmetry ~ dynamics, form ~ function (Kelso, 2008: 184). The forces of attraction that pull entities or different facets of circumstance into relationship are the affordances. The complementarities of action/reaction, opportunity/response, self/other, are all provided by affordances. The coordination dynamics of contrarieties are what affordances are all about.

Kelso further argues that the same sorts of things occur on a material as well as on a consciousness level. Metastability is not just a perceptual or action theory of affordance; it is also a concept of brain organization.

... in the metastable brain local segregative and global integrative processes coexist as complementary pairs, not as conflicting theories. Metastability, by reducing the strong hierarchical coupling between the parts of a complex

system while allowing them to retain their individuality, leads to a looser, more secure, more flexible form of functioning that promotes the creation of information. Too much autonomy of the component parts means no chance of their coordinating and communicating together. On the other hand, too much interdependence and the system gets stuck; global flexibility is lost (Kelso, 2008: 194).

Perceiving is an interactive event involving patterns of relationship between widely distributed neural ensembles, as well as between subjects and their contexts. Areas of the brain are multiple and diverse, just as social relationships are manifold and various. Kelso stresses that the laws of multiplicity and relationship are just as relevant for the brain as for the mind. Dynamic metastability affords brain—consciousness, as well as actant—world relationships.

We have argued that affordances follow what we, following Ricoeur (1992), call the logic of *idem* and *ipso* (Letiche, Lissack & Schultz, forthcoming). *Idem* focuses on continuity, stability and permanence; *ipse* centers on change, development and activity. Identity requires both continuity and change, stability and development. If identity moves too close to *idem*, self dies, experience dries up, and reification overwhelms. If there is an excess of *ipse*, identity disintegrates in chaos, and 'self' threatens to disintegrate. The 'metastability' of *idem* and *ipse* is what is required. Identity is not an either/or of *idem* and *ipse,* but an unstable both/and of them both. In Kelso's work the key question about 'affordances' is what determines the 'dwell ~ escape' dynamic(s) of relationship. 'Dwell' dominance leads to *idem* predominance; the supremacy of 'escape' turns into *ipse* chaos. Some instances of *idem* persist much longer than others, without the extremes of subject/object dualism arising. Likewise, the fluid and dynamic interrelations of *ipse* can move faster or slower, and more or less creatively away from past structures. Kelso, taking a metaphor from William James, says that the stream of consciousness is like the 'perchings' and 'flight' of a bird. *Idem* stability is juxtaposed to *ipse* movement. Both are needed to create a living, complexifying and emergent world of affordances.

Affordances reveal the human being coupled to its world in metastable relationships. Some connections dwell longer, sometimes escape is instant. The 'stickiness' of the affordances is crucial. How much attachment, relatedness, or attraction does an affordance exert? Affordances produce relational motion and the complementary relating of contrarieties. They replace mind / world; self / other; material / consciousness dualisms, with mind ~ world, self ~ other, material ~ consciousness relatedness.

Discussion

Organism and organization are mediated, complex, and they resonate. Traditional dualism sees organization as a rational effort wherein the human will overmasters chaos and makes it possible to realize rational goals. Thereafter creativity is needed to reverse the 'iron cage' of rules and order. First the rational intentional subject is all-important; and then some sort of destabilizing reversal is required.

With affordances or of a theory of complementary relating of contrarieties, neither side of the dualisms ever predominates. Organization is neither an intentional manifestation of human will, nor a product of entrepreneurial destruction. Affordances are a matter of mind and circumstance, and of the resonance between them. Relationships between world and consciousness manifest themselves in concrete networks of activity. There is no single determining logic to these dynamic and emergent relationships. Possibilities, dangers, and spiritual beliefs all resonate with circumstances, others, and innovative actions. The effort to reduce all affordances to a few causal combinations, amounts to reduction ad absurdum.

There are thus critical lessons here for the ways in which we train managers to act. The use of rule based checklists, and of Demming inspired statistical controls, asserts a stasis to the world, which we do not find. Such a stasis would allow for the affordances to be predictable, the context to be controllable, and emergence to non-existent. Instead, we find that emergence is pervasive, context is seldom controllable, ecologies are emergent, and few affordances are predictable, other than in the abstract world of theory (or of a management school case study).

The absence of stasis means that you cannot predict and control affordances. One affordance will be violent and destructive, and another creative and fulfilling; the one can open up a field of fear and aggression, and the other an opportunity-space for generativity. When we tell stories and share languaging, the changing context can bring us from raw experience to the possibilities and limits of shared consciousness.

Affordances can bring us from a possibility space to an activity. In the relationships between persons and situations, the move from activity to consciousness and back again, can be co-shared and co-experienced. Affordances are in effect 'complementary relating contrarieties', providing the non-dualist logic needed in social complexity studies. One will be drawn out by affordances, made to do things by affordances, and confronted by possibilities by affordances. The logic of affordances is a logic of relationship and possibility. Affordances are about the *could be* and not the *IS*.

Management studies are geared around the retrospective reduction of what was to a set of rules. By contrast, a complexity inspired organizational studies would develop a language to describe multifaceted possibilities, or affordances—that is, to map the potential, resonating, inter-relating, and interacting events of relationship. Affordances are about possible maps of relationship and where they might lead one. Rationalization after-the-fact of the path already taken is what organizational and management studies now focus on. They have an infamous inability to be predictive, because they are analyzing retrospectively, avoiding the complexity of affordances.

A social complexity theory of affordances, at present, will only be descriptive and it will be with difficulty that it would even be that. Trying to 'model' social affordances is down right silly. We do not have the means to enumerate, let alone describe social affordances at present. Affordance-analysis allows for complexity recognition, description and some (critical) reflection. Affordances we assert are the next frontier for organization studies and a critical arena for management education.

References

Bateson, G. (1972). *Steps to an Ecology of Mind*, ISBN 9780226039053.

Cilliers, P. (1998). *Complexity & Postmodernism: Understanding Complex Systems*, ISBN 9780415152860.

Costall, A,P (1995). "Socializing affordances," *Theory and Psychology*, ISSN 0959-3543, 5: 467-482.

Costall, A. and Still, A. (1989). "Gibson's theory of direct perception and the problem of cultural relativism" *Journal for the Theory of Social Behavior*, ISSN 1468-5914, 19(4): 433-441.

Gaver, W.W. (1993). "What in the world do we hear? An ecological approach to auditory event perception" *Ecological Psychology*, ISSN 1532-6969, 5(1): 1-29.

Gibson, J.J. (1966). *The Senses Considered as Perceptual Systems*, ISBN 9780313239618.

Gibson, J.J. (1977). "The theory of affordances," in R. Shaw and J. Bransford (eds.), *Perceiving, Acting and Knowing*, ISBN 9780470990148.

Gibson, J.J. (1986 [1979]). *The Ecological Approach to Visual Perception*, ISBN 9780898599596.

Gilbert, S. and Sarkar, S. (2000). "Embracing complexity," *Developmental Dynamics*, ISSN 1097-0177, 219: 1-9.

Gombrich, E.H. (1999). *The Image and the Eye: Further Studies in the Psychology of Pictorial Representation*, ISBN 9780714822457.

Jonsson, E. (2002). *Inner Navigation: Why We Get Lost in the World and How We Find Our Way*, ISBN 9780743222068.

Kelso, A.J.S. (2008). "An essay on understanding the mind" *Ecological Psychology*, ISSN 1532-6969, 20(2): 180-208.

Letiche, H., Lissack, M. and Schultz, R. (forthcoming)

McArthur, L.Z. and Baron, R.M. (1983). "Towards an ecological theory of social perception" *Psychological Review*, ISSN 1939-1471, 90: 215-238.

McGrenere, J. and Ho, W. (2000). "Affordances: Clarifying and evolving a concept" *Graphic Interface Proceedings 2000*, ISBN 9781568811901.

Norman, D (1988). *The Design of Everyday Things*, ISBN 9780465067107.

Reed, E. (1996). *Encountering the World: Toward an Ecological Psychology*, ISBN 9780195073010.

Ricoeur, P. (1992). *Oneself as Another*, ISBN 9780226713298.

Sanders, J.T. (1997). "An ontology of affordances," *Ecological Psychology*, ISSN 1532-6969, 9(1): 97-112.

Vygotsky, L.S. (1962 [1934]). *Thought and Language*, ISBN 9780262720106 (1986).

Weick, K. (1995). *Sensemaking in Organizations*, ISBN 9780803971776.

Zakorik, P. and R.L. Jenison (1998). "Presence as being-in-the-world," *Presence: Teleoperators and Virtual Environments*, ISSN 1531-3263, 7(1): 78-89.

Hugo Letiche is Professor of 'Meaning in Organization' at the Universiteit voor Humanistiek, Utrecht (NL), where he is the Director of the part-time PhD program. He has been part-time Professor at Lancaster University, Keele University and Bristol Business School, and visiting Professor at the University of California at Berkeley. His recent writings include *Making Healthcare Care* (Charlotte NC: IAP, 2008) and *Social Complexity Theory* (forthcoming). Recent research has focused on normative professionalism, social ecological systems theory versus social complexity theory, and the relevance of practitioner investigation. Recent articles have been on post-phenomenology (*Culture & Organization*), polyphony (*Organization Studies*) and art/creativity (*Society & Business Review*).

Michael Lissack is the director of the Institute for the Study of Coherence and Emergence (ISCE) and an Internet entrepreneur. Most recently he lectured on business and public policy at the Central European University in Budapest. Dr. Lissack is also the founder and chairman of Knowledge Ventures Inc. (an educational software and search company) and Market2Buyers.com (a real estate broker software company). From 1999 to 2004, Dr. Lissack served as the editor-in-chief of *Emergence: A Journal of Complexity Issues in Organizations and Management—E:CO*'s predecessor.

23. The Dark Side of Organizational Knowing

Hugo Letiche
University for Humanistics, HOL

Rigorous investigation of organizational epistemology, or what can an organization know and why, is a sadly underdeveloped field. Knowledge management as a field has suffered from naïve assumptions about what knowledge is and how it can(not) be shared. David Seidl in E:CO (2007) made a significant contribution to organizational epistemology, which I want to further problematize. Seidl made two assumptions: one ontological namely that organizations know things; and one epistemological namely that knowledge can be defined as perceptual complexity reduction. I wish to counter that persons and not organizations know things and that knowledge is more social than perceptual. I will argue that the problem of social knowing is not so much grounded in the epistemological question of knowledge / nonknowledge—that is, in the relations of foreground and background, facts and assumptions or knowledge and hermeneutics, as in the much more radical circularities of eternal return (duration) and the continual (re-)founding of social order. I will be inspired for the first point by Pierre Klossowski and for the second by Michel Serres.

Introduction

Social complexity theory assumes that the interactions, boundaries and limits to knowledge, power and organization are interrelated, nonlinear and emergent. Knowing is situated, circumstantial and context-driven. Acknowledgement of complexity leads a reduction in confidence in knowledge. By denying or repressing complexity, one can seemingly boost 'confidence' in knowing. The status of knowledge can be maintained by making the nitty-gritty of the processes of knowing invisible. If ontological, epistemological and hermeneutic assumptions are hidden, knowing can appear to be automatic, self-evident and faultless. David Seidl in *E:CO* (2007) argued for just such a process of complexity reduction in order to facilitate organizational knowing and knowledge management. I wish here to argue against it.

For David Seidl the dark side to knowledge is what in deconstruction goes by the name of *aporia* (Derrida, 1998). Perception or knowing is only possible thanks to an unseen means of observation. Either a person sees with her or his eyes or the person looks at her or his eyes (for instance, in a mirror)—but one cannot see (for instance, the landscape) and see the instrument of seeing (i.e., the eyes) at the same time. The means of perception have to be unseen in order for the seeing to occur. Likewise knowledge requires assumptions. To know requires that key assumptions remain hidden. A well-know practical illustration of this theme is the observation that one cannot all at once conceptualize riding a bicycle (for instance, the rules of physics involved), and actually ride a bicycle. Performing an activity requires hidden assumptions. Seidl calls this epistemological dilemma the 'dark side of knowing'. Knowledge exists thanks to its nonknowledge. For an empiricist, who does not want to acknowledge the hermeneutic nature of knowing and/or perception's inherent dependence on means of perception, the focus on the aporia is epistemologically challenging. And Seidl's illustrative paradoxes are deeply engaging—knowledge demands the ability to address nonknowledge, growth in knowledge always produces more nonknowledge, higher levels of awareness or metaknowledge simply produce higher levels or more abstract types of conceptual blind spots. But by putting his analysis on almost purely epistemological level, Seidl does not address the relationships between power and knowledge or the social nature of knowing. And I think that the dark side of knowing is at least as much a social issue as an epistemological one.

Merging the conceptual worlds of knowledge (epistemology) and power (organization) often in the name of Foucault (McKinlay & Starkey, 1998, Knights, 2002) has become commonplace. *Emperie* [empire] and *emperique* [empirical] (both Old French) are entangled in what I wish to call *emperi(qu)e* or following the etymological line to English: empire + empirical = *empir(e)ical*. *Emperi(qu)e / empir(e)ical* points to an asserted isomorphism, parallelism and/or dependency between, on the one hand economic and political rule and on the other hand, science and technology. *Emperi(qu)e / empir(e)ical* assumes that rules / law, truth / science, and methods / generalization are more important than the individual process of thinking or than experiential awareness.

Looking to etymology, *emperie* is old French {Cil doi seignor qui l'emperie government, Quant il i evident les vertuz si apertes, Il le receivent, sil portent e sil servent: ... *La vie de saint Alexis* mid- 11th century} and refers to that which is imperial, governed by an Emperor and/or considers itself heritor to the Roman empire (L. imperium). With the industrial revolution, the term was broadened to include colonial and commercial 'empires' and in the late 20th century its use became increasingly metaphorical. *Empirique* (empirical) has a dual derivation: it comes from the Greek word from which (via L. *experientia*) the word 'experience' is derived, and also from a Greek/Roman concept of medical knowledge. In the text of Celus (Roman, 25 BCE—50 CE) one finds the concept of the *empirici* who were a sect of medical physicians who chose not to think about the (possible) causes of sickness but to discover medicinal substances. They foreswore the study of anatomy and physiology to focus on observing illnesses and cures. The *empirici* were the forerunners of contemporary biomedicine, producing pharmacological knowledge blind to existential or experiential issues and in the main ignoring illness-experiences. In the 17th and 18th century *empirici* was a name in medicine and literary studies for quacksalvers and uninformed or insensitive readers. But in the same centuries the word gained its current positive status with the philosophies of knowledge and science of John Locke and David Hume.

Empiri(qu)e / empir(e)ical entails a political truth regime where 'matters of fact' are asserted to be truths inductively derived from experience. But as Hume already asserted and Seidl repeats, the principles of inductive reasoning cannot consistently be inductively derived because one needs the self same principles to start up any process of derivation. But Hume unlike Seidl asserts that infinite regress can only be blocked by assumptions, procedures or propositions that are maintained because of habit, custom, prejudice and/or power, i.e., via induction (empirical) + power (empire). *Emperi(qu)e / empir(e)ical* is of course here a play on words asserting that organization or the structures of power is complementary to the organized or generalizable knowledge of 'reality'.

Seidl, following Luhmann, asserts that organizations are mechanisms for complexity reduction and that their reduction principles, like the aporia in all knowledge, have to remain at least partially hidden in order to work. Complexity "can be conceptualized as a condition where one possesses more possibilities than can be actualized ... complexity thus means pressure to select, [where] in order to insure its continued reproduction [i.e., existence], [the] system has to reduce its possibilities; that is *reduce its complexity*" (Seidl, 2007: 21). But exactly who or what is the *one* who possesses and/or has to reduce the complexity? And is the reduction normatively neutral? Seidl chooses not to follow Luhmann and to differentiate between cognitively and normatively stabilized structures (Seidl, 2007: 22-23). Luhmann identifies knowledge with cognitively stabilized structures—i.e., with what in science is self-evidently accepted, such as 'gravity', and must fall by the wayside in the event of counter theoretical observations—i.e., upward 'falling' objects. Normatively stabilized structures (like the law) are not invalidated by counter examples (i.e., crime). I see little evidence in organizational studies that organizations are deeply cognitively stabilized—i.e., self-evidently known, described and theorized in a particular manner, and I believe that their form of normative stabilization is unsteady, temporary and often contested.

Pierre Klossowski and Michel Serres have studied human normative action from a process perspective. They have prioritized thinking above thought and rejected the reification of order. Klossowski and Serres's position implies that order / organization and knowledge / observation should not be collapsed into a single term. Thinking and organizing are on-going processes of volatility, which are complex, unstable and irreducible, and at most temporarily and fragilely stabilized. Their complexity entails the ever-repeating self-regeneration of emergent occurrence. Systems or series without end, where variance follows strange attractors, and sense-making is without resolution, are indeterminate and complex.

Seidl and Klossowski/Serres agree on an idea-centered concept of research; research needs to be about ideas or knowledge. Description grounded in the repetition of the same assumptions is not likely to add to knowledge. Serres and Klossowski contrast the individuality, particularity and immediacy of thinking, to the reified, structured and institutionalized nature of organization. For them, creative thought is emergent—i.e., thinking and innovating (re-)happen over and over again. If one identifies thinking with becoming—that is with change, alteration and emergence; one ends up equating thought with the *Being of becoming*. This is similar to trying to make nonknowledge knowable or to inducing induction. Thinking requires the circularity of re-beginnings and the dynamism of changing once again. As long as there is life, thinking starts in each moment once again. In the repetition of the (re-)beginning, thinking is circular. Klossowski and Serres identify complexity with this circularity—that is, with continuous motion, recommencement and repetition.

Klossowski and Serres assert that thought operates via the *eternal return* of questioning, doubting and wondering. Thinking has to break through the political and scientific artifices of method and truth, reification and repetition, to exist. For Klossowski, what counts in thinking is the moment of questioning and not the abstract conclusions. Putting thinking before the power / knowledge episteme has a radical effect on how *organization* can or ought to be thought about. Applied social science is itself a form of organization and organizational studies is a form of applied social science. They portray research that does not aspire to power + knowledge as primitive, unscientific, archaic and irrelevant. Exclusion is accomplished in the name of science. Planned change, the technological mindset and economic rationality de-legitimize free thought, associative exploration and normative subjectivity.

Lifeworld, poetic, ethnographic and phenomenological research, which is not-*empiri(qu)e / empir(e)ical*, has continually to be reinvented or it ceases to exist. The repetition of recognized theories and proven 'truths' involves little or no thinking, but merely the copying, reproducing or parroting of what has already been asserted. Repetition of facts, theories, and categories only requires duplication. The passion and enthusiasm needed for thought and for understanding is a different matter. For thought to be alive, ideas and texts have to be experienced. All ideas, from whatever origins or period, can be thought here and now, and as alive. If thoughts are questioned, reflected upon and investigated, then they are alive. All real questioning lives. Thinkers and thought live perennially—the time of thought is circular. Thought has no beginning or end. Thinking revolves around questioning and inveighing. Each time an idea is thought—i.e., experienced, questioned and pursued—it (re-)lives.

For Klossowski, the choice for thinking has perpetually to be re-made—it cannot be embedded, grounded, stabilized or fixed. Organizations deny thinking as a process by creating principles, check-lists, methods and techniques. Klossowski asserts that lived-thought is actually a *vicious circle* of unending becomings, which entail over and over again the choice for reflection, discussion and engagement. Because the choice to think has continuously to be made and re-made, it is at once always on-going and always repeating. But organization is another matter. Michel Serres asserts that *organization* or institutionalized social order is a product of violent social impositions of rules, order and restrictions. He asserts that social order may be desperately needed, but comes at a fairly high price.

To develop these ideas further, I will analyze two texts—Klossowski's *Nietzsche's Vicious Circle* and Serres's *Rome*. Klossowski argues that the eternal return of the present, or the always arriving of yet another instance of the 'now', demands that we question the value of purpose, intentionality and order as epistemological assumptions. Serres asserts that *organization* and especially the foundation-ary acts of institutionalized social order are sustained via scapegoat-ing and violence (Soila-Wadman, 2008). The dark side of thought is formed by repetition (Klossowski) and social violence (Serres)—both of which are socially ever-present in acts of knowing. I choose Klossowski and Serres as my sources because they are two of the most radical philosophers of the dark side of the social epistemology of knowing.

Living Thought

Klossowski's *Nietzsche and the Vicious Circle (1998)—i.e., Nietzsche et le cercle vicieux (1968)*—contains an epistemological argument that attacks 'truth' in the name of living particularity. In the book, what is Klossowski and what is Nietszche, and what is Klossowski playing Nietzsche, remains unclear; but this is not very important for this writing. Klossowski's book asserts that 'truth' in the social or natural sciences is episteme-ically bound to repetition and generalization. There is, thus, an epistemological divide between the specificity of existence and the regime of (scientific) 'truth'. Klossowski's Nietzsche identifies social control and conformity with the epistemology of empiricism, where truths of the external, impersonal or lifeless are asserted.

Science supposedly exists to restrict thought, behavior and to ensure security. The force of singular existence is made irrelevant by prioritizing the general. Lived existence and authentic consciousness are never general—they are always individual and specific. By prioritizing an abstract simulacra of consciousness above actual lived consciousness, science contributes to the mastery of fiction over life. Lived experience is partial, restricted and specific. Science is general and its truths are a-temporal. Scientific theory produces a mythic coherence—images of general, permanent and stable truth. But lived existence only knows the rise and fall of intensities and discontinuities of fluctuation in energy, force and power. In science, the force of actual existence is smothered with generalizations of permanence and truth. The living is pushed aside for the generalizable. But a generalizable consciousness or existence does not and cannot exist. There is only specific becoming and occurrence. In the substitution of being for becoming, abstract truth or permanence replaces living relationship, event and circumstance.

The substitution of the dead-letter for the living context occurs via the semiotics of the text. Language substitutes generalized terminology for specific circumstances. The lived world is defined out of existence—it is replaced by known and predetermined terminology. Naming the actual, specific, and living becomes impossible when the semantics of the naming process only allows for the repetition of already circumscribed terminologies. Language then speaks the speaker and not the other way around. The code dominates. Klossowski asserts that science with its assumptions and vocabulary de-individuates and standardizes consciousness. This is a semiotic argument—the naming process as we know it makes direct contact with the complexity of life-force impossible. Languaging destroys becoming by reducing emergence to preconceptions, whereby standard postulates and clichés predominate.

Klossowski explores how Nietzsche was personally physically confronted by this semiotic crisis. Recurrent severe migraine headaches brought Nietzsche to feel that his consciousness and body were in conflict with one another. But rather than taking consciousness' side and demanding: *Body why have you betrayed me?*, Nietzsche took the body's side and demanded: *Mind why are you with such destructive results, repressing my living substance?"* Normally the brain dominates the body, whereby embodiment is sacrificed to abstraction. This insures that the continuity of identity is maintained. Self and identity apparently stay the same, while the body constantly changes. The conceptualized or idealized self appears to be consistent, while the body is re-corporealized over and over again. Consciousness identifies itself with a code of signs embraced as 'self'. The excitations of concrete existence disappear in the abstractions of identity, self and meaning. The body as a concrete locus of impulses, comes and goes, changes and adapts, but the 'self' endures. Actual experience or the awareness of the present only lasts a few seconds. But the 'self' is abstracted into a simulacra of existence lasting +/- seventy years or more.

For Klossowski/Nietzsche, on the one hand, there is the possibility of identification with concrete living consciousness, and on the other hand there is the choice to identify with the abstractions of consciousness. The latter focuses on truth, meaning and intentionality; the former is nigh on unchartered. The latter entails complexity reduction, the former reflection on complexity. Complexity is grounded in the body. Aliveness most of the time is captured in systems of signs, wherein signifiers of self, identity, and mankind replaces becoming with simulacra. Instead of constantly altering intensities and flux, there is the 'truth' of complexity reduction. Impulses and multiplicity are thwarted; ersatz continuity, permanence and meaning are made to prevail. The succession of discontinuous states is sacrificed to the appearances of intentionality, order and purposefulness. Consciousness, as a verbal, languaged, and intentional construct is split off from the dynamic force of change and innovation.

In what sort of duration does existence exist? Kossowski emphasizes the continuous recurrence of the now—time is characterized by the endless repetition of the present. In the endless becoming of the present there is eternal recurrence. The 'now' is both absolute and it is characterized by total contingency. Time is "a condition of possibility" and "an actuality and a necessity" (Lukacher, 1998: 101). Temporality is profoundly paradoxical—it is pure chaos, potential and possibility, and it is concrete, real and substantial. Time establishes the eternal return of unending difference(s). Time or the 'now' is characterized by a succession of discontinuous states of becoming. But in Western thought, language and logic, 'Being' is prioritized and contact to becoming is closed off. Consciousness is lost in signs with no openings to thinking's processes. Klossowski's assertions, often made in Nietzsche's name, are those of a French semiotician. Nietz-sche certainly was out to dismantle the mentalist automaton that locks up becoming in a metaphysics of Being. Human will—seen as plans, concepts and strategies—supposedly is hidden or reified in its 'intentions'. Only if the self abandons intentionality can aliveness prevail above identity. Nietzsche wanted to 're-will' the self—that is, to assert fluctuation and intensity. The problem is that when intensity doubles back on itself and represents itself, it looses itself. The sign representing intensity replaces intensity with a reification or fetish. Real flux has no beginning or end and cannot be possessed, represented or slowed down. The claim that the basic stability required to establish text and language denies the complexity of aliveness, is very much in the critical French semiotic tradition.

Klossowski/Nietzsche asserts that in (most) text that the falsely permanent 'self' is prioritized and that the eternal circle of the now is discarded. Nietzsche attacks this strategy as an anti-existential weakness. It is an existential strategy wherein the general and ordered prevails above the specific and living. It is a solution of mediocrity wherein averages are more important than circumstances. Impulses and forces are sacrificed to results and norms. Aliveness disappears behind statistics, and impulses are smothered by the forms they produce. Abstractions—such as "mankind" or "humanity"—take precedence above living existence and concrete circumstance. The "non-sense" of self, identity and purpose, overwhelm the "sense" of lived awareness. Complexity is thwarted.

The eternal return of complexity is a becoming with no goal or purpose. It is a totally radical principle of emergence, which is utterly at odds with Western epistemological practices. Klossowski/Nietzsche asserts that all there is—is the rise and fall of intensities—or a continual disruption of equilibrium(s). Energy and intensity do not seek to endure—they just increase and decrease. Life is an intensity without telos. The ascending and descending movements of energy deny equilibrium—energy is its own goal. There is no beginning or end to the continual processes of recombination; energy disorganizes whatever it creates. Organization in the sense of willed and controlled social order is identified by Nietzsche with the oppressive regime of mediocrity. Signs destroy aliveness and false sociability kills authenticity. Nietzsche calls this principle of oppression, "gregariousness." Communication, relationship and sociability overwhelm authenticity, becoming and energy. Becoming, emergence and duration, or the acknowledgement of the eternal repetition of the 'now' are conceptualized in Klossowski/Nietzsche in terms of *being* versus *becoming*, truth versus process, intention versus thinking. The radical embrace of thinking or of consciousness versus doxa, or of cues versus codes, is crucial to thinking or knowing. While Klossowski/Nietzsche clarify what the becoming of thinking looks like, or what epistemological complexity is all about, they do not help us to get closer to organizing as activity or event.

Living Organization

Can organizing be understood as something more or different from rationalist repression or existential denial? Michel Serres in his book *Rome* (1991) sets out to address just this question[1]. Serres's logic is poetic—the leitmotifs of back versus white, defined versus emergent, and violent versus interactive, are repeated over and over again (Brown, 2005). But with each repetition, the relationships shift and regroup. While the argument of Klossowski/Nietzsche is fixed in semiotic conflict, Serres's thought moves back and forth, reinterpreting its key metaphors. His terms act as shifters, i.e., are open to multiple meanings and interpretations. The terminology keeps adapting—there is consistency in the processes of thought but not in the signifiers. Truths or meanings are not what count. Motion is crucial. Klossowksi/Nietzsche assert ideas as if they were objects. Language is 'this', 'self' is that; "gregariousness" is identified as one thing and intentionality as something else. Klossowski/Nietsche argues for emergence, change and inconsistency, but the text remains entirely consistent. The definitions, style and assumptions are unswerving. Ideas are reified—key concepts are stable and seemly possess powerful moral significance.

Serres puts flux into his signifiers. Multiplicity and the (semi-)indeterminacy of the concepts manifest themselves in shifting meanings. The poetics shift and double-back on themselves, permitting practices of *différance* to man-

1. Serres abhors traditional academic debate, which he sees as more oppositional than innovative. He seeks links, connections and interactions, and never writes criticisms of others whose positions are close to his own. Thus there are (almost) no texts by Serres on Deleuze, Foucault or Nietzsche.

ifest themselves. Serres addresses many of the same issues as Klossowski/Nietzsche—that is the relationship between definition, language and repression; the repetition of the 'now'; the role of flow and complexity. *Rome* is explicitly an analysis of the city's foundational myth, centering on Livy's history of Rome and Romulus and Remus. The question is: *What is entailed in the naissance (birth) of (social or human) order*?

To grasp Serres's reflection(s), it is necessary to examine his poetry, wherein the contrast between black and white forms a key element. As I've noted, it is performatively self-contradictory to champion impulse, energy and difference with repetitive or assertive argumentation. Variability and emergence demand language as shifters, *différance* and semantic instability. Serres's use of black and white in *Rome,* meets these demands. Early in the book {+/- page 35-75} 'white' is identified with blankness, emptiness and nothingness. White is nonexistent and dead. Substance, actuality, and circumstance are tangible and embodied. Stuff or matter is earthy and physical, i.e., black. Theory is ethereal and empty of hyle (material). Objects are enclosed in their substance and thereby black. Theories seek light and transparency. But experience, circumstance, and world are substantial and specific. Activity requires getting one's hands dirty. Blackness absorbs everything—it is the collection of all elements, dimensions and aspects. In the involvement of events and engagement, complexity and multiplicity, there is confusion, doubt and indeterminacy. The imagery of the tangible focuses on the world in all its concreteness. Light alone is unlivable—pure possibility is unthinkable. Pure meaning is totally restrictive and inhumanly absolute. Greece and Jerusalem are symbols of the will to light and to truth; Rome was alive, muddied, unreflective and substantial.

Later in the same book {+/- page 100}, Serres reverses the imagery in his discussion of the 'black box'. Specified inputs and outputs undergo a transformation whose process(es) remain indeterminate. The black box represents "the ensemble of passages from the multiple to the single and from the single to the multiple" (Serres, 1991: 98). What happens in the black box is blank, unknown and indeterminate; it is fragmentary, multiple and not explained. In the black box there is movement, circulation and event. The black box represents the logic of *systrophe*—literally the 'turning together'. In the *systrophe* many qualities, details and elements coexist without providing explicit definition(s). The *systrophe* flows from the many, or multiple elements of description to the single 'self' or identity and back again. It is an ebb and flow of meanings, constituent qualities, and specificities. The black is no longer concrete, earthly and substantial—it is transformative, more potential than concrete, and indeterminate.

Serres goes on to describe the dynamics of the black box in terms of 'quasi object's. Quasi objects facilitate social interaction by making circulation possible (Brown, 2002; Carr & Downs, 2005). Human groups can exist in so far as materials, meanings and actions circulate. Interaction is grounded in flow, exchange and distribution. Everything from building materials to rugby balls, money to religious icons functions as a quasi object. Quasi objects create relationships by changing hands, being shared and getting utilized. Their significance depends on things being done with and to them. As long as they enter into action and become process and event, they realize their potential. Quasi objects are defined in terms of the circumstances and actions that they generate. A quasi object is not an object in-and-of-itself; it is an object via circulation, activity and circumstance. Quasi objects tie persons together into groups—the quasi object has identity only in so far as it helps to form a group. A rugby ball is only a rugby ball in so far as it facilitates that rugby is played—the identity exists in the doing. If quasi objects collapse in their ability to marshal activity, they loose their meaning. For instance, money is valuable only as long as it is exchanged. If a society comes to mistrust its quasi objects, then that society collapses. Social existence can only exist in so far as it retains confidence in its quasi objects. It is by engaging with the quasi objects that loose elements of possibility and partial significance are smelted together into event, society and structure. Quasi objects are in the social black box. They make it possible to bring highly disparate elements together into processes, activities and actions. They are the glue that holds all the multiplicity and difference together without repressing or destroying it. The black box brings together but does not destroy. It organizes but does not (necessarily) terrorize. Serres initially used black to point to the material, concrete and earthly versus the idealized, conceptual and rational; but in the second instance, he made use of black to point to in-determinant processes of structuration and social activity.

In the next reversal, Serres redeploys {+/- pages 175-200} the metaphor of the box. In place of the black box and the quasi object, he introduces the distinction between boxes and sacks. The box looses all connotations of process, alteration and change, and becomes a metaphor for rigidity and fixed relationships. Boxes have hard edges and point to circumscribed relationships. Sacks are characterized by folds and flexibility. Sacks define fluid margins, heterogeneous spaces and unexpected proximities. A logic of burlap sacks would (metaphorically) be one of folding and flexibility.

Serres goes on to develop a concept of complexity in juxtaposition to the concept of the 'excluded third'. At issue is the assertion that *'X or not X' is true for any statement 'X'*—that is, any statement is either true or not true and there is no third possibility. Serres asserts that a quasi object is always open to multiple interpretations—a rugby ball is blown up cow hide, a potential source of conflict and a possibility for unity, identity and cooperation. The law of the excluded third contradicts the logic of social interaction, human event and the possibility of multiplicity. If one applies the law of the excluded middle to society, one gets violence, repression and totalitarianism. In so far as democratic or peaceful society is possible, it must deny the law of the excluded middle. This is why, according to Serres, that those societies that are the most rational and strive the strongest for truth; collapse in violence and anarchy (Hagemeijer, 2005). Greece and Jerusalem prioritized truth, clarity of purpose and knowledge, which led to their destruction. Complexity and truth are incommensurable—Rome accepted multiplicity and difference, and lived with contradiction.

In open thought, there is always another position, an alternative option and a strange or indeterminate attractor. The messiness of events, circumstances, and emergence is characterized by instability, vagueness and multiplicity. The will to define, clarify, and prescribe rejects multiplicity. Contemporary organization is defined in terms of complexity reduction, system order and goals. Rome was defined in terms of mixtures, flows and motion. Modern organization abhors fluidity, indistinctness and indeterminacy. Companies and governments want rules, distinctions and clarity. They determine boundaries between self and other, the desirable and excluded, the owned and not owned. Modern organization is all about boundary setting—about asserting the excluded middle or denying the third. But quasi objects are inclusive and horizontal—people and circumstances are drawn in. Their logic runs in the opposite direction to that of the excluded third.

Serres is convinced that a society based on rational clarity, purposeful systems, and social exclusion is inherently and perpetually violent. It exists by defining over and over again its boundaries of inclusion and exclusion. It continually legislates what is permitted and what is not permitted, which goals are to be pursued and which not, what is desirable and what is forbidden. Each definition creates yet another line of demarcation and principle of exclusion. Nonstandard multiplicity is an anathema to rationally planned organization. The excluded third leads to stability, repetition and structure. The fluctuating disorder of pure multitude is inhibited and denied by rational organization. Quasi objects do, undo, and redo order. One can map the movements of quasi objects; one can narrate their processes. One cannot legislate free interaction; genuine variability cannot be strategically managed. If one embraces the play of quasi objects, then one recognizes that flows of interaction will evolve and change, transmute and morph.

Events of quasi objects create interaction and flows of change. But modern organization and society are not fluid or participative. Organization determines boundaries and exclusion. Systems define how production is to proceed and what is to be done. The free flow of possibility is repressed. Activity is directed, structured and not free. Serres argues that all of this is achieved via 'scape-goating'. Social order can only be bought at the price of destroying the excluded middle—the third. Organization is an act of social violence and exclusion. It demands the destruction of multiple, complex and emergent. Groups exist by denying the random multiplicity and radical possibility of existence. Only by establishing the principle of 'this and only this' is organization made possible. All that is possible in the variety of flows and impulses has to be denied. This denial does not come gently. Difference is eliminated violently. According to Serres, what the myth of Romulus and Remus illustrates is the violence of the foundation of order.

Organization is grounded in the rejection of heterogeneity. Life's multiplicity is denied; specific definable objectives are prioritized above everything else. Possibility is murdered, complexity is destroyed and diversity is bridled. The choice not taken or the option not honored is honed out of existence. The humanity not followed is violently denigrated. The dominant 'this is' can support no alternatives, options or indeterminacy. The choice not chosen or the other that is not 'us' is ruthlessly killed off. In this sense, all organization is grounded in genocide—that is in the slaughter of some form of humanity. The principle(s) of inclusion and exclusion, differentiation and boundary setting, are investigated by Serres in *Rome* via the image of the one brother who survives and the other who is killed. One option lives and is prioritized, the other option dies and is repressed. Organization is grounded in the violence of the destruction of human possibility and the multiplicity denied.

Serres asserts that only in so far as the play of quasi objects can be accepted is sociability possible without continual foundational violence. If event and interaction are strong, people will need one another and will be so positively preoccupied that they do not need to kill. But as soon as the quasi objects fall still or loose their credibility, the process of goal-setting and defining, planning and restricting will recommence and scape-goating restarts. Momentary eventfulness and cooperative activity is possible; but the scape-goating needed to maintain the excluded third or middle is always looming just around the corner. Women who do not conform are burned as witches or stoned-to-death, homosexuals face violence, discrimination and ghetto-ization, etcetera. Organization requires foundational action(s). The primary assertion of 'this and not that'—the basic boundary setting inherent to the organization has to be maintained. Doctors who do not know their patients take decisions that nurses can better oversee, CEOs take product decisions that for instance only R and D really understands, politicians pronounce on matters that they do not oversee. Those in authority often refuse to acknowledge the expertise or value of others.

The act of exclusion, violence and repression is foundational. For order to be maintained, the choices for restriction, boundary and structure have to endlessly be repeated. The foundational act of social order defines the fundamental vicious circle of human society. Society is never permanently started—organization has to be continually re-started, re-done and re-enacted. The time of organization is temporary—initiatory boundaries and closures have to be repeated in all their violence, over and over again. Organization, once commenced, holds for a period. Difference is eliminated and order is asserted. But sooner or later, the power of the foundational violence looses its hold. The restrictions and borders stop to function and complexity, multiplicity and indeterminacy reappear.

Such moments of possibility are scarce and normally are re-repressed as soon as order is (re-)asserted. Organization is (re-)established for just as long as the new 'choices' or 'order' can be sustained.

The fundamental vicious circle of human society according to Serres is grounded in the destruction of complexity, multiplicity and human possibility. Scape-goating, or the violent rejection of some human possibilities in order to validate the 'truth' being adhered to, prevails. In the ex-Yugoslavia and the Caucasus, who does and does not control the state and have economic power has literally been fought out with the loss of life. The options not honored and the possibilities rejected are dismissed, abused and cursed in every possible way. Those in power label the 'road not taken' as negatively as they can.

Moments of quasi object activity define a liminal space between all the human violence, wherein constructive interaction can occur. Disgusted and tired, fearful and anxious humanity sticks to the results of the last scape-goating for as long as it can. Recreating organization is painful and violent, dangerous and threatening. The vicious circle will come round again and order will have to be re-started. The violence done to existence in the re-starting of the cycle of organization makes recommencement something to be dreaded.

Conclusion: Thinking, Organizing and Complexity

Organization and knowing have continuously to be (re-)established and (re-)founded. Organization entails repetition, renewal and recommencing. It is a 'vicious circle' in the sense that it entails continuous repetition, of the 'now' of the (re-)beginning, and of the recommencing. The dark side to organizing and knowing entails the human and epistemic violence of complexity reduction:

> *Organizations create their knowledge in the form of decision premises, which make decision-making possible by limiting the number of possible decisions. In other words, organizational knowledge makes observations possible, but at the same time it implies nonknowledge, which means unobservability. Regarding organizational observations, this means that they are at least as much based on nonknowledge as they are on knowledge. Because of the specific mode of reproduction, which can be referred to as uncertainty absorption or equivocality reduction, the demarcation between knowledge and nonknowledge, or observability and unobservability, is particularly strict* (Seidl, 2007: 24).

I obviously agree with almost all of this essentially epistemological argument, but assess its social and human import very differently than Seidl does. Organization is indeed characterized by complexity reduction or possibility elimination. But I refuse to call the repression of awareness and/or the banishment of the otherness of the other, 'knowledge'.

Organizational knowledge may indeed be the nonknowledge of human awareness or sociability. In the eternal return one instant follows upon the other in a succession of recommencements where existence emerges over and over again. But I must not fall into the paradoxical vicious circle of trying to define the Being of becoming. Serres avoids self-contradiction by making the vicious circle rigorous. Organization re-begins again and again. But each recommencement implies a moment of scape-goating wherein multiplicity is sacrificed to order. Organizations hold their participants in their grip, grounded in the death of chaos, possibility and aliveness, attached to Organization Klossowski/Nietzsche assert is life-denying in its repression of intensity, flow and emergence. And Serres's social logic of quasi objects only offers partial and temporary moments of respite. All the assumptions and rules needed for quasi objects to occur require social organization—that is, foundational acts of organizing. Organizing and organization cannot really be avoided.

The mania of current management practice(s) of re-initiating or re-starting organizations as often as possible via re-organizations, strategy development, mergers and acquisitions, etcetera, weakens the quasi objects. Current practice favors exclusionary violence. Scape-goating is being (re-)enacted in shorter and shorter cycles. Existence as thinking, creativity, and flow is in principle always possible. What is needed is the possibility to make links, favor interactions and to encourage relationships. But how to encourage multiplicity and flow, and to minimize violence and scape-goating, remains a little explored territory.

Complexity has been examined here as the continual restarting of thinking and (social) organization. Thinking and power, and ideas and social order, can all be collapsed into complexity reducing or denying hegemonious epistemes. But to do so one has to repress thinking's dynamism and emergence. One can argue that complexity reduction is the most efficient avenue to knowledge consensus, but complexity reduction requires the enforced inhibition of self-awareness. A simulacrum of agreement can be achieved by imposing assumptions, procedures and though patterns. But awareness of how the postulates are used has to be repressed, the role of theories left unexamined, and the mechanisms of perception ignored. The media of knowing has to be black-boxed so that 'the message is the message'. After forty years of an epistemology-in-use influenced by McLuhen (1967), wherein 'the medium is the message', can the processes of knowing really be ignored without coercion, bad faith or violence? Processes of knowing can be defined, prescribed and structured; but to do so one has to deny movement, change and the inherent instability of living relationships.

Thinking and social order share that they re-start, re-establish and recreate their activity constantly. In every moment of the 'now', thinking and social interaction are (potentially) recommencing. The complexity of thinking is unstable, troubling and makes the 'self' uncertain and indefinite. The complexity of the social is even more profoundly troubling. The relationships, exchanges and interactions of the social require boundary-setting and exclusion to exist; these are processes that victimize the excluded. Complexity describes the aliveness of thinking and society. But the cost to 'self', and in violence to humanity, of these processes has to be faced up to.

I disagree with Seidl's conclusion that:

> *...organizations are intelligent to the extent that they are able to re-enter nonknowledge into knowledge; in other words, to the extent that they are able to produce representations of their nonknowledge, which allows them to base their calculations on nonknowledge as if it were knowledge* (Seidl, 2007: 27).

Persons think and are conscious and live, not organizations. As Klossowski/Nietzsche insist, the ever-restarting process of complexity entails creativity and change, experiential duration and fragile, self-replicating cycles of order and disorder, often characterized by implicit violence, instability and tentativeness. Thinking and organizing entail repeated processes of motion and instability, whose flux and volatility need to be respected and cannot simply be controlled by 'complexity reduction'. Organizations as the holders or providers of quasi objects can make human social interaction possible. But I question how far organizations can 'represent' their own principles of exclusion and boundary-setting to themselves without destabilizing themselves. I know of no organization that openly acknowledges and deals with its own principles of inclusion and exclusion, partiality and limitation-setting, boundary imposition and restrictiveness. Only some are (de facto) permitted to buy the product, get the service, receive the promotion or profit from the success. The borders and confines of wealth, power and possibility form the determinate nonknowledge of organization, and this knowledge is not going to be opened up for participation, discussion or transparent negotiation.

References

Brown, S.D. (2002). "Michel Serres, science, translation and the logic of the parasite," *Theory Culture & Society*, ISSN 0263-2764, 19(3): 1-27.

Brown, S.D. (2005) "The theater of measurement: Michel Serres," *The Sociological Review*, ISSN 0038-0261, 53(1): 215-227.

Carr, A. and Downs, A. (2004). "Transitional and Quasi-objects in organizational studies," *Journal of Organizational Change Management*, ISSN 0953-4814, 17(4): 352-364.

Cook, S.A. (1928). *The Cambridge Ancient History*, ISBN 9780521044875.

Derrida, J. (1998). *Of Grammatology*, ISBN 9780801858307.

Haagemeijer, R. (2005). *The Unmasking of the Other*, ISBN 9058920976, http://publishing.eur.nl/ir/repub/asset/6963/EPS2005068ORG_9058920976_HAGEMEIJER.pdf.

Heidegger, M. (1991). *Nietzsche Volumes I & II*, ISBN 9780060638412.

Kellner, D. (2004). "Review Klossowski Nietzsche and the Vicious Circle & Waite Nietzche's Corpse/e," http://www.gseis.ucla.edu/faculty/kellner/Illumina%20Folder/kell23.htm.

Klossowski, P. (1968). *Nietzsche and the Vicious Circle*, ISBN 9780226443874.

Knights, D. (2002). "Writing organizational analysis into Foucault," *Organization*, ISSN 1350-5084, 9(4): 575-593.

Letiche, H. (2004). "Talk & Hermès," *Culture & Organization*, ISSN 1475-9551, 10(2): 143-161.

Lukacher, N. (1998). *Time-Fetishes*, ISBN 9780822322733.

McKinley, A. and Starkey, K. (eds.) (1998). *Foucault, Management and Organization Theory*, ISBN 9780803975477.

McLuhan, M. (1967). *The Medium is the Message*, ISBN 9781584230700 (2005).

Seidl, D. (2007). "The dark side of knowledge," *Emergence: Complexity & Organization*, ISSN 1521-3250, 9(3): 16-29.

Serres, M. (1991). *Rome: The Book of Found*ations, ISBN 9780804718677.

Serres, M. and Latour, B. (1995). Conversations on Science, Culture and Time, ISBN 9780472065486.

Serres, M. and Webb, D. (2003). "The science of relations: An Interview" Angelaki, ISSN 1084-8770, 8(2): 227-238.

Soila-Wasman (2008). "The role of scapegoats in organizations" paper presented at The Dark Side of Organization, June 25-27.

Walbank, F.B. (1984). The Cambridge Ancient History, ISBN 9780521234467.

White, D. (2003). "Review: Nietzsche and the Vicious Circle" The European Legacy, ISSN 1084-8770, 8(5): 635-639.

Zembylas, M. (2002). "Michel Serres: A troubadour for science, philosophy and education" Educational Philosophy & Theory, ISSN 0013-1857, 34(4): 477-502.

Hugo Letiche is Professor of 'Meaning in Organization' at the Universiteit voor Humanistiek, Utrecht (NL), where he is the Director of the part-time PhD program. He has been part-time Professor at Lancaster University, Keele University and Bristol Business School, and visiting Professor at the University of California at Berkeley. His recent writings include *Making Healthcare Care* (Charlotte NC: IAP, 2008) and *Social Complexity Theory* (forthcoming). Recent research has focused on normative professionalism, social ecological systems theory versus social complexity theory, and the relevance of practitioner investigation. Recent articles have been on post-phenomenology (*Culture & Organization*), polyphony (*Organization Studies*) and art/creativity (*Society & Business Review*).

Classical Papers Section

With a field so broad in both breadth and depth as is complexity thinking it is nigh on impossible to cover all our thematic bases in the development of well argued, robust and relevant understanding. In each issue of E:CO we include a Classical Papers section to help address some of the difficulties confronting not only complexity thinkers, but any thinker in modern times. These difficulties include:

Chinese Whispers: As past writings move even further into the past the author's ideas tend to be retranslated into the language of the time in which they are read. Often the intellectual landscape has evolved considerably; the political landscape has undoubtedly evolved, both of which prevent a true reading of the author's original text. Given the vastness of the material available, it is no surprise that many of us only know the works of previous authors through short quotations used by modern writers to support a particular point of view. The Chinese Whisper problem arises because quotes are always taken out of context in order to support the current author's position. If another author has only had access to the original author through quotations made by more recent authors it is no surprise that over time the meaning originally intended by the original author slowly shifts—sometimes to the point of not even being reconcilable with the original text. In politics this is all too often done in order to deceive, but in science it is an unfortunate by-product of managing the overwhelming complexity of past thoughts and writings (though of course it does also have a positive creative dimension). If you really want to know what a long gone author said on a particular subject then you really need to explore their original texts. Even then your modern cultural biases will prevent you from obtaining the original senses (researching the history surrounding the author often allows for a more genuine interpretation—though who has the time to perform such deep research into every author they take an interest in?)—you will at least remove the layers of reinterpretation and reframing imposed by many intermediate interpreters.

Reinventing the Wheel: With all that has gone before it is no surprise whatsoever that the wheel gets reinvented time and time again. Moving in and out of favor is part of the normal evolution of any idea, though each reinvention often changes the previous incarnation in subtle ways. My particular interest is philosophy, and the more I read the writings of past philosophers the more modern ideas I uncover that are supposedly original. It wouldn't surprise me at all if one could put together a modern text on complexity thinking using only Ancient Greek texts! Even with the best will in the world, and a never-ending budget, we can never take everything that has gone before into account. Very few new ideas are really original. Exploring past texts can reduce the effort expended in searching for novel solutions, but it can also drive the wheel of reinvention too. Claims of originality in the modern world where there is no basis for such claims seems to be an integral part of the Market driven economy. However, originality is not as important as one might think—which is rather fortunate given its rareness. An old idea brought to a new audience can be more valuable than a truly original idea. In absolute terms an old idea reframed for modern consumption really is a new idea, and so the notion of reinvention (like all notions) is problematic.

Empire Building: The development of any school of thought is often associated with a rush to differentiate itself from other schools of thought. This often means highlighting differences and subduing similarities. Empire building in science often leads to an 'ism', i.e., a collection of coherent thoughts and theories that seem self-contained and quite distinguishable from other 'isms'. Many complexity writers begin their treatises by expounding the shortcomings of so-called Newtonian thinking and developing a new way, Complexity thinking, built upon corrections to those prior limitations. The education theorist John Dewey noted exactly the same process in the development of progressive educational methods from traditional methods. Progressivism emerged in response to the perceived shortcomings of 'traditionalism', rather than the perceived needs of Education itself. Dewey says:

> *For in spite of itself any movement that thinks and acts in terms of an 'ism becomes so involved in reaction against other 'isms that it is unwittingly controlled by them. For it then forms its principles by reaction against them instead of by a comprehensive, constructive survey of actual needs, problems, and possibilities* (Dewey, 1938: 6).

I've lost count of the number of times I've read in one form or another that "linear thinking is dead—long live nonlinear thinking"—a rather pertinent illustration of reactionary-based theory development, which severely limits the possibility of realizing the full implications and extent of complexity.

An example of a literature that seems to have been arbitrarily (not necessarily intentionally) sidelined by the complexity community is that of *soft systems thinking* which evolved from the mathematically-focused General Systems Theory. Offspring literatures include *Critical Systems Heuristics*, *Critical Systems Thinking*, and *Boundary Critique* to name but a few. Through these different fields GST moved from the abstract world of mathematics (and some might say, un-appliable) to the concrete, though vague, world of useful application. Complexity Thinking has only just taken the first steps in the direction towards application. However, I think that in the effort to distinguish Complexity Thinking from General Systems Theory and its many outgrowths, Complexity Thinkers are severely limiting their subject's potential.

Of course *Empire Building*, like all seemingly negative concepts, has its more positive flip-side—it helps a community keep some degree of focus. This is essential if (partial) solutions to emerging problems/issues are to be found. The trick of course is to not reify one's Empire's boundaries in a politically motivated effort to differentiate and marginalize.

Historicity: Not all papers are created equal. Even though the obscurest of papers will have had some direct influence on the evolution of thought (although complexity places profound limitations on our ability to judge such influence objectively), some papers really do deserve the honor of being called *seminal*. Sometimes entire fields of endeavor can be traced back to a single paper (although many less known papers probably inspired the seminal one). Understanding the appearance and development of one's field of interest undoubtedly leads to a richer understanding of that particular field, which can open up avenues of exploration that might not have previously seemed legitimate.

In future volumes of *E:CO* we aim to reprint further classical papers. Each paper will be chosen to exemplify at least one if not all of the issues briefly discussed above, and will be introduced by whomever recommended the paper. We also aim to present the papers in their original formatting, rather than re-typesetting them. As such, the printing quality will not be up to our usual standards, but we think it is important to see the papers how they first appeared.

Kurt A. Richardson

Dewey, J. (1938). *Experience & Education*, New York, NY: Touchstone, ISBN 0684838281.

24. Thoughts on Organization Theory

Anatol Rapoport & William J. Horvath
(with an introduction by Jeffrey Goldstein)

Originally published as Rapoport A. and Horvath, W.J. (1959). "Thoughts on organization theory," *General Systems*, 4: 87-91. Reprinted with kind permission.

Rapoport's Career

This introduction will focus on the first author, Anatol Rapoport, because of his outstanding role in the development of systems sciences plus the fact that not much could be found out about William J. Horvath although he co-wrote with Rapoport on more than one occasion and was an esteemed scientist in his own right. Trained as a musician, mathematician, biologist, as well as psychologist, Anatol Rapoport was one of the most seminal pioneers in modern systems theory (see, Anatol Raport, nd; and Anatol Rapoport (*1911), n.d.). In that capacity, he cofounded the Society for General Systems Research at Stanford University in the mid nineteen fifties (I've found both '54 and '56 as the dates of the founding) along with other systems thinking luminaries at the time such as Ludwig von Bertalannfy, Ralph Gerard, and Kenneth Boulding. This society was later renamed The International Society for the Systems Sciences (ISSS) (Abraham, 2002; see also, Umpleby & Dent, 1999).

Born in 1911, Rapoport emigrated early from Russia to the United States, stopping along the way to study piano, composition, and conducting in Vienna. Eventually he settled at the University of Chicago where he received a Ph.D. under Nicholas Rashevsky who had almost single-handedly initiated the field of mathematical biology. Rashevsky had other students who went on to become preeminent in their fields, e.g., the very influential complexity-oriented biologist Robert Rosen.

Rapoport served during WWII in the US Army Air Corps. During his exceptionally long career, Rapoport was a professor at the University of Chicago, the University of Michigan, and the University of Toronto where he had gone to protest the involvement of the United States in the Vietnam War. He was also Director of the Institute for Advanced Studies in Vienna until 1983. He authored many books and over 500 research papers. His main areas of interest were in game theory, particularly regarding the latter's application in studies of conflict, cooperation, and peace studies, as well as highly original research into social networks, stochastic models of contagion, and mathematical models of parasitism and symbiosis. He had a truly illustrious career, receiving numerous awards and honorary doctorates, and was the president of and on the boards of numerous associations and journals. Like so many others of the early systems thinkers,

Rapoport was an authentic polymath, as at home in the abstract world of mathematical theorems, as in philosophy, social psychological experiments, computer simulations, music composition and conducting, as well as being totally engaged in the betterment of society and culture by promoting and initiating many vital currents in the study of peace.

Organized Complexity

Regarding Rapoport's and Horvath's paper in this Classic Paper Section of this issue of *E:CO*, it is first interesting to note that they quote Whitehead, in the latter role as a philosopher of science and metaphysician, in his call for new modes of thought which could break free of that type of analytical thinking circumscribed by placing physics as the prototype of how thinking ought to proceed. In particular, they followed Whitehead's case for an organismic approach in contrast to that type of analytical thinking which sought to grasp a "complexity" by examining its components alone, the idea being that because the parts were supposed to be simpler, an examination of these parts would more easily yield insight into the complexity. More specifically, Rapoport and Horvath point to that analytical method which believes it can comprehend a complexity by erecting a *superimposition* of the analytical knowledge of the parts, an explanatory strategy which has proved very fruitful in analyzing actual machines as superimposition of parts. By so doing, however, one has broken up the object under scrutiny into a temporal chain of causal events. Such a scheme works in celestial *mechanics* if perturbations to celestial orbits do indeed remain small enough. All of this implies, of course, that the whole is composed of an adding-up process—vector laws of addition, e.g., found in the theory of electromagnetism due to the linearity of the partial differential equations that Maxwell's formulation relied upon.

Rapoport and Horvath were quite prescient here about such *linear* assumptions. Because this additive approach had worked in those fields where linearity assumptions were all that was required, it was tempting to keep applying it to other "complexities." Again, they presaged, as an alternative, to the holistic claims of a more fertile "organismic" approach which would operate between the regions of "organized simplicity" and "chaotic complexity". Although agreeing with the tenor of such holistic approaches, Rapoport and Horvath pointed out how vague such holistic proposals tend to be. They turn to the use of the construct of an "organism" as an example and further probe into just what organismic thinkers have had particularly in mind when putting forth organisms as the holistic answer.

A more helpful approach, they contend, is to consider the differences between the biological and physical sciences, the former much more concerned about categorization and classifications, e.g., in taxonomies. One is reminded in this case of Walter Elssaser's essay as a Classical Paper in an earlier issue of

E:CO (Elsasser, 2006) in which the atoms studied by physics are all assumed to be identical and hence classification or taxonomy would be no use, whereas organisms as such are all different and hence taxonomies are not just a sound basis on which to start of build a theory, they are a necessary foundation for a viable biology. Furthermore, whereas the early biological observations were of the naturalist sort, physicists' observations depend on preconceived laboratory contexts. Nowadays, though, the biologist proceeds much like the physicist through experiments and quantification strategies as the pre-conceptual context of what then can be observed.

What Rapoport and Horvath wound-up suggesting by contrast was the incorporation of older biological dispositions which assume an "organized complexity" in the object of their study. They define an "organized complexity" as "a collection interconnected by a complex net of relations. It is neither simple in the sense of being amenable to linearization or addition, more specifically, a system that lacks "close loops in the causal chain." It is interesting to note that the first Macy Conference had closed feedback loops as its theme (Dupuy, 2000). Rapoport and Horvath again presciently point to the three body problem in celestial mechanics as what organized complexity is about, Poincaré, of course, having introduced many of the critical constructions of nonlinear dynamical systems in his study of such three body problems. But organized complexity is also to be distinguished from systems where the entities involved are vast, or ones relying on continuous gradations and thus require a statistical point of view, which indeed were among the innovations to the physicist's armamentarium at the end of the nineteenth and beginning of the twentieth century.

Then, trying to foresee what conceptual tools need to be added to provide a cogent approach to organized complexity, Rapoport and Horvath suggest teleology and taxonomy. The first suggestions may seem heretical given the fact that modern biology and the rest of modern science for that matter had been so adamant in ridding itself of teleological explanations—so why bring it back now? But what they mean by teleology is the mathematics of cybernetic machines which behave as it they had goals (for a cogate reinterpretation of teleology, see Asma, 1996). Such cybernetics focused on closed causal loops in the varied feedbacks whereby the output is affected by the detection of error between output and pre-set goals (see also, Juarerro, 1999). According to Rapoport and Horvath, such a theoretical move can free the biologist from having to be confined to unneeded reductionist strategies just by knowing it is possible in principle to be free from such a limited explanatory strategy.

Indeed, this kind of circularly causal feedback, introduced by these early cyberneticians, didn't disappear but became subject to increasing mathematical formulations as complexity science has proceeded. An example is Holland's (1998) scheme for the present day "organized complexity" of artificial life with his notion of "constrained generating procedures" which involve recursive transition functions acting on input combinations to create component mechanisms: $f_c: I_c \times S_c \rightarrow S_c$ (where f_c is defined so that: $f_c(t+1) = f_c(I_c(t), S_c(t))$; I = possible input combinations at time t; S = states of system at time t).

Iterating this equation at each new emergent level is one way to mathematically formulate the kinds of circular feedback loops to which Rapoport and Horvath were referring since the function ***f*** recursively builds new "component mechanisms" as the result of feedback of its previous state is pumped back into the full function in order to generate any present state.

On the other hand, taxonomy enters the picture by utilizing graph theoretical and topological notions of directed graphs, thus intuiting the eventual great importance that would later be made available by graph theory in social network theory. Purposeful dynamics is understood as directed graphs and closed causal loops as cyclic graphs whereas those free of such circularity in their causal nature are cycle-free or tree graphs. Indeed, Rapoport was also a pioneer tin social network theory.

Even the bringing in of topology is seen by Rapoport and Horvath as a taxonomic type of theoretical strategy since it concerns more the qualitative aspects, e.g., in topological approaches to the qualitative dynamics of phase space portraits. That topological theorems resemble classification, as indeed they do, e.g., in algebraic and differential geometry with all the classification theorems, and in group theory and use of algebraic topology in classifying all finite simple groups. In this context, it is quite startling to look at the huge atlas-type book that classifies finite simple groups, something I had accidentally come across one day in geography section of my university's library! I was surprised to see not continents and countries but long lists of different kinds of finite simply groups, a monumental project. Finally, it should be mentioned that Rapoport and Horvath were not finished with their suggestions for future methods including, e.g., decision theory and more particularly, game theory.

Tit-for-Tat

Indeed, Rapoport also made seminal contributions to game theory. In the early nineteen eighties, Rapoport's career evidently hadn't slowed down since he won a tournament sponsored by Robert Axelrod, well-known today for his agent based modeling of cooperation, that sought suggestions for repeated game theoretical models applicable to the evolution of cooperation. Axelrod (1984) was looking for a winning strategy promoting cooperation in repeated prisoner's dilemmas games. To do so, he sponsored a computer tournament in which strategies were to be submitted for playing 200 games of prisoner's dilemma. Although some of the submissions were highly intricate, it turned out that it was to be the simplest of all the submissions that achieved the highest score (on average), namely, the strategy submitted by Anatoal Rapoport which he called "Tit-for-Tat," requiring only four lines of computer code with two rules.

Axelrod has pointed out that the simplicity of Tit-for-Tat's approach stands in marked contrast to the very complex strategies found in computerized chess playing.

Tit-for-Tat is a variation of the iterated prisoner dilemma two-person game which has been used extensively to model situations of conflict and cooperation (see Tit_for_tat, n.d.). The prisoner's dilemma game is typically described in terms of two alleged perpetrators of a crime who are first arrested and then placed in separate interrogation rooms. The police give each of the perps the same choice, either confess to the crime (called "to defect" in game theory, a condition that automatically implicates the other as having in fact committed the crime) or to remain silent (called "to cooperate" in game theory since they both are not admitting to the crime and accordingly are also not implicating the other). During the interrogation, the police intentionally try to get each of them to implicate the other in the crime. The rule is that if one of the accomplices "defects" and the other "cooperates," then the defector gets off free but the cooperator receives a sentence of 10 years of jail time. Each of the perps are informed that a confession which implicates the other will lead to the confessor's release plus, as an added incentive, to a small reward (Meredith, 1998). If both confess, each one will be imprisoned. But if one individual implicates the other and not vice versa, then the implicated partner receives a harsher sentence than if each had implicated the other. However, if both cooperate (i.e., remain silent), then they both get six months in jail. Yet, if both accomplices defect, then they both receive six years of jail time. Also, it must be kept in mind that the interrogation process in each room does not leave that room, that is, neither of the suspects has true information about what choices, defection or cooperation, the other one makes.

One of the implications of the game is that if both alleged perpetrators are perfectly rational, then it should be a no brainer for each accomplice to conclude the best course is to implicate the other even though, in fact, they both would both be better off trusting each other, i.e., not confessing or implicating the other (see Grossman, 2004; Meredith, 1998). To get deeper into the type of thinking involved in the dilemma, imagine how one of the accomplices might think. If the partner does not implicate him/herself, then one would think that he/she should consequently implicate his or her partner and thereby received the most advantageous pay-off. But, the situation is even more complex since each of the players' decisions will be an important factor in future encounters. This means it is not just the immediate current situation faced by the two players that is relevant, anticipation about the future is also a significant prospect that needs to be taken into consideration. For example, if the two players know that they will never meet again, then defection, that is, confessing and thereby implicating the other, would seem to be the only *rational* choice. If that happens, then both will wind-up with a poor outcome. But if the prisoner's dilemma is repeated a number of times, then it may be advantageous to cooperate on the early moves and cheat only towards the end of the game. When people know the total number of games of prisoner's dilemma, they do indeed cheat more often in the final games.

Rapoport's Tit-for-Tat is a modified form of the iterated prisoner's dilemma in the following way. One player opens by cooperating with their opponent. Then, at the next play, the same player plays exactly as the other player did in the previous game. Thus, if the other side had defected in the previous game, the first player also defects but only for one game. However, if the other side had cooperated, the player continues to cooperate. In this manner, the players are punished for selfish behavior and rewarded for cooperation, but the punishment only lasts as long as the selfish behavior lasts. Tit-for-Tat proved to effectively demonstrate the value of cooperation.

The success of Titi-for-Tat took many by surprise because of its primarily cooperative orientation. Even when in successive competitions, various complex strategies were used to undercut it, Tit-for-Tat eventually prevailed. The theoretical hope remains that its cooperative results provide insight into how cooperation in animal and human social groupings have evolved during evolution

Axelrod published the results of the tournament leading to the inauguration of a second round although this time, to avoid special cheating strategies for the last game, there would only be a median of 200 games each of random length. Rapoport again submitted Tit-for-Tat and it won again. From an analysis of the 3-million choices made in the second competition, four features of Tit for Tat emerged (Meredith, 1998):

1. Never be the first to defect;
2. Retaliate only after your partner has defected;
3. Be prepared to forgive after carrying out just one act of retaliation;
4. Adopt this strategy only if the probability of meeting the same player again exceeds 2/3.

These results provide a model for the evolution of cooperative behavior.

Axelrod emphasizes that a formal theory for the evolution of cooperation needs to answer three questions:

1. How can a cooperative strategy get an initial foothold in an environment which is predominantly noncooperative?
2. What type of strategy can thrive in a varied environment composed of other individuals using a wide diversity of more or less sophisticated strategies?
3. Under what conditions can such a strategy, once fully established, resist invasion by mutant strategies (such as cheating)? According to Axelrod, Tit-for-

Tat is successful because it is 'nice', 'provokable' and 'forgiving'. A nice strategy is one which is never first to defect. In a match between two nice strategies, both do well. A provokable strategy responds by defecting at once in response to defection. A forgiving strategy is one which readily returns to cooperation if its opponent does so; unforgiving strategies are likely to produce isolation and end cooperative encounters. The most important theoretical result of those tournaments was that although the Tit-for-Tat-Strategy cannot possibly win the iterated Prisoners' Dilemma in an encounter with another single strategy, it is more likely to win in a "war of all against all" of different strategies (for details, see Axelrod. 1984).

References

Abraham, R. (2002). "The genesis of complexity," in G. Bocchi and M. Ceruti, *The Narrative Universe: Advances in Systems Theory, Complexity, and the Human Sciences*, ISBN 9781572732803.

Asma, S. (1996). *Following Form and Function: A Philosophical Archaeology of Life Science*, ISBN 9780810113985.

Axelrod, R. (1984). *The Evolution of Cooperation*, ISBN 9780465021222.

Dupuy, G. (2000). *The Mechanization of the Mind*, ISBN 9780691025742.

Elsasser, W. (2006). "A form of logic suited for biology (with an Introduction by J. Goldstein)," in K. Richardson, D. Snowden, P. Allen and J. Goldstein (eds.), *Emergence: Complexity and Organization 2005 Annual*, Volume 7, ISBN 9780976681472, pp. 423-466.

Grossman, W. (2004). "New tack wins Prisoner's Dilemma," http://www.wired.com/culture/lifestyle/news/2004/10/65317.

Holland, J. (1998). *Emergence: From Chaos to Order*, ISBN 9780738201429.

Juarerro, A. (2000). *Dynamics in Action: Intentional Behavior as a Complex System*, ISBN 9780262100816.

Meredith, C. (1998). "Tit for tat," *The Slab, Australian Broadcasting Company*, http://www2.owen.vanderbilt.edu/mike.shor/Courses/GTheory/docs/Axelrod.html.

Rapoport, A. (*1911) (nd). International Society for Systems Sciences, http://www.isss.org/lumrapo.htm.

Rapoport, A. (nd). http://en.wikipedia.org/wiki/Anatol_Rapoport.

Tit_for_tat. (nd). http://en.wikipedia.org/wiki/Tit_for_tat.

Umpleby, S. and Dent, E. (1999). "The origins and purposes of several traditions in systems theory and cybernetics," *Cybernetics and Systems: An International Journal*, ISSN 0196-9722, 30: 79-103.

8.
Thoughts on Organization Theory

ANATOL RAPOPORT AND WILLIAM J. HORVATH

THE CHARGE brought against his age (some 30 years ago) by Alfred North Whitehead was that the age was living on the intellectual capital accumulated in the seventeenth century. The implication was that if no new sources of "intellectual income" are discovered, the resources will be squandered. Elsewhere Whitehead said that the culture that cannot burst through the framework of its own concepts is doomed.

These remarks are metaphors, and if they remained unelucidated by a more concrete analysis, they would have to be viewed like so many other examples of poetically expressed discontent characteristic of pessimistic philosophers from Heraclitus to Schopenhauer. Fortunately Whitehead had a profound understanding of science and was able to spell out just what the philosophical limitations were which hemmed in *scientific* thought and just where the threat of impending sterility lay.

The constraining framework of thought, according to Whitehead, was the overwhelming predominance of analytic thinking in science, stemming from the hegemony of physics as *the* model of exact science. Analysis is an attempt to understand a complexity by examining its constituent parts. The parts being simpler, they are supposedly more amenable to understanding. The idea of analysis, then, is to understand the working of the parts. Note how in presenting the analytical view, it is natural to fall into the language appropriate for describing a mechanism —the "working of the parts." The implied hope is that it is possible to "build up" the understanding of a complexity by "superimposing" the workings of the various parts.

There is no denying that this approach is effective in specific areas of investigation. To understand how a complex machine works, it is often sufficient to understand the separate links in its construction: this lever transmits a force to effect a torque on this wheel; when the wheel has turned through a certain angle, this relay is closed, which allows a current to flow through this wire; the rotation of the wire, in turn, gives rise to a magnetic field, which . . ., etc. One has broken up the whole operation into a temporal chain of events, all connected by determinate "causal" relations.

Another good example is the explanation of the solar system (i.e., the motions of its components) in terms of classical mechanics. Having isolated certain measurable quantities (distance, time, mass) and having postulated certain laws of motion and a universal law of gravitation, the mathematical physicist deduces first the behavior in empty space of two point masses, subject to mutual gravitational attraction. Upon his solution he *super-imposes* perturbations due to the presence of other point masses. The scheme works because of the *additivity* of the fundamental effects if the perturbing influences are small compared with the interactions between the original pair. And of course underlying the whole scheme is the additivity of the fundamental quantities. Distances and forces are added according to the vector laws of addition; times and masses are added like scalars. In electro-magnetic theory too, additivity, stemming from the linearity of the basic partial differential equations, is the fundamental *tenable* assumption, which allows the super-position of

From Anatol Rapoport and William J. Horvath, "Thoughts on Organization Theory," *General Systems*, 4 (1959), 87–91. Reprinted by permission of the authors and the Society for General Systems Research.

vector fields (physical additivity) and the superposition of the many particular solutions of partial differential equations, multiplied by appropriate constants and added, to obtain the solution satisfying specific boundary conditions (mathematical additivity). Thus the classical picture "works" where it is applicable. And because of its success it has a terrible seductive power over the human mind. One is tempted to ask whether all complexities cannot be explained in this way. Is it not only a matter of super-posing more and more parts, each of which is understandable in itself, until the entire complexity is accounted for? Or, consequently, is not the understanding of any complexity to be gained by analyzing it further and further into its constituent parts? The essence of Whitehead's critique is to answer this question in the negative.

Whitehead's critique has been echoed during recent decades in many quarters. Most of the echoes have been solemn but vague. We are told that scientific thought needs to be spiked with "organismic" and "holistic" ideas. The holists even have a motto inscribed on their banner: "The whole is greater than the sum of its parts." In the writings of the proponents of a neo-organismic philosophy, we often read that Goethe had the right idea all along, since he thought "holistically" and "organismically." Some point to biology as the science to serve as a model for a new organismic orientation. Others proclaim the dawn of the new era as being ushered in by a preoccupation with "organized complexity," the region between the extremes of "organized simplicity" and "chaotic complexity," with which the exact scienitst has been hitherto exclusively concerned.

In spite of the vagueness of the language, much of this has a ring of truth. Still the ideas remain elusive, and they stand in constant danger of becoming clichés through repetition. Occasionally the arguments are supported by concrete examples, which can serve as points of anchorage for further exploration.

The notion of the organism, for example, is taken to be central in most holistic expositions of method. Well, one might ask, what about the organism? To begin with, what is it? This is a legitimate question, but it should not be pushed too hard if the "holist" is to be given a fair chance to explain his position. In order to avoid futile hair-splitting right at the start, we should not insist on rigor in the definition of "organism." An extensional definition will do. Organisms are exemplified by living things, most of which are sufficiently individuated to be recognized as separate entities. Next, what about the organism? How does placing the organism at the center of one's conceptual scheme help one get away from "limited mechanistic concepts"? The only answer that suggests itself is related to the well-known methodological differences between biological and the physical sciences.

First, the biological sciences depend to a far greater extent than the physical on categorization, and classification. Second, the biological sciences have often resorted to teleological explanations, which are taboo in physics (except for pseudo-teleological explanations to be dealt with below).

To illustrate the first distinction, we need only notice the importance of taxomony in biology: a classification of living things is its natural point of departure; important insights reflect themselves in reclassifications. In medicine, too, an art nurtured by biology, classification of diseases is a central problem. In short, the problem of *recognition* is an important one for the biologist. Before he begins to say anything about *events*, he must have a proper language to talk about *entities*, of which some can be directly observed and others must be specifically invented for the purpose.

Physics, as we understand it today, does *not* start with a classification of "entities." It starts with what have become to us intuitively evident abstractions from experience, which are *immediately* quantified. The development, then, proceeds by elaborate mathematical deductions from "first principles." Observations are not the wide-eyed explorations of the naturalist; the observations of physics are pre-arranged and carefully delineated: they are the observations of controlled experiment. They are recorded not as meticulous and intricate descriptions of a talented observer (the naturalist's art) but rather routinely, according to preconceived rules, practically always as measurements, often mechanically. Thus the physicist and the (old-time) biologist differ fundamentally in their approach to observing nature and describing what they observe, and this fundamental difference is reflected in the great importance of classification as a tool of cognition in biology and its relative unimportance in physics.

To illustrate the second difference, the early explanations given by biologists for observed events relevant to that discipline were in terms of the needs of the organism or, on the grand scale, in terms of the "strivings of the race" (as in Lamarckian evolutionary theory). Such explanations were also dominant in pre-Galilean physics. But with the mathematization of mechanics, such teleological causation was expelled from physics except when such "causation" could be shown logically equivalent to a mechanistic one. For example, the Hamiltonian Principle of Least Action *sounds* like a teleological explanation (trajectories are "chosen" so as to minimize certain integrals taken over the traversed path), but the Principle is mathematically derivable from Newtonian laws, in which no reference is made to future events as "causes" of present ones. The particles of Newtonian mechanics have no goals. The forces acting upon them act *here* and *now*.

It will be obvious to any contemporary biologist that in comparing the methods of biology and physics, we have not drawn a picture of modern biology at all. Natural history, taxonomy, collections, etc. are now only a small portion of the biologist's world. He is now quite as at home in the laboratory and even sometimes with mathematics as is the physicist. He certainly performs controlled experiments and does not confine himself to mere descriptions of what he has seen. Also the biologist is relying less and less on teleological explanations. He seeks to explain physiological processes in terms of physics and chemistry, which means that he is no longer satisfied with relating events to the needs of the organism but relies on the same laws of "here and now causality" as the physicist. Teleological ideas are no more welcome in evolution: genetic theories, based on random mutations and the theory of natural selection have replaced the Lamarckian outlook.

In short, the methods and outlook of biology approach more and more those of the physical sciences. What then can a philosopher mean when he suggests that the method of biology may have more to offer than physics to the science of the future, the science of organized complexity? Can he mean that the modern scientist should place more reliance on those aspects of biology which are carry-overs from a previous age? Such a position would be difficult to defend. Yet a defense of the "holist's" position can be made, provided we can spell out just how the older schemes of cognition (the "static" logic of classification and the teleological explanation) can fit into the new modes of analysis.

The key concept is "organized complexity" as exemplified by, say, a living organism or any "organized" collection of entities, that is, a collection interconnected by a complex net of relations, as distinguished from (1) organized simplicity and (2) chaotic complexity. We will try to clarify the difference by giving examples of (1) and (2), by showing how these have been treated by classical methods, and by assuming what is needed to extend those methods to the study of "organized complexity"—i.e., what is needed for a theory of organization.

The organization of a system is simple if the system is a serial or an additive complex of components, each of which is understood. We have already given examples of such systems. One is a machine whose working is a time-linear chain of events, each a determinate consequence of the preceding one; in other words, a system without closed loops in the causal chain. Another is, say, a propagated wave form which can be analyzed into additively super-imposed sinusoidal components.

As soon as strict sequential sequence or linear additivity is transcended, an "organized system" becomes rapidly more complex, usually too complex for detailed analysis into super-posable parts or effects. Already the three-body problem (three mass points of the same order of magnitude in space interacting gravitationally) becomes enormously complex for determinate analysis.

At the other extreme is "chaotic complexity" where the number of entities involved is so vast that the interactions can be described in terms of continuously distributed quantities or gradients, i.e., do not need to be specifically identified with regard to the individual entities. Such systems can be described by the methods of statistical mechanics which merge with those of classical mechanics when the collections of entities are treated as continuous. Thus it can be said that the class of events amenable to exact treatment has been extended from "organized simplicity" to "chaotic complexity" when a new conceptual scheme was added to the physicist's kit, namely the mathematics of probability (which does not enter classical mechanics at all). Hence it was not altogether true that in the first decades of this century physics still remained imprisoned in the conceptual framework of the seventeenth century as Whitehead implied. The new metaphysics of probability had been added to that of determinate causality and had been digested.

The question before us is whether we can foresee what conceptual tools must be further added in order to extend systematic rigorous theoretical methods to "organized complexity," with which the holists are presumably concerned. There are at least two such classes of concepts discernible, and both are reminiscent of the ways of thinking of the biologist. It is this resemblance that gives the proposal to give more attention to the ways of thinking derived from biology an aura of plausibility. Of course, in the process of being incorporated into modern scientific thought, these concepts lost their resemblance to their origins, but the origins remain recognizable, namely teleology and taxonomy.

When we say teleology, we mean teleology in its modern garb, made respectable by cybernetics. For cybernetics has concerned itself with a new phylum of machines, machines which behave as if they had goals. The mathematics of cybernetics has shown how, on the basis of physical laws alone *plus* certain principles of constructing networks of causal relations (networks including closed loops, which, recall, we have excluded from our simple-chain linkages of relations), "goal-seeking" behavior of mechanisms emerges.

Such machines can simulate goal-seeking behavior through the feedback principle, according to which the output of the system is modified by the error between its output and some pre-set goal. It is not unlikely that simple goal-seeking actions of living organisms are effected by similar organizational principles. If so, then the apparently teleologically determined behavior can be reduced to non-teleological principles (plus organizational principles hitherto not considered in the study of non-living systems). Once this possibility of reduction is established *in principle*, the biologist can rely on "teleological" (now pseudo-teleological) theories with an eased conscience. He can say with some justification that it is the cybernetician's job, not his, to show how a "teleological" theory is to be reduced to a non-teleological one; he assumes it can be done, and this frees him to construct theories "on a higher level" so to speak.

In discussing the rehabilitation of teleological principles, we have already touched on taxonomic ones. For example, we have distinguished between machines with and without feedback loops, only the first of which can exhibit "goal-seeking" behavior. The distinction is a topological one. If "causal relations" are indicated schematically by directed segments, the systems so described appear as directed graphs. It appears, then, that systems described by graphs containing cycles can exhibit forms of behavior which systems described by cycle-free graphs ("trees") cannot. This was recognized already in the early formulation of the isomorphism between symbolic logic and network theory by McCulloch and Pitts in 1943. The development of that theory begins with the explicit assumption of cycle-free networks. Many theorems do not hold if cycles are allowed, and even the conceptual repertoire becomes insufficient. For example, the introduction of cycles into the topology necessitates the introduction of an existential operator into the logic.

Come to think of it, topology can be viewed as a "taxonomic" branch of mathematics. It is qualitative rather than quantitative in spirit. Its findings resemble classification more than computation (witness the concept "genus" as applied to surfaces or to more general topological manifolds). The theorems of topology have an "all or none" rather than a "continuous" flavor. They frequently are assertions that something does or does not exist; that something is or is not possible, rather than expressions of functional relations among variables taking on a continuum of values.

Thus, we hope it is clear now what was meant by the statement that the "older" ways of thinking characteristic of biology, namely classification and teleological explanation, appear in modern development in a new garb. Cybernetics has shown that a teleological way of thinking is not incompatible with the metaphysics of physical determinism; topology has shown that a taxonomic way of thinking is not inconsistent with mathematical sophistication. The two disciplines are related: cybernetics is a dynamics superimposed on a topology.

It seems likely that these two disciplines will be at the foundation of that branch of science which deals with "organized complexity," i.e., organization theory.

There has also been a third line of development, quite unknown at the time Whitehead wrote his critique of the analytic method, but which has since come to be viewed as the third corner stone of organization theory. This is the theory of decisions, comprising both "games against nature" (including statistical inference) and game theory proper. Obviously this orientation is much more relevant to the "theory of organizations" than to organization theory. The distinction between the two we see as follows: We see organization theory as dealing with general and abstract organizational principles; it applies to any system exhibiting organized complexity. As such, organization theory is seen as an extension of mathematical physics or, even more generally, of mathematics designed to deal with organized systems. The theory of organizations, on the other hand, purports to be a social science. It puts real human organizations at the center of interest. It may study the social structure of organizations and so can be viewed as a branch of sociology; it can study the behavior of individuals or groups as members of organizations and so can be viewed as a part of social psychology; it can study power relations and principles of control in organizations and so fits into political science. The "decision," i.e., a choice based on examining a state of affairs *and* the range of possible outcomes is the fundamental event of decision theory. Such events are frankly based on teleological principles (I choose this action, because I hope to attain that goal). But we have already seen that what appear as teleological determinants can *in principle* be reduced to non-teleological ones; therefore it is not necessary to assume that a permanent gulf will persist between the methods of physical and social science. The decision-maker may one day be described as a complex goal-seeking system. Until then, the psychologist and the social scientist can simply take his goal-seeking behavior for granted.

To summarize, three lines of rigorous theoretical development have been intensively pursued in the past fifteen years or so, none of which had played an important part in classical physics and all of which show promise as the theoretical underpinnings of organization theory in general and perhaps in the theory of organizations specifically, namely:

(1) cybernetics—the theory of complex interlocking "chains of causation," from which goal-seeking and self-controlling forms of behavior emerge;

(2) topology or relational mathematics;

(3) decision theory, whose specific application potential is to the theory of human organizations.

The subdivisions of these disciplines of particular interest are:

(1) information theory, related to cybernetics;

(2) network theory, related to topology;

(3) game theory, i.e., that branch of decision theory in which decisions must be made not only in the face of uncertainty but also taking into account the presence of other decision makers, some or all of whose interests may be opposed to those of ego.

We cannot expect, however, that these developments will be rigorously applied in every serious study of real organizations. Indeed, we expect that premature application will be largely sterile. Information theory has been extensively developed only in connection with the problems of telecommunication, where only syntactic characteristics of language are relevant. Attempts to extend information-theoretical methods to semantic and pragmatic aspects of communication often degenerate into metaphorical speculations and contribute nothing to rigorous analysis. Attempts at rigorous application of game theory have fared no better. Network terminology has been useful in providing a descriptive language for communication patterns but the deductive powers of topology (specifically of graph theory) have been tapped only to a limited extent in application.

"Scientism" is still a predominant childhood disease of organization theory as it is applied in social science. Application of rigorous methods of analysis to real organizations is still largely a pious hope rather than a genuine opportunity. Therefore the impact of the new ideas can be expected to be at the time being only indirect. We expect that the study of human organizations will continue to depend on the insights and methods of the social scientist, i.e., the psychologist, the sociologist, the economist, and the political scientist. With the growth of interdisciplinary research we may expect the biologist to make additional contributions. There is some sense in considering a real organization as an organism, that is, there is reason to believe that this comparison need not be a sterile metaphorical analogy, such as was common in scholastic speculations about the body politic. Quasi-biological functions are demonstrable in organizations. They maintain themselves; they grow; they sometimes reproduce or metastasize; they respond to stresses; they age, and they die. Organizations have discernible anatomies and those at least which transform material inputs (like industries) have physiologies. All have "neural physiologies," since organization without internal communication, integration, and control is unthinkable.

In totality, then, we have today a variety of approaches to the study of organization (as an abstract principle) and a variety of approaches to the study of organizations (i.e., human aggregates with certain specified relations of interdependence among the members). The two developments are destined to travel along two separate roads for a while. Occasionally, a connecting path will be discerned, along which ideas can trickle from one stream to the other. Eventually, it is hoped, the two streams of ideas will actually merge.

25. The Doctrine of Levels

George P. Conger (with an introduction by Jeffrey Goldstein)

Originally published as Conger, G.P. (1925). "The doctrine of levels," *The Journal of Philosophy*, 22(12): 309-321. Reprinted with kind permission.

From Crisp to Confounded: An Introduction to George Conger's "Doctrine of Levels"

Metaphysical and Emergentist Levels

A professor of philosophy at the University of Minnesota, George Conger was loosely associated with the movement known as Emergent Evolutionism, the *proto*-phase of emergence-based thought in philosophy, the sciences, and even theology existing roughly from 1915 to 1940 (see, Blitz, 1992; Stephan, 1999; Goldstein, Forthcoming; and various Classical Papers published in *E:CO* since 2004). Conger's paper on the idea of emergent levels, published in the prestigious *Journal of Philosophy* in 1925, was a careful and thought-provoking examination of the idea coming right at the intellectual zenith of proto-emergentism. Conger explicated what he found conceptually helpful in the notion of emergent level, but did not fail in taking-on what he found conceptually troublesome about it as well. Conger's views on emergent levels are not only important for their insights into what Emergent Evolutionists in general held about levels, but also because of their display of striking prescience in foreseeing critical issues about levels that would arise later when the phenomena of emergence became a key construct within the sciences of complex systems. Conger's same prescience also provides perceptiveness into a central aspect of the notion of emergent level, a kind of paradox associated with the idea, a subject I'll return to in the conclusion.

Emergent Evolutionism held that the Darwinian theory of evolution needed to be supplemented by a series of saltations or jumps in novelty, the coming into being of a series of new emergent levels constituted by new emergent phenomena with new emergent properties. In the words of one of its chief proponents, the animal behaviorist and philosopher Conwy Lloyd Morgan (1923:15), "Each higher entity in the ascending series is an emergent 'complex' of many entities of lower grades, within which a new kind of relatedness gives integral unity...each higher com-*plex* takes on the role of a *com*-plex in virtue of its integral unity." Morgan emphasized the prefix "com-" because of its meaning of "together" in order to highlight how the emergent level was, in Morgan's words, *supervenient* upon whatever was coming together. An emergent level, accordingly, was a concept that called attention to the importance of considering the novel dynamics of a level operationally "higher" than that of the parts that were being newly related. As Conger understood it, each level was thought to be composed of structures sharing major attributes and interacting with one another. The beginning signs of the paradox of emergent level mentioned above can be gleaned in the simultaneous positing of a discontinuous *crispness* distinguishing one level from another and the need to attend to the *entanglements and interactions* of said levels in order to account for emergence.

The stratification of reality, nature or the cosmos into a hierarchy of levels, each representing a separate domain with unique entities, dynamics, and laws, is an ancient notion. Aristotle (cited in Lovejoy, 1936), for instance, proposed several different hierarchical schemes, one of the most influential being his vertical partitioning of the "powers of soul": the *nutritive* soul of plants; the *animated* soul of animals; and the *rational* soul of humans. Such layered cosmologies were typically tied closely with sequenced cosmogonies, e.g., the Neo-Platonic hierarchy of Plotinus (1991) which conceived the phenomenal world as a lower level in a series of "hypostases" emanating downward out of the highest, the One beyond Being Itself.

The idea of such metaphysical levels need not have had a religious basis as is shown in both modern philosophical, but non-theological, conceptions of metaphysical cosmic stratification such as those developed respectively by Nicolai Hartmann (see Werkmeister, 1990) and James Feibleman (1951) as well as the prevalent use of the construct of level found in various scientific theories. In the latter, different justifications have been offered for the particular allotment of levels, which entities exist on which levels, and so forth. Perhaps the most sophisticated and trenchant account of the promise and perils of appealing to the construct of theoretical levels in scientific thinking, especially that found in theoretical biology including its many recent complexity-based inspirations, is the one offered by the philosopher of biology William Wimsatt (many of whose papers were pulled together in his *magnum opus* of 2007), a name we will be returning to below.

What Emergent Evolutionists bestowed on antediluvian hierarchy scenarios was an emphasis on an ascension of novel strata, not as somehow pre-given and thereby built-into the very substratum of reality, but instead as *emerging* dynamically over time in a series of discontinuous *emergent* levels. This bestowal can be seen, for instance, in how the proto-emergentists reworked the philosopher Walter Marvin's (see Blitz, 1992) notion of "logical strata" that covered everything from ordinary things, to events, to constituents of these things, to constituents of these events, as well as logical, mathematical, physical, biological, and mental categories. Although in Marvin's layered reality, there were novel characteristics at each new level, proto-emergentists emphasized that the dynamical emergence of each new level would continuously supplement whatever antecedent levels were pre-existing. Accordingly, for proto-emergentists, the novelty of the new emergent levels should be considered a *radical*[1] type of novelty in the sense that the structures, patterns, dynamics, and properties of each new level composed of novel emergent phenomena were held to be neither predictable nor deducible from the lower level components alone, nor were they reducible to the latter.

The claim that nature is arranged according to a hierarchy of levels has been linked to various rationales for the existence of "special" sciences whose conceptual apparatuses are only deemed appropriate for "special" levels, e.g., biology aims at the level of living beings, chemistry at the level of the chemical reactions by which living beings function, physics at the level of atomic (as well as *supra*-atomic ensembles and *sub*-atomic particles) by which chemical reactions are determined, and so on. Conger's own commitment to such a scheme was along the lines of the Emergent Evolutionist take on the correspondence between levels and the "special sciences", a position which they had adopted and adapted from their immediate intellectual precursors, the English philosopher John Stuart Mills and the French philosopher and mentor of Henri Bergson, Emile Boutroux (Blitz, 1992). However, as we'll say more about below, Conger also insightfully queried just how many levels there would need to be in such a correspondence between natural strata and the domain specific to each special science.

In modern times, one of the most interesting versions of this kind of level correspondence was that offered by the late eminent biologist Sir Peter Medawar (1974) who formalized this correspondence according to the particular *geometry* operationally most appropriate for each level—"geometry" here referring to the famous classification founded on group theoretic invariances put

---

1. The term "radical emergence" has been used to describe the position taken by the philosopher Michael Silberstein (2001; Silberstein & McGeever, 1999; and comments on Silberstein's work in Freeman, 2001; and Feinberg, 2001) regarding his argument for considering consciousness as an emergent phenomenon. The word "radical" presumably connotes two related aspects of consciousness: 1. The claim that consciousness as an emergent phenomenon possesses an ontological and not mere epistemological status, i.e., it is non-deducible in a much stronger sense than the claim that only epistemic limitations on the part of observers restrict its deducibility; 2. Consciousness as emergent is completely irreducible to the dynamics and laws operative on its neuronal substrate in the brain. My use of the term "radical novelty" is a qualification about the kind of novelty an emergent phenomenon is thought to possess. Using "radical" to qualify emergent novelty affords insight into the specific nature of the type of novelty needed for a doctrine of emergence without claiming that some emergences are radical, while some are not. It seems to me that "radical emergence" is qualifying the wrong thing since, in my opinion, the various typologies of emergence that have been offered, e.g., so-called "weak" versus "strong" emergence and other classifications (see Bedau & Humphries, 2008) are not particularly illuminating. But this is a long story that I don't have the space here to get into—see Goldstein (Forthcoming).

forth by the great German mathematician Felix Klein in the last quarter of the nineteenth century[2]. The preceding footnote lays-out Medawar's main elements. In Medawar's hierarchy, for example, the "geometry" of *topology* corresponded to the level of physics because of this "geometry's" invariances of sidedness and ordinality on a line thereby enabling physics a wider capacity for abstraction while *affine geometry* matched the level of biology since shape as such was not invariant whereas living organisms do possess homographic invariance.

---

2. It should be noted that although Medawar's system of levels was not directly emergentist, it did include what could be called a kind of "conceptual emergence" in that new concepts could emerge in a level which have not appeared in the geometry above (Medawar's hierarchy goes vertically downwards from the most abstract to the least abstract). Medawar's hierarchy was not emergentist since it didn't provide an account of how new levels may come about besides appealing to a repeated restriction of possible transformations (Klein's transformation groups) which, again, was only a formalistic account of "conceptual emergence." It also should be recognized that Medawar's scheme did not consider what might take place between levels such as via the interactions of entities on different levels.

Here is Klein's classification of group theoretic invariances of different geometries:

Metrical geometry: a group of operations including translations of figures on a plane, rotations, inversions or any combination of these so that ordinary properties associated with our everyday meaning of shape do not change, e.g., distance remains unchanged; Euclidian geometry: operations allowing for symmetrical magnification in 3-dimensional space, e.g., shape is not changed but distance is not invariant; Affine geometry: new points or new co-ordinates are related to old points by linear integral functions; these affine transformations are similar to Euclidian transformations except that the degree of extension or diminution in all three dimensions are not the same; e.g., shape is not invariant but retains a homographic relationship; Projective transformations: the mapping functions are fractional linear functions; among the properties invariant with respect to projective transformation is linearity, i.e., straight lines stay straight lines whereas anharmonic ratio and conic section properties are invariant but parallelism is not; Topology: mapping functions are singled valued both ways and thus bring new points into a one-to-one correspondence with the old ones, plus they are continuous functions; e.g., figures are transformed but remain homeomorphic, i.e., general properties of the space remain such as insideness and outsidedness as well as the order of points on a line.

1. Topology	1. Physics
2. Projective Geometry	2. Chemistry
3. Affine Geometry	3. Biology
4. Euclidian-metrical geometry	4. Sociology/Ecology

Each geometry, and by implication, each "special" science is a special case of the one above (since the group of operations it defines is a subgroup of the group related to the one above it; the theorems then becoming more particular).

Conger's Prescient Probings

Whereas Medawar's hierarchical conceptual edifice was undoubtedly rigorous and sophisticated, it was also *metaphoric* in its speculating on the usefulness of group theoretically defined geometries. I find it quite interesting that Conger had called attention to what he thought of as the *metaphoric* status of the concept of an emergent level when this idea was employed in psychology or sociology or metaphysics (notice which disciplines he groups together!) and he did this way before all the modern squabbling about metaphor versus "hard science" among present-day complexity advocates.

Although agreeing in principle with the Emergent Evolutionist doctrine of levels, Conger pointed to several problematic aspects with how the notion of level was being employed, e.g., the issue of just how many emergent levels there were. Even among the more strident proponents of Emergent Evolutionism, one can find an uneasiness with how literally and precisely the idea of emergent level should be taken. The New Zealander/English philosopher and experimental psychologist Samuel Alexander (1966), for one, contended that to try to be complete or exhaustive about listing all the multifarious levels was to miss the point since there was a certain relativity to the apparent crispness of each level. The aforementioned Morgan (1923) had conceded it was beyond the "wit of man to number instances of emergences" and presumably this immeasurability extended to the number of different levels on which these innumerable instances dwelt.

Alexander also proposed there existed many intermediary stages such as the complex chemical complexes between matter and life, a proposal along the same vein as that of another Emergent Evolutionist, the American entomologist William Morton Wheeler (1927) who, by the way, was one of E. O. Wilson's professors and mentors. Wheeler argued that the assignment of any specific number of levels was arbitrary since emergence was observed in a manifold of spatio-temporal events. Wheeler went so far as to criticize the proto-emergentist emphasis on such weighty emergent level novelties as life and mind which he thought were conceived in such a wholesale and broad-brushed fashion, they represented only the final accumulative stages of very long series of more minimal emergences. He asked whether it really made that much sense to view life as a "higher" emergent level in respect to a "lower" level of inorganic matter, or, was it more accurate to speak about innumerous intermediary emergent levels. But then just how many intermediary levels were there? In such a context, the contemporary emergentist-oriented philosopher William Hasker (1999) has keenly suggested that the weaker the definition of emergence, the more emergent levels that would appear. This should have some bearing on the typological distinction between so-called "weak" and "strong" emergence offered by the contemporary complexity-oriented philosopher Mark Bedau (Bedau & Humphreys, 2008).

Conger himself speculated, after considering varied scientific sources in which the notion of level had been theoretically utilized, that there was indeed a precise number of levels, *viz.*, 25, each of which agglomerated together an unusual congeries (pun intended) of constituents ranging from electrons, spiral nebulae, families, perceptual reflexes, sentiments, and other oddly associated units within levels. For Conger, all 25 levels were distinguishable according to the way empirical investigation, by means of each of the "special" sciences was supposed to proceed on that level. For example, there were different investigational strategies for organic compounds, for unicellular organisms, for multi-cellular organisms, and so forth. Furthermore, Conger warned that all of the levels interact and therefore should be considered, simultaneously, as both independent as well as interdependent with the implication that the designation of a specific emergent level didn't indicate a sharp distinctiveness between them after all, anticipating, as we'll soon see, later remarks coming from within complexity theory and adding more hints about the paradox of emergent levels alluded to earlier.

Conger also made it a point to criticize what he believed were the misleading *spatial* and *valuational* implications involved in using the terms "higher" and "lower" to describe emergent levels in contrast to their substrate levels. In this context, Conger cited Morgan's use of the "higher-"connoted term "supervenient" (a term which has been revived in recent years by the philosopher Jaegwon Kim to describe the status of emergent levels—see Bedau & Humphreys, 2008). For Morgan, the emergent level of "new relatedness" is "supervenient" on the "lower" level but, unlike in Kim's disavowal of any causal efficacy on the part of the emergent level and its emergent entities upon the lower level and its entities, Morgan's "supervenience" did imply that the events at the lower level would run their course differently as a result, a preliminary statement of what would later be known under the appelation of *macro-determination* or *downward causation*, another subject to which we'll return in the conclusion.

Other, later emergentists have voiced similar concerns of discourse utilizing "higher" and "lower". For example, the mid-phase emergentist biologist and philosopher Paul Weiss (1969) attributed thinking in terms of spatially conceived level hierarchies as an undesirable vestige of the hard-to-change habit of visualizing a complex system as a whole on a "higher" plane and then dissecting this whole into components displayed on another "lower" plane. In a similar critique of such "higher/lower" descriptors, another mid-phase emergentist, the physicist and philosopher of science Mario Bunge (1979), who incidentally developed a rigorous formalism for emergence and emergent levels mostly unknown to current complexity theorists, pointed to the militaristic insinuation of rank and superiority by which "higher" levels *command* "lower" ones—the etymology of "hierarchy" includes the Greek "hieros-" meaning "sacred" and "-archein" "to rule."

Another insight into the less-than-salutatory connotations of spatial ascendance in describing emergent levels can be found in the formalism of emergence offered in a previous issue of *E:CO* by the mathematician and psychiatrist William Sulis (2004) who suggests that there is as much conceptual justification for understanding emergence in a *horizontal* as in a *vertical* fashion, e.g., two entities on the same *horizontal* level coming together to generate the conditions for the emergence of a third still on that same level such as is found in biological symbiogenesis (my example, not Sulis's, about which I'll say more below). Of course, "higher" can be taken in a non-spatial sense, for instance, of wider inclusiveness, although such a move could open the door to the misleading conclusion that emergence takes place by way of some mathematical/logical operation involving the erection of type hierarchies and similar ideas. Conger himself proposed interpreting "higher" in the sense of being further along a developmental pathway, emergent levels then understood from the perspective of a process of development (Salthe, 1993, being the prime representative of allying developmental with evolutionary scenarios in the determination of hierarchical schemes).

In regard to the question of how emergent levels arise, a subject about which Emergent Evolutionism was conspicuously reticent, Conger ventured forth with several possible scenarios. One involved processes of *integration* or *creative synthesis* which he understood as originating not only in the earlier work of Herbert Spencer, famous in the mid-nineteenth century for his pre-Darwinian speculations on evolution, but also in intimations on the part of the proto-emergentists E. G. Spaulding and R. W. Sellars on how parts were brought together into novel "organizing relations" exhibited on the new level. Another conjecture offered by Conger along these lines, although, in my opinion, not sufficiently different from the first to be considered an authentic alternative, concerned *combining relations* or *mutuality of relations* which referred to the critical importance of the conception of *inner relations*, a philosophical category somewhat typifying metaphysics in Britain and the US at the end of the nineteenth century, e.g., in the philosophies of F. H. Bradley and William James (see Sprigge, 1993) or what Morgan had identified as "new terms in new relations".

Interestingly, in this context, Conger cited the work of Mary Parker Follett, otherwise known as one of the main forerunners of modern organizational theory. Follett's quotation was very prescient itself in its dynamical systems, and Habermas-sounding analysis of human relations (e.g., see Habermas, 1985, where the influence of another proto-emergentist thinker, George Herbert Mead, is overtly evident; see also Mead, 2007, in *E:CO*): "...I never react to you but to you-plus-me; or to be more accurate... it is plus the-interweaving-between-you-and-me meeting you plus the-interweaving-between-you-and-me, etc. If we were doing it mathematically we should work it out to the *n*th power... We should obtain a differential equation or a set of differential equations..." In this context, Conger conjectured about universal processes whereby individuals in a group could adjust themselves according to mutual dependence, perhaps something like what happens in synchronization.

Conclusion—Crispness versus Interaction: Flirting with Paradox in the Idea of Emergent Levels

Conger also suggested, in another manifestation of foresight, that much of the "mystery" surrounding emergence could be cleared-up if the "cumulative interaction between units of various levels" and not just from "higher" to "lower" levels was investigated since the potentialities of the novel units on the novel levels were "unknown until its interactions reveal them". Conger further claimed, in an almost eerie prognostication, that entities at the same level could have different rates of interaction resulting in them being grouped together according to these differing rates and that, accordingly, boundaries would form with the units having slower rates inside the "inner portions" of the system and those with faster rates at the outer portions of the emergent level. An almost identical formulation can be found in one of the most sophisticated contemporary mathematical formalisms of emergence in the work of Ehresmann and Vanbremeersch (2007): "The evolution of the [emergent] system will mostly depend on the interactions between agents of various levels of complexity, acting with different time scales."

Similarly, Wimsatt (2007) has for many years now been pushing for a similar, more sophisticated treatment of the complexity of complex systems, in particular, the need to consider the crucial fact that as complex systems become more complex, the operative levels themselves get more complex in the sense of being more diffuse, overlapping, and capable of being decomposed in more numerous ways. Thus, a researcher might decompose a multi-cellular organism into various systems of parts making-up different levels, or into levels made-up of cells with common developmental fates, or according to phenotypic features determined by common sets of genes, or in accord with biochemical reactions, or... Similarly, cells can integrate into a multi-cellular organism, the latter being at a "higher" level yet chemical radicals can enter into the structure of some molecules without there being a difference between the level of the radical and the molecule into which it adheres. Consequently, the existence of an integrated entity neither guarantees the presence of crisp level distinctions nor there being just one sure way to analyze it. As the complexity-oriented biologist Richard Levins (cited in Wimsatt, 1972) put it, it's even harder to distinguish levels when their component subsystems have evolved together. And the mid-phase emergentist theoretical biologist Robert Rosen (1969) emphasized that biologists employ the term "hierarchy" only in an inexact, non-etymological sense since biological functions tended to be distributed over much if not all of a system.

In a similar respect, Emmech *et al.* (1998) pointed out that levels in complex systems possess *non-homomorphic* inter-level relations meaning that the nature of the relation between levels varies for each, a point that is not generally incorporated into grand schemes of emergent levels, each emerging sharply and consistently out of the one beneath, all towards the creation of ever increasing complexity in the universe. There's also the related issue so cogently put forward by Stanley Salthe (1993) that a "higher" emergent level is better thought of as a *boundary condition* within which "lower" emergent entities integrate so that, strictly speaking, at any one time the notion of an emergent level is an occurrence at level l_n out of substrate components at level l_{n-1} in the context of boundary conditions at level l_{n+1}.

The great importance of interactions in defining levels, particularly emergent levels, was also one of the main themes brilliantly articulated by Douglas Hofstadter (1979) in such works as *Gödel, Escher, Bach* and *I Am a Strange Loop*. Many readers may not have realized that, although these book dazzle with their mind-stretching *Gödelian* logic interposed with Escher's startling artworks and Bach's mathematically recursive music, one of Hofstadters main objectives all along was to build-up to his own unique take on emergence, specifically how the mind should be considered an emergent level of the brain, thesis which is spelled out in the last pages of *Gödel, Escher, Bach*. For Hofstadter, emergence is not a simple matter of crisply demarcated strata but, instead, is a far more interwoven, entangled, messier process and novel emergent levels express precisely this interwovenness, entanglement, and messiness, features which Hofstadter termed "tangled hierarchies" and "strange loops" since the "higher" and "lower" level constituted each other in a cross-referential fashion.

With this entangled/interactive view of an emergent level, a whole new slant became possible on such issues as the downward causation or macro-determination of the "higher" emergent level (e.g., of the mind) on the "lower" level (e.g., of the brain) and *vice versa* that the neuroscientist Roger Sperry (1986) had argued for in his emergentist theory of consciousness. Indeed, the very idea of downward causation, as it is typically formulated, presumes level distinctions crisp enough to make the idea of an influence of one on the other thinkable at all. In this regard, it perplexes me that in the face of so much prattle being written these days about downward causation in relation to emergence, few putatively complex systems-oriented thinkers are familiar with territory already plowed so rigorously and creatively as Hofstadter has accomplished.

To be sure, although hierarchical descriptions do tend to be of help to researchers in navigating through often impenetrable thickets of complex systems by isolating and then focusing on the dynamics at a crisply defined level, a focus made possible by heuristically ignoring details at lower or higher levels (see, e.g., Salthe, 1993; and O'Neill, et al., 1986), the above remarks of Conger and later complexity theorists reveal that things in nature are most often not so crisp. In this context, it is important to keep in mind an insight into theory-making in biology (and I would add: in complexity science as a whole) put forward the philosopher of biology Morton Beckner (1974), namely, that there is a critical difference between the hierarchies of levels constituting natural systems and the hierarchies of levels devised in our theories of systems.

I propose that the discrepancy between understanding the idea of a "higher" emergent level in the sense of a crisp, sharply defined demarcation and the theoretical cautions enumerated above concerning the arbitrariness and thicket of interactions involved in conceptions of an emergent level indicate that the very concept of an emergent level is one approaching a genuine paradox. On one horn of this paradox is the decided sense that an emergent level is a crisp category, much of this crispness resulting from what I (Goldstein, 2002) have elsewhere referred to as the "singular" or "privileged" status of emergent levels in distinction to the arbitrary designation of a particular scale. Scale is arbitrary in the sense that it depends on an arbitrarily set power of resolution used in observing a system, e.g., the power of resolution of different kinds of microscopes determines the specific scale at which an investigation utilizing such microscopes can proceed. However, an emergent level is not arbitrary since it is just that level at which, phenomenologically, emergent phenomena are observed ostensively.

It is the supposed obviousness of this crispness of emergent level that supports the artificial life researcher Andrew Wuensche (1998) in his contention that much of the promise of artificial life lies in our computational capability for tracking and measuring multiple, sharply-delineated levels simultaneously for the purpose of ascertaining the laws relating one level to another. At first impression, it does seem to be the case that the emergent levels of such artificial life standbys as the Game of Life are glaringly apparent, at least, in our quick perception that the movement of "gliders", for example, is on a "higher" emergent level than that of the substrate constituted by the on/off status of the cells. Yet, more recent research into artificial life, specifically the Game of Life, reveals that the issue of exactly which levels there are as well as the crispness of these emergent levels is certainly much more complicated and also more open for different interpretations than previously assumed (see, e.g., Griffeath & Moore, 2003). This implies, for me at least, that, if, even in such artificially devised scenarios of emergence as computational simulations where everything, in an important sense, is potentially measurable (if we only know, that is, where exactly to point our measuring instruments at and what exactly to use for appropriate metrics), the nature of emergent levels are not crisp but an admixture and commixture with other levels, then there is in general much more room for equivocations on how many levels there are, ambiguity about how these levels form by interacting among themselves, their transience, and so forth.

Although I think that Hofstadter with his labyrinthian tangled hierarchies and Wimsatt with his causal thickets have come the closest to exhibiting how the idea of an emergent level flirts with paradox, I don't think they have gone far enough for it is the case that entangled knots can be untangled (as I've seen methodical fisherman do with their long lines in a little fishing village north of Viña del Mar, Chile) and thickets of briar thatches can even be traversed (with sharp enough machetes and a willingness to be inflicted with superficial flesh wounds)! That is why, in contrast to tangles and thickets, I am proposing a more fundamental shift in our understanding of what it is that happens when an emergent level emerges: the lower level is so taken-up into the new emergent level by varied complexification operations that the former no longer exists as the *separate* echelon it had previously, but rather gets *confounded* up into the new emergent level according to the literally etymological sense of "confounded," that is, a "pouring together" so thoroughly, it is not possible even in principle to detect what was separate before the pouring together.

An image I have of this confounding is of traveling on a bridge which strangely curls -p behind itself as one traverses it, a bridge operating in just the opposite and reverse manner of the bridge of reductionist explanations which also curl-up behind itself as it traverses from the higher to the lower level. In the bridge of confounding, once the higher level is reached, the lower level has been subsumed into it so radically that lower level components are now not the same separate parts they were before but instead are themselves reconfigured within the confoundedness of the higher level coherence. Confounding, then, transgresses those level distinctions crisp enough to even allow for entanglement. The example of emergence that comes to mind as the most illustrative of confoundedness in a visualizable manner is that of symbiogenesis, one instance being the remarkable theory propounded by Lynn Margulis of the emergence of the first eukaryotic cells through endosymbiosis between primitive prokaryotes (see Reid, 2007). The first eukaryotic cells paradoxically were crisp in their boundaried constitution as separate entities yet their internal constitution was also a confoundedness of their pre-endosymbiotic parts.

Understanding an emergent level as confounded in this sense is definitely not meant to add greater mysteriousness to the construct of emergence such that Samuel Alexander's call for "natural piety" in the face of emergence should be taken as the most appropriate response to it. However, "confounded" does connote something obdurate to easy understanding as shown in the following quotation from Shakespeare's "The Phoenix and the Turtle":

Reason, in itself, confounded,

Saw division grow together,

to themselves yet either neither,

Simple were so well compounded...

By qualifying emergent level as *confounded* in this sense, I might be judged guilty of the logical imprudence of *obscurum per obscurius*, that is, trying to explain something obscure by appealing to something even more obscure, a conceptual ruse not infrequently found even more the most eminent of complexity thinkers! However, I believe that qualifying emergent levels as confounded can escape this charge by indicating how to follow through on another of Conger's prescient insights, namely, his contention that much of the discourse on emergence in his day and age was way too general and indefinite. His remedy was to call for testable hypotheses to initiate promising avenues for detecting processes of emergence. Similarly, by considering an emergent level as essentially confounded, I am also calling for greater definiteness in our probing of emergence, probing deeper and more perspicaciously into those processes and operations of nature and society which possess the potency of generating emergent outcomes that are confounded in the manner I have proposed. That is, confoundedness is being put forward as a goad for greater probity. I have suggested a new logic for emergence along such a line called the logic of *self-transcending constructions* (see Goldstein, 2002, 2003, 2004, 2006, Forthcoming). This new logic flirts with but does not embrace paradox, just as I am claiming the notion of an emergent level flirts with but does not embrace the paradox of simultaneous crispness and entanglement. But also like the place of emergence in a scientific or philosophical explanation, an appeal to which I (Goldstein, 1999) have suggested functions more as the beginning and not the terminus of an explanation, so the logic of self-transcending constructions is more like an indexical pointing to where explanatory strategies employing the notion of emergence could start looking.

References

Alexander, S. (1966). *Space, Time, and Deity: The Gifford Lectures at Glasgow 1916-1918* (Volume II), NY: Dover.

Beckner, M. (1974). "Reduction, hierarchies and organicism," in F. Ayala and T. Dobzhansky (eds.), *Studies in the Philosophy of Biology: Reduction and Related Problems*, ISBN 9780333148600, pp. 57-63.

Bechtel, W. and Richardson, R. (1993). *Discovering Complexity: Decomposition and Localization as Strategies in Scientific Research*, ISBN 9780691087627.

Bedau, M. and Humphreys, P. (eds.) (2008). *Emergence: Contemporary Readings in Philosophy and Science*, ISBN 9780262524759.

Blitz, D. (1992). *Emergent Evolution: Qualitative Novelty and the Levels of Reality*, ISBN 9780792316589.

Bunge, M. (1979). *Treatise on Basic Philosophy: Volume 4: Ontology II: A World of Systems*, ISBN 9789027709448.

Ehresmann, A. and Vanbremeersch, J.-P. (2007). *Memory Evolutive Systems: Hierarchy, Emergence, Cognition*, ISBN 9780444522443.

Emmech, C., Køppe, S. and Stjernfelt, F. (1998). "Explaining emergence: Towards an ontology of levels," *Journal for General Philosophy of Science*, ISSN 0925-4560, 28: 83-119.

Feibleman, J. (1951). *Ontology*, Baltimore: Johns Hopkins University Press.

Feinberg, T. (2001). "Why the mind is not a radically emergent feature of the brain," *Journal of Consciousness Studies*, ISSN 1355-8250, 8(9-10): 123-145.

Freeman, A. (2001). "Editorial introduction," *Journal of Consciousness Studies*, ISSN 1355-8250, 8(9-10): vii-ix.

Goldstein, J. (1999). "Emergence as a construct: History and issues," *Emergence: A Journal of Complexity Issues in Organization and Management*, ISSN 1521-3250, 1(1): 49-62.

Goldstein, J. (2002). "The singular nature of emergent levels: Suggestions for a theory of Emergence," *Nonlinear Dynamics, Psychology, and Life Sciences*, ISSN 1090-0578, 6(4): 293-309.

Goldstein, J. (2003). "The construction of emergence order, or how to resist the temptation of Hylozoism," *Nonlinear Dynamics, Psychology, and Life Sciences*, ISSN 1090-0578, 7(4): 295-314.

Goldstein, J. (2004). "Flirting with paradox: Emergence, creative process, and self-transcending constructions," *Conference: Capital Science 2004*-Washington Academy of Sciences, Washington, D.C., March 20, 2004.

Goldstein, J. (2006). "Emergence, creative process, and self-transcending constructions," in K.A. Richardson (ed.), *Managing Organizational Complexity: Philosophy, Theory, and Application*, ISBN 9781593113186, pp. 63-78.

Goldstein, J. (Forthcoming). *Flirting with Paradox: Emergence, Creative Process, and Self-transcending Constructions in Complex Systems*, Litchfield Park, AZ: ISCE Publishing.

Griffeath, D. and Moore, C. (2003). *New Constructions in Cellular Automata*, ISBN 9780195137170.

Habermas, J. (1985). *The Theory of Communicative Action* (Volumes 1 & 2), T. McCarthy (trans.), ISBN 9780807015070.

Hofstadter, D. (1979). *Gödel, Escher, Bach: An Eternal Golden Braid*, ISBN 9780465026562.

Lovejoy, A. (1936). *The Great Chain of Being: A Study of the History of an Idea*, ISBN 9780674361539.

Mead, G. H. (2007). "The philosophy of the present" (with an introduction by J. Goldstein), *Emergence: Complexity and Organization*, ISSN 1521-3250, 9(3):75-96.

Medawar, P. (1974)."A geometric model of reduction and emergence," in F. Ayala and T. Dobzhansky, (eds.), *Studies in the Philosophy of Biology*, ISBN 9780333148600, pp. 57-63.

Morgan, C.L. (1923). *Emergent Evolution: the Gifford Lectures Delivered in the University of St. Andrews in the Year 1922*, NY: Henry Holt.

O'Neill, R.V., Deangelis, D.L., Waide, J.B. and Allen, G.E. (1986). *A Hierarchical Concept of Ecosystems*, ISBN 9780691084367.

Plotinus (1991). *The Enneads*, J. Dillon (ed.), S. Mackenna (trans.), ISBN 9780140445206.

Reid, R. (2007). *Biological Emergences: Evolution by Natural Experiment*, ISBN 9780262182577.

Rosen, R. (1969). "Hierarchical systems in biology and comments on the use of the term hierarchy," in L.L. Whyte, A.G. Wilson, and D. Wilson (eds). *Hierarchical Structures*, ISBN 9780444000699, pp. 52-53.

Salthe, S. (1993). *Development and Evolution: Complexity and Change in Biology*, ISBN 9780262193351.

Silberstein, M. (2001). "Converging on emergence: Consciousness, causation, and emergence," *Journal of Consciousness Studies*, ISSN 1355-8250, 8(9-10): 61-98.

Silberstein, M. and McGeever, J. (1999). "The search for ontological emergence," *The Philosophical Quarterly*, ISSN 0031-8094, 49: 182-200.

Stephan, A. (1999). *Emergenz: Von der Unvorhersagbarkeit zur Selbstorganisation*, ISBN 9783933168092.

Sperry, R. W. (1986) "Discussion: Macro- versus micro-determinism," *Philosophy of Science*, ISSN 0031-8248, 53: 265-70.

Sprigge, T. (1993). *James and Bradley: American Truth and British Reality*, ISBN 9780812692273.

Sulis, W. (2004). "Archetypal dynamical systems and semantic frames in vertical and horizontal emergence," *Emergence: Complexity and Organization*, ISSN 1521-3250, 6(3): 52-64.

Weiss, P. (1969). "The living system," in. A. Koestler and J. R. Smythies (eds.), *Beyond Reductionism: The Alpbach Symposium 1968 - New Perspectives in the Life Sciences*, NY: Macmillan, pp. 3-55.

Werkmeister, W.H. (1990). *Nicolai Hartmann's New Ontology*, ISBN 9780813010083.

Wheeler, W.M. (1927). *Emergent Evolution and the Social*, London: Psyche Miniatures, Gen. Ser. No. 11.

Wimsatt, W. (1972). "Complexity and organization," *Philosophy of Science Addendum*, K.F. Schaffner and R.S. Cohen (eds), PSA-1972 (Boston Studies in the Philosophy of Science, Volume 20), Dordrecht: Reidel, pp. 67-86.

Wimsatt, W. (2007). *Re-Engineering Philosophy for Limited Beings: Piecewise Approximations to Reality*, ISBN 9780674015456.

Wuensche, A. (1998). "Classifying cellular automata automatically," Santa Fe Institute Preprint 98-02-018. Santa Fe, NM: Santa Fe Institute.

VOL. XXII., No. 12 JUNE 4, 1925

THE JOURNAL OF PHILOSOPHY

THE DOCTRINE OF LEVELS [1]

PHRASES denoting differences of level are familiar, and I think are becoming increasingly frequent, in the literature of several sciences. One might trace a development from the literal meanings encountered in geology and archæology to the various metaphorical meanings found in history, sociology, psychology, epistemology, and metaphysics. As used in metaphysics, it is evident that such phrases have the fascination of great generalizations, but also the dangers of loose metaphors. Without attempting anything like a complete treatment in this paper, I should like to consider briefly four questions bearing upon the subject: (1) What is a metaphysical level? (2) What levels does the world exhibit? (3) How do later levels develop from earlier levels? (4) What are some of the more important applications of the doctrine of levels to the classical problems of philosophy?

1. *What is a metaphysical level?* A level may be defined as a class of structures or processes which are distinguishable from others as being either higher or lower. The difficulty for metaphysics is in fixing the meanings of the terms "higher" and "lower." These terms are used variously, to denote differences which are spatial, valuational, logical, and developmental. As denoting spatial differences (*e.g.*, "above" and "below") they are obviously inadequate for the more elaborate requirements of metaphysics; if they have any importance here, it is secondary. As denoting valuational differences (*e.g.*, "better" and "worse"), the terms are important for idealists and pragmatists, according to whose metaphysics the world becomes eventually harmonious and intelligible, or increasingly amenable to human control. It is impossible to pass lightly by these strongly held positions; but in order to look back at them from a point of view to be reached later, I shall assume that the valuational differences which might be denoted by the terms "higher" and "lower," although they are indispensable for us, are not intrinsic to other portions of the world, and hence are not entirely adequate for metaphysics. As denoting logical differences (*e.g.*, "more inclusive" and "less inclusive"), the terms "higher" and "lower" are im-

[1] Read at the twenty-sixth meeting of the Western Division of the American Philosophical Association, April 9, 1925.

portant for the neo-realists and critical realists who ascribe metaphysical reality to subsistents and essences, and include among these the logical relations as subtending, if not characterizing, the world of existences. These positions also are difficult to pass by; but in order to look at them from a point of view to be reached later, I shall assume that the logical differences which might be denoted by the terms "higher" and "lower," although they too are indispensable for us, are refinements of epistemology rather than essentials of metaphysics.

Assuming, then, for the sake of the argument, that spatial, valuational, and logical meanings of the terms "higher" and "lower" are, as noted above, too limited to be adequate for metaphysics, I would pass to the developmental meaning of the terms. Lloyd Morgan has defined them in accordance with what he calls "involution and dependence." "When two or more kinds of events . . . *A*, *B*, and *C*, co-exist in one complex system in such wise that the *C* kind involves the co-existence of *B*, and *B* in like manner involves *A*, whereas the *A*-kind does not involve the co-existence of *B*, nor *B* that of *C*, we may speak of *C* as, in this sense, higher than *B*, and *B* than *A*. . . . I speak of events at any given level in the pyramid of emergent evolution as 'involving' concurrent events at lower levels. . . . But when some new kind of relatedness is supervenient (say at the level of life), the way in which the physical events which are involved run their course is different in virtue of its presence—different from what it would have been if life had been absent. . . . I shall say that this new manner in which lower events happen—this touch of novelty in evolutionary advance—*depends on* the new kind of relatedness."[2] It might be said of this construction that whether or not it is top-heavy, its point of view is decidedly that of the upper levels, and that the results of development rather than the process of development are conspicuous. In order to emphasize the process, one might define a metaphysical level as a class of structures existing in relative, but only in relative, independence; distinguishable from others as occurring earlier or later in time, and as being correspondingly simpler or more complex; and (at least within the limits of investigation) observed or inferred to have been produced from structures of earlier and simpler levels, and in turn to produce structures of later and more complex levels. The last point might be made briefly by saying that the various levels are observed or inferred to be in transitive and asymmetrical relations of producers and produced.

2. *What levels does the world exhibit?* When a level is defined in the sense just indicated, there appear to be differences of opinion

[2] C. Lloyd Morgan, *Emergent Evolution* (1923), p. 15 f.

as to what levels are actually found in the world. S. Alexander distinguishes six—Space-Time, primary qualities, matter, secondary qualities, life, and mind [3]—and recognizes the possibility of a seventh, or deity. Alexander's criterion for a difference of level is apparently the emergence, in us, of the quality of consciousness from a lower level of complexity which is vital. Not every difference of complexity is a difference of level. If matter, for example, is only a compound of electrons, the atom would not be on a different level of existence from the electron, but as compared with it might be like the more complex forms of life as compared with the unicellular organism, displaying greater complexity of structure, but not of such an order as to lead to the emergence of a new quality. On each level, in fact, there may be variations within that order of existence which exhibit secondary differences so great as to be called in common parlance differences of quality or kind.[4] But only those emergent qualities which are comparable to mind in us mark true differences of level. I am not able to say whether Lloyd Morgan regards each instance of emergence as marking a difference of level; at any rate he holds that salient examples of emergence are afforded in the advent of life, in the advent of mind, and in the advent of reflective thought, but that in the physical world emergence is no less exemplified in the advent of each new kind of atom and of each new kind of molecule. He says that it is beyond the wit of man to number the instances of emergence.[5]

It is noticeable that these writers, and with them R. W. Sellars,[6] whatever their differences of opinion concerning subordinate or secondary levels may be, agree in the widespread view that at least the realms of matter, life, and mind, respectively, constitute different levels in the universe. It is noticeable also that in these discussions of a universe of levels, there is marked emphasis upon problems of epistemology. Alexander, as we saw, calls primary qualities and also matter with secondary qualities levels, and distinguishes levels by a criterion for which the emergence of consciousness furnishes the clue. With frequent provisos against misinterpretation of his terminology, he maintains that at each level of the universe the new quality which emerges may be called the mind of that level.[7] Lloyd Morgan and Sellars both emphasize the process by which we know the developing world of naturalism, and link their theories of emergent evolution with their earlier works in the theory of knowledge. It

[3] S. Alexander, *Space, Time, and Deity* (1920), Vol. II, p. 335.

[4] *Ibid.*, Vol. II, pp. 45, 54.

[5] L. Morgan, *op. cit.*, p. 1.

[6] R. W. Sellars, *Evolutionary Naturalism* (1922), pp. 8, 332.

[7] Alexander, *op. cit.*, Vol. II, p. 68.

may be added that Alexander and Sellars present elaborate discussions of various ontological categories. One finds that, measured by the work of three of its chief protagonists, the doctrine of levels has been treated in close connection with the epistemological traditions.

I am not concerned with a discussion of these positions, so much as I am with an approach to the problem of levels from another angle. My point is not merely that metaphysics should be "emancipated from epistemology"; it is that sometimes even those whose naturalism would help to effect the emancipation have exaggerated the importance of epistemology. In order to win a new approach to the problem of levels, it may be advisable to go outside the literature of philosophy in the more technical sense and to appropriate some material from the sciences, where the categories of epistemology and ontology are, to say the least, not very prominent. As a sample of such material, may I take the following quotation from an article by A. P. Mathews on "The Mechanistic Conception of Life"?

> "It is generally believed that living matter originated from lifeless; that at first there was a void or formless universe composed of ether or electrons or whatever the primitive stuff may have been and still is; that some of these electrons got arranged somehow into systems, making what we call the atoms of the chemical elements; that those atoms combined slowly into more complex systems, the molecules; that these molecules condensed or aggregated into larger masses, meteors, comets, stars, suns, worlds; that these masses in their turn became arranged into systems, the solar system, great nebulæ, universes, and so on; that on the surface of one of these planets of one of these systems water condensed, amino acids and other organic and inorganic substances were formed, principally by the action of light; that these amino acids condensed or agglomerated into larger groups which may be called micellæ or colloid particles; and that finally these micellæ aggregated into still larger groups of cells and unicellular organisms, and these cells aggregated into larger multicellular animals and plants."[8]

I am interested in the possibilities which I think are open for attempts to enumerate the metaphysical levels in terms of such current scientific theories, and particularly in attempts to extend the enumeration, as Mathews does not extend it here, beyond the realm of biology and make it include the various levels of mind. It is not possible to be very definite in such an enumeration, or to be sure of the accuracy of some of the assumptions involved; but I think the scientific evidence is becoming clear enough so that one may tentatively consider a world of let us say about twenty-five levels, which I would enumerate somewhat as follows: (1) energies, (2) electrons,

[8] A. P. Mathews, "The Mechanistic Conception of Life," *Scientia*, Vol. 36, p. 243 (1924).

(3) atoms, (4) molecules, (5) astronomical masses, or bodies, such as planetesimals, (6) solar systems, (7) star clusters, (8) galaxies, or great spiral nebulæ, (9) possibly one or more groups of spiral nebulæ —that is, one or more astronomical "universes." Then, in the biological realm, let us say (10) organic compounds, (11) infra-cellular organisms, such as Mathews calls micellæ, (12) unicellular organisms, (13) multicellular organisms, (14) plant-and-animal groups, and then several levels exhibiting more complex types of social organization, such as (15) families or tribes, (16) nations or races, and, one might go on to say, (17) "The Great Society." Once more, in the neuropsychological realm, one might add levels for (18) specialized cells, (19) nervous areas where the conduction is indiscriminate or reversible, (20) reflex arcs, and then doubtless for several different levels of higher neural synthesis, providing for (21) complex reflexes such as are apparently operative in perception and language-reactions, (22) instinctive-emotional complexes, (23) sentiments, (24) values, and, as a final term, (25) personalities.

Of many points which might be considered in connection with this list of levels, let us mention three very briefly. First, the levels here enumerated are classes of structures, with their characteristic processes, which are observed and experimented upon in the sciences, and the problem of levels is thus stated in terms more amenable to empirical methods. In the second place, none of these levels now exists apart from the others, and doubtless they all interact. In the third place, the list, of course, shows sharp breaks in continuity, corresponding to the well-recognized distinctions between the so-called realms of matter, life, and mind. If each of the twenty-five or so classes of structures is to be called a level, there ought to be some other term to indicate the fact that each of these three realms or series of levels is so markedly different from the others. The term "realm" or "series" might be used to indicate this; but in view of considerations which I hope to develop later I would prefer the term "cycle."

3. *How do later levels develop from earlier levels?* This is, of course, the crux of the whole problem. There seem to be three slightly differing emphases in contemporary theories. We may distinguish them according as the emphasis is upon (1) integration, or creative synthesis, (2) combining relations, or mutuality of relations, and (3) emergence of new qualities. They are not mutually exclusive, and in fact usually overlap; but it is convenient to consider them separately.

The first emphasis, on integration or creative synthesis, recalls the work of Herbert Spencer and other older writers, and is represented among contemporary thinkers, for example, by E. G. Spaulding and Sellars. As Spaulding puts it, "in the physical world (and elsewhere) it is an established empirical fact that parts as non-additively organized form a whole which has characteristics that are qualitatively different from the characteristics of the parts. A simple and familiar illustration of this is the formation of water out of hydrogen and oxygen. . . . This process of the formation of new qualities through the organization of parts into wholes may be called creative synthesis. . . . The atom differs from the electrons that compose it, the molecule from its constituent atoms, the cell from its molecules and colloidal particles, and finally society from the human individuals that are its units. Indeed, no realm of fact, whether subsistent or existent, is exempt from this principle of creative synthesis, in accordance with which one or more specific organizing relations so relate parts that there are new qualities in the resulting whole, and whole and part belong to specifically different universes of discourse."[9] According to Sellars, "evolution means that there are levels in nature, that the higher is an outgrowth of the lower, that *A* and *B* integrated are more than *A* and *B* separated."[10]

The second emphasis, on combining relations or mutuality of relations, has just appeared in the quotation from Spaulding. It is noteworthy also in the work of Lloyd Morgan, who says that "what is supervenient at any emergent stage of evolutionary progress is a new kind of relatedness—new terms in new relations—hitherto not in being. In virtue of such new kinds of relatedness not only have natural entities new qualities within their own proper being, but new properties in relation to other entities."[11] Within the field of the social sciences M. P. Follett, developing Holt's work on circular response and Köhler's *Gestaltpsychologie,* has argued that social wholes are constituted by the progressively changing mutual relations of individuals. "In human relations . . . I never react to you but to you-plus-me; or to be more accurate, . . . it is I plus the-interweaving-between-you-and-me meeting you plus the-interweaving-between-you-and-me, etc., etc. If we were doing it mathematically we should work it out to the *n*th power. . . . We should obtain a differential equation or a set of differential equations to be solved by integration. . . . The principles of relation as given by differential calculus help us to a clear understanding of this fundamental principle of life."[12] Miss Follett thinks that we must avoid both the concepts of a super-whole and a static whole; "what the psychologist must do in his field and the social scientist in his is to study the whole a-making; this

[9] E. G. Spaulding, *The New Rationalism* (1918), pp. 447 f, 500.
[10] R. W. Sellars, *op. cit.,* p. 329.
[11] Lloyd Morgan, *op. cit.,* p. 19.
[12] M. P. Follett, *Creative Experience* (1924), pp. 62–64.

involves a study of whole and parts in their active and continuous relation to each other." [13]

The third emphasis, on the emergence of new qualities, is especially familiar in the work of Alexander and Lloyd Morgan. According to Alexander, "The emergence of a new quality from any level of existence means that at that level there comes into being a certain constellation or collocation of the motions belonging to that level, and possessing the quality appropriate to it, and this collocation possesses a new quality distinctive of the higher complex. The quality and the constellation to which it belongs are at once new and expressible without residue in terms of the processes proper to the level from which they emerge. . . . If, to borrow the language of Mr. Lloyd Morgan, the processes of a particular level are represented as *a* processes, a constellation of such processes is of such a kind as to be a new process *ab* with its quality *B*. That is, the thing which is based on that constellation of *a* processes has an emergent quality *B*, whose behavior consists in *ab* processes; and though *ab* processes are also *a* processes they are not merely such." [14] In more general terms, both Alexander and Lloyd Morgan interpret this emergence which they find to be characteristic of the universe in terms of what they call a "nisus," or pressing of the universe towards something unattained.[15]

It may be said of all these attempts to account for the various levels of the universe, and the development of later levels from earlier, that while they are perhaps true enough in their general terms, they still lack precision and definiteness. They all need to be tested, and if possible developed, by reference to the data and inferences of the sciences regarding such levels as those enumerated in Section 2, above. I think that there are promising possibilities of detecting processes which, if not operative at all the levels, are operative at levels typical enough so that the processes may be generalized and regarded as metaphysical principles.[16] As an indication of possibilities along this line, I might suggest some hypotheses. I am not yet altogether sure about these results; but I believe that sooner or later attempts like this will have to be made, if philosophy is to deal adequately with the questions presented in the amazing developments of the sciences.

We may regard it as commonly agreed upon that the process of evolution is marked by the integration of individuated units of a

13 *Ibid.*, p. 102.

14 Alexander, *op. cit.*, Vol. II, p. 45 f.

15 L. Morgan, *op. cit.*, p. 34.

16 *Cf.* Alexander, *op. cit.*, Vol. II, pp. 54–55. I have discussed the subject in Appendix B of *A Course in Philosophy* (1924).

lower level, let us say *n*, into individuated units of a higher level, let us say enumeratively $n+1$. Integration is, however, by no means the only process which is going on; each individuated unit at its own level *n* maintains itself by interactions with other units at level *n*, and with units of levels $n-1$, $n-2$, *etc.* Atoms, for example, are not merely integrated into molecules, but at their own level take on or give off electrons, and absorb and radiate energy. The material, the biological and neuropsychological cycles are so different that we call these processes of interaction by different names as the higher levels and cycles are reached. In the first, or lowest, cycle we have absorption and emission; in the second cycle, that of the biological organisms, we have ingestion and excretion, with their accompanying growth and decay; and in the third, or neuropsychological cycle, we have the processes which we call sensation and response. I think that these processes, as we pass from lower to higher levels, may be called continuous, but not for that reason identical. It seems to me that at least some of the mysteries connected with the advent of new qualities may become a little clearer if the appearance of new qualities is traced not merely to processes of integration, but also to processes of cumulative interaction between units of various levels. Let us take for example the wetness of H_2O. When molecules of H_2O interact with molecules of certain other substances, for example, those of a stone, within a certain range of temperature, the H_2O adheres to the stone. If instead of, or along with, its adhesion to the stone, the H_2O adheres to an apprehending organism, we say the water, or the stone, or the organism, is wet. If any one insists, he may maintain that the H_2O without the stone had the quality of adhesiveness, and without the organism had the quality of wetness; but if one must consider such abstractions as the H_2O without other bodies to act on and without organisms, it seems to me more suitable to the abstraction to say that the H_2O is potentially adhesive, or is potentially wet, or has potentialities of adhesiveness or wetness. Potentiality seems to be a category which is read into structures or processes at one level in the light of some other level. Integration is a realignment of parts, and a realignment which is potent. But just how it is potent is impossible to say *in abstracto*. The potentialities of the new unit are unknown until its interactions reveal them, and identify them as qualities of the new unit. Perhaps the secondary qualities and even the primary qualities could be interpreted in some similar way.

Besides the processes of interaction, and along with them, there are at least at some levels processes of repetition, or reproduction of units of a given level *n* from other units of the same level. Without going into details of these complicated subjects, let it suffice to say that other processes besides integration must be reckoned with as important for evolution from lower to higher levels. The process of integration, too, is particularly complicated by its relations with processes of differentiation. For example, molecules may be integrated from atoms, which we have called units of a lower level; but molecules may also arise as differentiations within a nebula or other astronomical body. Similarly, unicellular organisms may be integrated from infra-cellular units; but some cells certainly arise by processes of differentiation in multicellular organisms. Again, reflex arcs may be regarded as integrated from lower levels, or differentiated within units of higher levels. Apparently the process of integration of units of a level n into those of a level $n + 1$ is likely to be overlaid by later differentiations of units comparable to those of level n within the new units of level $n + 1$. In order to be sure whether there is real difference of level we must apparently look beyond the mere fact of togetherness or combining relations. It is for this reason, among others, that it is important to consider the other processes besides integration which as we have said are also at work in evolution. The test for difference of level seems to be the maintenance by the combination of units, which once severally belonged to level n, of a new, relatively independent individuated existence at the level $n + 1$, with processes of interaction, perhaps reproduction, and even in time perhaps new integration, proper to units of level $n + 1$, forming units of level $n + 2$, *etc.* And our immediate problem in connection with the doctrine of levels may be said to be to find characteristics or conditions of units at level n which promote their integration into units of level $n + 1$. It seems to me that examination of the data suggests two or three such characteristics.

In the first place, it is safe to say that all these individuated units, at all levels, exhibit what may be called individual variations, even if, in extreme cases, the variations pertain only to differences in spatio-temporal coördinates. And it is clear enough that integrations, when they take place among units of level n, take place among units which at that level differ somewhat from one another. Perhaps these individual differences, when large numbers of cases are taken, appear in the aggregate, in the material cycle, as residual attractions, which, according to several theories, are evident to us in integrations marked by the chemical affinity of atoms, the cohesion or adhesion of molecules, and the gravitation of larger masses. There seems to be something continuous, but not identical with such residual attraction in the biological realm, evident in the way in which individually varying organisms in a group adjust their individual insufficiencies and superfluities in what we call mutual dependence upon one another. Again, there seems to be something continuous, but not identical, with this residual attraction and mutual dependence, in the implications, references, and meanings of our individually varying single acts and ideas, which frequently point beyond themselves to larger systems of acts and ideas, usually purposive. From this point of view, one might define a meaning as the part which a relatively isolated act or idea has in a system, such as a sentiment or value, belonging to a higher level. We need to remember here also that processes of differentiation are at work, as well as processes of integration; but in any case, individual variation seems to be an important factor in the result.

The problem of integration can perhaps be simplified by asking what individual variations seem to be suited to the process of integration; and this question can perhaps be partially answered by calling attention to certain variations in dynamic properties. These variations in dynamic properties are not clearly made out for all the levels, but there seems to be a tendency for integrations among units which vary in the rate at which their interactions go on; units of level n which interact more slowly with units of level n, $n-1$, *etc.*, seem to integrate with other units of level n which react more rapidly, and those which interact more slowly and those which interact more rapidly seem to be locked together in the newly integrated units of level $n+1$. In several cases at least, as more and more complicated newly integrated structures of a level $n+1$ are developed, the n units whose processes are slower apparently tend toward positions in the inner portions of the new unit of $n+1$, while the n units whose processes are more rapid apparently tend toward positions in the outer portions. Possibly one may go a step further, and say that the dynamic variations which tend to bring about such distribution in the newly integrated units also make possible a regulation of the more rapidly interacting outer portions by the more slowly interacting inner portions. Examples of this may be seen in the "binding of electrons" by atoms, in the slowing down of the speed of revolution of a satellite to the speed of rotation of the primary body, in the relations between the nucleus and cytoplasm of a cell, between the digestive and the muscular systems of a multicellular organism, in processes of control in psychology, and perhaps in other cases.

It should be noted that, while the foregoing hypotheses may apply to differences of level in general, some special considerations may be necessary in order to do justice to the data and the inferences concerning the more striking differences which we called differences between series or cycles. I think that the most fruitful suggestion here is the hypothesis that the relation between material and biological cycles can be clarified by analogy with that between biological and neuropsychological cycles. Perhaps the latter member of each of these pairs is a means of coördination, equilibration, and speeding up for the former. At any rate it ought to be clear for philosophy that, along with the other ways of studying the living organisms, there ought to be more studies of them as a whole in their relation to the earth as a whole.

I am quite aware that none of these hypotheses looking toward more detailed descriptions of the process of integration and of passage from earlier to later levels needs to be regarded as any ultimate explanation of the process. Perhaps it will never be possible to demonstrate the emergence of novelty. One reason for this may be our own value-centric and persono-centric predicament. The very principle of novelty demands that we should be new at our level and different from the levels below us; and novelty, accordingly, must be recognized, rather than demonstrated or explained. Or, if there is to be explanation, other types of explanation, held to be more nearly ultimate, may be sought within the data, or outside and beyond the data. For every integration there may be an Integrator, for every level a Leveler. My point is that in some way such as I have indicated our theories can be empirically enriched by examination of the data. While we may need "natural piety" as much as ever, we can direct a larger portion of it toward familiar processes, and less of it toward processes which we give up as utterly obscure.

4. *What are some of the more important applications of the doctrine of levels to the classical problems of philosophy?* Merely in order to indicate the importance of the doctrine of levels, one should mention a number of its applications, although none of them can be developed at length within the necessary limits of this paper. (1) In terms of the doctrine of levels, evolution can be interpreted more precisely and accurately in terms of progress. According to Lloyd Morgan, progress is a "net result."[17] (2) According to Sellars, Lloyd Morgan, and others, the doctrine of levels affords a reconciliation of mechanism and vitalism in biology.[18] (3) According to Alexander, Sellars, and others, it affords a fruitful answer for the mind-body problem.[19] (4) In accordance with the work of Bergson and Alexander, it affords more definite ground for a concept of futurity, and a view that the future is essentially unpredictable.[20] (5) According to Spaulding, Alexander, and others, it affords a notable solution for the problem of freedom.[21] One may say that

17 Lloyd Morgan, *op. cit.*, p. 32.

18 Sellars, *op. cit.*, p. 329 f; Lloyd Morgan, *op. cit.*, pp. 8, 12.

19 Alexander, *op. cit.*, Vol. II, pp. 3, 7; Sellars, *op. cit.*, p. 334.

20 Bergson, *Creative Evolution*, translated by A. Mitchell (1911), p. 341; Alexander, *op. cit.*, Vol. II, Book IV.

21 Spaulding, *op. cit.*, p. 448 f; Alexander, *op. cit.*, Vol. II, p. 326 f.

fatalism is true at the level of reflexes, whereas determinism (in the sense of self-determination, involving personal responsibility) is at the same time true on the social and ethical level which we occupy. (6) Within the field of psychology, the doctrine may be made to interpret the significance of purposiveness. A purpose is a temporarily inhibited tendency which organizes or integrates activities with reference to it. Thus all purposiveness may be said to be organization, but all organization, which is characteristic of all levels, is not necessarily purposive. (7) The doctrine of levels reënforces the oft-desired and oft-proclaimed revision of our logical and ontological categories to make room for creativity and novelty, and to make clear that continuity does not necessarily mean identity. (8) With reference to the position of the idealists, personalists, and pragmatists, mentioned early in this paper, one may say that according to such a doctrine of levels as above developed, mind and personality are unique in the universe, but they are not universal. They are not fundamental, but apical. Yet, according to a comparative study of the structures and processes of the various levels such as I have been sketching, it is possible that the significance of mind and personality is not exhausted by placing them at the upper extremes of a long series of levels. It is possible that their structures and processes are not merely apical, but also typical of the universe as a whole; and although the world is not a Great Mind, the mind may turn out to be a little world, which at its own level gives us at least some light upon the nature of the whole. (9) With reference to the position of the realists, also mentioned early in this paper, one may say that the fact that the units of all levels exhibit cumulative interactions with units of lower levels goes far to indicate that the interactions of minds with bodies and environments, which interactions make up the subject-matter of epistemology, are special and highly complicated cases of the general types of interactions exhibited at other levels. In order to interpret these interactions at the level of mind it may be necessary to introduce subsistents, essences, floating adjectives, and similar concepts or realities; but it does not appear to be necessary to regard these as characteristic of the universe at all levels, or of the universe as a whole. One might say that if essences are really essential, they are essential to epistemology rather than to metaphysics. (10) Finally, in the field of philosophy of religion, one may say that the doctrine of levels portrays a universe which tends naturally to produce higher and higher levels. At present the highest level seems to be at or near the one which we occupy, or better, which we *are*. Perhaps, as Alexander suggests, in the future there will be levels higher than any which we have known. A philosophy of religion and psychology of religion which are moving away from the idea that matter and spirit are opposites, and which are sometimes referring to the universe, rather than anything outside or beyond the universe, as the object of religious devotion may, with the help of the doctrine of levels, measure, if they do not define, spirit as difference of level in matter, and spirituality as difference of level in ourselves.

GEORGE P. CONGER.

UNIVERSITY OF MINNESOTA.

26. The Role of Somatic Change in Evolution

Gregory Bateson (with an introduction by Hugo Letiche)

Originally published as Bateson, G. (1963). "The role of somatic change in evolution," Evolution, 17: 529-539. Reprinted with kind permission.

In the article reprinted here Bateson studies 'evolution' from the perspective of three factors: (i) fundamental (genotype) structural change, (ii) adaptation (somatic) response to environment/circumstances, and (iii) environmental change. He asserts that evolution is a matter of the interrelationships between these three factors in the perspective of natural selection. Bateson argues that adaptation is more surface-level and reversible than is structural change. But the economics of flexibility reveal that we can only do so much adapting before its possibilities and resources are exhausted. In principle, adaptation costs much more energy/effort than does mutation. Change must not be analyzed in isolation—the 'pre-giraffe' may have a longer neck (better to drink with), but its heart would be stressed and under pressure.

All change operates via complex interrelationships. Change is never local; it is always interlocked into complex fields of relatedness. Bateson stresses that 'fast change' is not a good thing. Control or identity and continuity lag behind what is controlled. In 'fast change', control will fail and the pressures on the organism will escalate. Too much flexibility or change on the individual level will destroy variability or the diversity of population. Individuals will be stressed to meet immediate life-threatening challenges leaving little or no room for serendipity or eccentricity. In a society that celebrates flexibility, it is important to understand Bateson's paradox. The more flexibility there is, the less real (new) difference the system can support.

Bateson rounds off his analysis with a comparison between 'regulators', 'adjustors' and 'extraregulators'. Adjustors let the environmental variables enter into the organism and the organism must then cope with the results. Regulators negotiate their relationships with their environment at their boundaries (such as at the skin) and keep their internal housekeeping (relatively) stable. According to Bateson the evolutionary trend seems to be towards 'extraregulators' who project their control mechanisms into their environment. Thus organizations and cultures are means of 'controlling' the self/environment relationship in the environment.

Bateson's concepts call up some very challenging questions. Is the current emphasis on 'flexibility' in organizations an evolutionary regression to 'adjustment' that weakens 'regulation' and endangers 'extraregulation'? If so, does this pose a fundamental danger to society and economy? Research, analysis and debate into the organization (organism)/environment relationship inspired by Bateson is important, challenging and innovative. Because of his communications/interactive/coevolving perspective, complexity studies is the right place for it to happen.

THE ROLE OF SOMATIC CHANGE IN EVOLUTION[1]

GREGORY BATESON
Veterans Administration Hospital, Palo Alto, California

Received November 8, 1962

All theories of biological evolution depend upon at least three sorts of change: (a) change of genotype, either by mutation or by redistribution of genes; (b) somatic change under pressure of environment; and (c) changes in environmental conditions. The problem for the evolutionist is to build a theory combining these types of change into an ongoing process which, under natural selection, will account for the phenomena of adaptation and phylogeny.

Certain conventional premises may be selected to govern such theory building:

a. *The theory shall not depend upon Lamarckian inheritance.* August Weismann's argument for this premise still stands. There is no reason to believe that either somatic change or changes in environment can, in principle, call (by physiological communication) for appropriate genotypic change. Indeed, the little that we know about communication within the multicellular[2] individual indicates that such communication from soma to gene script is likely to be rare and unlikely to be adaptive in effect. However, it is appropriate to attempt to spell out in this essay what this premise implies:

Whenever some characteristic of an organism is measurably modifiable under measurable environmental impact or under measurable impact of internal physiology, it is possible to write an equation in which the value of the characteristic in question is expressed as some function of the value of the impacting circumstance. "Human skin color is some function of exposure to sunlight," "respiration rate is some function of atmospheric pressure," etc. Such equations are constructed to be true for a variety of particular observations, and necessarily contain subsidiary propositions which are stable (i.e., continue to be true) over a wide range of values of impacting circumstance and somatic characteristic. These subsidiary propositions are of different logical type from the original observations in the laboratory and are, in fact, descriptive not of the data but of *our* equations. They are statements about the form of the particular equation and about the values of the parameters mentioned within it.

It would be simple, at this point, to draw the line between genotype and phenotype by saying that the *forms and parameters* of such equations are provided by genes, while the impacts of environment, etc. determine the actual event within this frame. This would amount to saying, e.g., that the *ability* to tan is genotypically determined, while the amount of tanning in a particular case depends upon exposure to sunlight.

In terms of this oversimplified approach to the overlapping roles of genotype and environment, the proposition excluding Lamarckian inheritance would read somewhat as follows: In the attempt to explain evolutionary process, there shall be no assumption that the achievement of a particular value of some variable under particular circumstances will affect, in the gametes produced by that individual, the form or parameters of the functional equation governing the relationship between that variable and its environmental circumstances.

---

[1] This essay is a by-product of research conducted at the Veterans Administration Hospital, Palo Alto, and sponsored by the Palo Alto Medical Research Foundation under the NIMH Grant OM-324.

[2] The problems of bacterial genetics are here deliberately excluded.

Such a view is oversimplified, and parentheses must be added to deal with more complex and extreme cases. First, it is important to recognize that the organism, considered as a communicational system, may itself operate at multiple levels of logical typing; i.e., that there will be instances in which what were above called "parameters" are subject to change. The individual organism might as a result of "training" change its ability to develop a tan under sunlight. And this type of change is certainly of very great importance in the field of animal behavior, where "learning to learn" can never be ignored.

Second, the oversimplfied view must be elaborated to cover *negative* effects. An environmental circumstance may have such impact upon an organism unable to adapt to it, that the individual in question will in fact produce *no* gametes.

Third, it is expectable that some of the parameters in one equation may be subject to change under impact from some environmental or physiologic circumstance other than the circumstance mentioned in that equation.

Be all that as it may, both Weismann's objection to Lamarckian theory and my own attempt to spell the matter out share a certain parsimony: an assumption that the principles which order phenomena shall not themselves be supposed changed by those phenomena which they order. William of Occam's razor might be reformulated: in any explanation, logical types shall not be multiplied beyond necessity.

b. *Somatic change is absolutely necessary for survival.* Any change of environment which requires adaptive change in the species will be lethal unless, by somatic change, the organisms (or some of them) are able to weather out a period of unpredictable duration, until either appropriate genotypic change occurs (whether by mutation or by redistribution of genes already available in the population), or because the environment returns to the previous normal. The premise is truistical, regardless of the magnitude of the time span involved.

c. *Somatic change is also necessary to cope with any changes of genotype which might aid the organism in its external struggle with the environment.* The individual organism is a complex organization of interdependent parts. A mutational or other genotypic change in any one of these (however externally valuable in terms of survival) is certain to require change in many others—which changes will probably not be specified or implicit in the single mutational change of the genes. A hypothetical pre-giraffe, which had the luck to carry a mutant gene "long neck," would have to adjust to this change by complex modifications of the heart and circulatory system. These collateral adjustments would have to be achieved at the somatic level. Only those pre-giraffes which are (genotypically) capable of these somatic modifications would survive.

d. In this essay, it is assumed that *the corpus of genotypic messages is preponderantly digital* in nature. In contrast, the soma is seen as a working system in which the genotypic recipes are tried out. Should it transpire that the genotypic corpus is also in some degree analogic—a working model of the soma—premise c (above) would be negated to that degree. It would then be conceivable that the mutant gene "long neck" might modify the message of those genes which affect the development of the heart. It is, of course, known that genes may have pleiotropic effect, but these phenomena are relevant in the present connection only if it can be shown, e.g., that the effect of gene A upon the phenotype and its effect upon the phenotypic expression of gene B are mutually appropriate in the overall integration and adaptation of the organism.

These considerations lead to a classifying of both genotypic and environmental changes in terms of the *price* which they exact of the flexibility of the somatic system. A lethal change in either environment or genotype is simply one which demands somatic modifications which the organism cannot achieve.

But the somatic price of a given change must depend, not absolutely upon the change in question, but upon the range of somatic flexibility available to the organism at the given time. This range, in turn, will depend upon how much of the organism's somatic flexibility is already being used up in adjusting to other mutations or environmental changes. We face an *economics* of flexibility which, like any other economics, will become determinative for the course of evolution if and only if the organism is operating close to the limits set by this economics.

However, this economics of somatic flexibility will differ in one important respect from the more familiar economics of money or available energy. In these latter, each new expenditure can simply be *added* to the preceding expenditures and the economics becomes coercive when the additive total approaches the limit of the budget. In contrast, the combined effect of multiple changes, each of which exacts a price in the soma, will be *multiplicative*. This point may be stated as follows: Let S be the finite set of all possible living states of the organism. Within S, let s_1 be the smaller set of all states compatible with a given mutation (m_1), and let s_2 be the set of states compatible with a second mutation (m_2). It follows that the two mutations in combination will limit the organism to the logical product of s_1 and s_2, i.e., to that usually smaller subset of states which is composed only of members common to both s_1 and s_2. In this way each successive mutation (or other genotypic change) will fractionate the possibilities for the somatic adjustment of the organism. And, should the one mutation require some somatic change, the exact opposite of a change required by the other, the possibilities for somatic adjustment may immediately be reduced to zero.

The same argument must surely apply to multiple environmental changes which demand somatic adjustments; and this will be true even of those changes in environment which might seem to benefit the organism. An improvement in diet, for example, will exclude from the organism's range of somatic adjustments those patterns of growth which we would call "stunted" and which might be required to meet some other exigency of the environment.

From these considerations it follows that if evolution proceeded in accordance with conventional theory, its process would be blocked. The finite nature of somatic change indicates that no ongoing process of evolution can result only from successive externally adaptive genotypic changes since these must, in combination, become lethal, demanding combinations of internal somatic adjustments of which the soma is incapable.

We turn therefore to a consideration of other classes of genotypic change. What is required to give a balanced theory of evolution is the occurrence of genotypic changes which shall *increase* the available range of somatic flexibility. When the internal organization of the organisms of a species has been limited by environmental or mutational pressure to some narrow subset of the total range of living states, further evolutionary progress will require some sort of genotypic change which will compensate for this limitation.

We note first that while the results of genotypic change are irreversible within the life of the individual organism, the opposite is usually true of changes which are achieved at the somatic level. When the latter are produced in response to special environmental conditions, a return of the environment to the previous norm is usually followed by a diminution or loss of the characteristic. (We may reasonably expect that the same would be true of those somatic adjustments which must accompany an externally adaptive mutation but, of course, it is impossible in this case to remove from the individual the impact of the mutational change.)

A further point regarding these reversible somatic changes is of special interest.

Among higher organisms it is not unusual to find that there is what we may call a "defense in depth" against environmental demands. If a man is moved from sea level to 10,000 feet, he may begin to pant and his heart may race. But these first changes are swiftly reversible: if he descends the same day, they will disappear immediately. If, however, he remains at the high altitude, a second line of defense appears. He will become slowly acclimated as a result of complex physiological changes. His heart will cease to race, and he will no longer pant unless he undertakes some special exertion. If now he returns to sea level, the characteristics of the second line of defense will disappear rather slowly and he may even experience some discomfort.

From the point of view of an economics of somatic flexibility, the first effect of high altitude is to reduce the organism to a limited set of states (s_1) characterized by the racing of the heart and the panting. The man can still survive, but only as a comparatively inflexible creature. The later acclimation has precisely this value: it corrects for the loss of flexibility. After the man is acclimated he can use his panting mechanisms to adjust to *other* emergencies which might otherwise be lethal.

A similar "defense in depth" is clearly recognizable in the field of behavior. When we encounter a new problem for the first time, we deal with it either by trial and error or possibly by insight. Later, and more or less gradually, we form the "habit" of acting in the way which earlier experience rewarded. To continue to use insight or trial and error upon this class of problem would be wasteful. These mechanisms can now be saved for *other* problems (Bateson, 1960).

Both in acclimation and in habit formation the economy of flexibility is achieved by substituting a deeper and more enduring change for a more superficial and more reversible one. In the terms used above in discussing the anti-Lamarckian premise, a change has occurred in the parameters of the functional equation linking rate of respiration to external atmospheric pressure. Here it seems that the organism is behaving as we may expect any ultrastable system to behave. Ashby (1945, 1952) has shown that it is a general, formal characteristic of such systems that those circuits controlling the more rapidly fluctuating variables act as balancing mechanisms to protect the ongoing constancy of those variables in which change is normally slow and of small amplitude; and that any interference which fixes the values of the changeful variables must have a disturbing effect upon the constancy of the normally steady components of the system. For the man who must constantly pant at high altitudes, the respiration rate can no longer be used as a changeable quantity in the maintaining of physiological balance. Conversely, if the respiration rate is to become available again as a rapidly fluctuating variable, some change must occur among the more stable components of the system. Such a change will, in the nature of the case, be achieved comparatively slowly and be comparatively irreversible.

Even acclimation and habit formation are, however, still reversible within the life of the individual, and this very reversibility indicates a lack of communicational economy in these adaptive mechanisms. Reversibility implies that the changed value of some variable is achieved by means of homeostatic, error-activated circuits. There must be a means of detecting an undesirable or threatening change in some variable, and there must be a train of cause and effect whereby corrective action is initiated. Moreover, this entire circuit must, in some degree, be available for this purpose for the entire time during which the reversible change is maintained —a considerable using up of available message pathways.

The matter of communicational economics becomes still more serious when we note that the homeostatic circuits of an organism are not separate but complexly interlocked, e.g., hormonal messengers which play a part in the homeostatic control of organ A will also affect the states of organs B, C, and D. Any special ongoing loading of the circuit controlling A will therefore diminish the organism's freedom to control B, C, and D.

In contrast, the changes brought about by mutation or other genotypic change are presumably of a totally different nature. Every cell contains a copy of the new genotypic corpus and therefore will (when appropriate) behave in the changed manner, without any change in the messages which it receives from surrounding tissues or organs. If the hypothetical pre-giraffes carrying the mutant gene "long neck" could also get the gene "big heart," their hearts would be enlarged without the necessity of using the homeostatic pathways of the body to achieve and maintain this enlargement. Such a mutation will have survival value not because it enables the pre-giraffe to supply its elevated head with sufficient blood, since this was already achieved by somatic change—but because it increases the overall flexibility of the organism, enabling it to survive *other* demands which may be placed upon it either by environmental or genotypic change.

It appears, then, that the process of biological evolution could be continuous if there were a class of mutations or other genotypic changes which would simulate Lamarckian inheritance. The function of these changes would be to achieve by genotypic fiat those characteristics which the organism at the given time is already achieving by the uneconomical method of somatic change.

Such a hypothesis, I believe, conflicts in no way with conventional theories of genetics and natural selection. It does, however, somewhat alter the current conventional picture of evolution as a whole, though related ideas were put forward over sixty years ago. Baldwin (1897) suggested that we consider not only the operation of the external environment in natural selection but also what he called "organic selection" in which the fate of a given variation would depend upon its physiologic viability. In the same article, Baldwin attributes to Lloyd Morgan the suggestion that there might exist "coincident variations" which would simulate Lamarckian inheritance (the so-called "Baldwin effect").

According to such a hypothesis, genotypic change in an organism becomes comparable to legislative change in a society. The wise legislator will only rarely initiate a new rule of behavior; more usually he will confine himself to affirming in law that which has already become the custom of the people. An innovative rule can be introduced only at the price of activating and perhaps overloading a large number of homeostatic circuits in the society.

It is interesting to ask how a hypothetical process of evolution would work *if* Lamarckian inheritance were the rule, i.e., if characteristics achieved by somatic homeostasis were inherited. The answer is simple: *it would not work,* for the following reasons:

1. The question turns upon the concept of economy in the use of homeostatic circuits, and it would be the reverse of economical to fix by genotypic change *all* the variables which accompany a given desirable and homeostatically achieved characteristic. Every such characteristic is achieved by ancillary homeostatic changes all around the circuits, and it is most undesirable that these ancillary changes should be fixed by inheritance, as would logically happen according to any theory involving an indiscriminate Lamarckian inheritance. Those who would defend a Lamarckian theory must be prepared to suggest how in the genotype an appropriate selection can be achieved. Without such a selection, the inheritance of acquired characteristics would merely increase the proportion of non-viable genotypic changes.

2. Lamarckian inheritance would disturb the relative timing of the processes upon which evolution must—according to the present hypothesis—depend. It is es-

sential that there be a time lag between the uneconomical but reversible somatic achievement of a given characteristic and the economical but more enduring alterations of the genotype. If we look upon every soma as a working model which can be modified in various ways in the workshop, it is clear that sufficient but not infinite time must be given for these workshop trials before the results of these trials are incorporated into the final blueprint for mass production. This delay is provided by the indirection of stochastic process. It would be unduly shortened by Lamarckian inheritance.

The principle involved here is general and by no means trivial. It obtains in all homeostatic systems in which a given effect can be brought about by means of a homeostatic circuit, which circuit can, in turn, be modified in its characteristics by some higher system of control. In all such systems (ranging from the house thermostat to systems of government and administration) it is important that the higher system of control *lag behind* the event sequences in the peripheral homeostatic circuit.

In evolution two control systems are present: the homeostases of the body which deal with tolerable internal stress, and the action of natural selection upon the (genetically) non-viable members of the population. From an engineering point of view, the problem is to *limit* communication from the lower, reversible somatic system to the higher irreversible genotypic system.

Another aspect of the proposed hypothesis about which we can only speculate is the probable relative frequency of the two classes of genotypic change: those which initiate something new and those which affirm some homeostatically achieved characteristic. In the Metazoa and multicellular plants, we face complex networks of multiple interlocking homeostatic circuits, and any given mutation or gene recombination which initiates change will probably require very various and multiple somatic characteristics to be achieved by homeostasis. The hypothetical pre-giraffe with the mutant gene "long neck" will need to modify not only its heart and circulatory system but also perhaps its semicircular canals, its intervertebral discs, its postural reflexes, the ratio of length and thickness of many muscles, its evasive tactics *vis-à-vis* predators, etc. This suggests that in such complex organisms, the merely affirmative genotypic changes must far outnumber those which initiate change, if the species is to avoid that *cul de sac* in which the flexibility of the soma approaches zero.

Conversely, this picture suggests that most organisms, at any given time, are probably in such a state that there are multiple possibilities for affirmative genotypic change. If, as seems probable, both mutation and gene redistribution are in some sense random phenomena, at least the chances are considerable that one or other of these multiple possibilities will be met.

Finally, it is appropriate to discuss what evidence is available or might be sought to support or disprove such a hypothesis. It is clear at the outset that such a testing will be difficult. The affirmative mutations upon which the hypothesis depends will usually be *invisible*. From among the many members of a population which are achieving a given adjustment to environmental circumstances by somatic change, it will not be possible immediately to pick out those few in which the same adjustment is provided by the genotypic method. In such a case, the genotypically changed individuals will have to be identified by breeding and raising the offspring under more normal conditions.

A still greater difficulty arises in cases where we would investigate those homeostatically acquired characteristics which are achieved in response to some innovative genotypic change. It will often be impossible, by mere inspection of the organism, to tell which of its characteristics are the primary results of genotypic change and which are secondary somatic adjustments to these. In the imaginary case of the pre-giraffe with a somewhat elongated neck and an enlarged heart, it may be easy to *guess* that the modification of the neck is genotypic while that of the heart is somatic. But all such guesses will depend upon the very imperfect present knowledge of what an organism can achieve in way of somatic adjustment.

It is a major tragedy that the Lamarckian controversy has deflected the attention of geneticists away from the phenomenon of somatic adaptability. After all, the mechanisms, thresholds, and maxima of individual phenotypic change under stress must surely be genotypically determined.

Another difficulty, of rather similar nature, arises at the population level, where we encounter another "economics" of potential change, theoretically distinguishable from that which operates within the individual. The population of a wild species is today conventionally regarded as genotypically heterogeneous in spite of the high degree of superficial resemblance between the individual phenotypes. Such a population expectably functions as a storehouse of genotypic possibilities. The economic aspect of this storehouse of possibilities has, for example, been stressed by Simmonds (1962). He points out that farmers and breeders who demand 100% phenotypic uniformity in a highly select crop are in fact throwing away most of the multiple genetic possibilities accumulated through hundreds of generations in the wild population. From this Simmonds argues that there is urgent need for institutions which shall "conserve" this storehouse of variability by maintaining unselected populations.

Lerner (1954) has argued that self-corrective or buffering mechanisms operate to hold constant the composition of these mixtures of wild genotypes and to resist the effects of artificial selection. There is therefore at least a presumption that this economics of variability within the population will turn out to be of the multiplicative kind.

Now, the difficulty of discriminating between a characteristic achieved by somatic homeostasis and the same characteristic achieved (more economically) by a genotypic shortcut is clearly going to be compounded when we come to consider populations instead of physiologic individuals. All actual experimentation in the field will inevitably work with populations, and, in this work, it will be necessary to discriminate the effects of that economics of *flexibility* which operates inside the individuals from the effects of the economics of *variability* which operates at the population level. These two orders of economics may be easy to separate in theory, but to separate them in experimentation will surely be difficult.

Be all that as it may, let us consider what evidential support may be available for some of the propositions which are crucial to the hypothesis:

1. *That the phenomena of somatic adjustment are appropriately described in terms of an economics of flexibility.* In general, we believe that the presence of stress A may reduce an organism's ability to respond to stress B and, guided by this opinion, we commonly protect the sick from the weather. Those who have adjusted to the office life may have difficulty in climbing mountains, and trained mountain climbers may have difficulty when confined to offices; the stresses of retirement from business may be lethal; and so on. But scientific knowledge of these matters, in man or other organisms, is very slight.

2. *That this economics of flexibility has the logical structure described above—each successive demand upon flexibility fractionating the set of available possibilities.* The proposition is expectable, but so far as I know there is no evidence for it. It is, however, worthwhile to examine the criteria which determine whether a given "economic" system is more appropriately described in additive or multiplicative

terms. There would seem to be two such criteria:

(a) A system will be additive insofar as the units of its currency are mutually interchangeable and, therefore, cannot meaningfully be classified into sets such as were used earlier in this paper to show that the economics of flexibility must surely be multiplicative. Calories in the economics of energy are completely interchangeable and unclassifiable, as are dollars in the individual budget. Both these systems are therefore additive. The permutations and combinations of variables which define the states of an organism are classifiable and—to this extent—non-interchangeable. The system is therefore multiplicative. Its mathematics will resemble that of information theory or negative entropy rather than that of money or energy conservation.

(b) A system will be additive insofar as the units of its currency are mutually independent. Here there would seem to be a difference between the economic system of the individual, whose budgetary problems are additive (or subtractive) and those of society at large, where the overall distribution or flow of wealth is governed by complex (and perhaps imperfect) homeostatic systems. Is there, perhaps, an economics of economic flexibility (a meta-economics) which is multiplicative and so resembles the economics of physiological flexibility discussed above? Notice, however, that the units of this wider economics will be not dollars but patterns of distribution of wealth. Similarly, Lerner's "genetic homeostasis," insofar as it is truly homeostatic, will have multiplicative character.

The matter is, however, not simple and we cannot expect that every system will be either totally multiplicative or totally additive. There will be intermediate cases which combine the two characteristics. Specifically, where several *independent* alternative homeostatic circuits control a single variable, it is clear that the system may show additive characteristics—and even that it may pay to incorporate such alternative pathways in the system provided they can be effectively insulated from each other. Such systems of multiple alternative controls may give survival advantage insofar as the mathematics of addition and subtraction will pay better than the mathematics of logical fractionation.

3. *That innovative genotypic change commonly makes demands upon the adjustive ability of the soma.* This proposition is orthodoxly believed by biologists but cannot in the nature of the case be verified by direct evidence.

4. *That successive genotypic innovations make multiplicative demands upon the soma.* This proposition (which involves *both* the notion of multiplicative economics of flexibility and the notion that each innovative genotypic change has its somatic price) has several interesting and perhaps verifiable implications.

(a) We may expect that organisms in which numerous recent genotypic changes have accumulated (e.g., as a result of selection, or planned breeding) will be delicate, i.e., will need to be protected from environmental stress. This sensitivity to stress is to be expected in new breeds of domesticated animals and plants and experimentally produced organisms carrying either several mutant genes or unusual (i.e., recently achieved) genotypic combinations.

(b) We may expect that for such organisms further genotypic innovation (of any kind other than the affirmative changes discussed above) will be progressively deleterious.

(c) Such new and special breeds should become more resistant both to environmental stress and to genotypic change, as selection works upon successive generations to favor those individuals in which "genetic assimilation of acquired characteristics" (Waddington) is achieved (Proposition 5).

5. *That environmentally induced acquired characteristics may, under appropriate conditions of selection, be replaced*

by similar characteristics which are genetically determined. This phenomenon has been demonstrated by Waddington (1953, 1957) for the *bithorax* phenotypes of *Drosophila*. He calls it the "genetic assimilation of acquired characteristics." Similar phenomena have also probably occurred in various experiments when the experimenters set out to prove the inheritance of acquired characteristics but did not achieve this proof through failure to control the conditions of selection. We have, however, no evidence at all as to the frequency of this phenomenon of genetic assimilation. It is worth noting, however, that, according to the arguments of this essay, it may be impossible, in principle, to exclude the factor of selection from experiments which would test "the inheritance of acquired characteristics." It is precisely my thesis that the *simulation* of Lamarckian inheritance will have survival value under circumstance of *undefined* or multiple stress.

6. *That it is, in general, more economical of flexibility to achieve a given characteristic by genotypic than by somatic change.* Here the Waddington experiments do not throw any light, because it was the experimenter who did the selecting. To test this proposition, we need experiments in which the population of organisms is placed under double stress: (a) that stress which will induce the characteristic in which we are interested, and (b) a second stress which will selectively decimate the population, favoring, we hope, the survival of those individuals whose flexibility is more able to meet this second stress after adjusting to the first. According to the hypothesis, such a system should favor those individuals which achieve their adjustment to the first stress by genotypic process.

7. Finally, it is interesting to consider a corollary which is the converse of the thesis of this essay. It has been argued here that simulated Lamarckian inheritance will have survival value when the population must adjust to a stress which remains constant over successive generations. This case is in fact the one which has been examined by those who would demonstrate an inheritance of acquired characteristics. A converse problem is presented by those cases in which a population faces a stress which changes its intensity unpredictably and rather often—perhaps every two or three generations. Such situations are perhaps very rare in nature, but could be produced in the laboratory.

Under such variable circumstances, it might pay the organisms in survival terms to achieve the *converse* of the genetic assimilation of acquired characteristics. That is, they might profitably hand over to somatic homeostatic mechanisms the control of some characteristic which had previously been more rigidly controlled by the genotype.

It is evident, however, that such experimentation would be very difficult. Merely to establish the genetic assimilation of such characteristics as *bithorax* requires selection on an astronomical scale, the final population in which the genetically determined *bithorax* individuals can be found being a selected sample from a potential population of something like 10^{50} or 10^{60} individuals. It is very doubtful whether, after this selective process, there would still exist in the sample enough genetic heterogeneity to undergo a further converse selection favoring those individuals which still achieve their *bithorax* phenotype by somatic means.

Nevertheless, though this converse corollary is possibly not demonstrable in the laboratory, something of the sort seems to operate in the broad picture of evolution. The matter may be presented in dramatic form by considering the dichotomy between "regulators" and "adjusters" (Prosser, 1955). Prosser proposes that where internal physiology contains some variable of the same dimensions as some external environmental variable, it is convenient to classify organisms according to the degree to which they hold the internal variable constant in spite of changes in the external variable. Thus, the homoiothermic animals are classified as "regulators" in regard to temperature while the poikilothermic are "adjusters." The same dichotomy can be applied to aquatic animals according to how they handle internal and external osmotic pressure.

We usually think of regulators as being in some broad evolutionary sense "higher" than adjusters. Let us now consider what this might mean. If there is a broad evolutionary trend in favor of regulators, is this trend consistent with what has been said above about the survival benefits which accrue when control is transferred to genotypic mechanisms?

Clearly, not only the regulators but also the adjusters must rely upon homeostatic mechanisms. If life is to go on, a large number of essential physiological variables must be held within narrow limits. If the internal osmotic pressure, for example, is allowed to change, there must be mechanisms which will defend these essential variables. It follows that the difference between adjusters and regulators is a matter of *where,* in the complex network of physiologic causes and effects, homeostatic process operates.

In the regulators, the homeostatic processes operate at or close to the input and output points of that network which is the individual organism. In the adjusters, the environmental variables are permitted to enter the body and the organism must then cope with their effects, using mechanisms which will involve deeper loops of the total network.

In terms of this analysis, the polarity between adjusters and regulators can be extrapolated another step to include what we may call "extraregulators" which achieve homeostatic controls *outside* the body by changing and controlling the environment—man being the most conspicuous example of this class.

In the earlier part of this essay, it was argued that in adjusting to high altitude there is a benefit to be obtained, in terms of an economics of flexibility, by shifting from, e.g., panting to the more profound and less reversible changes of acclimation; that habit is more economical than trial and error; and that genotypic control may be more economical than acclimation. These are all *centripetal* changes in the location of control.

In the broad picture of evolution, however, it seems that the trend is in the opposite direction: that natural selection, in the long run, favors regulators more than adjusters, and extraregulators more than regulators. This seems to indicate that there is a long time evolutionary advantage to be gained by *centrifugal* shifts in the locus of control.

To speculate about problems so vast is perhaps romantic, but it is worth noting that this contrast between the overall evolutionary trend and the trend in a population faced with constant stress is what we might expect from the converse corollary here being considered. If constant stress favors centripetal shift in the locus of control, and variable stress favors centrifugal shift, then it should follow that in the vast spans of time and change which determine the broad evolutionary picture, centrifugal shift of control will be favored.

Summary

In this essay the author uses a deductive approach. Starting from premises of conventional physiology and evolutionary theory and applying to these the arguments of cybernetics, he shows that there must be an *economics of somatic flexibility* and that this economics must, in the long run, be coercive upon the evolutionary process. External adaptation by mutation or genotypic reshuffling, as ordinarily thought of, will inevitably use up the available somatic flexibility. It follows—if evolution is to be continuous—that there must also be a class of genotypic changes which will confer a bonus of somatic flexibility.

In general, the somatic achievement of change is uneconomical because the process depends upon homeostasis, i.e., upon whole circuits of interdependent variables. It follows that inheritance of acquired characteristics would be lethal to the evolutionary system because it would *fix* the values of these variables all around the circuits. The organism or species would, however, benefit (in survival terms) by genotypic change which would *simulate* Lamarckian inheritance, i.e., would bring about the adaptive component of somatic homeostasis without involving the whole homeostatic circuit. Such a genotypic change (erroneously called the "Baldwin effect") would confer a bonus of somatic flexibility and would therefore have marked survival value.

Finally, it is suggested that a contrary argument can be applied in those cases where a population must acclimate to *variable* stress. Here natural selection should favor an anti-Baldwin effect.

Literature Cited

ASHBY, W. R. 1945. The effect of controls on stability. Nature, **155**: 242.

——. 1952. Design for a brain. Wiley, New York.

BALDWIN, J. M. 1897. Organic selection. Science, **5**: 634.

BATESON, G. 1960. Minimal requirements for a theory of schizophrenia. AMA Arch. Gen. Psychiatry, **2**: 447.

LERNER, I. M. 1954. Genetic homeostasis. Oliver and Boyd, Edinburgh.

PROSSER, C. L. 1955. Physiological variation in animals. Biol. Rev., **30**: 22–262.

SIMMONDS, N. W. 1962. Variability in crop plants, its use and conservation. Biol. Rev., **37**: 422–462.

WADDINGTON, C. H. 1953. Genetic assimilation of an acquired character. EVOLUTION, **7**: 118.

——. 1957. The strategy of the genes. Allen and Unwin, London.

27. The Status of Emergence

Paul Henle (with an introduction by Jeffrey A. Goldstein)

Originally published as Henle, P. (1942). "The status of emergence," *Journal of Philosophy*, 39(18): 486-93. Reprinted with kind permission.

The Hard Problem of Emergence is its Property of Self-transcendence

I first read this classic paper on emergence by the American philosopher Paul Henle over ten years ago. Rereading it, I am surprised by several themes which did not strike me so the first time around. First is Henle's early avowal—this article is after all from the early nineteen forties and was published in the prestigious *Journal of Philosophy*—that the idea of emergence had become generally accepted as a corollary to the theory of evolution "and there can be little doubt as to its truth." Clearly the idea of emergence was not just a fly-by-night notion that was awaiting for complexity science to give it substance. On the contrary, Henle's paper and others like it around that time touched on many issues concerning emergence that were hotly debated, and fostered important insights which complexity science can learn much from and need not waste time by continually reinventing the wheel. It appears to me that the divide between the two cultures, science and the humanities, that C. P. Snow so strongly derided is not only still with us, but may have grown wider, much to the detriment of both sides.

One particularly salient topic Henle took up was emergent *novelty* since he believed the soundness of the concept of emergence depended in large measure on whether the novelty it proclaimed was plausibly construed. This issue of novelty was closely connected to the way Henle thought emergence defied explanation although he, at the same time, somewhat paradoxically insisted that emergent phenomena necessarily retained a causal linkage with the substrate out of which they emerged. This is very much pertinent to current complexity science inspired disputes placing the idea of emergence against reductionist strategies of explanation.

Looking over these themes now, I suggest we view Henle's argument as his unique grappling with the central conceptual crux of the idea of emergence, what I'll call here the "hard" problem of emergence, borrowing the phrase (but altering the meaning a bit as I shall say more about below) from the Australian philosopher David Chalmers's famous characterization of the refractoriness of consciousness to scientific explanation (see, e.g., Chalmers, 1995-1997; see also Rosenberg, 2004, for a intriguing answer to Chalmers's hard problem by incorporating emergence into the fundamental nature of nature). Shifting from Chalmers's views on consciousness to the idea of emergence is not a far reach given two relevant facts. The first is that Chalmers studied for his Ph.D. under Douglas Hofstadter whose magnum opus, *Gödel, Escher, Bach* (1979) is a very long (742 pages!) speculation aiming to understand consciousness as an emergent phenomena, a proposal that comes to a head in the last several pages when the fertile notions of Gödelian "strange loops" and "tangled hierarchies" are employed in order to account for how consciousness could emerge from the brain. The relevant second fact has to do with a growing trend in consciousness studies that seeks to explain consciousness as emergent in ways apart from Hofstadter's meta-mathematical approach (see Freeman, 2001; and Clayton, 2006).

According to Chalmers, the "easy" problem of consciousness lies in coming up with credible scientific explanations, couched in terms of computational or neural mechanisms, of such phenomena as the ability to react to environmental stimuli, the integration of information by a cognitive system, the focusing of attention, and similar cognitive functions. The easy problem aims to explain cognitive or mental functioning according to constructs having to do with brain dynamics, the latter which I'll generalize as the *substrate* or component level. The "hard" problem of consciousness, though, concerns a far tougher challenge, *viz.*, how experience itself, i.e., the subjective awareness of immediate "raw feels" or "qualia", can be explained by the same brain substrate. Thus, whereas for any act of perception there may an information processing explanation, there occurs at the same time the undeniably subjective aspect of perception which does not seem accessible to scientific explanation—at least in the same way.

There are two things making the hard problem hard which can be transferred over to emergence. The first is the radical disjunction in qualities between the kinds of things that are to be linked in an explanation, that is, the qualities observed on the emergent level (not to be taken necessarily as vertical or hierarchical) and the qualities of the substrate components. In the case of emergence *per se*, however, unless one is considering an emergence based theory of consciousness, the qualities of emergent phenomena do not include qualia or awareness as such but instead include the properties of being: integrated or coordinated as in collective phenomena; dynamical or arising over time and not pre-given; unpredictable or ostensive in the sense of not able to be known before actually showing themselves as in a simulation; non-deducible from the substrate components; and irreducible to this same substrate (see Goldstein, 1999). It is the claimed incommensurability of these two different domains of qualities which renders explanations utilizing the idea of emergence baffling.

The second component of the hard problem which is pertinent to emergence has to do with the very call for an explanation itself, which presumes that these two discrepant kinds of things can indeed be somehow causally linked. If there were no call for an explanation, then the discrepancy or incommensurability could just stand as a brute fact. By not having to account for subjectivity of first person awareness, it might seem that the hard problem of emergence would prove to be less difficult than the hard problem of consciousness. Nevertheless, the singular qualities of emergent phenomena have shown themselves to be similarly obdurate to explanation. However, it should be noted that emergence, by covering so many different fields of study and therefore types of complex systems, does provide for a much greater variety of causal mechanisms and the other accouterments of explanation.

The main reasons why I interpret Henle as grappling with the hard problem of emergence involve both his discussion of the special novelty of emergent phenomena as well as his severing of causality from explanation. Henle wanted to retain causation in how emergents emerge from their substrate since, otherwise, their novelty could be understood as the sort of thing one sees in those types of random processes which lack a causal connectivity between antecedent and consequent states of affairs. But the genuine novelty of emergent phenomena implied for Henle that this necessary causal connection was not explainable as such, e.g., in terms of Aristotle's famous four types of causes. At this juncture, Henle proffered such cryptic utterances as: "... emergence occurs where the formal cause of the efficient cause is inadequate to explain the formal cause of the effect" (p. 488). We'll see that Henle was not content to rely on such enigmatic sayings, but tried to explicate how it could be both that causality was at work with emergence yet emergents at the same time surpassed explanation, at least how explanation was typically understood.

In order to indicate what is specifically problematic in the hard problem of emergence, I am suggesting that we phrase this conundrum in terms of the "self-transcending" property of emergence (Goldstein, 2002; 2004): emergents emerge out of a substrate, i.e., are caused somehow by this substrate, while simultaneously are novel with respect to or, in other words, transcend their substrate roots. The "self-" of "self-transcending" is unlike the "self-" of "self-organization" since the phrase is not trying to say anything about how emergence supposedly happens by itself or out of its own "self" resources. Instead, the "self-" of "self-transcending" is the substrate, and the explanatory models that are sufficient for this substrate, which are transcended during emergence by operations which bring forth novel emergent phenomena. It should be noted that "self-transcending" is not meant as some highfalutin phrase but simply as a concise description of this dual nature of emergence: *continuous with* the substrate in the sense of being causally related to it, but at the same time *discontinuous with* it in the sense of being novel with respect to the substrate.

The Nature of Emergent Novelty: The Unexplainable and the Non-deducible

Henle distinguished emergent novelty from that of a subjective experience such as seeing a certain color for the first time, taking an opposite direction than contemporary advocates of understanding consciousness as emergent. He wanted to make it clear that emergent novelty, by not being constituted by subjective experiences, did not possess just epistemological status. Yet, as we'll see later on, Henle left off his paper suggesting that emergents were more epistemological than ontological. Although attending to the role of the observer, proto-emergentists such as the English philosophers C. D. Broad, Samuel Alexander, and Conwy Lloyd Morgan considered emergents as mostly of monumental, that is, ontological status, e.g., life as emerging out of the inert or new species taking place via evolutionary saltations (Goldstein, 1999).

Henle also averred that the novelty of emergent phenomena was not adequately captured by the coming into being of a new copy of something already existing, for instance, the very next Intel Quad processor, say, the 5432th chip. Henle here applied an Aristotelian interpretation of *form* imposed on *matter* as indicating what the emergentists had in mind: emergent novelty was of a *novel form* and not simply a copy of the previous 5431 microprocessors. Samuel Alexander (1939, 1966; Brettschneider, 1964; see also Gillette, 2006), one of the key proto-emergentist philosophers, had in fact appealed to Aristotelian form to characterize the new *qualities* demonstrating emergent novelty, borrowing this from Morgan's earlier usage (Passmore, 1957).

Furthermore, for Alexander, a novel emergent form was more than just the *organization* of the substrate components, organization as such has become a staple of how emergence has been conceived by complexity inspired philosophers as well as scientists (see Bechtel & Richardson, 1993). For example, Alexander's form included the idea of self-regulation (presaging later concepts of homeostasis) which is certainly not presumed in organization *per se*. Indeed, for Alexander, it was novel forms which could attain the "requisite complexity of integration is attained" [sic] (Alexander, 1939). The positing of emergent novelty as a new form can be appreciated by considering how the notion of form had been appropriated by Aristotle from earlier ancient Greek philosophers. For instance, in the cosmogony of Philolaus which was influential on both Plato and Aristotle, the very beginning of the world (what could be a grander novelty than that?) marked the imposition of intelligible form on unintelligible matter (Barnes, 1979). And, Timaeus, the philosopher who synthesized many of the pre-Socratic philosophical positions in the famous dialogue of Plato by that name, offered a cosmogony whereby originary forms established the natural kinds of things which distinguish one kind from another (Zuckert, 2009). Emergent novelty as novel forms consequently is to be understood more along the lines of a difference in kind, not just degree, but kinds as understood in the primordial

Aristotelian sense which Alexander had appropriated and which it seems that Henle had done as well.

For Henle, the emergent novelty of emergent phenomena was also understood as a function of their not being accessible to explanation which, as we noted above, is problematic since Henle maintained at the same time a necessary causal linkage between an emergent and its substrate. Non-deducibility, accordingly, must amount to something different than excluding causal connectives. Indeed, the establishing of emergent novelty on the basis of non-deducibility has served as a dominant theme since proto-emergentist days, and a very brief review of some trends in this direction can be helpful in developing a sense of what Henle was aiming at:

1. The *trans*ordinal nature of the laws operative on the emergent domain, that such laws *transcend* the laws operative in the domain of the substrate components (see Broad, 1925; Stephan, 1999; O'Connor & Hong Yu Wong. 2002). The philosopher James Van Cleve (1990) understood transordinal laws with the following analogy: a body *A* in relation to Body *B* and *C*: if *B* was the only present object *A* would be attracted to it, if *C*, then *A* would be attracted to *C*, but if *B* and *C* were both together *A* would be attracted to point between *B* and *C*; but this last fact could not be deduced from *A* and *B* alone or *A* and *C* alone. Here we could also place Ryan's (2007) topological classification of a Mobius strip which, as a one-sided, one-edged, and non-orientable surface with boundary, possesses a topological property not present in the substrate which in this case are the subsets of disjoint simplicial triangles;
2. In mathematical logical terms as:
    a. "Observational" emergence to which the Danish mathematical physicist Nils Baas and the Danish philosopher Claus Emmeche (Baas & Emmeche, 1997) classify 'Gödel sentences' in a formal system (reminiscent of Hofstadter's meta-mathematical approach) as well as the property of membership of the Mandelbrot set and most Julia sets, the latter similar to the philosopher David Newman's (1996) description of being in the basin of a strange attractor in a dynamical system as an instance of emergence;
    b. The Existence Theorem for Emergent Phenomena proposed by Ian Jack Cohen and Ian Stewart (1994) which uses a Gödel/Turing-like argument to informally demonstrate that in any sufficiently rule based system there exist simple true statements (emergents) whose deduction from the rules is necessarily enormously long and complicated (i.e., deductions via substrate);
3. In the abstract algebraic terms of category theory, the non-existence of an isomorphism between specific diagrams or specific categories given some functor connecting them (i.e., particular mapping functions) (Ehresmann & Vanbremeersch, 2007; and Guitart, 2009; see also Baas and Emmeche's "hy-

per-structure" model of emergence which borrows from the work of Ehresmann and Vanbremeersch);

4. In terms of the laws of physics, the Constructionist Hypothesis put forward by the Nobel Laureate physicist Philip Anderson (1972) which helped get the whole neo-emergentist program started: although it might be possible to reduce nature to certain simple, fundamental laws, this did not entail a similar ability for re-*constructing* the universe from these simple laws since each new level of complexity requires the consolidation of entirely new properties and laws not appearing at the lower levels.

Again, what seems to me be the problematic central issue regarding the claim of the non-deducibility of phenomena is what I term above as the "self-transcending" nature of emergence, that emergents are causally related to but non-deductive (explainable) from their substrate. This was even more evident in Henle's characterization of emergent novelty as *logical novelty*, which was closely associated to the unpredictability of emergents (in complexity theory unpredictability is more likely expressed as the ostensive property of not being able to be foreseen except through the actually occurring run of a simulation—see Darley, 1995). The logical novelty of emergent novelty implied unpredictability but in explicating what he meant, Henle offered a strange digression that I believe is fundamentally unsound. Specifically, Henle contended that it was theoretically possible that Newton could have anticipated Einstein by predicting the Michelson-Morley experiments on the constancy of the speed of light. The only thing amiss would be that Newton's theory would not have been the "simplest" explanation which Henle demanded for a credible explanation. Yet, in my estimation, while it is indeed true that Newton's theory can be considered a limiting case of Einstein's theory of special relativity this is only for non-relativistic velocities, and it simply doesn't make much sense to hold Newton capable of this feat, particularly so in the context of what Henle offered as an early statement of what later were called "bridge laws" to facilitate explanation. So a short digression on bridge laws is in order.

In the usual model of reductionist explanations, the context is taken to be an *inter-theoretic* reduction whereby a reducing theory reduces a reduced theory. However, the domains of both the reducing and the reduced theory may be so disparate that there needs to be some way to bridge over them; hence the need for bridge laws, the basic idea of which was formulated by the eminent philosopher of science Ernst Nagel (cited in Batterman, 2007) twenty years after Henle's article. By uniting the two domains, bridge laws are supposed to help make predictions by supplementing the explanation with boundary conditions, limiting assumptions, approximations, and other addenda (Sklar, 1967). As the noted philosopher of science Lawrence Sklar has pointed out, if in order to move forward in a reductive fashion, a scientific explanation must have recourse to bridge laws at all implies there must be a great deal of preliminary groundwork to effectuate the needed fit between levels. This preparatory spadework has been described by the philosopher Jaegwon Kim (1993), otherwise known for his analysis of the emergence-like notion of supervenience, as a strong enough "connectability" or "co-extensivity" between the levels. Sklar points out how the ultimate goal of a strict reductionist explanatory strategy aims at complete identification of the two domains that eliminates the higher emergent level altogether.

Bridge laws, then, can be seen as a way to buttress a reducing theory and thus render the system under examination more predictable. Henle though wanted to keep unpredictability by pointing out how it was always relative to the knowledge gained about a system, that is, to available evidence. Thus, to strengthen his view on emergent novelty he posited logical novelty as that which would transcend any available knowledge or scheme of explanation or specific system of logic, even any supplementation by way of bridge laws. Indeed, bridge laws don't come without a whole slew of additional problems as the philosopher of science Robert Batterman (2007) points to: Which bridge laws are suitable, e.g., are they "established by linguistic investigation alone? Are they factual discoveries? If the latter, what sort of necessity do they involve? ..."

Henle's logical novelty as the criterion for emergent novelty can be interpreted as another instance in emergentist thought (see other examples in the classical papers on emergence and their introductory essays in Richardson & Goldstein, 2008) where a bar is set as to the kind of novelty demanded by emergent phenomena. If the novelty can surpass this hurdle, then we have a case of emergence. For Henle the bar was quite high: "Imagine the simplest possible system of the world, or theory comprehensive enough to link together all of the characteristics exemplified by nature to date. An emergent then is a characteristic whose appearance calls for a revision of this all-embracing system...an emergent halts reason in its attempt to map the cosmos" (p. 491). Note that Henle suggested that this "halting" of reason may not be permanent. This appears to imply that Henle had accepted that the view of emergence being a provisional matter until a "re-drafting" occurred coincident to an advance in knowledge (see Goldstein, 2000).

However, because of this dependency of emergent novelty on some particular conceptual system, Henle was led to conclude that emergence only rises to the level of epistemological and not metaphysical (or ontological) significance. The reason is that any conceptual system works on the basis of hypotheses and the criterion for selecting one hypothesis as more true than another rests on its "logical simplicity" which comes down to an epistemologically based judgment: "Emergence can be given a metaphysical status only if a similar status can be given to logical simplicity. Indications are that this is not the case and that emergence has to do with our knowledge only..." (p. 493).

Logical Novelty and Self-Transcending Constructions

Henle's final conclusion about emergence's epistemological status seems to go against the grain of most of his thinking in this paper, which tended more to establishing the credibility of emergence as having to do with the actual "furniture of the world". However, if we take his idea of logical novelty as a hurdle over which emergent novelty had to pass, there can be a way around the seeming self-contradictory nature of Henle's argument: by considering this hurdle as hinging on what we are able to know as of *yet* so that, if emergent novelty in a particular case passes over the epistemological hurdle of emergent novelty, then it can be taken as ontological although this latter status may be in a condition of ongoing transition back into epistemology *per se*. In other words, the provisionality of granting ontological significance to any particular manifestation of emergence becomes just another way that emergence involves an ongoing transcendence of whatever models have been devised up to the point of the emergents showing themselves. Then, if we interpret emergent novelty as bringing about the appearance of new forms or natural kinds, then these new forms or kinds exhibit the ontological import of emergence. Moreover, a similar fate would befall certain already existing natural kinds, e.g., new species during the course of evolution. An example is Lyn Margulis's idea of *symbiogenesis* (see Reid, 2009) in which a kind of biological emergence has been used to reinterpret what has already occurred in primordial times in the origin of eukaryotic cells by way of ingesting protokaryotic and cells.

Moreover, if we take Anderson's Constructionist Hypothesis and link it up to the complexifying operations described in terms of category theory on the part of Ehresmann and Vanbremeersch as well as Baas and Emmeche, then we can understand emergence as a set of constructional operations on already existing natural kind building blocks, more specifically, a self-transcending kind of construction (Goldstein, 2002, 2004) which, at the same time, includes the role of the observer and subjective experience. Understanding emergence in terms of self-transcendence has the benefit of transcending the way that the hard problem of emergence has become a major stumbling block getting in the way of a wider appreciation of the concept of emergence. On the one hand, this hard problem has led those of a more reductionist bent to simply reject the whole notion of emergence in order to dispose of the troublesome aspect of its self-transcending nature. On the other hand, this same hard problem of emergence has led others to see in emergence the work of a supra-natural, vitalist, or holist impetus precisely because the emergent phenomena are not amenable to explanation as in Henle's sense. The hard thing about the hard problem of emergence is how to somehow keep both poles, the continuous and discontinuous, in conceptual tension. This is the road emergentists need to walk, a fact even the early emergentists recognized when they saw what they were doing as steering between, on one side, vitalism, and, on the other side, mechanism (their word for reductionism). The perspective of self-transcending constructions doesn't

Original Ref—E:CO Vol. 11 No. 4 2009 pp. 73-87

lead to easy answers whether of the reductionist or holistic variety but it does try to justice to the claims of emergentist thinkers for the past century.

From the perspective of self-transcending constructions, Henle's and the other descriptions of the non-deducibility of emergents can be read as sources of insight into what these self-transcending constructions must consist in so that they have the potency of bringing about phenomena with the quality of emergent novelty. So far, much of the uneasiness with the idea of emergence can be traced to a failure of imagination in conceiving the possibility of naturalistic processes with the power of self-transcending constructions in generating emergent phenomena.

References

Alexander, S. (1939). *Philosophical and Literary Pieces*, J. Laird (ed.), ISBN 9781843713135.

Alexander, S. (1966) *Space, Time, and Deity: The Gifford Lectures at Glasgow: 1916-1918 in Two Volumes*, ISBN 9780766187016.

Anderson, P. (1972). "More is different: Broken symmetry and the nature of the hierarchical structure of science," *Science*, ISSN 0036-8075, 177(4047): 393-396.

Baas, N. and Emmeche, C. (1997). "On emergence and explanation," *Intellectica*, ISSN 0769-4113, 2(25): 67-83.

Barnes, J. (1979). *The Presocratic Philosophers*, ISBN 9780415050791.

Batterman, R. (2007). "Intertheory relations in physics," Stanford Encyclopedia of Philosophy, http://plato.stanford.edu/entries/physics-interrelate/.

Bechtel, W. and Richardson, R. (1993). *Discovering Complexity: Decomposition and Localization as Strategies in Scientific Research*, ISBN 9780691087627.

Bettschneider, B. (1964). *The Philosophy of Samuel Alexander*, NY: Humanities Press.

Broad, C.D. (1925). *The Mind and its Place in Nature*, ISBN 9780415488259 (2008).

Clatyon, P. (2006). *Mind and Emergence: From Quantum to Consciousness*, ISBN 9780199291434.

Cohen, J. and Stewart, I. (1994). *The Collapse Of Chaos: Discovering Simplicity in a Complex World*, ISBN 9780140291254 (2000).

Darley, V. (1995). "Emergent phenomena and complexity," in R. Brooks and P. Maes (eds.), *Artificial Life IV: Proceedings of the Fourth International Workshop on the Synthesis and Simulation of Living Systems*, ISBN 9780262521901, pp. 411-416.

Ehresmann, A. C. and Vanbremeersch, J.P. (2007). *Memory Evolutive Systems: Hierarchy, Emergence, Cognition*, ISBN 9780444522443.

Freeman, A. (ed.) (2001). *The Emergence of Consciousness*, ISBN 9780907845188.

Gillett, Carl (2006). "Samuel Alexander's emergentism," *Synthese*, ISSN 0039-7857, 153(2): 261-296.

Goldstein, J. (1999). "Emergence as a construct: History and issues," *Emergence*, ISSN 1521-3250, 1(1): 49-62.

Goldstein, J. (2000). Emergence: A construct amid a thicket of conceptual snares, *Emergence*, ISSN 1521-3250, 2(1): 5-22

Goldstein, J. (2001). "Emergence, radical novelty, and the philosophy of mathematics," in W. Sulis and I. Trofimova (eds.), *Nonlinear Dynamics in the Life and Social Sciences*, ISBN 9784274904080.

Goldstein, J. (2002). "The singular nature of emergent levels: suggestions for a theory of emergence," *Nonlinear Dynamics, Psychology, and Life Sciences*, ISSN 1090-0578, 6(4): 293-309.

Goldstein, J. (2003). "The construction of emergence order, or how to resist the temptation of hylozoism," *Nonlinear Dynamics, Psychology, and Life Sciences*, ISSN 1090-0578, 7(4): 295-314.

Goldstein, J. (2004). "How emergence flirts with paradox: Emergence, creative process, and self-transcending constructions," *Proceedings of Capitol Science 2004*, Washington Evolutionary Systems Society, Washington, DC.

Guitart, R. (2009). "Cohomological emergence of sense in discourses (as living systems following Ehresmann and Vanbremeersch)," *Axiomathes*, ISSN 1572-8390, 19(3).

Hofstadter, D. (1979). *Gödel, Escher, Bach: An Eternal Golden Braid*, ISBN 9780465026852.

Kim, J. (1993). *Supervenience and Mind: Selected Philosophical Essays*, ISBN 9780521439961.

Newman, D. (1996) "Emergence and strange attractors," *Philosophy of Science*, ISSN 0031-8248, 63: 245-61.

O'Connor, T. and Wong, H. (2002). "Emergent properties," *Stanford Encyclopedia of Philosophy*, http://plato.stanford.edu/entries/properties-emergent/.

Passmore, J. (1957). *A Hundred Years of Philosophy*, ISBN 9780140136715 (1994).

Prokopenko, M., Boschetti, F. and Ryan, A. (2009). "An information-theoretic primer on complexity, self-organization and emergence," *Complexity*, ISSN 1076-2787, 15(1): 11-28.

Reid, R. (2009). *Biological Emergences: Evolution by Natural Experiment*, ISBN 9780262513401.

Richardson, K. and Goldstein, J. (eds.) (2007). *Classic Complexity: From the Abstract to the Concrete*, ISBN 9780979168833.

Rosenberg, G. (2004). *A Place for Consciousness: Rethinking the Deep Structure of the Natural World*, ISBN 9780195168143.

Ryan, A. (2007). "Emergence is coupled to scope, not level," *Complexity*, ISSN 1076-2787, 13(2): 67-77.

Sklar, L. (1967). "Types of inter-theoretic reduction," *The British Journal for the Philosophy of Science*, ISSN 0007-0882, 18: 109-124.

Stephan, A. (1999). *Emergenz: Von der Unvorhersagbarkeit zur Selbstorganisation*, ISBN 9783933168092.

Van Cleve, J. (1990). "Mind: Dust or magic? Panpsychism versus emergence," *Philosophical Perspectives*, ISSN 1520-8583, 4: 215-226.

Zuckert, C. (2009).Plato's Philosophers: The Coherence of the Dialogues, ISBN 9780226993355.

THE STATUS OF EMERGENCE [1]

THE doctrine of emergence, having passed through the stage of acute philosophic controversy, has become to a large extent a generally accepted view. Unfortunately it has not become an unambiguous view. In one sense it seems to amount to no more than the insistence that there are novelties in the world. If this is all that is intended, the doctrine is a corollary of evolutionary theory and there can be little doubt as to its truth. To take a simple case, color vision can have appeared in the world only with the development of animal life and there must have been a first animal capable of distinguishing, say, the color violet. There might be variations in the clarity with which violet is distinguished, but, allowing for such differences, the statement stands. This color of violet, in its purely qualitative aspect as it appeared to the primitive animal, must have been something new under the sun.

Color names are, of course, used in a variety of senses. They may indicate the quite subjective, sensuously perceived aspect of vision, the sense in which a philosopher might say that his present sense datum was red. Again, they may be used to indicate any subjective manifestation, regardless of its qualitative character, of a type of act of perception, as in the case in which one might define green as the color of grass. In this second sense, two people might have qualitatively different sensations and still both agree that they were seeing something green. Finally, color names

[1] Presented before the Eastern Division of the American Philosophical Association, Vassar College, Dec. 30, 1941, and revised.

may refer to physical correlates of color, as if one were to define violet as a vibration of about 0.0004 mm. wave length impinging on the retina. It is only in the first of these senses that novelty is clearly indubitable.

If belief in novelty were confined to this subjective aspect of experience, then there could be no question of its existence. In the hands of many writers, Alexander, Broad, and Morgan, to mention only a few, the belief goes much farther. Thus life itself, not merely what it feels like to be alive, is taken as a novelty. Novelties, moreover, are assumed to arise not everywhere, but merely at certain stages in the evolution of the world, and are designated as emergents, in contrast to resultants which involve no novelty. It is this more complex and fully developed theory of emergent evolution which we proceed to examine.

We may begin by noticing the sense in which the term "novelty" is employed in discussions of emergent evolution. It is clear, I think, that the sense of the term is not the sense in which automobile number 439,752 is new as it rolls off the assembly line. It is true that in one sense we are dealing with a new car, but, in the common phrase, since thousands of similar cars have been manufactured, the car is "old stuff." In more Aristotelian terminology, the form and matter of the car have only recently been conjoined and in this sense the car is new, but the form has often been found in other cars and in this sense there is no novelty. The latter sense, the sense in which forms are new, is the sense intended by emergent evolutionists. A Platonist might well object here that *per se* no forms or qualities are new and that the only novelty arises in the appearance of those forms in the sensible world. This contention need neither be affirmed nor denied, for the sense in which a Platonist might admit novelty is the only sense required for the doctrine of emergence.

Novelty, then, pertains primarily to the qualities of things and a quality may be said to be new if it was never before embodied. We may, of course, define a derivative sense in which things are new, namely, if they embody new qualities, but this is not the primary sense.

Not every occurrence of a new form is, however, a case of emergence. If a new universe, replete with qualities never before exemplified, were to appear *ex nihilo,* this would hardly be counted as a case of emergence. Again, to take a less fantastic example, imagine a universe which contained irreflexive relations and transitive relations, but, up to a given time, no relations which exemplified both characteristics. Imagine further that there were no asymmetrical relations. Suppose now a relation were to arise which were both irreflexive and transitive. Since all irreflexive and transitive relations are of necessity asymmetrical, a new quality would be born into this universe; but even the most sanguine philosopher would hesitate to consider this a case of emergent evolution.

The crux of these exceptions seems to be that emergent evolution has to do with a causal situation. If one claims that life is a quality emergent from the inorganic, this is to claim that living organisms are brought into being by inorganic processes, though they are not explained by these processes. Again, to claim that mind is an emergent from living organisms, takes these organisms as the cause of mind, although in this case as well, the effect is inexplicable in terms of the cause. Aristotle has remarked that a chance event is one which is not the final cause of its efficient cause, even though the efficient cause does have a final cause.[2] In a rough parallel, one might say that emergence occurs where the formal cause of the efficient cause is inadequate to explain the formal cause of the effect. Again roughly, we have emergence where a new form appears and where the causes of the appearance are unable to explain the form.

This principle would explain why the cases cited above are not instances of emergence. In the universe arising *ex nihilo,* there is no causal relationship. In the case of the appearance of asymmetry, whatever causes explain the conjunction of transitivity and irreflexiveness are adequate to explain asymmetry, since it follows deductively from the other two.

It is because of the failure of one group of qualities to explain another that unpredictability in advance of the fact can be used as the criterion of emergence. If there is an adequate basis in one group of characteristics for the explanation of the second, the latter could be predicted in advance of occurrence. Again, after the occurrence of one group of qualities with another, their joint appearance can be generalized and so be made the basis of future predictions; hence, in general, where there is an emergent, there is unpredictability. The converse, however, is not true and not every unpredictable event or characteristic marks an emergent. Suppose a new planet were to appear in the solar system, and to appear either for no reason at all, through sheer chance, or to appear for causes which had been hitherto undiscoverable. Such an occurrence would surely be unpredictable, but it would not constitute a case of emergence, unless the planet presented novelty in the sense discussed previously.

[2] *Physics,* 196b, 30 ff.

With the exception just noted, however, unpredictability has been taken as the criterion of emergence, and here discussions of emergent evolution have been at their weakest, for unpredictability has been treated as a simple quality of events. Thus, we are told by Lloyd Morgan, among others, that when carbon and sulphur are combined, the weight of the carbon-bisulphide formed is predictable, but many of its other properties are not.[3] In view of the fact that the principle of the conservation of matter has been on a scientific footing for less than two centuries, it must be apparent that the weight of the compound, far from being inherently predictable, could have been predicted only recently. Since, moreover, the boiling points of hydrocarbons, which are *prima facie* just as mysterious properties as the characteristics of carbon-bisulphide, have been successfully predicted in advance of the synthesis of the compounds, predictability is seen once more to be relative to knowledge. Indeed, a moment's reflection must convince one that predictability, far from being a mere predicate, is at least a tetradic relation and we must say, not that quality *A* is predictable, but rather that quality *A* is predictable by person *B* on evidence *C* with degree of assurance *D*.

Once the dependence of predictability upon evidence is admitted, it becomes clear that unpredictability is not peculiarly the mark of the novel, but merely of what may be loosely called the logically unrelated. Any characteristic past, present, or future, not related to a given body of evidence would be unpredictable on that evidence. This independence of predictability and futurity, far from undermining the concept of emergence, however, rather permits a more exact formulation of it. We will indicate a kind of novelty which may be termed logical novelty and which is independent of any notion of time. In terms of this, we will be able to formulate the idea of emergence more precisely.

We may begin by noticing the way in which any characteristics are predictable on scientific evidence. This evidence consists in particular observations of relations of characters and these observations do not entail the prediction, but rather give rise, by inductive reasoning, to a theory which entails the prediction. For example, to return to the illustration used before, the combination of sulphur and carbon may be the cause of the occurrence of carbon-bisulphide, but a statement of the relevant properties of the elements, their atomic weights, molecular structure, and combining ratios by themselves afford no basis for any conclusions about the weight of the compound formed. Before any conclusion can be reached, the assumption of the conservation of mass is

[3] *Emergent Evolution*, p. 3.

required. Again, to take an example which appears to display emergence, the explosiveness of nitrogen chloride could hardly be predicted in advance of the fact on the basis of the properties of nitrogen and chlorine alone. Yet the explosiveness could easily be predicted from a knowledge of the properties of nitrogen bromide, nitrogen iodide, and a general knowledge of the halogens. More generally, the situation would appear to be this: only when we are dealing with the properties of relations as in logic and pure mathematics do qualities seem to be so related that the presence of certain ones of them guarantees and explains the presence of others. In physical cases the conjunction of characteristics is always empirical, *i.e.*, characteristics do not entail one another, but at most are found conjoined. Hence predictions can be made not on the basis of the characters themselves, but only when they are supplemented by hypotheses—and this applies equally to resultants and emergents.

Thus, the situation always involves a group of data, a, b, c, *etc.*, a system of logic, L, a context of accepted theory, T, an hypothesis, H,[4] and an occurrence of a characteristic, x, whose predictability is in question. In order that H may be an admissible hypothesis, it is requisite that, by the standards of L, H be compatible with T and that a, b, c, *etc.*, be consequences of H and T. In order for x to be predictable, it as well must be a consequence of a, b, c, H, and T. Assuming the logic, L, and the accepted theory, T, to be fixed, we may say that under these conditions, x is predictable on the basis of a, b, c, and H.

Unfortunately, the H which satisfies the conditions above is not, as a rule, uniquely determined. There may be several such, H_1, H_2, *etc.*, some allowing for the prediction of x and others not. If we wish to speak of the predictability of x on the basis of the data a, b, c alone, we must introduce some restriction on the hypotheses involved. Since the alternative hypotheses are assumed to be similar in respect to their relationship to other knowledge, some internal means of differentiating them must be found and the only such means seems to be the characteristic of logical simplicity.

We must introduce the assumption, then, that among all the alternative hypotheses, one is simpler than the others and we may then say that x is predictable on the basis of a, b, c, provided it is predictable on the basis of a, b, c, and the simplest H which will explain a, b, c.

With this much by way of preliminary, we may say that x is

[4] The term "hypothesis" must be understood throughout this discussion in a sense which makes prediction possible.

logically novel with respect to *a, b, c,* provided it is not predictable on the basis of *a, b, c.* No notion of temporal priority is implied in this concept of logical novelty; thus, relative to a description of points on a line in terms of between-ness, the characteristics of the line in terms of serial order are logically novel. Or, relative to Euclidean geometry, classical mechanics is logically novel. It should be observed also that there is no absurdity in supposing something logically novel actually to have been predicted. It would be agreed, I believe, that the outcome of the Michelson-Morley experiment is logically novel, relative to Newtonian physics. Yet it is conceivable that Newton might have explained his data by relativity physics, anticipating Einstein, and have predicted the outcome of this experiment. Such explanation would have been theoretically possible, but it would not have provided the simplest explanation of Newton's data, hence even though actually predicted, the results of the experiment would have remained logically novel.

In terms of this conception of logical novelty we may now formulate the concept of emergence more accurately. A characteristic is an emergent provided (1) it is logically novel relative to all previously exemplified characteristics, and (2) it is not merely a spatio-temporal characteristic. This latter requirement is inserted merely to rule out cases like that of the previously mentioned unexpected planet.

This more or less formal definition may be made clearer by considerations of the following sort: Imagine the simplest possible system of the world, or theory comprehensive enough to link together all the characteristics exemplified by nature to date. An emergent then is a characteristic whose appearance calls for a revision of this all-embracing system. More figuratively, an emergent halts reason in its attempt to map the cosmos. This halt need not be permanent, but may merely call for some re-drafting; yet in a universe in which emergence was the rule or in which every event involved emergence, rational thought could at best give a partial and inadequate account.

If the preceding account is correct and no characteristic is *per se* unpredictable, but is unpredictable only relative to some system, there arises the question: Is emergence a metaphysical concept or does it have epistemological significance only? We have argued that prediction occurs only in a nexus of explanation and that a variety of explanations of the same data is possible, on some of which the occurrence of a given quality would be predictable, on others not. In order to secure uniqueness of explanatory framework, it was necessary to have recourse to the hypothesis of a simplest explanation in terms of which prediction might proceed. The problem resolves itself into the problem of the nature of logical simplicity. If simplicity is merely simplicity for a given individual, a given society, or even the human mind generally, then emergence would seem to be primarily of epistemological significance; if on the other hand, simplicity can be given some higher ontological status, then emergence takes on a metaphysical importance as well. Unfortunately, the notion of simplicity has been largely neglected in logical analyses and the following considerations do not pretend to be entirely conclusive.

In general, it seems, we call one hypothesis simpler than another when it involves fewer independent assumptions than the other but has the same explanatory power. Thus, in the *Almagest*, Ptolemy notices that he can explain the motion of the sun either by assigning to it motion on an eccentric circle, or by means of a deferent concentric with the earth and a single epicycle, the deferent and epicycle moving in opposite directions with equal angular velocities.[5] Ptolemy unhesitatingly chooses the former mode of explanation.[6] Since he recognizes that the two are mathematically equivalent, any choice between them must be based on grounds of simplicity. Presumably Ptolemy's reasoning was as follows: In the first explanation the only arbitrary elements are the diameter of the circle and the amount of its eccentricity. In the second, the size of the circles, the directions of their motions, and the coincidence of their velocities are all arbitrary. The first theory involving fewer arbitrary assumptions is therefore simpler and for the same reason is more likely to be true.

There seem to be two reasons, however, for not taking this notion of objective simplicity at its face value. First, the view presupposes that theories are literally true. It is, of course, correct that if one were fashioning an orrery on a cosmic scale and attaching the sun to it, it would be more likely that the sun's motion would come out as it has if the orrery were constructed according to the view that Ptolemy accepts. A more sophisticated view of theory, however, would attach literal truth only to verifiable consequences and regard theories merely as ways of obtaining this set of consequences. The number of assumptions going into the formulation of the theory would then be irrelevant. It should be noticed that the acceptance of this view of the function of theory does not involve acceptance of the entire pragmatic theory of meaning.

The other reason which makes it difficult to accept the notion

[5] Bk. III, sect. 3; Heiberg 219.

[6] Bk. III, sect. 4; Heiberg 232.

of an objective simplicity is harder to formulate. To illustrate the point, however, it is common in logical systems to find that a single postulate may be equivalent to two or more others without on the surface appearing any more complex. Since it does the work of a pair, it must somehow be a broader assumption than either of the pair whose place it takes. We are thus led to the notion of something which we might adumbrate as magnitude of assumption which seems to vary inversely with the number of assumptions. It is hard to see how there is any gain in simplicity by diminishing one of these factors at the expense of increasing the other.

Certainly no one can claim that the concept of logical simplicity is clear, and it may be that this entire analysis misses the point, but, on the basis of it, logical simplicity would appear to apply to the formulation of theories rather than to the theories themselves.

To sum up: Novelty clearly exists in the universe and may, indeed, characterize every event, but this does not entail a doctrine of emergence. Predictability is not a simple predicate and can not be used as a criterion of emergence except on the basis of some hypothesis, presumably the simplest hypothesis which accounts for previous characteristics. Emergence can be given a metaphysical status only if a similar status can be given to logical simplicity. Indications are that this is not the case and that emergence has to do with our knowledge only, but, in any case, the focus of discussions of emergence must be the concept of logical simplicity.

PAUL HENLE.

UNIVERSITY OF MICHIGAN.

Forum section

28. An Alternative Sports Metaphor for Understanding Teamwork as Complex: Soccer

R. Michael Bokeno
Murray State University, USA

As critiques of and dislike for organizational teamwork increase, alternatives must be sought for both pedagogy and practice. Competitive sports metaphors are often used in management practice and teaching; unfortunately, these tend to reflect distinctly American values of zero-sum competition, cybernetic, error-correcting efficiency, individualistic success, therefore de-emphasizing what (American) organizational teamwork needs most: creativity, innovation, genuine autonomy and inventiveness. This is precisely what makes both the pedagogy and practice ineffective. This essay proposes the game of soccer as an alternative metaphor and heuristic device. I contend that both organizational teamwork and soccer are quantum phenomena. Specifically, I demonstrate how soccer teamwork is non-linear, holonic, emergent and engaged, and articulate those concepts with extant, conventional understandings of teamwork in organizations. My hope for the essay is that the soccer metaphor will inspire a more complex understanding of organizational teamwork as a collaborative (rather than simply cooperative or coordinated) activity.

Want Better Teamwork? Watch More Soccer!

We need to ground the reality of "we" in a new conceptual structure.

Danah Zohar, *The Quantum Self*

In the past few years, both the principle and practice of "teamwork" has come under increased scrutiny as one of those organizational practices that in wide variety of ways, has not lived up to its expectation for organizational effectiveness. While there has been some success for project teams—designed for a specific purpose and then no longer necessary—other kinds of teamwork, such as that needed by management and executive decision-making teams, seems much more difficult to achieve.

It's hard to find workplaces that exemplify teamwork. Susan Heathfield (2008) contends, for example, that:

We have miles to go before valuing teams and teamwork will be the norm.

In America, our institutions such as schools, our family structures, and our pastimes emphasize winning, being the best, and coming out on top. Workers are rarely raised in environments that emphasize true teamwork and collaboration.

As we typically understand teamwork, it is not a natural process. Our prototypes for teamwork are derived primarily from collectives of talented individuals who manage to coordinate their individual talents for large sums of money. "Team" is the name given to these groups of individually talented people.

Second, we may not be disposed to teamwork. Chris Argyris's famous thesis about Model I and Model II governing assumptions (e.g., Argyris, 1994) suggests that we are almost anthropologically predisposed to compete for zero-sum outcomes (Model I) rather than collaborate for positive sum outcomes (Model II). We do, according to Argyris, have an idea of what teamwork is, in the sense of genuine collaboration, but we seldom activate that idea.

Third, as conventionally practiced in organizational and business contexts, teamwork is a managed, "performed" process; it has goals, roles, expectations, objectives, reporting procedures and cybernetic mechanisms to insure efficient progress along a stipulated path toward goal accomplishment. This makes it simply coordinated or cooperative activity, rather than collaborative activity.

In a nutshell, these three reasons would account for its failure.

Genuine teamwork requires a level of participation and collaboration not usually requisite to specific project accomplishment. Importantly, it is this "next level" of knowledge, skill and ability that distinguishes genuine collaborative teamwork from simply cooperative or coordinated activity. Unfortunately, we only have a vague idea of what this next level is or entails.

The concepts most often used to capture or frame—teach and understand—this next level, seem almost magical, celestial, or otherwise unreal, unattainable, or out of our control: *synergy*, *jazz*, *improvisation*, or *flow*. While enchanting, in graduate management classes or training seminars I conduct, there is usually no one who can articulate how, for example, "synergy" means anything more than "*everyone working together*," or how it is different from cooperative activity or coordinated activity. "Yes, the whole is greater than the sum of the parts, *but what does THAT mean?"*

My professor-ish frustration with this subsides when I take a minute to reflect on the rationale. Not many practicing executives, managers, team leaders and MBA's are afforded a deep conceptual understanding of genuine teamwork; many have not been exposed to alternative pedagogies or heuristic devices that would drive an applied understanding of genuine teamwork or collaboration.

In fact, *contra* recent defenses of management education (see *BizEd*, Sept/Oct 2007), what B-schoolers have been very much exposed to instead are foundational principles of individual success, zero-sum interdependence, task-focused efficiency, a cybernetic understanding of team development, and the fiduciary gain that results from doing this well within a fragmented, competitive world view. Individualism, competition, control (error-correction), efficiency and economics are in fact the driving metaphors of business and management, in practice and pedagogy. Most companies and cultures still struggle with ways of awarding collective recognition for collective accomplishments. Little wonder that "teamwork" in the celestial, or quantum, sense of synergy or flow, remains merely a neat, ideal, unattainable concept, and that coordination or cooperation often suffices.

Consequently, most if not all the prescriptive advice for improving teamwork—decentralize control and authority, trust each other, have shared vision, appreciate differences, learn continually, support others—requires a mindset wholly foreign to conventional thinking. Certainly a decade or two of more new-agey ideas has made recent B-schoolers a little more malleable. But there is little evidence that much has changed in organizational practice. And alternative mindsets aren't simply inserted as new software—they must learned.

So the question becomes one of how to cultivate an alternative understanding of teamwork such that collaboration becomes more realistic or at least more realizable. Different or new ways of thinking about things are often aided by metaphor.

An Alternative Metaphor

Competitive metaphors from sports are used often in management and organization theory and education, as well as office and business practice. Intriguingly, one sport that is rarely if ever used is soccer. That's probably because Americans understand soccer about as much as they understand teamwork: as I will contend in the paper, this is due, in turn, to the fact that both are quantum phenomena.

In my decade plus of coaching, studying and observing the game of soccer, I have derived some interesting lessons about playing together that help cultivate an alternative understanding of teamwork to those furnished by American football, baseball or even basketball. More pointedly, I have concluded that soccer is a more effective metaphor for the pedagogy and practice of teamwork in organizations.[1]

1. Two things worth mentioning here. First, I have no coaching license. I stopped coaching at about U12, leaving it to those who had played the game. My observations are drawn from watching and studying it as my three sons progressed through the ranks, developed by increasingly better coaches for about ten years. Second, as such, coaches with more experi-

Soccer is Nonlinear

First and foremost, there is something nonlinear or noncybernetic about soccer that distinguishes it from most sports Americans are familiar with. Scientifically speaking, a nonlinear system is one whose outcomes cannot be predicted from initial conditions along a straight or stable trajectory. Soccer allows a very wide latitude of acceptable play, and is not necessarily error-correcting. It is dynamic in the sense of continuous play (stoppage only for serious injury or to award penalties) on the field, rather than turn-taking. It never seems to visibly, obviously progress, in the way a 3 minute 16 point run would change a basketball game, or a three-run homer would change a baseball game, or a couple of consecutive deep route completions would alter a football game. In soccer, the ball can go from one side of the field to another in a single play, and sometimes backward (see: the drop), in a single play. In basketball, by contrast, a the ball cannot go back to the other side of the mid-court line; a third-baseman would never think about throwing the ball into left field to effect a force play at second; and in American football, if we are going backward, we have a long day ahead of us. And that is only on offense—there are also eleven defenders to contend with, each with their own set of skills, moves and tricks, whose job it is to take the ball from you.

In this sense, we might also say soccer is "chaordic": it is as much chaos as it is order. It is "orderly, in that each team intends to try to do something in a designed way. Certainly every team seeks to move the ball down the field (efficiently) with the intent of a accomplishing a goal. But it is chaotic in that all players know there are hundreds of things that interfere with that objective; that is, the nature of the game as played is that it while it might be predictable in theory or principle, it is unpredictable in practice. Thus, the intent of moving down the field to score a goal might better be understood as *collaborating to create the conditions to finish a play* (score). And again, the game is inherently dynamic as it is constantly moving, hence changing.

But the important point is that its randomness or chaos is accepted and featured at least as much as the ordered part. Every misstep is not seen as deviation or correctable error; off-target passes are necessary and acceptable risks; missed shots and missed tackles are part of the game. There are very wide parameters for *effective* play and development, yet few rules for *correct* play. There are constants, such as "positions" or "areas" for which players are responsible (wing, midfielder, forward, sweeper, etc), and there are more or less standardized "runs" or executions that are tactically fundamental; also, certainly there are scripts at certain times (marking for set pieces like free kicks and corners). But the vast majority of the game is played within these very wide parameters. It is complex and creative—in fact once you are about sixteen years old and are assumed to have mastered the technical skills to play the game, most of your de-

---

ence and credentials may take issue with some of the soccer nuances in the essay.

velopment is about creativity—about the multiplicity of choices you can make with what you can do.

Teamwork Lessons from Nonlinearity

1. Control Less

As we know from other sports, the player with the ball controls the field. The American football quarterback, as well as the point guard in basketball, "controls" how the play will develop; in baseball, the pitcher "controls" the batter. However, and excepting extremely rare circumstances (L. Messi, for instance) in soccer, *the player with the ball is the slowest player in the field.* The player with the ball "controls" the field but only for a very short time, because carrying (dribbling with your feet)—technical speed—is very much slower than non-technical speed (unencumbered running). Hence, the carrying player attracts defenders very quickly who will in all likelihood, take the ball off their foot. So the *only* way the team progresses down the field, is by passing the ball to someone else.

2. The Drop

Often, if a carrying player realizes his or her productivity with the ball is stymied somehow you will see them kick the ball backward—in the opposite direction of their goal, to a field teammate or even all the way back to the goalkeeper. To the linear thinker, this seems counterintuitive and even dangerous as far as play development goes. However, to go "backward" is not a loss—it is to gain the advantage of more time, better field vision, for better execution of the next pass. It is an understanding of immediate personal unsuccess as an *opportunity* for someone else to do something for team or holistic success.

By contrast, how often has your team "dropped" or gone slower—taken the time to reexamine unchallenged assumptions in their conclusions, ask unasked questions, or challenge the goal and path that the project leader assumes to be correct? How often have you passed off a task that you wanted to do, but someone else was better qualified for? In conventional understandings of teamwork, aren't these characteristics of weakness, inefficiency or being "in trouble"?

Nonlinearity counters conventional and ineffective understandings of team leadership and control. The most common advice offered in hundreds of books and documents is that team leaders "control" the team. Allegedly, this is accomplished via clear communication of a clear singular vision, designing and sticking to deliberate plans, setting specific expectations, clarifying roles and procedures, and dozens of other prescriptive adages.

The consequence of such actions is to implicitly invite—and I imagine invoke sanctions for—unmet expectations, blurred role activity, procedural deviations, etc. In other words, the more prescriptive emphasis on specific leadership behaviors and control functions, the more opportunity for error-correction. In fact, look at any of those hundreds of documents and these will be nearly identical to the reasons given for team failure. The evidence is that the top leaders seldom create the teams they want other groups in the company to be. The top, after all, is a place you typically get to through competition and self-reliance, or at best, by building tight networks of power and affiliation.

Consequently, in some sense can we not say that teams fail or are ineffective *because* they are (over)controlled, and (over)managed, rather than undercontrolled or unmanaged?

In a way, yes. Georgia Institute of Technology's John Vallas (2008) found that top-down, expertise-driven error correction actually limited a firms ability to improvise and lowered morale; that a logic of standardization was pitted against a logic of participation, and, intriguingly that:

> *...successful implementation of [team participation] may depend of the ability of corporate executives to demonstrate the very capacity for flexibility they demand of their employees.*

John Allpress, National Player Development Coach of the Football Association in London, confirms this by recognizing that there are differences in error-correction. Certainly, there is good reason to spend a great deal of time honing skills and role expertise by correcting mistakes. But that alone leads to rigidity and staleness, an uninspired predictability of role performance, and most importantly, *a hesitation to experiment*. Crucially:

> *Experimenting develops flexibility and the ability to be unpredictable, the perception to see tiny differences [in possibilities for play] ... Experimenting helps develop creativity, innovation and inventiveness—hallmarks of the best and most effective footballers* (Allpress, 2006: 7).

With just a little extrapolation, what is the hallmark of an effective team member, or team leader for that matter? It could be their precision, their sharp expertise, their clear communication, their eager adherence to their role. But I think it could also be the creativity and inventiveness, the unpredictability, which only a quantum mindset could appreciate. And this is a mindset that is rarely if ever cultivated.

As nonlinearity counters conventional advice regarding leadership and control, it also cultivates its opposite: decentralization, equity and empowerment. Everyone a leader or no one a leader. Both work, keeping in mind that leadership and followership are both interdependent and situational.

In terms of the former, in the 2006 World Cup, Argentina's Esteban Cambiasso scored. He was happy; the team celebrated. But what actually happened was that Cambiasso "finished" a play comprised of an almost unheard of *twenty-four passes* up, back and across the field. Described by blogger Roger Cohen as *part ballet, part advanced mathematics, part instinct and part genius,* this is truly a pinnacle of team play. Could any of your team projects at work be described that way?

Now, certainly work teams in organizations celebrate upon projects well-done, and the feeling that comes from such accomplishment. But doesn't this usually mean that they haven't rained on someone else's parade, that their conclusions or plans were not unacceptable, politically or otherwise, to some else? Hardly the transcendent experience described above.

A second example supports the same point: Greece wins 2004 European Cup with a team of players virtually unknown to premiership soccer in Europe, let alone the rest of the world—no stars, no *prima donnas.* Much of Greece's effectiveness was attributable to flexible, non-rigid team play, facilitated, for example, by *defender* Katsouranis making offensive plays and runs, therefore blurring his stipulated role as a defender. By contrast, most of the literature about "why teams fail" cite precisely this blurring of roles and expectations as a reason.

Soccer is Holonic

Scientist Arthur Koestler (1978) coined the "holon" as the fundamental unit of the universe: everything is dyadic, simultaneously both whole and part, demonstrating both self-assertive tendencies and integrative tendencies. That is, tendencies for individual expression/prominence, as well as tendencies to be part of the larger whole.

Soccer exhibits this in its constant push-pull dynamic of movement, where a player or team is only rarely certain whether they are attacking or defending, and that could change in an instant. Moreover, such instances *emerge* from the play. More experienced coaches identify a certain phenomenon of the game as the "transition game"—where, for a minute, it is clear that one side has to quickly move from offense to defense. Those more experienced may disagree with this, but I view the entire game as a "transition" game—a constant, continual dynamic full of creative tension where anything can happen anywhere on the pitch (field). Indeed, soccer is a game where offensive skills are ALSO used at the defensive end of the field, and defensive skills are ALSO used at the offensive end of the field.

Teamwork Lessons from Holism

1. Technical AND Tactical Preparation

Know your role, be prepared and on top of your game (expertise and skills). But more important, work to understand how what you have fits into the scheme of the entire team, and how you can contribute it productively. One of my sons has exquisite footwork that would look stunning on a video. However, he usually does so much of it that he forgets he has teammates there to pass to. Better if he uses his skill to beat one defender to create space and then pass the ball off to someone in a better position to do something else with it.

2. Everyone Must see the Field...

(as well as the grass) and the possibilities. This means understanding the big picture. Here, we can draw from some of the conventional prescriptive wisdom about teamwork: team members should know why they are doing what they are doing, and what its outcome will be; correspondingly, project leaders must understand the management or executive rationale for the project. Intriguingly, in any given instant, soccer play may be understood as operating in what is scientifically called *phase space*: the mathematical space of all possibilities in a given situation, or the collection of all possible configurations. If everyone knows the big picture, or when everyone sees the field, there exist countless opportunities for collaboration that may be unseen and unpredictable from initial project conditions or direction.

3. Perform-Support-Perform-Support ad infinitum

One of the earliest tactical lessons a developing soccer player is exposed to is "pass and go". What this means more precisely is "Never pass and watch."

Holonically speaking, performing or executing your job or task is only half the holon, the self-assertive tendency; the other half is supporting others as they execute theirs, the integrative tendency. Along the same lines, your job on the team isn't perform OR support, it is perform-support constantly. This speaks somewhat to roles, contributions, tasks and their place in team accomplishment. More directly, it speaks to the intentional blurring of these roles for a more collaborative endeavor.

Soccer is Emergent

Again, in scientific language, emergence refers to the arising of new and unexpected structures and phenomena that could not have been expected to arise from the initial conditions of their dynamic. Emergence is a product of self-organizing systems. Soccer players understand emergence in three ways:

First, it is usually the reason that play seems slow (for Americans). A pass here, a header there, drop it backward 20 yards, switch it across the field, *for godssake when is someone going to knock it out of the park?!* But while this sort of fundamentally sound tactical play seems boring, slow and ineffective, understand that it is *within these initial conditions* that players are watching for openings to exploit in the defense, waiting for the right time for a support run, seeing where the space is that a teammate can get to. Some of it may be predictable, but much of it is not. Rather, those moments we would recognize as "progress" (Americans recognize this by the crown cheering), arise spontaneously from the existence of momental field dynamics.

Second, thus, it is what happens when a particular offensive configuration appears about the mid-line or anywhere really, when passes and drops and heads and switches and ball tricks have been received and reciprocated, when they know that what they are doing now is clearly working, and all of the possible conditions have collapsed into this zone of thrilling potential (effectiveness). The momentum is up, the spirit is ungaugeable, and the light in their eyes tells all of the exhausting unmitigated fun they have at this play.

Third, they understand it from the defensive perspective also. They realize that what the other team is doing clearly has transcended their own constants or structure; they are working harder and faster, not smarter, relying on the gut work of those hours of honing individual expertise and fundamental skills, trying desperately to be able to see a field in which particle is not becoming wave right in front of them. Indeed, emergence is rarely countered with managed skill or protocol.

Teamwork Lessons from Emergence

Form Follows Function

Nothing of surprising significance can emerge from tightly prescribed functions and conditions, assigned roles, managerially "necessary" team members, controlled expectations or status quo team structure.

From personal experience: Several years ago, as a function of large scale university restructuring, my department merged into the business school. In July or August, the B-school organized a "retreat" out at the lake to welcome its new members, to get to know each other, integrate, and start to organize the business of the coming year. Part of this was the assignment/organization of the yearly B-school committees. I was assigned to a new committee for the B-school—the "Innovation Committee." Our new, particularly foresighted Dean gave this committee one charge: "Think out of the box!" (At the time, the phrase was not the cliché it is now.) What are the possibilities here with these new members, what could we conceivably accomplish, given this influx of new and different talent? I was excited.

For a minute. Thinking out of the box began with a disheartening discussion about an agenda and a Chair. After several minutes of fumbling with that structural albatross, it became apparent, that as a new committee, we were in uncharted territory—agenda and Chair became sort of interdependent, well, because there had to be a goal and, well then a path to the goal, and certainly some measures of progress, and who's going to report this and so on *ad nause-*

um. It was evident to me that none of the accountants, management professors, marketing or economics folks that surrounded me had thought about innovation much. Though their default mechanism was "structure", how could there possibly be an agenda?

Sheepishly, I offered: "We are the *innovation* committee. The only charge we have is "think out of the box." How 'bout if we just go without an agenda for awhile?" Stunned silence—for several long agonizing moments. Like I had launched a tornado across their desks of nice neat papers. Then, some nervous chuckles, and the idea was dropped. I never spoke again in that meeting, and the Innovation Committee has never met since that retreat seven years ago. Evidently because there was no extant structure for discussing innovation or possibility.

The concepts I mentioned earlier in the essay—flow, synergy, jazz, improvisation, possess the same pedagogical and practical difficulties as does 'emergence". However, what is usually left out of those prior concepts is the distinctly quantum explanation: all derive from a technical deftness and expertise that strives to imagine the unimaginable breadth of possibility within formal, structured or accepted limits. Such striving often sounds/seems complex and disorderly—much like soccer.

Soccer is Engaged

Soccer players recognize the importance of engagement with their teammates on the field, as their actions are moment to moment. By engagement I mean things like talk, trust, communication. Developing soccer players are always communicating (or should)—where they are, where to put the ball, who to play to, who to mark (guard). By the time they have played together for a couple of seasons they know each other very well; on and off the field relationships are for the most part solid. Importantly, what ensues from this is sort of a telepathic self-organization on the field—where they all know each other's assets and liabilities, their tendencies. They trust each other to do their best with every pass, every reception, every tackle, every shot. And if ineffective at a particular moment, ""unlucky, we'll get it next time."

Teamwork Lessons From Engagement

1. Understand each other, as well as the project

Appreciate the diversity of talent on your team. Value your own talent for exactly want it is, no more, but no less.

2. Build relationships of connection and support

In organizational teamwork, are we afforded the opportunity to develop relationships that help us in the long-run? Do we communicate in supportive ways that help the team or each other, or mostly defensive ways that protect turf, are self-interested and don't rain outshine anyone else?

In conclusion, the implications and lessons I have drawn here are admittedly idealistic. They are derived from an amateurish, perhaps overdetermined infatuation with the beauty and perfection of the game. Undoubtedly, the sport of soccer is not absent those maladies characteristic of sports teams in general. There are the "superstars", the grumpy, the disenfranchised, the overpaid; certainly, there are team dynamics that don't work, personality conflicts and all sorts of mismanaged phenomena. And even in youth soccer, though strongly discouraged, there is always the most competitive coach who will put the 11 year old who has matured about two years too early "up top" at forward, where he can simply outrun and out muscle everyone. Despite that, the way the game is designed to be played is my emphasis here—collaborative, creative.

Finally, it may be worth mentioning that the popularity of soccer among American youth is tremendous. According to a 2003 report by Gary Davidson of *SoccerTimes*, there were 7.783 million avid participants in 2002; by that measure, that makes soccer the second most popular U.S. team sport behind basketball (19.982 million) and ahead of softball (5.438 million) in 2002. Moreover, the large majority of players in the country are youngsters and an increasing number are women. Three-quarters of those playing are aged between 6 and 17 while 37 percent are women.

Speculating just a little ... in 2008, these players should be at or near working or college age—perhaps entering B-schools or business/management education. I think the soccer metaphor would work for them; already possessed of the "alternative" quantum mindset that soccer entails, they may be the ones ready for the change to genuine collaborative teamwork.

References

Allpress, J. (2006). "Managing mistakes to the players' advantage: Lessons from coaching in football (soccer)," *Development and Learning in Organizations*, ISSN 1477-7282, 20(4): 6-9.

Argyris, C. (1994). *On Organizational Learning*, ISBN 9780631213093.

Heathfield, S. (2008). "How to build a teamwork culture: Do the hard stuff for teams," http://humanresources.about.com/od/involvementteams/a/team_culture.htm.

Koestler, A. (1978). *Janus: A Summing Up*, ISBN 9780091321000.

Vallas, J. (2003). "Why teamwork fails: Obstacles to workplace change in four manufacturing plants," *American Sociological Review*, ISSN 0003-1224, 68(April): 223-250.

Zohar, D. (1990). *The Quantum Self*, ISBN 9780688107369.

R. Michael Bokeno (Ph.D. Purdue University, 1985) is Professor of Organizational Communication BB&T Fellow in the College of Business and Public Affairs Murray State University, Murray Kentucky, USA. His research interests lie at the intersection of critical management studies and complexity/chaos theory, and his published work has appeared in *Journal of Organizational Change Management, Academy of Management Review, Management Communication Quarterly*, and *Development and Learning in Organizations*, among other national and international venues. He has been involved in youth soccer for twelve years as a coach and manager of premiere level teams, and held administrative offices in several local associations.

29. Emergence Is Not Always 'Good'

Robert Weetman
Independent Consultant, UK

The concepts of emergence and collective intelligence are fascinating, and from their study might come good things. But neither is 'good' by definition and we ought to be careful not to let our enthusiasm and interest lead to us into speaking too casually about the benefits of 'encouraging emergence' or 'developing collective intelligence'. We can find ourselves battling the emergent properties of a system, and working against its collective intelligence. This article explores an example to illustrate this from the field of social care. It also discusses some tentative 'laws' and some issues resulting from the positive nature of popular perspectives on emergence.

Introduction

The ideas of emergence and collective intelligence seem to be inherently attractive ones. When we read about ants solving a problem, or people being wiser as a group, we think in positive terms. That Wikipedia can even exist, never mind that it can sometimes be the best source of information on a subject, is surprising and wonderful. The thought that collective intelligence might be useful in finding a way forward on global warming is something worthy of detailed study.

From here it takes only one small step to a place where we are talking about how to encourage emergence and how to develop collective intelligence. We find ourselves thinking about these things as being inherently positive attributes of a system, particularly of a human one. Two of the four introductory paragraphs in Wikipedia's entry on collective intelligence (20 August 2008), to which I'm referring for obvious reasons, specifically present this positive slant (the other two are neutral).

But to take this step is, I think, a huge mistake. The fact is that emergence and collective intelligence aren't 'good' by definition. If good things can emerge, so can bad; and intelligence can be put to beneficial or to detrimental use. This article explores these points in more detail, examining some specific examples, suggesting some general conclusions, and discussing the consequences that arise from it being popular to take a positive view of emergence.

I'm aware that some readers might resent any implication that they didn't already know that emergence and collective intelligence can result in 'bad' as well as 'good'—so I should be clear that I'm not necessarily presenting new knowledge here. We've known about the awkward ways in which systems work for a long time. I'm simply reacting to the manner in which emergence and collective intelligence tend to be discussed.

A Simple Example

I find this much simplified example helpful as an introduction to this discussion.

An advice centre is staffed by passionate specialist workers. They individually reply quickly and efficiently to telephone queries. If they can answer the query directly they do so, and if not they immediately say so and refer the matter quickly to their colleagues. They each care deeply about getting the right replies sent to people.

We might expect that the emergent properties of this system—which is made up of passionate and efficient workers—would be positive. Unfortunately we all know that this isn't how emergence works. Putting a group of efficient and passionate people together doesn't necessarily create an organization which, in our dealings with it from outside, is efficient and passionate. When we look at our interaction with individual workers, we find we are dealt with efficiently and the worker's passion is clear. But we may also find that we are passed repeatedly around the system, that our query is never actually answered, and that it takes a long time for us to work out that the centre does not have the expertise we need.

The property of being inefficient at replying to queries is an emergent one. It is one that arises at, and is best observed at, the organizational level. It's clearly not a 'good' property.

A More Informative Example

A more in-depth example is required if we are to look at this properly—and I'll refer to the area of work in which I specialize, which is in supporting change within 'care' organizations. I believe that this example has much to teach us about collective intelligence and emergence more generally, and about what we miss if we tend to view these as inherently positive.

On the whole people who on the receiving end of care and support work need this because of some great difficulty in their lives. For a few, the provision of particular expertise (for instance pain relief), or help with the accomplishment of a physical task (for instance getting out of bed), is sufficient. Solving this technical matter, on its own, is sufficient to free them to have a full and rich life. But for many people the key issues aren't these technical matters. Rather, they have problems like a lack of people in their life, low self-esteem, or narrow life experience. These things might have been caused by the specific technical problem (e.g., disability, illness, or addiction), but 'solving' this technical matter has little lasting effect on their quality of life. The effects of exclusion, devaluation and disempowerment are at least as debilitating as the original issues ever were—and the person becomes trapped in a situation where exclusion, disempowerment and devaluation positively reinforce each other.

Fortunately it is quite possible to solve this problem for individual people. Much has been written about how to do this, and on the whole the problem isn't a particularly technical one. There are obvious difficulties to overcome—as in all work with people—and the work may well require creativity, skill, and tenacity, but even people who have never worked in this field can think of a host of solutions.

Unfortunately, when we look at how to make sure that our care and support systems do this for everyone, we come across much greater challenges. I want now to describe the way in which emergence and collective intelligence sit at the heart of these challenges.

On the whole our care and support services base everything they do on the technical issues a person faces—their disability, illness, or addiction for example. For as long as anyone can remember discussion has been taking place about how to change this. There have been repeated changes of legislation and government policy, and within organizations many rounds of consultation, policy change, structural change, training, and so on. Somehow it seems that the problem is intractable. There are, as there have always been, examples of good practise to be found if we look hard, and always the promise of real change on the horizon. Words change, structures change, stated values change, there are overall small improvements, but on the whole everything works in pretty much the same way as it has for decades.

Inevitably a system/emergence approach is useful to understanding what is going on. In particular, we notice what for many is a startling idea; that it isn't really in the interests of a care organization to tackle devaluation, exclusion and disempowerment, and that this fact influences how such an organization works. In simple and bold terms; the organization needs people to be lonely, devalued, and disempowered. These problems are why it exists, and the more people who have these problems the better. Of course at this point it is important that I acknowledge that individual staff and managers within an organization, and even individual projects or departments, don't think this way. They are often hugely dedicated, and they dream of making a real difference, and they do their very best within the system in which they work. However, it should be clear to readers who understand systems that what is in the interests of the system will be influential none the less.

Many people might believe that the individual goodwill of staff within these organizations will enable them to overcome the interests of the system. It is important, therefore, to spend a moment considering quite how powerful the interests of the system actually are.

Some of the main external pressures on care organizations arise from the public image of those people who are supported. If 'people with mental health problems' are generally seen firstly in terms of being a threat to society, then it is in the organization's interest to respond to this. If people with Downs Syndrome are generally understood to be a bit like children, then organizations will find it difficult to work in such a way as to allow those they support to take real risks or have sexual relationships. The existence of the organization will be more secure if it works in a way which doesn't 'frighten the horses'—and this will be true even if society's perceptions of the people it supports are utterly incorrect. Whatever other pressures or policies influence a care/support organization we can't get away from the fact that most of their income comes from society at large, through taxes (or an equivalent), and is dependent on political decisions (or an equivalent). They are, in a very direct way, working on behalf of society at large. It is unlikely that competing pressures—for instance an official policy demanding that people are included—will be anywhere near as powerful as the fundamental fact that it exists to do whatever society as a whole is expecting of it.

Unsurprisingly, in real world terms, the result is often an organization which 'talks the talk' of inclusion and empowerment, staffed by people doing their best to work on these issues, but which somehow manages to completely undermine its own efforts.

The point of providing this example should be clear. The system, the organization, works in a way which results in disempowerment, devaluation, and exclusion. This is an emergent property of the system. We can go further I think. The organization must have a collective intelligence. Why should we assume that this intelligence would be directed towards anything other than the interests of the system? To me it is clear that the collective intelligence of a care organization has this focus—and that it can do a very good job in working on the system's behalf. Many of those who write about changing these organizations mention the need to work 'under the system's radar' (or something equivalent). Some of these writers may be visualizing 'the system' as synonymous with 'those in charge' but I think that what they are responding to is the way in which such a system seems to be particularly creative and adaptive in fighting change. Personally, when I write in this context about 'the system' I'm thinking in terms of the collective intelligence of the whole system.

It is clear, at least in terms of particular desirable outcomes, that emergence and collective intelligence are not 'good' attributes of organizations like this.

Original Ref—E:CO Vol. 11 No. 2 2009 pp. 87-91

Generalizing

I find that thinking about emergence and collective intelligence in this way is hugely helpful. It also bridges a gap between what I understand about complexity and the work on 'change' I have encountered in systems theory. It becomes particularly powerful once also connected to ideas like those about the 'dangers' inherent in attempting to lead 'adaptive' change. These are outlined by Ronald A. Heifetz and Marty Linsky (2002) in their book 'Leadership on the Line,' which provides a comprehensive list of the different ways in which a system can neutralise a change effort.

While these ideas are very helpful for me, they are much more difficult to convey in a short time to those with whom I work. One approach that can help is to try to summarize what we know, almost as a set of rules:

- Emergence, and collective intelligence, can be powerful and are very interesting to study, but are not inherently 'good' or 'bad';
- In human systems the system's interests, and therefore its behavior and collective intelligence, may well be against us. This is particularly true in situations where we are facing an 'adaptive' challenge in trying to lead system change;
- It seems that the emergent behavior of a human system (or systems) tends to maintain power imbalances between different groups of people. It is worth noting that it is in the interests of a group of powerful people to maintain that power, and that the group's behavior (viewed as separate from the behavior of individuals) will respond to that.

Wider Issues

I have written this article because I have struggled to find other people writing in an easily accessible way on these specific matters. It may be that they are debated deep within the 'complexity' community—but from the outside I haven't been able to detect this. That is unfortunate for my specific work—I imagine that at least to some extent I've been re-inventing the wheel—but I think that other people must also be excluded and that there are probably wider consequences of this.

The simplest problem is that people who would find ideas around change, emergence, and collective intelligence to be of immediate practical use are much less likely to encounter them—or that they only do so in a fleeting way when with someone like me for an hour or two. It is difficult to find any appropriate writing to recommend on this subject to a middle manager busy with day-to-day issues.

I believe that there are also deeper problems resulting particularly from the positive bias from which much of the discussion views emergence and collective intelligence. If we repeatedly use one particular set of examples to speak about collective intelligence, such as Wikipedia or prediction markets, these problems will be inevitable. We find ourselves creating theories based only on the perspective these provide. James Surowiecki, in The Wisdom of Crowds (2004), discusses the way in which diversity leads to a greater level of crowd wisdom. Others interested in collective intelligence make a leap to state that collective intelligence depends for its existence on diversity.

To some extent language begins to fail us here, but my understanding of collective intelligence is very different. Diversity may well add to the *wisdom* of a collective intelligence—but, in somewhat clumsy language, un-wise collective intelligence can still exist in a group without diversity.

In a similar vein, I recently came across an influential academic discussing how emergence is visible on the Internet specifically because of the large number of people involved. I see this as another example of such confusion. While it may be true that Google or Wikipedia may depend for their *accuracy* on the number of people involved, that is not proof that emergence only appears when many people are involved in a human system. Even an extremely small human system can have emergent properties. For instance if I put only three 'shy' people in a room together it is impossible for me to predict the nature of the character of the group which will emerge. As a consultant I am very familiar with the way in which a small group can exhibit a character very different from that which one would predict from the characters of the individuals who are members of it.

Perhaps we should state another 'rule' as follows:

- The *wisdom* of the collective intelligence of a human system can depend on the large number of diverse of views within it.

Conclusions

The first key point I have made in this article is that both emergence and collective intelligence seem, at least to me, to always be interesting attributes of a system, but not always to be positive ones. The second point is that it is a mistake for us to forget this. I have also laid out a number of 'rules' which seem to me to be useful to those working with a certain set of problems caused by emergence and collective intelligence.

To conclude, I would like to suggest that the closer involvement of practitioners like myself in discussions about emergence and collective intelligence would be a useful step forward. We have very practical experience of dealing with human systems at a system level, and useful ideas to contribute toward theoretical discussion as a result.

I would be delighted to receive any correspondence on this article (at enquiryRW@capacitythinking.org.uk or see www.capacitythinking.org.uk).

Original Ref—E:CO Vol. 11 No. 2 2009 pp. 87-91

Readers might also consider contributing accessible writing on these themes to the 'change' section of the collection of articles at www.isja.org.uk/.

References

Heifetz, R.A. and Linsky, M. (2002). *Leadership on the Line: Staying Alive through the Dangers of Leading*, ISBN 9781578514373.

Surowiecki, J. (2004). *The Wisdom of Crowds*, ISBN 9780385503860.

Wikipedia (20 August 2008) "Collective Intelligence," http://en.wikipedia.org/wiki/Collective_intelligence.

Robert Weetman is an independent consultant and trainer working in the UK on social and organizational change. He has a BSc (Hons) in Physics and an MSc in 'Work Design and Ergonomics'. Of more relevance is that he has 17 years of experience with issues around 'change' in care and support organizations, and in society at large.

30. Networks, Emergence, Iteration and Evolution

William Chamberlin
Texas Tech University Health Sciences Center, US

Recognition that the reductionist approach to science leaves great gaps in our understanding has led to the synthesis approach to further explain the world around us. The synthetic approach examines the inter-relationships of individual entities as they interact to create complex networks. This approach spawned the creation of a new science—the study of Complex Systems. This article takes the concepts of Complexity Theory and hypothesizes a process of simple steps iterated many times over that explains the emergence of new entities and the evolution of our Universe. The concept of systems, emergence, iteration and evolution is proposed to explain the process underlying our evolving Universe. This process would be expected to leave fractal patterns in its wake. The fractal patterns are related to the shared tendencies for self-organization found in complex networks. The principles apply to all networks irrespective of their component parts and include both inanimate and living systems.

Introduction

What good is a worldview if it does not explain the events around us in a meaningful and internally consistent manner that helps us to understand and resolve the problems that we continually face? This work proposes a worldview that conceives of everything as a process of interacting systems acting on hierarchical, evolving levels. This process is theorized to be fundamental to the evolution of the Universe. Thus, the Universe is seen as being built out of layers of complex dynamic systems, each layer consisting of the 'emergent properties' that are generated from the complex interactions of the systems in the level below it.

The Universe is everything we know and think—and probably more. It definitely includes particles, forces, and energy. The subject matter of physics, chemistry and biology as well as the abstract notions of forces, ideas and information are all aspects of the Universe. A truly universal 'law' describing the Universe must apply to them all. This article discusses a process that seems to be universal and accounts for the development and evolution of all that was—and will be—'new' in our Universe.

Hopefully, most of the ideas expressed in this article are not new. They have already been developed, discussed and published in the scientific literature by others. The uniqueness of this article lies in the emphasis on the relationship of emergence and the iterative process with the fractal patterns observed throughout Nature, and proposes that this relationship underlies the evolution of the Universe.

Systems, Networks, Emergent Phenomena, Iteration and Evolution

A. Single entities (E) interact to form complex interacting systems. New behavior traits emerge (emergent phenomena) from the dynamic interplay of the components of the system...

B. The newly emerged properties define the system. The original system may no longer be viewed as a system, but rather as a single new unique entity (E1).

C. A detailed description of the characteristics of this new entity (E1) is a description of the emergent properties that define it.

A1. Single entities (E1) interact to form complex interacting systems. New behavior traits emerge (emergent phenomena) from the dynamic interplay of the components of the system.

B1. The newly emerged properties define the system. The original system may no longer be viewed as a system, but rather as a single new unique entity (E2).

C1. A detailed description of the characteristics of this new entity (E2) is a description of the emergent properties that define it.

A2. Single entities (E2) interact to form complex, etc., etc., etc.

The process of smaller systems interacting to create new 'emergent' properties that define new entities—and then these entities forming complex networks out of which newer 'emergent phenomena' arise is hypothesized to be the process by which the Universe evolved. It is an iterative process. The critical concept is that all single entities can be viewed as complex systems composed of their component parts, and these parts may be seen as smaller systems themselves. Since all single entities are systems, the identifiable characteristics of any single entity are the emergent properties that arise out of the interactions of the 'lesser systems' that comprise it.

In summary, a single entity on one level can be viewed as being the 'emergent properties' of a complex dynamic system of smaller entities existing at a lower level. Looking forward, we can predict that as our current systems form modules and interact in complex dynamic ways, new 'emergent phenomena' will arise and we will evolve, we just cannot predict what these emergent properties will be.

Systems/Emergence Theory is best explained by example. A very brief description of the evolution of the Universe will be given touching on the beginning of the eras of physics, chemistry and biology.

Era of Particle Physics

Based on the Standard Theory of Particle Physics it is theorized that by one trillionth of a second after the Big Bang fundamental particles were already interacting with each other in dynamic systems to create the building blocks of the Universe.

As conceptualized today, there are two major classes of matter, hadrons and leptons. Hadrons and leptons are distinguished by their behavior characteristics. They interact through the four Forces via the exchange of energy quanta. These packets of energy are best conceived as agents of interaction. The forces and their carriers of interaction are listed below:

Strong Force (gluons), *Weak* Force (bosons), *Electro-magnetic* Force (photons) and *Gravity* (gravitons?).

Hadrons are particles that participate in all four known types of interactions—strong, weak, electro-magnetic and gravitational.

Leptons are particles that do not participate in strong interactions.

Charged leptons participate in 3 interactions—weak, electromagnetic and gravitational.

Neutral leptons participate in 2 interactions—weak and gravitational.

Leptons are considered to be fundamental particles.

Hadrons are composed of the fundamental particles called quarks and gluons. Combinations of the six different 'flavors' of quarks produce hadrons with different properties (protons, anti-protons, neutrons, pi-mesons, k-mesons etc.).

> *The quarks inside a baryon or meson (types of hadrons) are continually interacting with one another via the strong force field. At any instant in time, they may contain many virtual particles: gluons and additional quark/anti-quark pairs. The picture of a proton as made of three quarks is thus a gross simplification. For example, we know from measurements that in a high momentum proton only about half the momentum is carried by quarks, the rest is carried by gluons* (www2.slac.stanford.edu).

Paraphrasing the above in systems/emergence terminology: the six different flavors of quarks and gluons interact to form dynamic systems of interacting fundamental particles. These dynamic networks develop behavior characteristics that we recognize as the various types of hadrons—protons, anti-protons, neutrons, anti-neutrons etc. The emergent characteristics define and become the particles of the Universe.

As the Universe expanded and the temperature cooled, hadrons (protons) and leptons (electrons) formed systems that displayed the emergent properties that we recognize as hydrogen atoms. Under the influence of gravity, hydrogen atoms coalesced to form stars. Conditions were suitable in these stars for hydrogen atoms to interact with each other, and out of these interactions came new systems of fused nuclei with the emergent qualities of the heavier elements.

In short, out of the interacting systems of quarks and gluons emerged protons, from the interactions of protons and electrons emerged hydrogen atoms, and in the furnaces of stars the hydrogen atoms formed systems displaying the emergent properties of the heavier atoms found in the Periodic Table of the Elements.

The process—of fundamental particles interacting to form systems from which new behavior characteristics emerged that in turn defined new entities—was an ongoing process from the very beginning.

Era of Chemistry

Once the formation of atoms has occurred, it is easy to trace the formation of systems of atoms as they form molecules. Each individual molecule may be viewed as a system composed of its constituent atoms. And of course, the myriad molecules that are formed interact with each other to form still larger molecular networks. The molecular systems manifest emergent properties that possess the attributes of the physical Universe. Thus the elements and compounds that form the earth and stars are created. Needless to say, the underpinnings of the chemical systems are based on the micro-systems of subatomic particles.

Era of Biology

A very special branch of chemical evolution involves the complex, adaptive dynamic systems that are formed from the chemistry of carbon, hydrogen, oxygen, nitrogen and other trace elements. At least at one site in the Universe C, H, O and N formed molecules that interacted with one another in a very special way. Some of the molecules that were products of chemical reactions in the network not only served as substrates for additional reactions, but they also served as catalysts for reactions that were involved in their original synthesis. In this manner reactions became 'autocatalytic' and the systems, when provided with sufficient substrates, developed the ability to grow and replicate. As these autocatalytic, self-replicating systems increased in number, size and complexity they developed emergent characteristics that we now associate with Life. Thus, out of 'emergent chemistry' came 'evolutionary biology'.

Under the evolutionary pressures of a coevolving environment, the systems became increasingly complex. Molecular pathways within simple bacteria

Original Ref—E:CO Vol. 11 No. 4 2009 pp. 93-100

increased in complexity and altered the phenotypes of the cells in which the systems were found. Simple prokaryotes developed complex relationships and became eukaryotes. Single cell eukaryotes formed systems with the emergent properties of colonies (sponges). Complex systems built upon underlying complex systems and the process continued until multi-cellular plants and animals with specialized organs evolved.

Skipping forward, it appears that arguably the highest level of systems/emergent integration found in Nature is the intelligent self-awareness emerging out of the network of trillions of interacting neurons in the human brain. This process continues and the future will predictably take us to the next level.

The key point is that everything has evolved by the same process—component parts interact to form complex systems that display new characteristics as a result of their complex interactions. The new and possibly unique emergent properties define new entities. These new entities may form complex systems performing on the next 'higher' evolutionary level. An iteration of this process explains the evolution of the Universe from its earliest moments, when all that existed was energy and a few subatomic particles, to the present.

Evidence for Iterative Emergence

The proposed iterative process is conceptually possible, but is there any evidence that the Universe really is built out of layers of interacting networks? Certainly the emergence of new entities occurs (the reader is referred to the numerous publications on this subject, e.g., Zukav, 1979; Casti 1994; Kauffman, 1995, 2000; Johnson, 2001; Morowitz, 2002; Laughlin, 2005) but does a repetition of this process account for the construction of everything in the Universe? Evidence for this hypothesis may be indirect and somewhat soft but it is nevertheless intriguing. The iterative process of interacting networks creating new entities on the smallest scales of space and time to the largest may be reflected in the fractal patterns found throughout Nature involving different time and space scales. Somehow these fractal patterns reflect similar tendencies for self-organization found in networks of all sizes as theorized in complexity theory.

Complex Systems as a Science

Reviewing the essential lessons of the science of Complexity will be helpful. There are several key observations that may not be intuitively obvious.

1. Multi-variable complex adaptive dynamic systems display similar behavior characteristics. These similarities exist even though the systems may be composed of different components. This observation implies that the interactions of the components possess an importance independent of the nature of the components. The study of these similarities is the study of complexity.

2. At a threshold of energy and complexity, systems develop new behavior characteristics that we call 'emergent phenomena'. The emergent traits may be totally unpredictable from what is known about the component parts. This observation is summarized in the phrase "The whole is more than the sum of its parts."
3. Emergent phenomena appear at phase transitions occurring at the interface of conditions of extreme stability and conditions of excessive instability. This interface has been colorfully named the 'edge of chaos'. Mathematical models predict bifurcation patterns—or a doubling- of the possible semi-stable states that can occur at these phase transitions.
4. These observations are seen in systems composed of atoms, or molecules, or cells, or animals, or neurons, or mixtures of any independent 'nodes' all interacting with each other.
5. Complex adaptive systems exist in different size scales and interact in different time scales. This brings us to one last interesting concept—the concept of fractals.

The science of Physics is built on the use of mathematical concepts and equations that construct models of the Universe. The mathematical concept of 'fractals' serves as a model for the systems/emergence theory. Fractals are entities that exhibit similar, but not necessarily identical, traits when examined on different measurement scales. They are said to display self-similarity and be invariant under length scale transformation. The wooden Babushka dolls that open up to reveal similar dolls within may serve as a concrete analogy. If our Universe is really built from emergent traits springing forth from layers of complex interacting systems and since complex systems all possess similarities, then the concept of fractals may model the most fundamental creative process in the Universe.

Mathematical Model for Complex Systems Behavior and Fractal Geometry

Song, Havlin and Makse published a letter in Nature that is particularly relevant to the premise put forth in this paper. They describe a mathematical process by which several different naturally occurring complex systems are shown to display self-similarity or invariance under a length scale transformation. Their work provides mathematical support for the idea that the self-similar fractal patterns found in nature may be related to the processes of networks and emergence described previously.

Interested readers must read their article but the abstract is as follows:

Complex networks have been studied extensively owing to their relevance to many real systems such as the world-wide web, the Internet, energy landscapes and biological and social networks. A large number of real networks are referred to as 'scale-free' because they show a power-law distribution of the number of links per node. However, it is widely believed that complex networks are not invariant or self-similar under a length-scale transformation. This conclusion originates from the 'small-world' property of these networks, which implies that the number of nodes increases exponentially with the 'diameter' of the network, rather than the power-law relation expected for a self-similar structure. Here we analyze a variety of real complex networks and find that, on the contrary, they consist of self-repeating patterns on all length scales. This result is achieved by the application of a renormalization procedure that coarse-grains the system into boxes containing nodes within a given 'size'. We identify a power-law relation between the number of boxes needed to cover the network and the size of the box, defining a finite self-similar exponent. These fundamental properties help to explain the scale-free nature of complex networks and suggest a common self-organization dynamics (Song *et al.*, 2005: 392).

The renormalization procedure that they use in which large boxes are drawn around numerous interacting nodes that are then considered to be a single new node represents the natural phenomena of the creation of a new entity out of the emergent qualities of an underlying complex system. The next step in their renormalization procedure involves connecting the larger boxes with one another to create a new network. This step represents the process of new networks that form using the newly created larger entities. In both cases the resultant relationship patterns are self-similar and independent of length-time transformation. Fractal patterns, by definition, display self-similarity and are independent of length-time transformation.

Complex Adaptive Dynamic Systems in Biology

Manifestations of Complexity Theory may be found throughout biology. The biosphere itself is a large complex adaptive dynamic system composed of all the interacting elements that comprise it. There are many smaller networks participating in different time and space scales found within the biosphere. Fractal patterns amongst these networks are self-evident. Examples are found in the inter-relationships of species within the global animal kingdom, the relationships of individual animals interacting in a defined ecosystem, the relationship of the bacterial microflora of the human colon and, if our hypothesis is correct, the relationship of key molecules with each other found in the metabolic pathways of human molecular biology. Similarities in the interactions of the components of these complex adaptive systems reflect fundamental attributes of Emergence and self-organization of complex adaptive dynamic systems.

Applying Complexity Theory to the field of human Medicine promises to provide new insights that may increase our ability to understand complex disease processes. Biologic networks exist at different levels. Molecules and organelles are the components of cells, cells are the components of the major organs, and the major organs form a physiologic network with the emergent properties manifested at the biologic level of the intact animal (humans). And, of course, individual humans form sociologic networks.

Health can be viewed as a state where there is the proper 'homeostatic' balance of the myriad biochemical networks, cellular networks and large organ networks that comprise the human body. All the involved complex adaptive dynamic systems are properly functioning around their correct attractors.

Disease, on the other hand, can be conceptualized as occurring when there is some disruptive influence that deviates the networks away from their homeostatic attractors. The clinical manifestations of a disease would be the altered emergent traits generated at the different biological levels.

Infection with Hepatitis A Virus (HAV) serves as a good example. HAV is ingested, absorbed through the intestinal mucosa and travels via the blood to the liver where it adheres to and invades the liver cell (hepatocytes). The virus synthesizes molecules that reproduce itself and also subvert molecular pathways of the hepatocyte. Homeostasis of the hepatocyte's molecular systems is perturbed resulting in 'disease' at the molecular level resulting in the emergence of disease traits at the cellular level. The human immune system has evolved mechanisms to sense and respond to microbial pathogens. Pattern Recognition Receptors sense the presence of HAV molecules and alter the host cell's genetic networks such that molecular signals (cytokines) are synthesized and sent to activate the Innate Immune System. The host's immune system (organ) is activated to respond to the viral infection. The immune system is another complex adaptive dynamic system. Immune activation can be conceptualized as a perturbation from its normal 'healthy' state to one associated with disease. The manufacture of myriad pro-inflammatory cytokines produces effects on the body that are associated with disease, i.e., fever, muscle aches, fatigue etc.

Thinking in terms of Complexity Theory can provide insight and help us to understand diseases that are currently poorly understood. Crohn's Disease (CD) is a good example. CD is an inflammatory disorder of the gastrointestinal system associated with dysregulated immunity. The etiology of CD is highly controversial. Perhaps the controversy stems as much from our current linear approach to thinking about disease causality as from anything else. Applying Complexity Theory to the way we think about the multiple factors known to be associated with CD may simplify the problem and resolve the controversy.

Linear logic about causality implies that A causes B causes C etc. and is represented in Figure 1.

Non-linear approaches are much more complex and include positive and negative feedback and feed-forward loops that incorporate multiple variables (see Figure 2). The Emergent characteristics of a 'Complexity Causal Network' (etiology) may produce several potential outcomes that Complexity Theory regards as semi-stable states. Such an approach allows us to conceptualize and comprehend the interacting influences of many different variables. Minor variations in any given variable could potentially drastically alter the resultant (semi-stable state) outcome. Thus, an etiologic agent when introduced into the mix may result in a disease state in one instance but may not cause any changes if the conditions are slightly different.

Factors thought to be associated with CD include the following: 32 known gene polymorphisms, prior exposure to other microbes, age, smoking, immune deficiency, use of corticosteroids, infection with Adhesive Invasive E. coli, infection with *Mycobacterium avium paratuberculosis*, other infections, unknown environmental factors.

The insight gained from applying the principles of Complexity Theory to contemplating the etiology of CD (and other diseases) is that no single factor is the causative agent; there must be a critical blend of factors that result in the emergence of a disease that is labeled as Crohn's Disease.

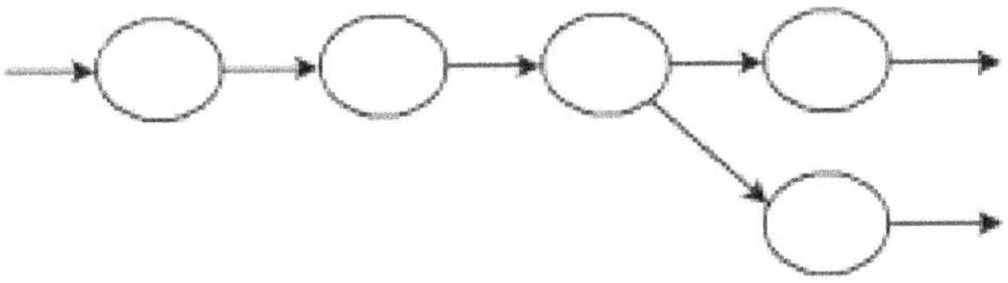

Figure 1 *Current Linear approach to Causality*

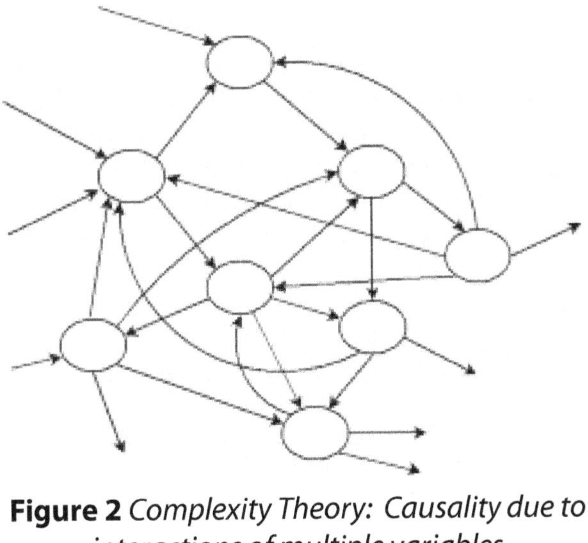

Figure 2 *Complexity Theory: Causality due to interactions of multiple variables*

Hierarchical Control Networks

"Nothing makes sense in biology except in the light of evolution". When thinking about the evolution of multi-cellular organisms such as Man it is apparent that a system must have also evolved that can coordinate and integrate all the different molecular, cellular and organ activities that occur during life. This control system must be complex, adaptive and dynamic. It must respond to changes in the internal and external environment to modify the organisms global response in order to survive. This system must be hierarchical in that it must coordinate activities at the local as well as the global level. Our known evolved hierarchical control systems are fundamentally molecular. The nervous system functions via molecular neurotransmitters, the endocrine system utilizes peptide and steroid hormones to modulate cellular activity throughout the body, and individual cells communicate on a local basis with an array of peptide, carbohydrate, lipid and other molecules that bind to specific receptors or directly modulate enzyme function.

Using the Complexity paradigm to think about the command and control mechanisms may allow early recognition of some exciting new research developments. Given that all evolution is essentially molecular evolution, it is intriguing to think about the family of steroid molecules as playing a key role in the evolution of an integrating and coordinating hierarchical control system that allowed the development of higher multi-cellular organisms. Since we have sequenced the human genome we know that 46 steroid receptors are encoded within our DNA. Since these receptors have been preserved they must be there for a reason. We only know what the ligands are for a small number of these receptors (androgens, estrogens, corticosteroids, mineral corticoids, Vitamin D and possibly some others). Thus the ligands and function of the majority of these steroid receptors are unknown. Changing an OH group from any one of the carbon atoms on the steroid molecule can alter the binding to a receptor—and elicit profound down-stream biologic effects. An example would be the one hydroxyl group that differentiates testosterone from estrogen—and we all recognize the biologic effects that occur. Most importantly, the addition of an OH group may take only one metabolic step. It is intriguing to think that a mutation in a receptor may find a possible ligand only one metabolic step (mutation) away that opens up potential new pathways for evolution to explore.

If this hypothesis is in fact true (and it is only a hypothesis), then steroid molecules may have played a fundamental role in evolution and the current endocrine, paracrine and autocrine activities of the naturally occurring steroids may be key to the integration of the different metabolic pathways in our bodies.

Assuming that this hypothesis is correct, then research and development of steroids may provide mechanisms to modulate our metabolic pathways in an advantageous manner. Such research is currently being performed by a California biotech company that has molecules in clinical trials. If successful, their en-

Original Ref—E:CO Vol. 11 No. 4 2009 pp. 93-100

deavors may bring a sea-change to the field of Medicine. Already their research is showing promising activity in the areas of immunology, inflammation, cancer, ageing, diabetes and others. The impact on medical care may surpass that of the entire antibiotic era—which of course only addressed infectious agents.

Applicability

So, even if this theory on the evolution of the Universe is correct, what good is it for us? It does not change anything; things are what they are regardless of how they got there. How does it help? There are two immediate answers to these questions. *First*, we have always sought to understand what things are, how they came to be and why they are what they are. Perhaps Systems-Emergence Theory can help answer these questions. *Secondly*, employing the systems-emergence approach may be a practical method for resolving problems. Perceiving things as being constructed of layers of interacting systems may provide insight into problems that still confuse us. Confusion often arises because it is unclear on which 'level' a given variable is functioning. Insight may be achieved by realizing that 'reality' is constructed from layers of complex systems. The concept of modules and new entities are related.

Summary

Systems/emergence theory emphasizes the importance of networks of interacting entities out of which new properties (entities) emerge. This process is fundamental to the creation and evolution of all things in the Universe. It is consistent with current theories of physics, chemistry and biology.

Acceptance of systems/emergence theory does not change a thing; things are what they are. But, an appreciation of the systems/emergence perspective may provide insight into the World around us and help us to resolve paradoxes and problems that we continually encounter. Utilization of this paradigm may be particularly useful when contemplating the relationships of the many networks found in the pyramidal levels of biological systems. This model allows us to more easily integrate our knowledge of molecular phenomena with cellular, organ and global function. It also allows medical scientists and physicians to better conceptualize the dynamic interactive relationships of the multiple factors that are involved in complex disease processes. Phenotypes are simply 'emergent traits' of underlying genetic, biochemical and cellular networks. Disturbances in these networks by whatever mechanism at any level may disturb the homeostatic balance, alter the emergent characteristics, and manifest as disease.

References

Casti, J. (1994). *Complexification*, ISBN 9780060925871 (1995).

Johnson, S. (2001). *Emergence: The Connected Lives of Ants, Brains, Cities and Software*, ISBN 9780684868769 (2002).

Kauffman, S. (1995). *At Home in the Universe: The Search for the Laws of Self-Organization and Complexity*, ISBN 9780195111309 (1996).

Kauffman, S. (2000). *Investigations*, ISBN 9780195121056 (2002).

Laughlin R.B. (2005). *A Different Universe: Reinventing Physics from the Bottom Down*, ISBN 9780465038299 (2006).

Morowitz, H.J. (2002). *The Emergence of Everything: How the World Became Complex*, ISBN 9780195173314 (2004).

Song, C., Havlin, S. and Makse, H. (2005). "Self-similarity of complex networks," *Nature*, ISSN 0028-0836, 433, 392-395.

Zukav, G. (1979). *The Dancing Wu Li Masters: An Overview of the New Physics*, ISBN 9780060959685 (2001).

William Chamberlin M.D., FACG is currently a member of the medical faculty at Texas Tech University Health Sciences Center in El Paso, Texas. He received his undergraduate education at Dartmouth College and Medical degree from Tufts Medical School. He is currently involved in an international project addressing and refining the concept of Causality in Complex Adaptive Systems.

31. Adjacent Opportunities: Engaged Emergence—Getting Beyond Enterprise Fortune Telling

Ron Schultz
The Center for Social Profit Leadership, US

Cause, condition, and effect characterize the process of emergence, but often times, rather than focusing on this, enterprises would rather spend billions trying to predict and divine what is going to happen in the future. How effective has this fortune telling been? The answer is costly and not very accurate. However, that fact hasn't seemed to stop anyone from investing more resources into enterprise soothsayers as a means of getting a leg up, no matter how inaccurate they prove to be.

If enterprises better understood what drives and influences business action, they might be a bit more successful in their prognostication by narrowing their sooth and recognizing the cues, as my colleague Michael Lissack would call them, that present themselves to those who, in fact can, see them as cues. This ability is what I am calling *Engaged Emergence*.

Engaged Emergence is attending to the resonance of the world, of which we are a part, and being authentically present to witness the auspicious coincidence of what emerges. In some cultures, this is also called mindfulness, our ability to focus our minds without distraction on what presents itself. The opposite of mindfulness is mindfilledness, a quality that is in constant distraction, and by its nature, will miss any but the most blatant cues that become present. The process begins with theory.

The Theory of Enterprise Action is based primarily on two core ingredients; 1) What we want—that which informs all our products and services, and can then be referred to as information, and 2) What we build to support getting what we want, or infrastructure. When the ratio between information and infrastructure is right, according to the given situation of the enterprise, the action takes place, successfully. When the ratio is not correct, and there is too much or too little infrastructure to support the acquisition of information, we do not get a successful action. Ah, if it were as simple as just that. Unfortunately, the actions we take in our enterprises are also influenced by two other forces: 1) The culture of the enterprise – a culture of fear automatically creates greater infrastructure to protect oneself from what is feared. A culture that is too loosey-goosey creates action that is too unbounded so that no organization, self or otherwise, can take place, and efforts dissipate. And then 2) Outside of the internal cultural influence there is the impact of the world at large that operates outside the control of the enterprise. This is the stuff of meltdowns.

The ability to recognize the available cues before a meltdown emerges can save valuable resources. If we are distracted by the causes and conditions before the meltdown, we will be unable to incorporate what presents itself in a fashion that will avoid disaster. Engaged Emergence is about being present enough to see what is emerging as it emerges.

Which is less costly—predicting what doesn't happen and/or missing what does, or seeing what's happening as it emerges and being able to incorporate it into present action? Engaged Emergence is attending to the cause and condition and then being able to witness the effect and respond accordingly. Mindfilled action is getting caught in the cause and condition and becoming a victim of the effect because it cannot be seen until after it has happened.

What is possible when we can make decisions by facing what emerges as it becomes known? What benefit arises from that ability to be present? For one, when we operate in the world as it happens, the possible is actual. There is no separation. This is also not a mystical occurrence. It is the meeting place where resources not only afford the emergence of the new and novel, but also the same place where new and novel resources can also emerge.

Now it may seem that to be able to recognize these events as they transpire requires a certain clairvoyant capability, but as my colleague Dr. Les Forman explained, that is not the case. Much like a psychotherapist, it does require us to be able to listen from a place that psychodynamic therapists call *free floating attention*. And, we're not going to operate effectively within this possibility space if we are already mindfilled. At its core, every emergent action takes place within the context of our ability to articulate it, so in order to deliver the message of what we see, we must not only witness it we must also be able to hear how it manifests itself so we can express it. If we can't listen with the appropriate capacity to grasp the message, if our minds are too distracted, it may mean that we miss the opportunity.

In honing this faculty to become engaged observers of the world in front of us and spot the cues that point us in the direction of new possibilities, we can also miss the opportunity if we limit our observations to strategic issues or the next hot item. Enterprise leaders often think they know what outcomes they want to take place and so they keep their focus and approach purely strategic. However, if we are actively opening our perceptions to the emergent opportunities that may make themselves available, we must also be ready to broaden our vision to include new outcomes that may arise and be required to satisfy the real need of the enterprise. This is a significant departure from the thinking that, as a matter of course, we define outcomes and goals and then create strategies to meet those outcomes. It compels us to rework our attachments to traditional outcomes as our only choices, and transform them, to say, focus on development rather than growth. The notion of growth, while certainly appealing to a short-term vision of exit strategies and market caps, may not be the best strategy for an enterprise that is looking for sustainability and longevity over the long haul. An example of this refocusing on new outcomes can be found in the emergence of green economies, where new answers to issues of environmental sustainability and financial opportunity are forcing us to look differently at our traditionally desired strategies.

Recognizing emergent opportunities as they make themselves known rather than after the fact might be the difference between creating a revolutionary market shift and following a safer and more linear progressed change that may actually lead an enterprise toward a less successful outcome. But we can only realize such shift if we can develop and trust the skills to see what is about to emerge and then allow for the listening, reflection and conversations out of which these emergent opportunities can properly manifest.

32. Adjacent Opportunities: Social Networks Emerge—You Have Nothing to Lose but Your Loneliness!

Ron Schultz
The Center for Social Profit Leadership, US

Social networks aren't becoming a part of our culture—they are our culture. And as never before, as societies and organizations are more connected, they are finding that they also have to be more responsive to the emergence of this new collective empowerment.

It is the underlying resonance within social networks that keeps them vital, connected and which provides their members access to emergent opportunities that in turn deepen the participant's own connectedness. Once that penetration space has been achieved, the interaction possible within the social network no longer needs to be influenced from outside, and the experience becomes addictive and self-generating. However, if the shared resonance were to end, the system would also decay or need to regroup under a new connective banner.

Identifying this resonance is not just about listening to language or conversation, although it is certainly expressed and manifest in those terms. The connective tissue of social networks is something more basic than patterns on a page. It has do with identity and not the kind that is easily stolen, but in this case willingly shared. Social networks are the antithesis of the image of the rugged individualist. They are more an appeal to recognize our interdependence, our collective need for each other to survive. The complexity of the information web we are weaving has increased to a level that it is actually pushing us closer rather than isolating us further. And within that drawing together the normally background and inaudible resonance has found its voice.

The question I am asking, is will the product of our interaction within social networks—our conversations—experience a similar emergent propensity, out of which innovation can arise? Or will these networked conversations produce only a continued state of ordinary novelty?

I find it quite amazing that with the advent of Twitter, an extraordinary marketing tool gone haywire, we also find a proliferation of face-to-face networking opportunities, often documented in real time via Twitter. Go to a conference or gathering arranged by Millennials or anyone under the age of 35 for that matter, and three-fourths of the time is allotted to networking and one-quarter to more traditional forms of presentation. And 100 percent of the time tweating to those not there about what they are missing.

It will be interesting to see if this continual state of social network connection can deliver the kind of emergent opportunities that are truly innovative. Oded Noy, the founder of a company that has produced a rather remarkable software product, called Social Approach that can identify what he calls "bonfire" conversations taking place within social networks, and then recognize the patterns in those conversations so that companies can use that information and market directly to that conversation, thinks that this new public discussion is filled with ample emergent opportunities.

The uniting force of these "bonfire" conversations for Noy, is the underlying resonance of the topic at hand. He believes that "the synchronicity of the individuals interacting within the network conversation provides a much broader group dynamic and opportunity for emergence, which ultimately will have a much greater impact on the world in which we are living." The evidence he sites to support this idea was the emergence of the Iranian opposition, and the use of Twitter and other social devices and the rapid coming of age of the real-time web. This particular story is still in process so it's too early to tell if the governmental forces will allow this conversation that is growing in resonance among the Iranian people to bring about real change. The change that could emerge is greater repression.

Still the social network was the means by which a movement was struck and couldn't be stopped. It continues in spite of the bans and the bullets. The technology removed the friction allowing the insurgent and emergent population to discover where the conversation was taking place and where innovation was broiling.

What Social Approach, itself, has done to help harness the resonance is figure out how to better align the values of the conversation with what is being produced, whether those are manufactured goods or revolutions. And in doing so they are shifting the prevailing perception that organizations have traditionally used to put their message out into the world, commonly referred to as, Public Relations, into the far more interactive notion of *Public Relationships*. When an organization engages in building public relationships through these social networks they open themselves up to a whole new realm of opportunities that could never have emerged previously.

This is not a prediction of things to come. This is what is currently taking place in any enterprise where the average age of an employee is 35 and under. They are already connected and in being so connected, those within the once solidly secure borders of corporate structures have turned this bastion of impenetrability into a porous mesh that has blurred the boundaries of many of today's leading enterprises. Their now public relationship means that they must be ready to interact more directly with the world in which they operate, because like it or not, they already do. They must also learn to build the relationships that are critically important to the enterprise, or if they fail to do so, run the risk of being out of step with the world around them.

The bonfire conversations that will last the night are those where the common resonance unites the participants and engages them at a level of active participation. If the resonance is lost, the search goes on for a new bonfire and a new set of interactions. Participating in these conversations alone won't create an emergent event that is both new, and now that may produce an innovative game changing event. But when the Tweet-up goes out and the crowds gather, the readiness of the public discussion to take up what is resonating within the network, and the synchronicity of the agents involved, will invariably bring about a new level of collective novelty and emergent opportunity that we have never seen before. And by connecting into the network, we also increase our awareness of our natural interdependence, which as we see in times of crisis is critical to our mutual development and survival.

33. Adjacent Opportunities: Engaged Emergence—Part 2

Ron Schultz
The Center for Social Profit Leadership, US

A couple of issues ago I wrote a column called "Engaged Emergence," about the presence required to witness emergence as it was happening. In that piece, I defined emergence as cause, condition, and effect. For those familiar with my particular Buddhist affliction—they say it should only last a lifetime or two—it might have been evident that the definition I used was also the common definition in Buddhist circles for Karma.

Karma is simply the effect that our actions create in the world we encounter. There is actually not too much here on the woo-woo-ometer that anyone would disagree with. What we do in the world, what we bring to each interaction, within a given environment, allows new things to emerge. What makes this connection between emergence and Karma interesting is that Karma isn't an event that just happens, it is a causal chain.

Within Karma, all our previous actions build on one another. Every actualization within our individual possibility spaces is carried with us into our next interaction, and they all influence the possibilities that make themselves known to us and their next iteration. We tend to think of emergent interactions as singular events—two or more agents interact and what can emerge is greater than the sum of the parts—rather than seeing it as two fully-developed causal chains interacting, with the history of both influencing what emerges. When looking at the process of the emergent event, we have to take into account that both agents bring all of this into each encounter. That's complexity.

Now for those long-time complexity arhats who have toiled and dallied in the world as it appears to show up, you may have thought in the back of your mind that delving into this soup of complex interactions would one day lead to enlightenment. Well, you just might be right.

In order to see emergence, as it emerges, requires presence. And to be present means that we aren't distracted by iPods, twitter, text messages, Netflix, the baseball playoffs or the looming deadline that is rapidly pressing down on us because we have been watching the baseball playoffs. It is about being mindful. Now mindfulness is not a spiritual or religious term, it is simply our ability to focus our mind on the landscape in front of us. Whenever we find ourselves distracted by the pheromones arising next to us, the blue, cloud-spotted skies out the window, or the piece of tape stuck to the top of the table, it is our ability to bring our minds back to our focus that we call mindfulness. When the alluring scent catches us and pulls us away from the landscape before us, we immediately become mind-filled and our original focus is lost.

Now I can hear the call arising, "well sometimes distraction is needed, and must be paid its due attention, especially when the olfactory nerve is so stimulated." And at that moment, it is our causal chain of experience that will ultimately determine if we follow that distraction, and then attempt to reapply our presence so that we might recognize what could be truly emerging from this new landscape, or return to our original focus.

It is at this instant, when the possible and actual are about to meet, that our ability to either recognize, or miss, the signposts telling us that something new and emergent is about to happen takes place. If we are present we see what is happening. If not, we miss it. Neither experience could have occurred independently of our causal chain of experience. However, we should be clear that Karma, like emergence, is not fate. Each of our possibility spaces is visible only to what we are able to see. How many of us have found ourselves in this position—that the possible was on the verge of actualizing and we didn't see what was actually emerging until after the fact? "Wow, I didn't see that coming." If we had only been present enough to read the signs correctly, rather than being distracted and completely mind-filled by other things, we might have seen what was taking place right in front of us.

Suffice it to say, to be fully conscious of the impact of our causal chain of emergent experiences on our next interaction, requires a level of Buddhahood most of us have yet to attain. However, our ability to recognize this notion of 'Engaged Emergence'—witnessing emergence as it is happening—can be developed through practice. Within this mindful practice, we are, in essence, influencing what we bring to the table. If we are simply building on our habitual causal chain of experience, we need to learn to change the pattern of the experiences. That is a learned operation and one that can be practiced. And like any such undertaking, whether it's learning to play a guitar without either Rock Band or Guitar Hero, or painting a pastoral landscape, our first attempts usually suck. But if we continue to practice, we gain a level of mastery over that to which we are applying our focus.

In practicing this level of engagement, we discover that something rather unexpected happens. When we operate from a mindful image of the world we encounter, we see things we might previously have missed. We still bring our causal chain of experience with us, but as we break through the patterned behavior and habitual responses that have influenced it in the past, and add a new and more present way of looking at the world we encounter, the possibility space surrounding us enlarges and we see more within it with greater clarity, so that when something does appear that we hadn't seen before, we recognize it before it hits us on the side of the head or passes us by.

Original Ref—E:CO Vol. 11 No. 3 2009 pp. 85-86

It is at every moment of interaction between two or more agents that the causal chain of experience of each agent surfaces and their individual thinking becomes biased by it. We tend to discount this, attempting to sanitize it out of the equation for the sake of simplicity, but we cannot do that. How or if we react is always based on the complex mix of our joined experiences, and unless we can do things differently that will break this pattern of reactive thinking, we will invariably miss the opportunity to see what new might emerge as it emerges. In some cases, this may not be as critical. We have been looking back at what has emerged for much of our lives, and often late getting to the party because of it. But now, when our ability to act quickly may be essential to moving forward successfully, engaging in an interaction, with focused and mindful intent, aware of what we bring to the table, can be the difference between seeing the next great emergent opportunity or becoming an also-ran.

We can redirect our causal chain of experience, our karma if you will, by acting differently and being more present when we do. The first step in this emergent process, however, is recognizing that it is influencing us before we can do anything really differently. As the Buddhists say, "Karma, neh?"

34. Adjacent Opportunities: Cultivating the Emergent Constituency

Ron Schultz
The Center for Social Profit Leadership, US

Engaging an audience to buy-in to what is being offered has been the goal of advertisers, scientists, salespeople, politicians, hate mongers, social activists, professors, all the world's religions and entrepreneurs trying to fill every enterprising nook they might possibly exploit, to name just a few. And to go along with those selling their ideas and products, there are a myriad of experts who want to sell what they know about how to make that sale.

In my test bed for 'getting the word out' known as the *International Social Action Film Festival*, I have watched hours of films about social activists, charities and the work of non-profits and NGOs. They share a number of common issues. First, the majority have no idea how to tell a compelling story about social change, and second they have little notion of the constituency required to develop an audience that will support them. The prevailing perception is that everyone will want to know. But when engaged in engaging others, "All the World is *Not* a Stage." If you think your audience is everyone, you may find yourself sitting alone.

The comedian, Steven Wright, once made the observation, "I don't want everything. Where would I put it?" And as we begin to look at how to engage the emergent process of developing an audience for our ideas, we have to start looking at the fact that what we think is not necessarily for everyone. Because, of course, if everyone showed up, where would we put them?

The emergent aspect of building any audience rests on its foundation. Successful self-organization is always bounded in some fashion, and unless there is a clear and resilient boundary within which to amass those who will show up, and the emergent properties that will invariably be unleashed, people simply won't stick around. The foundation of any audience is what I call the *constituency of need.* If there are not people who have a strong need for what you are offering, no matter how well you build it, they won't show up.

Providing that those who make up this constituency of need aren't too neurotic, the interactions of this group, as long as it is has become informed as a group around the offering, will lead to a level of emergence that will attract the next and broader level of constituency—that of interest. The difference here between constituency and the idea of community, is that a constituency is a body that is being served, while a community may provide services to, and by, its members for its members. A constituency is more dependent on what is being offered to it, rather than what it might do for others. If perhaps our political representatives were truly up to their founding ideals, they might view those they serve more as a community than a constituency. This would also require a level of participatory citizenship and responsibility from that community, rather than being a group that abdicates to the server delivering to the served. However, we'll leave that particular discussion to another column.

So how do we, who are proffering our intellectual wares and property, build an emergent constituency who will help spread the word we want to get out? Lisa Schneiderman, a Hollywood-trained, socially responsible brand builder and publicity coach, whose own work is about the intersection and interaction between pop culture and social action, believes that it begins with what she calls the "Fan Base." And the way she recommends building this base for new ideas and products is by, first, establishing the demographics of those that want or need your thinking—the constituency of need. Then once that group has been activated, we can identify which organizations and/or enterprises are similar to the ideas or products being put forth and then we can ally or partner with those that already have a healthy pre-existing fan base—a broader constituency of interest

Lying at the heart of growing and developing this audience, Schneiderman believes, is the creation of a compelling call to action around which these constituencies can rally and create a connection to the message, and as Schneiderman suggests, be "inspired to associate with the cause or product." This is where communities begin to emerge from interacting constituencies of interest and the word begins to spread. Just putting the word out and hoping that it's heard because it's the right formulation of words is not enough. We must set up opportunities for constituencies to interact around a clear brand-strategy that can, in turn, initiate networks that build communities formulated around the call to action. It is when networked constituencies begin to interact together that galvanized communities form and a message can be launched on an emergent and viral trajectory, all of which was triggered by the originating constituency of need—the fan base—that first resonated with the concept, cause or product.

What compelled and attracted that founding organization was their need. It wasn't the need of the entrepreneur or the academician, but rather the need of those who were being served. We often think that new ideas or products will spread because people will find them interesting or at least should. Eventually this may be so, but if we are planning on building an audience that will stay beyond that first look, it must be built by tapping into the "fan base" who finds their need satisfied to the extent that they are willing to engage with others who have felt likewise. Without the generation of that interaction, nothing new can emerge beyond its founding.

 Original Ref—E:CO Vol. 11 No. 4 2009 pp. 88-89

How many great discoveries, products or causes have died because they were unable to identify and then compel the constituency who really needed what they were offering? The historic field is littered with successful formulations like this that fail to satisfy the fundamental need of anyone other than its creator. Surprisingly, that compelling moment that encourages action is not about the benefit that might arise from the solution being offered, but the resonance shared by those who find it intolerable that the need exists. When people interact at that intersection a loyal audience begins its formulation and within the bounded self-organization of that constituency the development of a new and successful movement emerges.

35. Empowerment: Chapter 5 (of 10) Expectation & Nonresistance

Michael Shenker
Human Performance Technologies, US

Before I left last week, Myron told me that today's lesson would both lighten my load as well as expand my horizons which would be welcome information to say the least. Although I have tremendous responsibilities and most of the time I'm under a lot of stress, all-in-all I have a great life. My company is successful, my wife and I have two beautiful, happy children, I have great friends and I have good health. So I have to ask myself, why is my relationship with Myron such a boon to my existence?

Over and above trying to understand these new concepts, I'm fifty years old and although I've been extremely fortunate in my life, I'm starting to realize that there is more to life than what I previously believed. I attended a seminar recently and the person leading it had us do an exercise whereby we closed our eyes and pretended we were lying on a hospital bed. The lights were turned down low; we were to pretend we were all alone and had only three minutes to live. We were asked to quickly review our life and the question was then posed, "how do you feel about it?" And you know what? I didn't feel so hot. In fact, tears welled up in my eyes when I thought about all the things I wanted to do but never got around to it. I thought about all the places I wanted to take my family but was too busy building my business and I thought about the dreams I once had when I was younger which seemed to fall by the wayside as I had gotten older.

My experience at that seminar has made me realize that my meeting Myron was very fortuitous. As I pulled in front of his house, I saw him trimming some bushes which was pretty impressive for a man of his years.

As I approached him he greeted me warmly "Good morning Robert, how are you doing this fine day?"

"I'm doing great, sir. I'm surprised you haven't hired somebody to do your yard work."

"Oh hell, I've never found anybody that could do the job nearly as well as me. I do all the yard work myself; can't you see how young it makes me look?" Myron gave me a big smile and let out one of his trademark laughs. I realized that this was the perfect opportunity to ask him something I've wanted to know since I met him but felt uncomfortable asking. "Exactly how old are you, Myron?"

He lowered the clippers and looked at me, "How *young* do you think I am?"

OK, I asked for it. How do I tactfully get myself out of this one? "I'm going to say... seventy." It came out sounding like a question.

"Humph. Not even close," he said and went back to trimming the hedge.

"Eighty?" I asked, now certain if I done the right thing.

He lowered the clippers once again and turned to me. "You're getting warmer. I'm going to be ninety-seven years young next month," he bowed as he said it.

My mouth dropped open in astonishment and all I could say was, "Wow."

"What are we jawing out here for? Let's get some tea and go to work."

As I sank down into that couch which I'd grown to love, Myron poured me a nice hot cup of our traditional green tea.

"Robert, today we're going to talk about a couple of things that will definitely lighten the load. As a warrior, you might think that life is all about fighting reality, but nothing could be farther from the truth. In fact, a warrior's life is easier than most because of their perspective."

Myron then began searching through the hundreds of books that occupied his bookshelf, floor and desk. "Aha, there it is." Finding what he was looking for he removed an old-looking leather bound book from one of the shelves.

"Robert this little book is called the *Tao Te Ching* which was written by Taoist sage by the name of Lao Tzu in the 6th century BC and translated by a brilliant author by the name of Stephen Mitchell. Lao Tzu had never intended to document his wisdom, however, when he was making his way to the mountains to die, he was held captive at the border by the ruler of China until he wrote all of what he knew of life." He opened the book to a marked section. "His writings contained 81 precepts. In his 2nd precept, Lao Tzu writes," as he reads aloud:

> *When people see some things as beautiful, other things become ugly. When people see some things as good, other things become bad.*
>
> *Being and non-being create each other. Difficult and easy support each other. Long and short define each other. High and low depend on each other. Before and after follow each other.*
>
> *Therefore the Master acts without doing anything and teaches without saying anything. Things arise and he lets them come; things disappear and he lets them go. He has but doesn't possess, acts but doesn't expect. When his work is done, he forgets it. That is why it lasts forever.*

He closed the book and looked up at me, "I have read this to you Robert because of two words, *'expectation'* and *'non-resistance,'* which are the bane of our existence. Unlike many western motivational speakers who preach that one should *'expect what you want,'* I can tell you with 100% certainty that expectation

Original Ref—E:CO Vol. 11 No. 1 2009 pp. 106-107

will get you nothing but disappointment and heartache. The truth of the matter is that without expectations your life will be much happier and joyful. Now this is not to say that you don't have goals, because you do, but it's all about creating them, not expecting them because they will come at their perfect time and you don't know when that is."

Myron began to move closer as he continued speaking, "The point here is that instead of arguing with reality or *what is*, you can utilize another Universal Law, the *Law of Polarity*. The Law of Polarity which is discussed by Lau Tzu states that everything in life is like a coin with two sides; a 'good' side and a 'bad' side and in many instances you don't know which is which. It's kind of like the story about the sparrow and the cat. Once upon a time, there was a nonconforming sparrow who decided not to fly south for the winter. However, soon the weather turned so cold that he reluctantly started to fly south. In a short time ice began to form on his wings and he fell to earth in a barnyard, almost frozen. A cow passed by and crapped on the little sparrow. The sparrow thought that it was the end. But, the manure warmed him and defrosted his wings. Warm and happy, able to breath, he started to sing. Just then a large cat came by and heard the chirping, investigated the sounds. The cat cleared away the manure, found the chirping bird and promptly ate him. The moral of the story:

1. Everyone who shits on you is not necessarily your enemy.
2. Everyone who gets you out of the shit is not necessarily your friend.
3. If you're warm and happy in a pile of shit, keep your mouth shut."

With that, Myron let out a great pirate laugh, slapped his knee and said, "I love that story."

He then winked at me, patted me on the shoulder and said, "See you next week, Robert."

36. Empowerment: Chapter 6 (of 10) Personal Freedom

Michael Shenker
Human Performance Technologies, US

On the drive over to Myron's house I was tempted to pull out my cell phone and cancel our meeting due to the foul mood I was in, but that little voice inside kept nudging me to keep my appointment. I find it funny how difficult I can be at times. I swear, I fought myself all the way up the brick walkway to Myron's front porch until I finally called it a truce and knocked, once again, at precisely 8 AM. The door opened and Myron greeted me with a big friendly smile.

"Good morning Robert, how are you doing today?" and gestured for me to come in.

As I walked past him and into the living room I replied, "Well not so hot." We took our seats as I relayed to him the circumstances over the last couple of days. Myron poured the tea with a wry smile on his face and I wondered what the little elf was up to.

"Robert, one of the subjects I wanted to cover with you today is the works of Don Miguel Ruiz entitled the *Four Agreements*. Based upon what you just shared with me, it would appear that my timing is excellent."

I took my cup and settled back, prepared to hear some new pearls of wisdom I had become so used to experiencing. Myron also settled back in his chair and continued, "Don Miguel Ruiz comes from a long line of Toltec healers and teachers. Although his family anticipated that he would continue on with the family's centuries-old legacy, Don Miguel followed a different path and went to medical school and became a surgeon. After a near-death car accident, Don Miguel switched gears and now devotes his life to sharing the Toltec wisdom. In the interest of teaching mankind a path to personal freedom and not being controlled by your environment, he wrote his first book, *"The Four Agreements."* Although I am only going to discuss the highlights, I highly recommend that you get the book and study it in depth."

From the table next to him he picked up the book and held it up to show me, then opened it. "The first Agreement is BE IMPECCABLE WITH YOUR WORD." He looked up from the book and said, "... Which means to speak with integrity. This sounds somewhat simple, but it is actually very, very powerful. Depending on how you use it, your words can change lives for the better which is demonstrated by people like Lao Tzu, Socrates and Jesus or they can destroy lives which is demonstrated by Hitler, Stalin and Saddam Hussein. If you do what you say, you can create a heaven on earth. If you spend your time gossiping or not being true to your word, you render your speech weak and powerless." He took a sip of tea and looked closely at me. "Is this making any sense?"

"Yes, I replied. "I get it. If we don't have our word we don't have much at all."

"Good!" Myron smiled and returned to the book. "The second Agreement is DON'T TAKE ANYTHING PERSONALLY." He stops to think for a moment, looks at me and says, "The reality is that whatever someone says or does to you, it has nothing to do with you and everything to do with them. When you become immune to the opinion of others, you become very powerful and enjoy a level of independence enjoyed by few. Just look back on your life and see how often your words and actions were based upon how others would perceive them. Now think how your life would be if you had the freedom to act independently and didn't take anything personally, would that improve your life?"

That was a powerful question and I had to stop and think about my answer. "Every time I was ever involved in a confrontation whether it was verbal or otherwise, was a result of my reacting to what the other person said or did."

"Excellent, that's getting it!" He again read from the book, "The third Agreement is DON'T MAKE ASSUMPTIONS." He slowly put down the book and looked at me. "Most people spend their entire lives making assumptions of what others say and do based upon their personnel view of the world, not how the world really is. So the vast majority of what we believe to be true is false. By asking clarifying questions and not making assumptions, you can get to the actual truth as opposed to your *version* of it. With this one agreement, you can completely transform your life. Are you ready to give it a try?"

"I believe so." I reconsidered my words, "Yes, I'm ready."

Myron nodded affirmatively then went back to the book. "The fourth and last Agreement is ALWAYS DO YOUR BEST." He smiled, closed the book and looked at me once more. "By following this Agreement you will experience life with a positive intensity which will produce a high self esteem level and a lot of self fulfillment. By doing your best, which will vary, depending upon how you feel, you will be highly productive while at the same time being good to yourself. The main idea behind doing your best is that your actions are motivated by self love and respect—not for some reward at the end. Actions based upon rewards are hollow and not truly enjoyed."

He placed the book back on the table and picked up his tea and sipped. "What do you think of the Four Agreements?"

"I love them. If I would have incorporated them into my life prior to yesterday, the outcome of my day would have been entirely different. Up until now,

Original Ref—E:CO Vol. 11 No. 2 2009 pp. 92-94

I never even considered the concept of personal freedom and the tremendous impact it would have for me and every person in my life."

"Well Robert, today is your lucky day. In addition to the *Four Agreements*, I'm going to show you an exercise which will give you even more personal freedom."

"What have I done to deserve all this?" I said jovially.

"Like I said, 'when the student is ready the teacher or teaching will appear.' Now I want you to un-button your top button, un-tuck your shirt, loosen your belt a notch or two and lie down on the floor."

"Are you kidding me?"

"Do I look like I'm kidding, let's go."

I followed Myron's instructions feeling a little bit out of my comfort zone but then, I had to admit, my entire relationship with him has been about getting me out of my comfort zone. I proceeded to lie down on the floor.

"OK, now put this glass of water beside you and I'll give you little background on Transformational Breathing or TB, which was co-created by Dr. Judith Kravitz. TB is a unique form of breath work that facilitates the natural healing process for all types of trauma and is beneficial in gaining greater physical, mental and spiritual health. It has a number of associated benefits including:

- Higher energy level;
- Reduces worry and anxiety;
- Enhances the awareness of self-sabotaging pattern;
- Clears past traumas and dramas;
- Relieves depressive and negative emotions;
- Improves self-esteem.

In addition to all of this, I do it every day and can personally vouch for the many benefits it has to offer."

"Wow, that's pretty impressive. Is it painful?"

"Well," Myron said with a chuckle, "When I first started doing it I was a little concerned about maintaining my pace for forty five minutes."

"Forty five minutes, are you kidding me?"

"No, and the beauty is, Dr. Kravitz will guide you through the entire process. So let's get started."

Myron handed me a pillow for my head and one for my knees and then a set of head phones and turned on his CD player. As the music played and I breathed, Myron coached me on breathing deeper and deeper. Forty five minutes later, I was about as calm as I have ever been while at the same time feeling rejuvenated. What was weird was it didn't feel as if I was doing it for forty-five minutes. It felt more like fifteen or twenty. And, I felt myself actually looking forward to doing it again.

When I opened my eyes, Myron was peering down at me with that big smile,

"How'd it go, are you gonna live?"

"Wow that was really something."

"Are you up for doing it again sometime?"

"Absolutely!"

"OK," He then slipped a piece of paper in my hand with the TB website scribbled on it. "You better gather up your things and I'll see you next week."

Still reeling from my Transformational Breathing experience and feeling more relaxed than I can imagine, I waived a faint good bye as I made my way out the door.

Original Ref—E:CO Vol. 11 No. 2 2009 pp. 92-94

37. Empowerment: Chapter 7 (of 10) Sharing It

Michael Shenker
Human Performance Technologies, US

After six weeks of Myron, six weeks of *The Science of Getting Rich*, seven days of TB and a comprehensive study of the *Four Agreements*, I was really ready to rock. My wife, kids and coworkers have all commented on my calm demeanor and upbeat attitude. What's really cool is that I'm starting to work out again and instead of getting up a couple of times a night I'm sleeping straight through until the alarm goes off.

As I walked through Myron's door he could tell that things were working for me. While he poured us some tea and we got comfortable, I could see by his smile that he was pleased with me and it made me feel a sense of pride.

"So Robert, how's life treating you?" He sat back in his chair eager to hear my response.

I felt myself sit up straighter as I raised my chin, "It's more like how am I treating *it*. Our work together is really starting to make a huge difference; in my business, with my family and especially in me.

"Excellent, so now what?"

"I'm not sure I understand what you mean," I said as I leaned back into the sofa.

"Well, what are you going to do with all of this new found knowledge?" Myron took a slow sip of his tea and then looked at me probingly.

My confidence was suddenly evaporating as I tried to determine an appropriate response. "I plan to just keep improving and getting better."

"Isn't your Worthy Ideal to make a positive difference and be of service to as many people as possible before you die?"

"Yes."

"Then share it."

"What do you mean, share it?"

"Well, in addition to your immediate family, don't you have over two hundred people in your company?"

"Yes."

"What would be the impact to your company if your employees had the same knowledge and experiences you've had over the last seven weeks?"

"That would be a dream come true."

"Would you like for that to happen?"

"Absolutely, but do you actually have the bandwidth to implement such changes in my company?" After all, this little guy was in his late nineties and two hundred people would be extremely taxing on anyone let alone a person of Myron's age.

"No, nor would I want to." He took another sip of tea. "That's *your* job."

I practically fell off the sofa and spilled my tea. "Me?" I said incredulously. "I wouldn't know where to start."

"Well then, let's start with you," as he handed me a napkin and refilled my cup.

"Business is essentially about profitability and a strong bottom line – that is a given. However, building a sustainable, growing and profitable business requires much more than just financial acumen and having people turning out the work. Business leaders like yourself are faced with constant and unrelenting pressures. These pressures come from a wide array of sources including maintaining market share, interpersonal, cross cultural governance and maintaining profitability just to name a few. There's no time for developing your people… right?...Wrong!" He set the teapot down and held his cup as if getting ready to sip.

"Great Leaders are constantly looking for newer and better ways to empower and leverage the passion of their employees. They know that irrespective of the circumstances, increasing the return on investment of their human capital can make a *substantial* difference in their bottom line profits over and above the products and services they sell."

"I never looked at it that way."

"Well, I would suggest that you start. But don't feel bad because very few leaders have this *empowered* perspective and consequently struggle because of it. Studies have shown that highly motivated employees are up to 127% more productive than your average employees in highly complex jobs. I would also suggest that motivated employees turn up to work because they *want* to, not because they *have* to. I would even go so far as to say that motivated employees are more cost conscious and have less sick days. That all adds up to a lot at the end of the year." He takes a sip of tea and continues.

"Research has found on average that 70% of the employees are looking for more meaning in their work. That means that only 30% are relatively happy in their job. To me that says that there is a lot of underutilized human capital in

Original Ref—E:CO Vol. 11 No. 3 2009 pp. 87-90

the workplace. It makes sense to me Robert, that if you want to break away from the pack of the *unempowered* to the *empowered*, you had better implement modalities that create passion in your employees."

I felt a rush of energy; a wonderful feeling of lightness slowly flowed through me as I began playing "what if" scenarios in my imagination. I knew that a change of this manner was possible if I wanted it to be so and I was excited about the possibilities as Myron continued to elaborate.

"And I'm not talking about a short lived blip of motivation typically provided by books, tapes or an off-site meeting at the beach. The passion I'm referring to is a conscious choice to live a life of meaning where at the end of the day you know you have made a contribution, not only to your own life but in the lives of your employees and customers. How does that hold up to your Worthy Ideal?"

"It's dead on." I looked at him and then down to my cup, "But I have no idea where to begin."

"Again, Robert, it all starts with you. First you must know how to *feel*."

"Feel what?" I asked, slightly frustrated.

"If you have to ask then you better really take a deep breath, step outside the vortex of normality, and step into the world of *leadership from the heart*. Feeling is another way of listening to that small voice within. That voice, or as I like to call it, the 'Gollum,' a character in the Lord of the Rings, is the balance between good and evil, right and wrong. It's the voice that most people don't listen to, the voice that gives you that 'gut feeling.' And when you follow this voice, you take ownership and responsibility for every thought, decision and action that arises both personally and professionally."

"You mean I can't blame other people, I can't delegate the fallout?" I said jokingly.

"The answer is that great leaders, passionate leaders, take full responsibility for the outcome of their leadership. Great leaders constantly interact with their employees. They get out from behind their desks, and ivory towers and interact with those who are in the trenches."

"But I'm constantly asking my people about what's going on."

"No one is ever going to tell you the truth if *you* ask them how things are going. But if you're intermingling with your employees directly, you'll be able to see the truth first hand."

"I guess I see what you mean. To be honest, maybe I didn't really want to know the truth. I just wanted to hear that everything was working well."

"If you want to be a great leader, you must learn to build trust. Francois De La Rochefoucauld once wrote, 'The trust that we put in ourselves makes us feel trust in others.' To build true and meaningful trust you must first learn to trust *yourself*, then and only then will others learn to break down the barriers of skepticism and trust you."

"Wow, I never looked at it that way. You know, this is all a little overwhelming but I know I can do it. Where do I start?"

"Robert, a change in your company starts with you. The desire to change is paramount due to the intestinal fortitude you'll need in order to withstand the many blows from critics who'll ridicule and try to put you down. Remember though, the majority are only willing to make a difference if it benefits them and their bank accounts. Working to the contrary is the stuff from which legends rise. Just ask yourself, how do I want to be remembered, what legacy do I wish to leave and how do I want to feel when I look back on my life at the end of it all?"

Myron picks up a hand written piece of paper next to the teapot. "The following is a general checklist I put together for beginning your journey of becoming a passionate, heart-based leader," as he reads aloud:

"Take a good look at yourself—the good the bad and the ugly. Understand and learn from the many benefits and lessons that lay within any pain, hurt or suppressed experience. An executive coach or mentor can be of tremendous help in this self evaluation process.

Promote personal responsibility and ownership within your company. Always starting with *you* first.

Pursue an impactful purpose for your organization that results in a positive sustainable difference within the company while still creating a strong bottom line.

Promote personal growth, organizational leadership development and identify, nurture, recognize and reward your people and emerging leaders and those people with hidden talents.

Open up the positive feedback channels within your company, your family and your friends."

He then handed me the paper and said, "Now, do a quick mental audit of how many of the above steps you are actively planning, encouraging and pursuing within your company?"

I looked at the page and quickly accessed it. "I'm embarrassed to say that I'm not doing any of these things. Up until now, I thought I had a pretty good company."

"And Robert, I'm sure you do, but how would you like to have a *world class* company that really makes a contribution?"

"I'd love it."

"Well it all starts with a single thought that develops into an idea that transforms into a plan that gives birth to a proposal that with care and commitment will mature into a fully blown embraceable cultural shift."

"Wow, that was nicely stated," I said in awe.

"Now Robert, the question is, do you think you have what it takes? Is your company ready for such an incredible challenge? What have you got to lose… your job? Forget about the college tuitions, the mortgage and all of your other financial responsibilities just for a moment. When you have a Worthy Ideal that is warmly embraced and your primary aim is to be 'of service' you won't have to worry about where the money comes from, it will just manifest and flow to support you on your leadership journey."

"Thanks to you, I believe that I am ready, Myron. In fact, I can't wait to begin."

"That's what I want to hear. You have a lot of information to assimilate so let's call it a day. Next week we'll pick up where we left off."

As Myron walked me to the door I felt that a new page was turning in the evolution of my company and more importantly, my life.

38. Empowerment: Chapter 8 (of 10) Empowering Others

Michael Shenker
Human Performance Technologies, US

"Good morning Myron, how's it going today?" I said as I approached his front porch. Myron sat on an antique rocker, smiling back at me.

"Robert, it's always going the same. In fact, that reminds me of a story my grandfather used to tell me when I was a young boy. Several thousand years ago in Lebanon, there lived a wise man by the name of Useff who, throughout his life was always happy. It didn't matter if he had enough to eat, sufficient clothes to keep him warm or even a place to sleep, he was happy. People in the town would always ask him why he was always so happy but he would never tell them. On the last day of his life with just minutes to live, one of his oldest and dearest friends knelt down beside his bed and whispered in his ear, 'Useff, before you leave this earth you must share with me your secret of how you have remained so happy all these years.' With a smile, Useff looked up at his friend and told him that each day when he would awaken, God would ask him, 'Useff, today you can be happy or you can be sad, which will it be?' And very calmly, Useff told his friend that he always chose to be happy."

"That's a great story. I guess, all in all it's really as simple as that," I admitted.

Myron looked inquisitively at me and asked, "So Robert, how are you doing today?"

"Myron, I'm here with you on this beautiful, sunny Saturday, how else could I possibly be but excellent?"

"On that note," and a slight chuckle Myron replied, "Then let's go to work."

He then rose smoothly from the chair and ushered me into the house. "Today I would like to discuss a critical step in our work which is the process of empowering others."

I took my traditional place on the sofa as Myron poured us both a cup of tea. He then walked to his book case and began searching the stacks as he had done so many times before. As I watched him, I was overcome by a feeling of gratefulness for having the opportunity to study with such a learned man. Myron, with all of his foibles truly set a great example of how to live a quality life. Finally, with an old leather bound book in hand he sat back down and removed a weathered-looking sheet of paper. "I do not know the author of this marvelous document. When I purchased this first edition years ago, it was folded between the pages. I have tried many times to identify the author but to no avail. As it is too frail to carry around, I would like you to copy it down in your notebook. This will be a good way for you to acquaint yourself with the Thirty Six Steps to Empowerment. Then, pass it on to every person you value with the instruction of reading it every day for a minimum of thirty six days—or ideally longer should they so desire. On day one of their first reading have them write down how they rate themselves on a scale from one to five with one being the lowest and five the highest, on their *observance* of each of the thirty concepts. Then, add up the numbers and that will be their base-level score. Once completed, have them put their ratings away in a safe place. Then, on day thirty six they rate themselves again. Once they have totaled their score, have them subtract their score from day one from their new score on day thirty six and the difference will show them how much progress they have made. Of course, I would highly recommend that you be the first to do this."

He then carefully handed the sheet of paper to me as I began to read the neatly handwritten text.

Thirty Six Steps to Empowerment

1. I am grateful and give thanks to the difficult aspects of my life as I know that adversity is my greatest teacher.
2. I know that I am always at the right place at the right time and that each person, place and thing that comes my way is perfect.
3. I have the courage to be different as I no longer need the approval of others.
4. I manage my expectations accordingly as I know that all pain and agony occur when there is a contradiction between the ego's expectations and the reality of the moment.
5. I trust that there is an Infinite Self and I am less vulnerable to the manipulating fear of the ego.
6. I open my heart to hear the answers instead of depending upon the intellect.
7. I conform to what I feel from my Infinite Self as opposed to the norms and expectations of society.
8. I do not compare my unique self with anyone else.
9. I accept what I have and where I am today and don't confuse my current reality with goals for the future. [note: if you do not have any goals, create some]
10. I am satisfied as I know that all is well with the world right now.
11. I continue to expose myself to the present moment and come closer to my Infinite Self.
12. The less I care about the future the freer I become.
13. I need less and less to be happy.

Original Ref—E:CO Vol. 11 No. 4 2009 pp. 90-92

14. As I maintain balance I create energy and power.
15. I continue to give thanks for what I have and know that I am blessed.
16. I maintain discipline by having a greater desire for wisdom than my need to accommodate the ego.
17. I maintain harmony by not judging, criticizing or complaining about other people or circumstances.
18. I am conscious and respectful of the needs of others and they are conscious and respectful of my needs.
19. I maintain objectivity by being an unemotional observer.
20. I attract what I want by seeing, feeling and experiencing what I want as if I already have it.
21. I attract what I want by being centered and having a clear intention.
22. I don't talk about myself nor explain what I know or do.
23. When I feel angry, I stop and emotionally go within to figure out what I've lost and agree to loose it.
24. By respecting other living things, I am respected.
25. I practice non-action by watching the ebb and flow of events and getting underneath life.
26. I attract what I want by effortlessly allowing it to come to me.
27. I know that my job on this planet is to experience, express myself, create, grow and transcend.
28. I know that nothing is permanent and all change is a "gift."
29. I serve humanity through my silence.
30. I easily change as I have no fear of *not knowing.*
31. By changing I create energy. By creating energy I affirm life.
32. I am light hearted and laugh a lot. I flee from seriousness as it is a disease of the ego.
33. I accept being here and not being here as equal in value.
34. When things don't feel right, I do nothing as my actions are initiated by my certainty.
35. I know that life is a sacred journey and feel humility and gratitude.
36. I fear not death as it is my friend.

After carefully compiling my score I looked at Myron feeling humbled and excited at the same time, "Wow, that's quite a process. I can see how working on these concepts each day could have a major impact on myself, my family and my company. I'm almost embarrassed to say that I scored less than 40 out of a possible 180 points."

"I think you're starting to see the vast untapped resources and potential you have hiding in plain sight. I'm reminded of a story of a beggar in ancient China who sat on a great chest outside the gates of his village. Anyone who entered or left the village was approached by the beggar for food, clothing or some form of sustenance. One day a very wise man came to his village and was approached by the beggar for something to eat. The wise man asked him how long he had been begging. The beggar replied that both his father and his grandfather before him had sat on this very same chest. The wise man then asked if he'd ever opened the chest that he, his father and his grandfather had been sitting on for all these generations and the beggar said 'of course not, it was just a place for them to rest.' The wise man then said that nothing should ever be taken for granted and bid the beggar good bye without offering him a single crust of bread. Although the beggar was disgruntled at his lack of compassion, he became unsettled with the wise man's comments. All these years he had never considered looking inside the chest, so he opened it up and low and behold he found that it was full of gold." A smile warmed his features as he put down his teacup and continued.

"This story is similar to many of your contemporaries. Some will continue to run their companies with the idea that profits can only be made from external means and never consider the vast treasure they have in their employees and some, like you, will have the presence of mind to *open the chest and find the gold.*"

Myron rose from his chair and said, "In the meantime Robert, let's call it a day and go out and enjoy the sunshine."

As I was heading for my car I realized that it wasn't going to be easy but I knew I was about to open the chest and find the gold.

Appendices

A1. Editorial—Issue 11.1: Chaos and Complexity, Branches and Trees, and the Advantages of Vague Concepts

Jeffrey A. Goldstein
Coeditor-in-chief
Adelphi University, USA

E:CO has a New Subtitle

This issue signifies the start of volume number 11, a feat showing not just how far *E:CO* has come in a relatively short amount of time, but gives striking evidence that it is still only building steam. Just look at the rich contents of this issue: articles hailing from diverse countries around the globe and covering a wide variety of fields, and all done in a rigorous, thought-provoking fashion. This issue also inaugurates a new subtitled for *E:CO*: "An International Transdisciplinary Journal of Complex Social Systems." That's quite a mouthful but it more accurately reflects what *E:CO* is about and is evolving more and more into.

Three phrases in our new subtitle stand out. First, there is the emphasis on "International." Of course, internationality was essential in the journal's mission right from the start, and even before that with the precursor journal *Emergence*. But we're now including it in the title to reaffirm our commitment to a truly international focus in terms of our authors, our readers, and our article topics. Indeed, this very issue is like a mini United Nations with authors hailing from Australia, Italy, Egypt and Israel (our own little peace initiative although it wasn't planned that way!), South Africa, China, and the US. Then there's the term "Transdisciplinary". Again that was an indispensable aspect of our initial mission but here it is spelled out overtly. This current issue is a vivid demonstration of transdisciplinarity as can be seen by considering the fields of research of the authors: cognitive science, urban planning, environmental studies, architecture, knowledge management, social networks, organizational theory, operations research, information technology and computer science, and sociology. Moreover, the methods and constructs utilized herein similarly cut-across a wide range of scientific, mathematical, cultural, and philosophical disciplines.

The third new addition to the subtitle is the intentional mention of "Social Systems." We are in fact the only existent journal specializing in complexity applications to social systems. But of course we also supplement this specialization with numerous phalanges into philosophy, mathematics, education, physics, biology, semiotics, and many other fields related in one way or another to social systems.

In spite of the long and still unwinding road that *E:CO* and complexity studies in general have been on, it comes as a surprise to me how vexing it can be to pin down, into a concise definition, just what is it that the field of complexity theory or complexity science covers. I'm sure many of the readers have experienced a version of the following situation. I was recently at a party where I met some friends that I hadn't seen in 30 or more years. To catch up on all those years, we asked the customary questions of what we each had been up to personally and professionally. When my turn came, I again, as I always do in this situation, became tongued-tied as to how to describe just what it is that complexity theory is about. This elusiveness of a crisp definition of complexity always reminds me of that famous retort by one of the United States Supreme Court Judges when asked how he defined obscenity: he said he couldn't define it precisely in a few words, but he knew it when he saw it! That's how I see the study of complex systems, I can't precisely define it but I'm pretty sure I can recognize it when I see it.

Of course, this kind of response is not particularly enlightening and it brings with it all the vagueness accompanying ostensive, contextual, and indexical references, a topic I'll return to below. Moreover, although one tactic could have been to start listing the varied types of systems considered to be complex, I realized that this would not be of much help. Since, not only were there too many of them—with the result I would be hard pressed to say what they had in common besides such unilluminating vague properties as interconnectedness—it would be questionable whether many of the systems I mentioned were in fact complex, or at least complex all the time, i.e., some systems at certain conditions such as at equilibrium appear quite simple, but may not be when parameters change.

So, getting back to the situation of trying to explicate for my old friends what complexity theory was about, this time around I came up with a new strategy (for me at least) of asking my well meaning friends if they had ever heard of "chaos theory" and most of them indeed had. Chaos theory, of course, has entered the popular imagination in many forms, from Jeff Goldblum's famous allegory of it in the movie "Jurassic Park", when he dropped water on the hand of the character played by Laura Dern and told her to watch its unpredictable flow, to the more recent film "The Butterfly Effect" with Ashton Kutcher, to last year's surprisingly good romantic comedy "Chaos Theory". As a matter of fact, each of these films does include plot elements having to do with the property of sensitive dependence on initial conditions that characterizes chaotic systems.

So, when the heads of my friends shook "yes" that they had indeed heard of chaos theory as well as the butterfly effect, I then added that chaos theory was like a branch of a wider and taller tree called "complexity science." But immediately after offering that analogy, I thought to myself, uh oh, that doesn't really help at all for at least two strong reasons. First, it leaves the harder question of what this wider and taller tree is comprised of yet to be explained, and second, and even more troubling, it dawned on me that this analogy was simply wrong and misleading in a variety of ways.

First of all, there is a decided sense that chaos theory, at least in the form of its progenitor field of nonlinear dynamical systems theory (NDS), predates much of modern complexity theory, going back over a century to Poincaré and to the great Russian mathematicians Chebyshev, Lyapunov, and others. Branches can hardly be said to predate their trees! Second, a great deal of the mathematical apparatuses used by complexity theory, indeed, taken for granted by complexity aficionados, comes right out of NDS, e.g., phase space, attractors, bifurcations, qualitative dynamics, and so on. Wolfram first classified the dynamics of cellular automata using NDS types of categories. Furthermore, the continual updating of cellular automata, which underlies the apparent motion of artificial life, recapitulates the iterations of discrete difference equations as found in the logistic map which is so emblematic of NDS.

Also, there are the very important differences between chaotic and complex systems, e.g., it is not clear to me at least that complex systems generally possess sensitive dependence on initial conditions like chaotic systems do. In addition, chaotic systems can form out of very simple functions whereas it seems complex systems require relatively large networks of interacting agents where the mathematics of their interactions is typically much more complicated than that of logistic-type maps. On the other hand, Shaw (1981) made a convincing case how chaotic systems can be understood as information generators which it seems to me is cognate to what Peter Allen talks about as happening in complex systems as micro-level diversity generates the seeds of emergent order with new properties.

Furthermore, complexity science utilizes many more mathematical tools than what NDS can supply, e.g., there's all the graph theory of social networks, the group theory of the renormalization group used to explain phase transitional types of the emergence of new order, all the computational complexity stuff hovering around, the talk of Turing machines, algorithmic complexity theory, Bennett's logical depth, Crutchfield's computational mechanics, and so forth—a whole host of new mathematical objects and constructs which include but also exceed chaos theory.

Complexity is Vague and that is a Good Thing

So relating complexity to chaos as tree trunk to branch simply won't do. And, again, I was still left with the arduous task of trying to adequately describe the tree. That brings me back to the idea of vagueness. At first impression, it might seem problematic that an entire field of study appears to be perhaps hopelessly vague. "Vague" though doesn't necessarily entail being nebulous, ethereal, or inherently muddled. Indeed, in philosophical circles there has recently been a great deal of interest in exploring the philosophical ramifications of vagueness. In this regard, the English philosopher M. S. Sainsbury (1997) has made a good case for considering much of our ordinary discourse as composed of words that are fundamentally vague. Thus, we find vagueness attached to the ordinary idea of "small," demonstrated in what is called Wang's paradox after the mathematical logician Hao Wang, a protegé of Kurt Gödel:

* By mathematical induction,

0 is small,

If n is small, $n + 1$ is small;

Therefore, every number is small.

As the British philosopher of mathematics Michael Dummett (1997:101) explained about Wang's paradox, "...since every natural number is larger than only finitely many natural numbers, and smaller that infinitely many, every natural number is small, i.e., smaller than most natural numbers." This kind of vagueness is typically linked with the ancient conundrum of the *heap* known as the Sorites paradox (Sorensen, 2006):

1 grain of wheat does not make a heap.

If 1 grain of wheat does not make a heap then 2 grains of wheat do not.

If 2 grains of wheat do not make a heap then 3 grains do not.

... *If 9,999 grains of wheat do not make a heap then 10,000 do not.*

In the context of Sorites's types of paradox, Sainsbury points to the word and concept "red"—there is no crisply defined set of referents that could be considered red without also including colors that blend off into yellow. For instance, we might find on a shelf in an art store, a row of red oil paint tubes that go from a strong, bright, blood red all the way to something approaching orange yellow. There is a continuum of possible colors between red and yellow depending on the scale of resolution, although from a distance we can discern, in a prism, or example, a band separating red from yellow. Ralph Stacey has his well-known diagram distinguishing three zones of a coordinate plane into simple, random, and complex in relation to the degree of certainty on the horizontal axis and agreement on the vertical one. As helpful as this diagram can be, in my experience it turns out that the complexity region is itself highly complex, containing mini-regions, fractally embedded, of more random and more simple regions, like the pockets of order within the famous bifurcation diagram of the logistic map (ah ha! We're back to chaos theory after all that!).

As a prototypical example of vagueness, the philosopher Achille Varzi (2001) cites a wild passage from the many populating Saul Bellow's great novel *Herzog*:

> *Remember the story of the most-most? It's the story of that club in New York where people are the most of every type. There is the hairiest bald man and the baldest hairy man; the shortest giant and the tallest dwarf; the smartest idiot and the stupidest wise man. They are all there, including honest thieves and crippled acrobats. On Saturday night they have a party, eat, drink, dance. Then they have a contest. And if you can tell the hairiest bald man from the baldest hairy man—we are told—you get a prize.*

Varzi points out how vagueness runs through this passage because not only is there no sharp boundary exactly demarcating the category of bald men and no precise number of hairs separating hairy from the bald, it also makes no sense to think one can identify the hairiest bald man. Even though some men are obviously bald and some are clearly hairy, between the two there are a host of borderline cases including baldish guys, men with toupees, hirsute beatniks with a shiny bald spot on the top of their heads, and so on and on. Accordingly, the concepts of bald and hairy are vague with ranges of application having vague boundaries.

As another philosopher Max Black (1997) put it, vagueness does not necessarily indicate a conceptual defect just as an impressionist painting is not adequately understood as a flaw of precision in technique:

> *The impressionist painting of a London street in a fog is not a vague representation of what the artist sees, since his [sic] skill consists in the accuracy with which the visual impression is transcribed. But the picture is called vague in relation to a hypothetical laboratory record of the wave lengths and positions of the various objects in the street, while it is forgotten that that record, in supplying additional detail, obliterates just those large scale relations in which the artist or another observer may be interested* (71).

Related to Black's contention, the computer scientist Mark Changizi (1999) has made a good argument that vagueness is intimately related to the issues of undecidability and noncomputability in mathematical logic and computational complexity theory. Changizi interprets vagueness as resulting from bounds on computability, and not from human fallibility for it would exist even with the computer HAL from *2001: A Space Odyssey*. Instead, vagueness is what we are ultimately left with in the case of all natural languages. According to Changizi, "given that you are computationally bound, avoiding vagueness brings in greater costs than accepting it." (345).

Hence, I believe we can presume that complexity and complex system are in fact vague notions, but those designations have a silver lining. To see what the advantages can be to vagueness, it is helpful to distinguish it from the ideas of either ambiguity or generality. Concerning, ambiguity, the term "complexity" is definitely a candidate. Seth Lloyd at MIT several years back had identified at least 30 different definitions of "complexity." This could be taken as indicating there is something fishy about the whole field if it can mean so many things. But, vagueness is not the same as ambiguity. Whereas ambiguity is about words possessing many equivocal meanings, vagueness has more to do with the referents to which the word applies. Thus, vagueness makes a concept more inclusive, allowing all sorts of borderline cases, in effect, manifesting the fact that it deals with systems whose boundaries are themselves essentially vague, permeable, fractally bounded, and so forth. Moreover, as a vague notion, in this philosophical sense, complexity continually resists formulation in terms of this or that discrete mathematical formalism such as set theory, or graph theory, or whatever theory. This of course doesn't imply that we abandon such formalisms. On the contrary, it suggests instead that we go the whole length in utilizing such formalisms, but with the expectation that at some point we will need to transcend each formalism and devise new constructs and terms and methods to further our study of complex systems. This also means that vagueness doesn't imply that anything goes in describing and studying complex systems, i.e., vagueness is not some kind of a license for unbridled post-modernist (misunderstood post-modernism) claptrap. Rather, it is marker that context is a crucial factor in determining the scope of inquiry.

Vagueness is also not to be conflated with the idea of generality. Thus, although there are a multitude of types of complex systems, trying to get at what is complex about them by extracting some common element that could then be put into the form of a generalization would miss the boat by a country mile, since the highly important role that context plays in understanding complex systems would be sacrificed. Complex systems are related to other complex systems, but this relation is better thought of in terms of a Wittgensteinian family resemblance than some common essence that they share in general.

For a moment, let's assume the opposite, i.e., that the complexity of a complex system, indeed, the realm of complex systems in general are not vague notions but instead can be exactingly and precisely defined. This would entail that it would always be *decidable* whether a system was or was not complex, no borderline cases allowed. But this would paradoxically turn complexity into a simple notion for surely a crucial aspect of simplicity is the presence of crisp, uncontroversially so, boundaries. This, though, would in turn imply that complexity as simplicity must run up against Rice's theorem in mathematical logic and computatability theory, which suggests that only trivial problems are decidable or computable, i.e., there are algorithms to solve them (Rice's Theorem, Weisstein).

"Trivial" in this sense is closely related to properties that are always true or always false, i.e., tautologically demonstrable properties such as that bachelors possess the property of being non-married. More formally, Rice's theorem holds that for any non-trivial property of partial functions (we'll put aside for the sake of brevity the technical definition of such functions), there is no general and effective method to decide whether an algorithm computes a partial function with that property (Rice's Theorem, Wikipedia). Turing's famous noncomputability results turn out to be a special case of Rice's Theorem which states the more general case that every nontrivial decision problem regarding the function computed by a given Turing machine has no algorithmic solution (Goldreich, 2008). Rice's theorem is an important result for computer science because it sets up boundaries for research in that area.

If complexity and complex were in actuality nonvague, crisp ideas, then according to Changizi's linkage of vagueness to undecidability described above, most matters involving complex systems, by involving only nonvague, crisp areas of delineation, would therefore be decidable. But according to Rice's Theorem, this would suggest that most matters involving complex systems would be trivial. But clearly they're not, as our readers and authors and anyone else for that matter researching complex systems clearly knows. Rather, complex systems are difficult to research, difficult to measure, difficult to construct viable theories of, indeed difficult to determine if or if not they are in fact complex! A case could be made, I believe, that the more difficult it is to define a specific field of study, ultimately the more interesting it is precisely because it eludes preconceived constructs, frameworks, and pat answers.

Ultimately, the best strategy I could have followed in trying to describe to my friends what complexity theory was about was to simply show them a copy of this issue of *E:CO* and gone through it, weaving around the more arcane mathematical constructs, and pointing to the kinds of systems and the issues being investigated and the conclusions being drawn. I can think of no better way to describe what makes up complexity theory than these examples.

References

Black, M. (1997). "Vagueness: An exercise in logical analysis," R. Keefe and P. Smith (eds.), *Vagueness: A Reader*, ISBN 9780262112253, pp. 69-81.

Changizi, M. (1999). "Vagueness, rationality and undecidability: A theory of why there is vagueness," *Synthese*, ISSN 0039-7857, 120: 345-374.

Dummett, M. (1997). "Wang's paradox," in R. Keefe and P. Smith (eds.), *Vagueness: A Reader*, ISBN 9780262112253, pp. 99-118.

Goldreich, O. (2008). *Computational Complexity: A Conceptual Perspective*, ISBN 9780521884730.

Hyde, D. (2005). "Sorites paradox," Stanford Encyclopedia of Philosophy, http://plato.stanford.edu/entries/sorites-paradox/.

Rice's Theorem (Wikipedia). http://en.wikipedia.org/wiki/Rice%27s_theorem.

Sainsbury, M.S. (1997). "Concepts without boundaries," in R. Keefe and P. Smith (eds.), *Vagueness: A Reader*, ISBN 9780262112253, pp. 251-264.

Shaw, R. (1981). Strange attractors, chaotic behavior, and information theory," *Zeitschrift für Naturforschung*, ISSN 0932-0784, 36a: 80-112.

Sorensen, R. (2006). "Vagueness," Stanford Encyclopedia of Philosophy, http://plato.stanford.edu/entries/vagueness/.

Varzi, A.C. (2001). "Vagueness, logic, and ontology," *The Dialogue: Yearbooks for Philosophical Hermeneutics*, 1: 135-154.

A2. Editorial—Issue 11.2

Peter M. Allen
Coeditor-in-chief
Cranfield University, UK

Recently I started to supervise a Ph.D. student who was interested in the question, How do people deal, subjectively, with uncertainty? Here she was referring to the questions raised by economists such as Frank Knight and George Shackle which had really been somewhat overlooked by their disciplines. As a good 'complexity' person I immediately realized that much of what complexity science tells us is that the world is evolving and that prediction is never really possible, and therefore that uncertainty is in reality the norm and not the exception. However, as a cynical person of a certain age, I also thought that most people don't want to accept this message and prefer to cocoon themselves in routines and trivial pursuits of various kinds, perhaps even in studying complexity science! But of course, as my student pointed out, some people only want to be working in a realm where uncertainty abounds, since only there can the new and the innovative really flourish, and because nothing interesting can happen anywhere else. So people differ in their willingness to accept uncertainty, and many flee as fast as possible in the opposite direction.

Clearly complexity science reinforces the message of an evolving world and a world of freedom. Freedom can only flourish where there is uncertainty—about the facts, the system, the issues, the possibilities. Ignorance is also an important facilitator or experiment. Not knowing that one is performing a risky experiment by leaving 'physics'—a known subject and discipline—to work on 'complexity'—involving urban systems, ecologies, organizations etc.—an unknown subject—with unknown job opportunities and great career risks, is something that, fortunately, I have only realized in retrospect. Fundamentally then, as the world changes, many people try desperately not to change what they do, but of course they are actually performing experiments in what they continue to do; others have no idea of the benefits or risks of pursuing a certain path, and therefore pursue it in ignorance of the potential consequences, and others do genuinely seek to work in a recently opened domain where little is known. These ideas have been found by various authors before—a successful population must contain both explorers and exploiters if it is to survive, since it must push into the unknown and hopefully exploit some of what it finds there.

This is an interesting idea and indeed we may think that complexity shows us how evolution will lead to successful populations that have this dual behavior of facing or fleeing uncertainty. However, my wife then told me about the book she was reading by Erich Fromm who seems to build greatly on these ideas before they even existed. In 1941 he wrote *Escape from Freedom* or *Fear of Freedom* as it was variously known, and talked about the consequences of people faced with freedom and hence with uncertainty. He thought that embracing freedom and facing uncertainty was healthy, whereas trying to deny or escape from it was the root cause of psychological conflicts. Fromm outlined three main escape mechanisms: automaton conformity, authoritarianism, and destructiveness. Automaton conformity is about suppressing your own identity in favour of what is perceived as the preferred type of personality of society. This pushes the burden of choice from the self onto society. Authoritarianism is allowing oneself to be controlled by others, perhaps political or religious leaders another, removing freedom of choice almost entirely and submitting that freedom to someone else. Lastly, the mechanism of destructiveness is a process that attempts to eliminate others or the world as a whole to escape freedom.

In many ways his views provide an explanation not only of the rise of Fascism and Communism, but also the prevalence of feudalism and religions, as people run to avoid making choices and decisions themselves about the many different issues that arise in our world. Some MBA students may consider that, having paid their fees, they are entitled to receive the Recipe Book—with the answers in. Democracies become sufficiently opaque for governing and decision making to be made by relatively few people, and the adoption of the idea of a free market economy means that people simply get on with consuming increasingly unneeded things in vast quantities, thereby avoiding the difficult thoughts and judgements that life produces. The idea that markets operate above moral frameworks or ethical standards, in that the 'invisible hand' can't be wrong, not only flies in the face of what Adam Smith himself believed, but as recent events have shown, can potentially crash the global system of interconnected dishonesty, ruining and dispossessing real people. But these disasters are just part of the price we pay when we delegate power to others, and console ourselves by watching and reading superficial rubbish as a substitute for living and spending our efforts on acquiring more material goods to try to fill the emptiness that we feel.

So I think that Erich Fromm was quite clairvoyant in his writings and said some very important things way before their time. Complexity science and thinking tells us about emergence, creativity and freedom as a positive message, although it might still tell us that it is impossible to have only explorers—bold individuals with their own thoughts, views and ideas, innovating and creating new structures and organizations with improved performance. In fact, evolution will lead to a society in which there have to be a range of cognitive types, so that teams can be formed, can be lead and can also operate. In some ways then, Fromm considered the acceptance of freedom to be healthy and its non-acceptance to be sick, but since society needs organizations and structures if it is to deliver the things we need, then we require an emergence of harmonious cooperation between different kinds of people. However, this does suggest why evolution and complexity will have no end, but travelling hopefully is all we can do and in any case is better than its opposite.

Let me finish with some quotes of Erich Fromm that seem particularly relevant to *E:CO*:

> *Creativity requires the courage to let go of certainties.*
>
> *In the nineteenth century the problem was that God is dead. In the twentieth century the problem is that man is dead.*
>
> *Just as modern mass production requires the standardization of commodities, so the social process requires standardization of man, and this standardization is called equality.*
>
> *The capacity to be puzzled is the premise of all creation, be it in art or in science.*
>
> *The quest for certainty blocks the search for meaning. Uncertainty is the very condition to impel man to unfold his powers.*
>
> *The task we must set for ourselves is not to feel secure, but to be able to tolerate insecurity.*
>
> *There can be no real freedom without the freedom to fail.*

A3. Guest Editorial—11.3: Complexity and Environment

Kevin J. Dooley[1] *& Hugo Letiche*[2]
1 Arizona State University, USA
2 Universiteit voor Humanistiek, HOL

Introduction to Special Issue on Complexity and Environment

Welcome to the *E:CO* special issue on *Complexity and Environment*. The genesis of this collection of papers was the Fourth Annual Organization Studies Conference, "Embracing Complexity," during July 2008 in Pissouri, Cyprus. The call for papers for the seminar was grounded in Gregory Bateson's complex understanding of environment, change and evolution.

Appropriately the "classic" article chosen for this special number was written by Bateson (1963). Bateson believed that the social / environmental degradation of justice, milieu and culture, is co-derived from cultural ideas of humans and that restoration is best accomplished by concentrating on these cultural ideas. The environment that we can see is the one that our ideas can produce; the human is always implicated in what we claim to 'know'. Bateson was a theoretician of communication—thus, he saw the environment that we study as a communicative artifact. Thanks to how we communicate with ourselves and surroundings we see understand and manipulate those surroundings in particular ways and not in other ways. Bateson's epistemic mutuality does not prioritize the knower or the known and thereby is a rich source for developing complexity insights.

Luhmann's "Social Systems" (1995) has provided us with a useful complexity (autopoesis) inspired introduction to the concept of *environment*. From a systems theoretic perspective, an environment must exist for a system to exist, for by definition environment is everything outside the boundary of a system. Environment is the dual of system; expanding the boundary of a system will decrease what is considered environment, and decreasing the boundary of a system will expand what is considered environment. Any system will differentiate itself from its environment in a number of ways, including behaviorally and temporally.

According to Luhmann, an environment contains many other complex systems itself, thus any environment is always more complex than any system embedded in it. In a social system, this causes tension in the system. For control, a controller must be more complex than the system being controlled (Ashby,

1958). Thus if the environment of a social system is more complex than the social system itself, this implies that an environment is beyond the control of a social system. Environments must be grappled with: influenced and adapted to, but not controlled.

As no social system is closed, it must interact with its environment. The social system in turn treats its environment and its complexity as a contingency. If the social system seeks resources from its environment, resource dependencies emerge, whereby the social system and its environment coevolve. If the social system seeks information from its environment, information uncertainty emerges, whereby the social system seeks to either reduce uncertainty or mitigate its negative consequences (Luhmann, 1995).

Thompson's "Organizations in Action" (1967) views the organization in relation to its environment from the perspective of input-core-output. The environment provides input into the organization (e.g., raw material, requirements information, demand) and the organization creates outputs (e.g., product, service, information, waste) that go back into the environment. For Thompson, uncertainty associated with both resources and information is key. Thompson suggested that organizations purposefully buffer their core activities from uncertainty by (e.g.) adhering to master schedules or vertically integrating to secure resources. If buffering is not possible, organizations use smoothing, forecasting, and rationing to attempt to mitigate losses from uncertainty.

Lawrence and Lorsch's "Organization and Environment" (1967) took the notion of contingency one step further by suggesting that the manner in which one organized should be contingent upon the nature of the environment. Their work and others that followed, posited and found evidence to support that hierarchical organizational structure was fitter for companies in stable environments while organic organizational structure was fitter for companies in turbulent environments.

As complexity science emerged, different types of models treated environment differently. Computational and dynamical models, such as Kauffman's NK model (Kauffman, 1993) or nonlinear dynamical systems models (Guastello, 2002) model the environment as a source of exogenous (often random) shocks. Agent-based models, such as Sugarscape (Epstein & Axtell, 1996), treated environment both as a source of exogenous shocks (i.e., a source of information uncertainty) and as a repository of resources that agents seek to acquire from some functional purpose. Self-organizing models such as autopoiesis (Maturana & Varela, 1987) and far-from-equilibrium models (Lichtenstein, 2000) however have incorporated environment as a key theoretical construct. And via consciousness studies, fundamental reflection about the relationships between life and structure, self and organization are advancing.

More recently, environment has taken on a more literal meaning for organizations. As both market and regulatory forces push organizations to operate and create products and services with less negative environmental impact, the concept of environment has been used more literally. Life cycle assessment, which seeks to estimate the environmental impact of a product or system across its life cycle, treats environment as a source of resources (extractions) and as a receiver of emissions. As extractions indicate resource dependency, so too do emissions indicate information uncertainty, in terms of increased risk to future human or eco-system health.

In this issue we have compiled a strong set of papers that discuss organizations, environments, and complexity. The papers were invited by us, the issue's coeditors, and went through the review process in order to improve.

Tom Eide explores organization/environmental definitions via analysis of an Ibsen play wherein the individual ethical, social-economic and political levels are explored as interlinked and contradictory. He is interested in whistle-blowing and ethics, but also in multiple-perspectives. The situation analyzed is different for the actants than for the researcher. Contemporary social science methods often stress that the researched must agree to the data presented, Eide shows us a situation where this would never lead to honesty or insight. And he asks us if literary methods that can reveal such conflicts and paradoxes are not needed in the study of complexity.

The next three papers use the physical concept of the environment. In "Sustainable Supply Networks: A Complex Systems Perspective," Liz Varga *et al.*, use a case study of the evolution of supply networks in the UK aerospace industry to demonstrate how supply networks coevolve with their environment. They find empirical support that links environmental drivers to change in supply network structure and behavior. Their work highlights the importance of path dependency and coevolution with the environment.

"Social and Ecological Transitions: Winemaking in California", by Greig Tor Guthey and Gail Whiteman, uses a case study of the northern California wine industry to show how environmental pressures affected regional planning and industrial practice. They use the concept of "ecology" to unify the physical environment and the organizations embedded with it. Their study emphasizes the need for an interdisciplinary approach when addressing these phenomena.

Kathryn Pavlovich's "A Fractal Approach to Sustainable Networks" uses the concept of fractals to discuss the structure of sustainable inter-organizational networks. In studying ethnographic data concerning a tourist destination, she examines self-similarity in the networks that emerged, and finds evidence of volume-filling, reciprocity, and enfoldment, characteristics of a complex adaptive system. The paper demonstrates how the concept of fractals can provide a useful lens to understand the coevolution of environment and organizational system.

Our Complexity and Philosophy paper is "Making Room for Affordances" by Hugo Letiche and Michael Lissack. If the relationship between the entity (organization, organism, person) and its environment is as mutual as Bateson claims, how can we still describe and understand it? If we are always relationally implied in what we study, how can we still 'study' it? James Gibson's concept of 'affordances' potentially cuts the Gordian knot of self/other, researcher/researched and organism/environment by focusing on the activity between the two instead of on the dualisms. Scott Kelso's 'complementary relating of contrarieties' is a contemporary conceptualization of the activity of the in-between—i.e., of cognition that is situated and contextual, dynamic and relational. In this article these tools of complexity studies are examined and developed.

This issue of *E:CO* offers a post-Bateson(ian) exploration of thinking the identity/environment relationship. Neither 'self' nor 'environment' is an adequate concept any more—thinking their in-between and mutuality is a crucial challenge of our times.

References

Ashby, W.R. (1958). "Requisite variety and its implications for the control of complex systems," *Cybernetica*, ISSN 0011-4227, 1(2): 83-99.

Bateson, G. (1963). "The role of somatic change in evolution," *Evolution*, ISSN 0014-3820, 17(Dec): 529-539.

Epstein, J.M. and Axtell, R. (1996). *Growing Artificial Societies: Social Science from the Bottom Up*, ISBN 9780262550253.

Guastello, S.J. (2002). *Managing Emergent Phenomena: Nonlinear Dynamics in Work Organizations*, ISBN 9780805831634.

Kauffman, S. (1993). *Origins of Order: Self-Organization and Selection in Evolution*, ISBN 9780195079517.

Lawrence, P. and Lorsch, J. (1967). "Differentiation and integration in complex organizations," *Administrative Science Quarterly*, ISSN 0001-8392, 12:1-30.

Lichtenstein, B. (2000). "Self-organized transitions: A pattern amidst the 'chaos' of transformative change," Academy of Management Executive, ISSN 0896-3789, 14(4): 128-141.

Luhmann, N. (1995). *Social Systems*, ISBN 9780804726252.

Maturana, H.R. and Varela, F.J. (1987). The Tree of Knowledge: The Biological Roots of Human Understanding, ISBN 9780877734031.

Thompson, J. (1967). *Organizations in Action: Social Science Bases of Administrative Theory*, ISBN 9780765809919.

Kevin J. Dooley is a Professor of Supply Chain Management at Arizona State University. Dr. Dooley is a world-known expert in the application of complexity science to organizations. He is a Dean's Council of 100 *Distinguished Scholar*, and he has published over 100 research articles and coauthored an award winning book, *Organizational Change and Innovation Processes*. He is currently Director of Sustainability Initiatives in the W. P. Carey School of Business, and Director of Life Cycle Research in The Sustainability Consortium. He has coauthored two patents concerning Centering Resonance Analysis, a novel form of network text analysis, and is cofounder and CEO of Crawdad Technologies, LLC. He has a Ph.D. in Mechanical Engineering from the University of Illinois.

Hugo Letiche is Professor of 'Meaning in Organization' at the Universiteit voor Humanistiek, Utrecht (NL), where he is the Director of the part-time PhD program. He has been part-time Professor at Lancaster University, Keele University and Bristol Business School, and visiting Professor at the University of California at Berkeley. His recent writings include *Making Healthcare Care* (Charlotte NC: IAP, 2008) and *Social Complexity Theory* (forthcoming). Recent research has focused on normative professionalism, social ecological systems theory versus social complexity theory, and the relevance of practitioner investigation. Recent articles have been on post-phenomenology (*Culture & Organization*), polyphony (*Organization Studies*) and art/creativity (*Society & Business Review*).

A4. Editorial—Issue 11.4

Peter M. Allen
Cranfield University, ENG

This journal, like many others, is written and read by people who presumably consider that the words within it can provide some knowledge that should increase their ability to deal with situations that they may be called upon to deal with. In other words, we live in the hope that reading the results of the experiences, experiments and reflections of others can improve our interpretive framework and lead to better responses to given problems or an improved ability to the recognition of problem types. In this way, the 'information' carried within the communication reduces uncertainty and improves outcomes. And information can be acquired either simply from experience—ordinary pragmatism, or as the result of deliberate experimentation and acquisition of data which is really the scientific method.

It seems probable that most people, firms and organizations actually function on the basis of simple pragmatism and the learning obtained by practical experiment and the rejection of unsuccessful initiatives. The question then is whether a 'scientific approach' to problems can really reduce inherent uncertainties and lead to higher performance? If knowledge is supposed to reduce uncertainties then how do we get knowledge—and can we separate it from belief?

The answer is that we obtain knowledge by making successive assumptions that constrain what can happen—and so reduce the freedom of the elements present until the future is forced into a single trajectory into the future—and prediction appears possible.

The first assumption concerning a situation is to say that there is a 'boundary'—and that some things will be considered to be inside and others will be outside in the environment. In other words this corresponds to saying that the inside can be understood in its own terms, in the context of the outside. This assumes that we know *enough* about the links between the inside and the outside to be able to get understanding just from the elements within. The second assumption concerns our knowledge of those elements within—the components of which the inside is made. This assumes that we have sufficient ability to be able to 'classify' the elements within into the different types of objects and their characteristic behaviors. This might be biological species and perhaps their age cohorts, or in social systems people classified according to their ethnic, politic or philosophical beliefs, or their skills, professional activities etc. When we do this for any system of interest we find that over time the constituent types have changed. New types and activities have emerged and others have left. Classification shows us that over time qualitative evolution occurs and the system is not structurally stable in that the variables, and therefore the equations describing the mechanisms and processes at work within it must change over time. This is the key territory and realization of complexity thinking. It recognizes that 'systems' do not simply run according to fixed mechanisms but undergo qualitative evolution through a series of instabilities as new things emerge and others disappear. Although this is illuminating it remains a situation of real 'uncertainty' since we know that the current system we perceive will change qualitatively—but we don't know when and we don't know in what way. However, merely stating that things change qualitatively at times and in ways that we cannot predict does not, in our society, "bring home the bacon"! Our society demands more than this—it demands prediction.

Further assumptions have to be made in order to appear to achieve this. Firstly, one can take the current entities of the system and include their mechanisms of interaction. Here the micro-events - of birth, death, production, sales, migration etc.—can take place according to some probability distribution. If we allow for events to occur with their actual probabilities then we arrive at a dynamical description that includes fluctuations and generates endogenously both average behavior and the deviations around it. These are stochastic differential equations—known also as the Chapman-Kolmogorov equations, and allow the study of system resilience the effectiveness of different possible contingency plans. However, it does assume the absence of new behaviors and types within the system, as well as of learning behavior on the part of participating agents. So, it is now better able to 'predict'—providing that the internal elements are considered as fixed, mechanical objects incapable of experiment or learning.

In our quest for understanding and prediction we find two choices for our next assumption:

1. That the probabilistic dynamics in fact moves to a stationary solution rapidly giving us 'self-organized criticality';
2. That the probabilistic dynamics can be approximated by retaining only the most probable events leading to 'System Dynamics'.

The first of these leads particular probability distribution functions shaped by the mechanisms contained in the stochastic equations. For particularly simple mechanisms this can lead to the emergence of power-law distributions and rank-size rules for earthquakes, city sizes, firm sizes etc. In reality though the charts of city and firm size shows a great deal of dynamic change as cities and firms grow and decline, and it seems by no means obvious that the probability distribution must be stationary. In other words, although the model results in a probability distribution, it pushes the 'uncertainty' into that of the assumptions on which the result rests.

Original Ref—E:CO Vol. 11 No. 4 2009 pp. vi-viii

If we take the second route to greater 'prediction, we assume that only the most probable events occur. This lead us to 'system dynamics' which is in general a non-linear set of dynamical equations that appear to be predictive and deterministic. In other words, they seem to allow the future trajectory of the system to be calculated, and to provide a basis for policy and strategy analysis by seeing the differences made over time by one intervention as opposed to another. This is a very tempting picture for any decision or policy maker or indeed manager. It appears to offer a way to test different decisions and allow their advantages and disadvantages to be compared.

But the important point that we need to reflect on is that such apparent power of prediction is only real if, and only if, the assumptions made in achieving it are in fact true. In other words the real uncertainty that we know to characterize the long term evolution of an ecology, economy, market or firm is only banished by assumption. In this light therefore, we must admit that understanding and predictions will only hold until things change and our expectations are confounded. Our methods therefore do not scientifically eradicate the uncertainty of an evolving world, but instead mask it and tell us that providing the system doesn't change then we can predict what it will do. But clearly the uncertainty is now as to whether the system will change or not.

While it may be reasonable to believe that the system may hold its structure for short times, then this becomes increasingly unlikely for longer times, since history has shown us that over longer time periods everything of interest seems to change as new entities and types appear in the system and others become extinct. Ultimately, the uncertainty remains. All that we can say is that for short times we may have a coherent structure in which we can talk of prediction and calculate the 'risk' involved in fluctuations around the average values of variables. However, the system will, over longer times possess uncertainties associated with the emergence of new variables, mechanisms and features in the future, where new dimensions of will be turned on and new factors become relevant.

Ultimately then, we may say that if our assumptions are true then our predictive models, either probabilistic or deterministic will hold. In reality though we have exchanged one ill-defined uncertainty for another, that is the uncertainty as to whether the system is structurally stable or not. If we own and control firmly the situation in question, then our current interpretive framework will probably be a good indicator of what will happen. But often our very first assumption about how the system connects to the environment can still trip us up and invalidate our predictions. Essentially we push uncertainty into the question of whether or not the situation, system or organization is structurally stable, and in general do not attempt to answer the question. The first difficulty being—unstable to what disturbance? If we want to study this, then we must examine in more detail both the internal levels of detail within our system and the external world in which our system is located, and without knowing what to look for, this is a difficult task.

In summary then, complexity really recognizes these limits to knowledge and shows us that our quest to banish uncertainty and replace it with knowledge is in fact bound to fail eventually. We only get 'certainty' by closing our minds and pretending that the situation we are faced with is also 'closed'—however, it is indeed wisdom to know that 'knowledge' concerning particularly social, economic and cultural systems is really only current 'belief', and that surprises and uncertainty will always be present. In an evolutionary universe, we are destined to continue a perpetual learning process without end, as we develop, correct and adapt our interpretive frameworks to gradual fossilization which, fortunately, we then fail to pass on to the next generation.

Index

A

B

C

D

E

F

G

H

I

K

L

M

N

O

P

Q

R

S

T

U

V

W

www.ingramcontent.com/pod-product-compliance
Lightning Source LLC
LaVergne TN
LVHW062029101126
852251LV00005B/62
9780984216567